Environment, Energy and Sustainable Development

Environment, Energy and Sustainable Development

Edited by **George Thomson**

SYRAWOOD
PUBLISHING HOUSE
New York

Published by Syrawood Publishing House,
750 Third Avenue, 9th Floor,
New York, NY 10017, USA
www.syrawoodpublishinghouse.com

Environment, Energy and Sustainable Development
Edited by George Thomson

International Standard Book Number: 978-1-68286-300-8 (Hardback)

Printed in China.

Contents

Preface

Energy is indispensable for development. There is critical need for a cleaner energy technology for sustainable growth in the modern times. This book discusses the sources of energy and the impact they have on the environment in depth with a special focus on renewable sources of energy like hydroelectricity, solar energy, wind energy, etc. It consists of researches contributed by experts from the field of environment and energy sciences. The aim of this text is to acquaint the readers with concepts like sustainability and effective environmental protection. It is ideal for environmentalists and researchers.

This book is a result of research of several months to collate the most relevant data in the field.

When I was approached with the idea of this book and the proposal to edit it, I was overwhelmed. It gave me an opportunity to reach out to all those who share a common interest with me in this field. I had 3 main parameters for editing this text:

1. Accuracy - The data and information provided in this book should be up-to-date and valuable to the readers.
2. Structure - The data must be presented in a structured format for easy understanding and better grasping of the readers.
3. Universal Approach - This book not only targets students but also experts and innovators in the field, thus my aim was to present topics which are of use to all.

Thus, it took me a couple of months to finish the editing of this book.

I would like to make a special mention of my publisher who considered me worthy of this opportunity and also supported me throughout the editing process. I would also like to thank the editing team at the back-end who extended their help whenever required.

Editor

1

Diverse configurations of the boiler feed pump drive for the ultra-supercritical 900-MW steam plant

Katarzyna Stępczyńska*, Henryk Łukowicz and Sławomir Dykas

Abstract

Coal-based electric power generation remains the basic source of obtaining energy. With increasing pressure to reduce CO_2 emissions, improving power unit efficiency has become an issue of utmost significance. Surely, one of the possibilities to improve the efficiency of new power units is raising the steam parameters. With improved power plant efficiency, there is a lower demand for power of almost all auxiliary equipment except the boiler feed pump. The reason for this is that the power needed to drive the feed pump is an almost linear function of the steam pressure. This means that, even though the steam mass flow (and, consequently, the feed water mass flow) is reduced and the efficiency of feed pumps is improved, their power increases. For this reason, it is very important to find the optimum drive of the boiler feed pump. The main aim of the conducted analysis was to compare various drive options of the boiler feed pump for a conceptual ultra-supercritical 900-MW steam power unit. The following drive configurations of the boiler feed pump were presented and compared:

- A frequency-controlled electric motor
- A condensing turbine fed with steam extraction from the immediate-pressure turbine
- An extraction-backpressure turbine fed with steam from a cold reheat steam line with bleeds shifted from the low-pressure turbine
- A backpressure turbine fed with steam from a hot reheat steam line operating in parallel with the intermediate-pressure turbine
- An extraction-backpressure turbine fed with steam from a cold reheat steam line with bleeds shifted from the intermediate-pressure turbine (the master cycle idea).

The analysis of the operation of the 900-MW unit with various configurations of the feed pump drive was carried out for three load levels: for the nominal mass flow of live steam and for the partial mass flow of 75% and 50%.

Keywords: Ultra-supercritical coal-fired power units, Boiler feed pump drive, Master cycle, Net efficiency of electric power generation

Background

The perspective of ever more stringent restrictions on emissions and of the rising coal consumption pose a challenge to the professional power engineering and necessitate the search for effective and economically acceptable methods of reducing CO_2 emissions. In the next years to come, condensing coal-fired power units will still be the pillar of the Polish electric power system. In the present situation, it is, therefore, necessary to achieve better and better efficiency of electric energy generation which will have a critical significance in meeting the emission-related requirements. Undoubtedly, raising the steam parameters is the main driving factor in the efforts to increase efficiency. Thus, the construction of new generation coal-fired units for ultra-supercritical parameters is a matter of the next few years to come. The efficiency of power generation in conventional coal-fired units has increased in the last 10 years by approximately 7% points. The power units which are now being built in the pulverized coal-fired (PF) technologies reach a capacity of the order of 1,000 MW. The pacemakers in the

* Correspondence: katarzyna.stepczynska@polsl.pl
Institute of Power Engineering and Turbomachinery, Silesian University of Technology, Konarskiego 18, Gliwice, 44-100, Poland

electric power generation net efficiency are now: Nordjyland in Denmark (47% for hard coal) and the BoA power unit in Niederaussem, Germany (45.5% for brown coal) [1].

Improving a power unit efficiency results in a reduction in fuel and steam consumption per unit which involves a reduction in the mass flow of burned fuel, feed water and other combustion products: slag and ash. The expansion of regenerative systems and the raising of feed water temperature additionally result in reducing the mass flow of the main condensate and of the mass flow of the water cooling the condenser. Generally, an effect of an increase in the steam parameters is a decreased demand for power to satisfy the needs of most facilities. However, there is a greater need in terms of the power with which the feed water is pumped. Since the power of the boiler feed pump is an almost linear function of the live steam pressure despite a reduction in the unit steam consumption and the feed water consumption, as well as an improvement in the efficiency of the pump itself, its share in the total demand for power to satisfy own needs increases substantially.

Boiler feed pump drives currently in use

The feed water system is one of the critical components of a power unit. Its properties decide about many parameters of the thermal cycle but also about a failure-free and safe operation of the power plant. Therefore, the problem of the selection of the boiler feed pump drive is one of the most important issues in designing power units.

Due to the requirements, each power unit should be equipped with at least two boiler feed pumps whose nominal capacity should ensure the maximum capacity of the once-through boiler. The value of the pumping effective pressure is included in the range of 1.1-1.4 of the steam pressure rating in the boiler. The efficiency of modern feed pumps exceeds 85%. The demand for power while pumping feed water is from 2% to 4.2% of the power generated by the unit [2]. Consequently, the driving motor of the feed pump is usually the principal receiver of electricity in the power plant.

In high power units, an extra low speed booster pump is used to prevent cavitation. The pump, which features an inlet designed specially to avoid cavitation, feeds water into the main pump, thus, protecting it entirely from this phenomenon. The booster pump may have a separate drive, or it can be driven by the main pump driving motor through a speed reduction gear [3].

To drive boiler feed pumps in power units with a capacity of up to approximately 250 MW, induction squirrel-cage motors and a fluid coupling to adjust rotation speed are used almost exclusively. However, as the power capacity of newly constructed units grew, the lack of appropriate electric motors (EM) became a problem. For this reason, for power units with a capacity exceeding 300 MW, steam drives were applied by means of a separate turbine incorporated into the thermal cycle. The steam turbine starts to drive the feed pump only after a sufficient amount of steam has become available. Due to that, it is necessary to use an additional start-up pump with a partial capacity of 30% to 50% with an electric drive. The driving turbine of the boiler feed pump can be incorporated into the thermal cycle in two ways [2]:

- As a condensing turbine (CT), fed with steam from the intermediate-pressure turbine (outlet steam is condensed in a separate condenser or it is introduced into the main condenser)
- As a backpressure turbine (BPT) or extraction-backpressure turbine (EBPT), fed with steam from the high-pressure turbine (outlet steam and steam bleeds, if any, are condensed in low-pressure regenerative heaters).

Because condensing turbines used to drive boiler feed pumps have a limited effect on the thermal cycle of the unit, they have found a much wider application in the professional power engineering. In Poland, this configuration was used in the units of the power plants in Bełchatów, Opole, Pątnów and Łagisza.

Another possible option is the feed pump driven by the shaft of the main turbine. The pump can be joined to the turbine shaft either at the high- or low-pressure turbine side. Both solutions were used in the past, but both increased costs and the turbo set complexity. A gear had to be applied because of the difference in the speed of the turbo set and the pump. As these solutions involved a bigger investment outlay and higher operational costs, they are no longer used. Their substantial drawback was also the forced and extremely unfavorable location of the feed pump in the powerhouse which caused additional construction problems.

Condensing turbines to drive boiler feed pumps are usually fed with the steam from the outlet of the intermediate-pressure turbine or steam from the intermediate-pressure turbine bleeds, and they give steam to the main or additional condenser. The amount of steam needed to drive the condensing turbine reduces the mass flow through the low-pressure turbine and can contribute to a reduction in the total discharge loss and, consequently, reduce the heat consumption per unit. Therefore, an application of a condensing turbine to drive the feed pump is more effective if the low-pressure turbine is heavily loaded.

Under a low load of the unit, the steam mass flow through the low-pressure turbine is usually not big enough which causes the danger of a substantial increase in the temperature of the blades. As a rule, the problem is solved by increasing the steam mass flow from the intermediate- to the low-pressure turbine. Under low loads of the unit, there may then be not enough steam fed into the turbine driving the feed pump which necessitates an introduction of additional steam into it. This may cause mechanical problems because the turbine driving the boiler feed pump was designed for lower parameters of the incoming steam. In consequence, additional facilities are required for the system, which increase the investment costs of the plant. Another option in this situation is a transition to an electric drive of the boiler feed pump.

The large mass flow of the steam needed to satisfy the pump's demand for the driving power causes the volume of the outlet steam mass flow to increase significantly, which results in a large annulus area and, consequently, in high blades of the last turbine stage. In the case of high power units, the relatively high rotation speeds of the boiler feed pump (up to as much as 7, 300 rotations per minute) are not suitable for the condensing steam turbine. Due to the high rotation speed required by the pump, the peripheral velocities of the last stages differ considerably from the optimum values that ensure the highest efficiency. Combining high peripheral velocities with high blades often causes mechanical problems. To avoid them, some manufacturers offer turbines with the flow separated in the last stage into two stages with a smaller diameter. This sometimes solves mechanical problems with high blades, but due to the flow losses resulting from the separation of the steam mass flow, no improvement in efficiency is achieved. In fact, the efficiency of condensing turbines which drive boiler feed pumps is often by 6% to 7% points lower than that of the main steam turbine [4]. Some manufacturers offer reaction-type condensing turbines with a fully separated flow with the steam inlet in the middle and symmetrical outlets. This leads to the rotor elongation, which is not a problem in the case of reaction constructions because they solve the mechanical problems related to long blades and ensure an efficiency level which is 2% to 3% points, different from the efficiency of the main turbine [4]. However, such machines are very expensive.

In recent years, both the power capacity of available electric motors and the efficiency of low-pressure turbines and drive control have increased substantially. Table 1 presents example efficiency values for turbine and electric drives. The efficiency of modern electric motors is much better than the one of motors coming from earlier batches which is still used in electric power stations. The efficiency of the motors offered today exceeds 97%, while older motors reached the efficiency level in the range of 91% to 94%. Moreover, the characteristic curves of the efficiency of modern motors have a flat course, which is of utmost importance for operation under a partial load [5]. Hence, the overall efficiency related to electric motors driving boiler feed pumps has improved. On the other hand, the efficiency of turbines driving BFPs has not changed much. Therefore, in the latest constructions of high capacity power units for supercritical parameters, especially German ones, the system which is most often used is the 2 × 60% or 2 × 50% one, with an electric drive and rotation speed controlled by frequency, as this kind of speed control has a fairly flat efficiency characteristic in a wide range of loads compared to the fluid coupling. The adjustment of the pump efficiency is carried out through the control of the frequency and voltage of the current powering the electric motor in order to obtain the required rotation speed. The pump is connected to the driving motor through an elastic coupling to reduce the torsional harmonic forces transmitted by the pump. Changing rotation speed is the most effective and economical way of adjusting the pump efficiency. Adjustments made to rotation speed cause only a slight decrease in the pump efficiency compared to operation with fixed revolutions. A change in rotation speed involves a change in the pump characteristic. This type of control allows a reduction in electricity consumption in a wide range of loads. Usually, frequency-controlled synchronous motors are used for such solutions. The rotation speed of a synchronous motor can be reduced from the nominal value to very low speeds by reducing the frequency and voltage of the electric current powering the motor [5].

The improvement in the reliability and availability of the pumping engine components, which is a part of the general technological progress, has caused that the availability of the 2 × 60% (2 × 50%) systems driven with electric motors is only slightly different from the values achieved by the EM 3 x 50% or BPT 100% + EM 30% configurations. On the other hand, the investment expenditures for the EM 2 × 60% (EM 2 × 50%) systems are much lower compared to the other two [6]. For these reasons, in many recent constructions of high capacity power units, it was the pump driven by a frequency-controlled electric motor that was selected, not the pump driven by the turbine. Despite the electric power losses in the generator, transformer, supply network, frequency converter and motor, this

Table 1 Comparison of boiler feed pump drive efficiency [6]

Efficiency	Condensing turbine	Electric drive
Low-pressure turbine	-	88 - 92
Generator	-	98.5
Transformer	-	99.5
Motor	-	97
Drive control	-	95 - 96
TOTAL	83	74 - 84

system is sometimes more efficient than the feed pump driven by a condensing turbine.

Alternative options of the boiler feed pump drive

The assessment of the possibility of a further improvement in the net efficiency of electricity generation in conventional coal-fired units for ultra-supercritical parameters calls for consideration of options of the boiler feed pump drive other than the most often used ones, namely those powered by the electric motor and by the condensing turbine.

One of them is an application of the aforementioned backpressure turbine or extraction-backpressure turbine fed with outlet steam from the high-pressure turbine from the cold reheat steam line. The outlet steam together with the driving turbine bleeds (if any) would condense in low-pressure regenerative exchangers. Owing to the fact that outlet steam also gives up heat in the regenerative system, the extraction-backpressure turbine does not require a costly system of the condenser and cooling water which usually is the case with the condensing turbine driving the main feed pump.

An interesting option could be the implementation of the 'master cycle' (MC) solution patented by the company Elsam Engineering (Fredericia, Denmark), which consists in shifting all the intermediate-pressure turbine bleeds to an auxiliary extraction-backpressure turbine referred to as 'tuning turbine' which is also fed with the steam from the high-pressure turbine outlet [7-9]. The auxiliary turbine drives the main boiler feed pump and, optionally, an auxiliary generator which could carry away the power surplus generated by the system to the unit transformer through separate, additional primary coils. During the start-up or holdup of the unit, the 'synchro-self-shifting' clutch is open and the auxiliary turbine does not work. The auxiliary generator is switched to motor operation and drives the main feed pump, which renders the start-up feed pump driven by an electric motor unnecessary. The use of an additional generator on a single shaft together with the pump and the turbine would require revolution control to maintain a steady speed on the part of the generator.

A solution like this has many advantages because a substantial rise in the temperature of steam, especially reheated steam, increases the problem of the difference between the bleed steam and the saturation temperatures in regenerative heaters fed with bleeds from the intermediate-pressure turbine, which are situated right after the steam reheater. This causes material and thermodynamic problems because, with the rise in the mean difference in temperature of the heat-exchanging agents, the exergy losses rise too which leads to a reduction in the process effectiveness and to an increase in the heating steam consumption. By keeping pressures in the bleeds of the auxiliary turbine at the same level as in the intermediate-pressure turbine, the temperature difference in the exchangers will be much lower (the outlet of this turbine is even in the area of saturated steam). Another advantage of the master cycle is that the construction of an intermediate-pressure turbine without bleeds is much simpler and cheaper. This is essential in view of the need to use expensive construction materials. Lower temperatures of the steam in the bleeds from the driving turbine, compared to the intermediate-pressure turbine, result in an increase in the mass flow from them. This causes a reduction in the steam mass flow to the reheater, i.e., lower costs of the reheated steam pipelines and of the reheater in the boiler. The steam mass flow from the main turbine into the condenser is also smaller which decreases heat losses of the cycle.

Yet, another solution is the use of a backpressure turbine working in parallel with the intermediate-pressure turbine. The driving turbine is fed with steam from a hot reheat steam line. Outlet steam from the turbine is in turn introduced into the inlet of the low-pressure turbine. Such a configuration limits the problem of too little steam fed to the low-pressure turbine at low loads of the unit. What can be a problem, however, are the high parameters of the inlet steam (identical to those at the inlet to the intermediate-pressure turbine), which is related to the use of expensive construction materials.

An analysis of selected configurations of the boiler feed pump drive

In order to assess the impact of the boiler feed pump drive configuration on the net efficiency of electric power generation, balance calculations for a unit for ultra-supercritical parameters were carried out. The subject of the study was a conceptual coal-fired unit with a net electric power capacity of 900 MW. The basic parameters of the unit are listed in Table 2. The system has a single inter-stage steam reheat, a four-stage low pressure regeneration and a three-stage high pressure regeneration, plus a steam desuperheater.

Five configurations of the boiler feed pump drive were analyzed:

- A frequency-controlled EM (Figure 1)
- A CT fed with steam extraction from the immediate-pressure turbine (Figure 2)

Table 2 Basic parameters of the conceptual 900-MW unit

Live steam pressure	30 MPa
Live steam temperature	650°C
Reheated steam pressure	6 MPa
Reheated steam temperature	670°C
Pressure in the condenser	0.005 MPa
Feed water temperature	310°C
Net electric power	approx. 900 MW

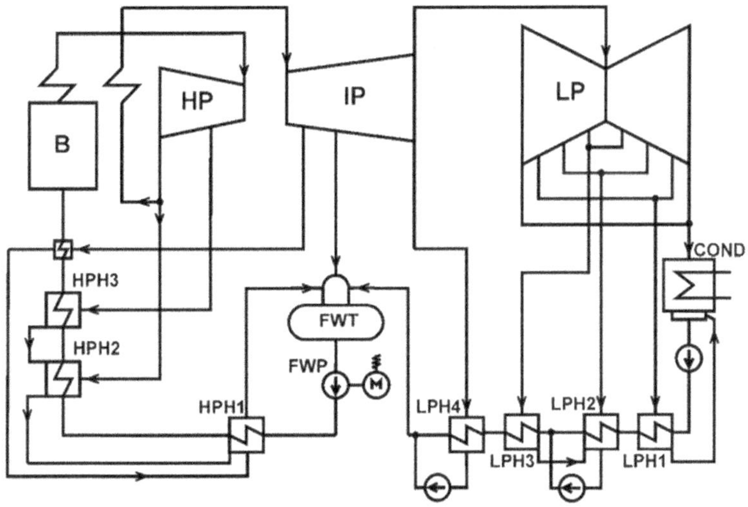

Figure 1 Diagram of the thermal cycle of a 900-MW unit with feed pump driven by EM. B, boiler; HP, high-pressure turbine; IP, immediate-pressure turbine; LP, low-pressure turbine; HPH3, HPH2 and HPH1, high-pressure feed water heaters; FWP, feed water pump; FWT, feed water tank + deaerator; LPH4, LPH3, LPH2 and LPH1, low-pressure feed water heaters; COND, condenser.

- An EBPT fed with steam from a cold reheat steam line with one bleed and an outlet directed to low pressure regenerative exchangers (Figure 3)
- A BPT fed with steam from a hot reheat steam line operating in parallel with the intermediate-pressure turbine (Figure 4)
- An application of the idea of the MC: an extraction-backpressure turbine fed with outlet steam from the high-pressure turbine with bleeds and an outlet directed to regenerative exchangers which, in a classic system, are fed from the intermediate-pressure turbine (Figure 5).

Figures 1, 2, 3, 4 and 5 present simplified diagrams of the thermal cycle of a 900-MW unit for the boiler feed pump drive configurations under consideration.

Methods and assumptions in the cycle analysis

The calculations of the cycle for selected configurations of the boiler feed pump drive ware conducted with the software Gate Cycle v. 5.40 (GE Company, Billerica, MA, USA). The software is used for balance calculations of electric power systems and makes it possible to model their operation under various loads.

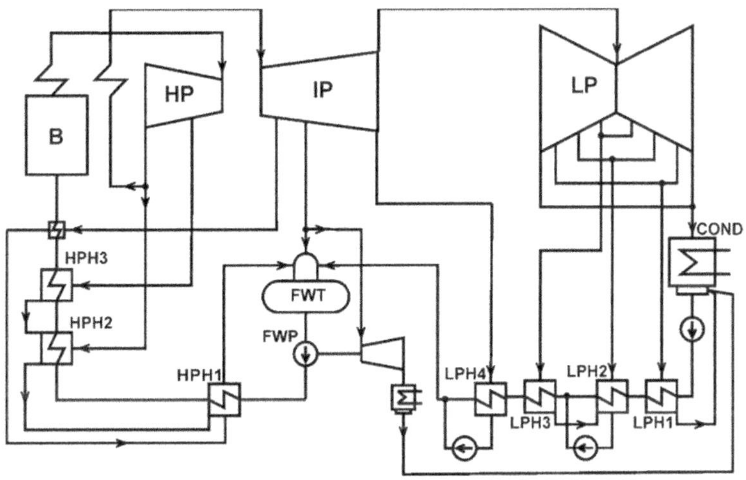

Figure 2 Diagram of the thermal cycle of a 900-MW unit with feed pump driven by CT. B, boiler; HP, high-pressure turbine; IP, immediate-pressure turbine; LP, low-pressure turbine; HPH3, HPH2 and HPH1, high-pressure feed water heaters; FWP, feed water pump; FWT, feed water tank + deaerator; LPH4, LPH3, LPH2 and LPH1, low-pressure feed water heaters; COND, condenser.

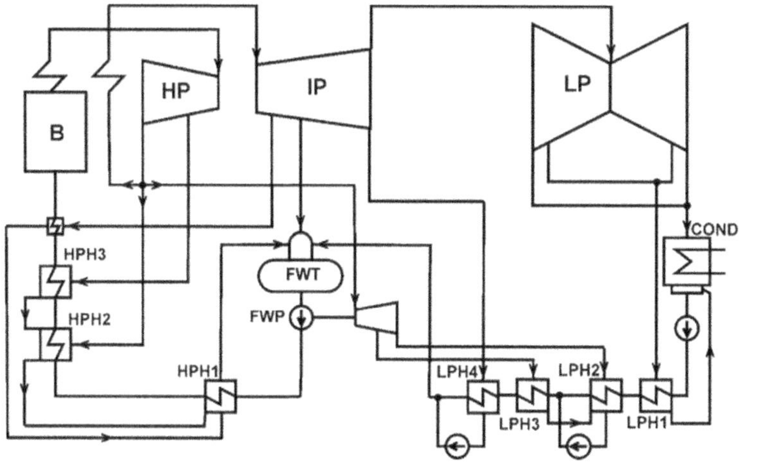

Figure 3 Diagram of the thermal cycle of a 900-MW unit with feed pump driven by EBPT. B, boiler; HP, high-pressure turbine; IP, immediate-pressure turbine; LP, low-pressure turbine; HPH3, HPH2 and HPH1, high-pressure feed water heaters; FWP, feed water pump; FWT, feed water tank + deaerator; LPH4, LPH3, LPH2 and LPH1, low-pressure feed water heaters; COND, condenser.

The analysis of the cycle was conducted with an assumption that for each unit configuration the nominal mass flow of live steam was identical. The efficiencies of the groups of turbine stages as well as the pressures in corresponding bleeds in individual unit configurations have the same values so do all the live steam parameters, the pressure in the condenser, the temperature of the water feeding the boiler and the boiler efficiency.

The calculations were made for the nominal mass flow of live steam produced by the boiler and for the partial mass flow of 75% and 50%. For the needs of the analysis, it was also assumed that, for each configuration, the system had the same boiler feed pump that ensured a 100% efficiency of the steam boiler, and the booster pump had a separate electric drive. In the case of the variants with the turbine drive of the feed pump, the turbine generates power needed only to drive the pump and does not generate extra electricity.

In order to calculate the net electric power and the net efficiency of the unit, an assessment was made of the demand for electric power of all the basic auxiliary equipment: the boiler feed pumps: both booster and main (for the EM variant), the condensate pumps, the pump of the water cooling the condenser, the air and flue gas fans,

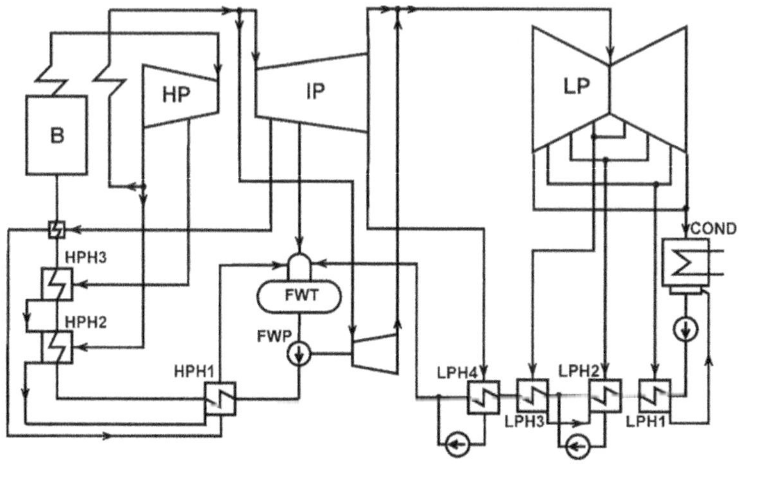

Figure 4 Diagram of the thermal cycle of a 900-MW unit with feed pump driven by BPT. B, boiler; HP, high-pressure turbine; IP, immediate-pressure turbine; LP, low-pressure turbine; HPH3, HPH2 and HPH1, high-pressure feed water heaters; FWP, feed water pump; FWT, feed water tank + deaerator; LPH4, LPH3, LPH2 and LPH1, low-pressure feed water heaters; COND, condenser.

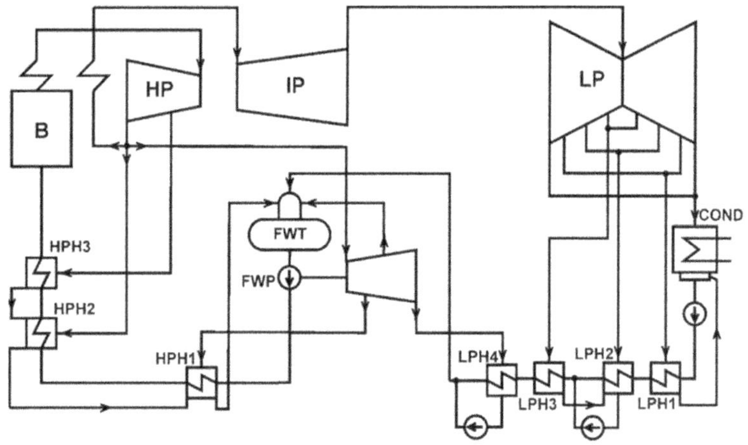

Figure 5 Diagram of the thermal cycle of a 900-MW unit. With feed pump driven by extraction-backpressure turbine: the idea of the MC. B, boiler; HP, high-pressure turbine; IP, immediate-pressure turbine; LP, low-pressure turbine; HPH3, HPH2 and HPH1, high-pressure feed water heaters; FWP, feed water pump; FWT, feed water tank + deaerator; LPH4, LPH3, LPH2 and LPH1, low-pressure feed water heaters; COND, condenser.

and the coal pulverizers. In each variant, identical operation and efficiency characteristics were adopted for corresponding facilities.

Results and discussion

The paper presents a comparison of the variants of the boiler feed pump drive for a conceptual 900-MW coal-fired unit for ultra-supercritical parameters. Table 3 shows the values of electric power needed to drive the feed pump for individual options depending on the live steam mass flow. In the electric motor option, the demand for electric power accounts for energy losses in the transformer, the supply network, the electric motor and in the frequency converter. For the remaining options with the feed pump turbine drive, the demand for electric power to drive the pump was calculated as the difference between the unit gross electric power for the option with an electric motor and the unit gross electric power for the other variants. The variant with the BPT shows the lowest demand for power for the 100% and 75% mass flow of live steam; for a smaller load of 50%, the CT requires the least power. However, for the EM, CT and BPT variants, the values are similar. A considerable reduction in the generated gross electric power occurred for the EBPT and the MC variants. This is because, in both of these cases, a relatively large steam mass flow is directed to the turbine driving the feed pump (for the MC option, it is more than 15% of the live steam mass flow), and it does not work in the main turbine. As the driving steam is taken into the turbine from a cold reheat line, it is not reheated in the boiler. Consequently, the amount of heat delivered into the cycle is reduced in comparison with the other variants.

Table 4 presents the unit consumption of the fuel chemical energy per unit of the generated net energy and the unit net electric power for three values of the live steam mass flow: 100%, 75% and 50% of the nominal mass flow. For the nominal load of the unit, the lowest unit consumption of the fuel chemical energy, i.e., the highest net efficiency of electric power generation was achieved for the MC variant, whereas the least efficient was the EM option. Figure 6 presents in the graph form the changes in the values of the unit net efficiency depending on the variant of the boiler feed pump drive.

Table 3 Demand for electric power to drive the feed pump

Drive variant of the feed water pump	Live steam mass flow/ nominal live steam mass flow		
	50%	75%	100%
Electric motor (EM)	9,844 kW_e	18,364 kW_e	32,660 kW_e
Condensing turbine (CT)	9,261 kW_e	18,239 kW_e	32,947 kW_e
Extraction-backpressure turbine (EBPT)	16,322 kW_e	28,723 kW_e	47,703 kW_e
Backpressure turbine (BPT)	10,041 kW_e	17,821 kW_e	30,169 kW_e
The idea of the Master Cycle (MC)	26,674 kW_e	44,441 kW_e	66,756 kW_e

Table 4 Unit net electric power (N_{elN}) and unit fuel consumption per unit of generated net energy (q_{enchN})

Drive variant of the feed water pump	Live steam mass flow/ nominal live steam mass flow					
	50%		75%		100%	
	N_{elN} (kW$_e$)	q_{enchN} (kJ/kWh)	N_{elN} (kW$_e$)	q_{enchN} (kJ/kWh)	N_{elN} (kW$_e$)	q_{enchN} (kJ/kWh)
Electric motor	478,980	7,847	701,798	7,684	908,255	7,636
Condensing turbine	479,719	7,835	702,200	7,679	908,461	7,632
Extraction-backpressure turbine	472,875	7,849	692,001	7,677	894,126	7,620
Backpressure turbine	479,001	7,819	702,686	7,656	911,273	7,608
The idea of the Master Cycle	462,999	7,854	676,888	7,666	875,774	7,581

The lines in the graph correspond to the set values of the live steam mass flow produced by the boiler. For lower values of the live steam mass flow, the CT variant turned out to be much better. For the MC and EBPT configurations, there is a bigger fall in the unit efficiency as the load decreases than for the others. This is probably due to the fact that turbines in this configuration have a large impact on the operation of regenerative exchangers where the steam from their bleeds and outlets is directed. Whenever the mass flow and the pressure of the generated steam decrease, the demand for power of the boiler feed pump falls substantially. Owing to that, the mass flow of the steam reaching the driving turbine also decreases significantly, and as a result, the pressures and steam mass flow at its bleeds and outlet decrease as well. Less steam flows into regenerative exchangers, so less heat is given up to the water that is being heated. The situation would be different if the steam mass flow feeding the turbine was increased, and the resulting surplus of power was given up in the auxiliary generator. However, as it was mentioned before, there is a problem with the difference in rotation speed between the feed pump and the generator. The pump operates at high speeds which change according to load, whereas the generator has to maintain a steady synchronous rotation speed.

Conclusions

The analysis showed that for the assumptions adopted for the calculations of a 900-MW ultra-supercritical unit, the highest net efficiency of electric power generation with a nominal steam mass flow had been achieved for the MC variant. It is by 0.34% point higher than the one obtained for the feed pump driven by an electric motor. According to Blum et al. [7], the application of such solutions gives much better results if they are combined with the use of dual steam reheat, where the problem of temperature accumulation in regenerative exchangers is more significant.

For smaller mass flows of generated steam, the cycle with a BPT features the highest efficiency. This type of

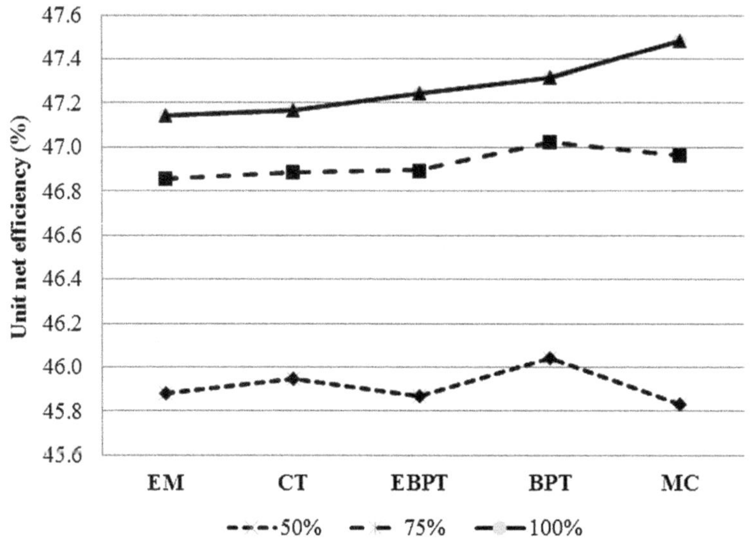

Figure 6 Net efficiency of a 900-MW unit electric power generation. For individual feed pump drive options depending on the live steam mass flow. EM, electric motors; CT, condensing turbine; EBPT, extraction-backpressure turbine; MC, master cycle.

solution might, however, entail higher investment expenditures due to very high parameters of the inlet steam (reheated steam). The differences in the unit net efficiency for variants using typical drive configurations, EM or CT, are very slight.

The load schedule of the unit should also have a considerable influence on the selection of the boiler feed pump drive. Assuming a longer operation of the unit under lower loads would exclude the application of the EBPT and MC variants due to a substantial decrease in the unit efficiency with load.

The choice of the optimum option of the boiler feed pump drive has a significant meaning from the point of view of reliability and availability of a power unit, the amount of investment expenditures and the net efficiency of electric power generation. Due to that, for each new investment, a detailed technical-economic analysis of the possible solutions has to be conducted.

Competing interests
The authors declare that they have no competing interests.

Acknowledgments
The results presented in this paper were obtained from research work co-financed by the National Centre for Research and Development within the framework of contract SP/E/1/67484/10—Strategic Research Program—Advanced Technologies for Energy Generation: Development of a technology for highly efficient zero-emission coal-fired power units integrated with CO_2 capture.

Authors' contributions
KS carried out the calculations for all 900-MW power unit configurations, participated in the sequence alignment and drafted the manuscript. SD conceived of the study and participated in its design and coordination, and helped draft the manuscript. HŁ participated in the design of the study and helped draft the manuscript. All authors read and approved the final manuscript.

References
1. Buchta, J, Pawlik, M, Oziemski, A, Kotlicki, T, Wawszczak, A: Potrzeby własne bloku nadkrytycznego. In: Kosman, G., Rusin, A., Taler, J., Pawlik, M. (eds.) Zagadnienia projektowania i eksploatacji kotłów i turbin do nadkrytycznych bloków węglowych, pp. 390–394. Politechniki Śląskiej, Gliwice (2010)
2. Laudyn, D., Pawlik, M., Strzelczyk, F.: Elektrownie. Naukowo-Techniczne, Warszawa (2005)
3. Westhuizen, W, Cattaert, T: Power station pump selection: part 2. Worlds Pump. **1**, 14–17 (2010)
4. Enery-Tech.Consider dual-driven boiler feed pumps for coal plants with capacity greater than 500 MW. http://www.energy-tech.com/article.cfm?id=18376&PageNum=1 (2008). Accessed 10 January 2011
5. Szulc, Z: Regulowany napęd elektryczny pompy wody zasilającej o podwyższonej pewności zasilania. (2008). Elektro.info. http://www.elektro.info.pl/artykul/id2534,regulowany-naped-elektryczny-pompy-wody-zasilajacej-o-podwyzszonej-pewnosci-zasilania. Accessed 10 January 2011
6. www.voilth.pl. Zastosowanie przekładni hydrodynamicznych Voith w napędach pomp wody zasilającej nowych bloków energetycznych. http://www.voith.pl/index.php?id_strony=302&id=241&idmain=26. Accessed 10 January 2011
7. Blum, R, Bugge, J, Kjaer, S: USC 700°C power technology—a European success story. VGB PowerTech. **26**, 4 (2009)
8. Blum, R, Bugge, J, Kjaer, S: Development of a PF fired high efficiency power plant (AD700). In: (ed.) Proceedings of the Riso International Energy Conference "Energy Solutions for Sustainable Development. Riso R-1608 (EN), Denmark (2007). http://130.226.56.153/rispubl/reports/ris-r-1608_69-80.pdf. Accessed 10 January 2011
9. Pawlik, M: Elektrownie opalane węglem w świetle polityki ograniczania emisji CO_2. Biuletyn Techniczno-Informacyjny Zarządu Oddziału Łódzkiego Stowarzyszenie Elektryków Polskich **2**(45), 1–36 (2009). http://sep.p.lodz.pl/biuletyn/sep_2_2009.pdf

2

Thermal comfort design of traditional houses in hot dry region of Algeria

Maatouk Khoukhi[1*] and Naïma Fezzioui[2]

Abstract

The new architecture produced recently in the south of Algeria known as 'modern construction' following the trend of the northern cities, which have different climate, is completely non-adapted to the harsh climate of the south of Algeria and, therefore, has high-energy consumption. The Ksar of Kenadsa is considered among the most important old cities of the south-west region of Algeria by its cultural and religious dimensions. In this area, the traditional architecture has been built to achieve the comfort in hot season. The new constructed houses have been built following the north architectural design neglecting the very harsh climate of the south region of Algeria characterized by very hot and dry climate. To evaluate the thermal comfort of this modern housing, a comparative analysis for the existing traditional housing is carried out using TRNSYS software. The simulation results show that the modern typical house seems to be inappropriate for the desert climate. Indeed, except the use of air-conditioning in summer, there is no other solution which can ensure thermal comfort.

Keywords: Energy, Modern house, TRNSYS, Thermal comfort, Simulation

Background

Passive methods of achieving thermal comfort in buildings in various countries have been widely investigated [1-7]. Fanger has calculated the thermal comfort by introducing a basic comfort equation [8]. Indeed, those passive methods are the best solution to provide a healthy and energy-efficient indoor environment [9-12].

The new houses built during these last decades in Algeria particularly in the south, called modern construction, are high-energy consumer and are completely not adapted for the hot climate. Indeed, the high demand of housing combined with the poor economic management of the public authorities made that the qualitative aspect of the built houses was sacrificed to the profit of the quantitative aspect to solve the enormous deficit in housing. It was urgent to find immediate solutions, resulting in the importation of new housing models, completely unsuitable for the climatic, cultural, and social contexts of the region [13].

Algeria has a traditional architectural heritage which can be noticed through the knowledge in the construction field based on vernacular architecture integrating climatic data and local resources. In the south west of Algeria (hot and dry climate), the traditional architecture had developed an empirical knowledge, particularly oriented to the achievement of the comfort in hot season [14]. The users' comfort was ensured by a combination of several passive strategies of thermal control, which are the results of a thorough knowledge of the climatic conditions.

The comfort requirements have changed, but the current architectural design does not answer these new needs. These climates are characterized by long and severe hot seasons. For this reason, the buildings must be designed according to the summer requirements. Those of the winter will be satisfied consequently [15].

The city of Kenadsa is characterized by long periods of overheating where discomfort is strongly felt. The aim of this study is to evaluate the thermal comfort of the modern constructed houses. A comparison analysis within the existing traditional housing is carried out using simulation by means of TRNSYS software.

Geographical location and climatic data

Kenadsa is located in the south west of Algeria about 20-km west of Bechar city and from 950-km south west of Algiers, at the altitude of 806 m, latitude of 31°38′N, and longitude of 2°15′W [16]. It lies at the bottom of a valley by

* Correspondence: mkhoukhi@squ.edu.om
[1]Civil and Architectural Engineering Department, College of Engineering, University of Sultan Qaboos, P.O. Box 33, Al-Khoudh, Muscat 123, Oman
Full list of author information is available at the end of the article

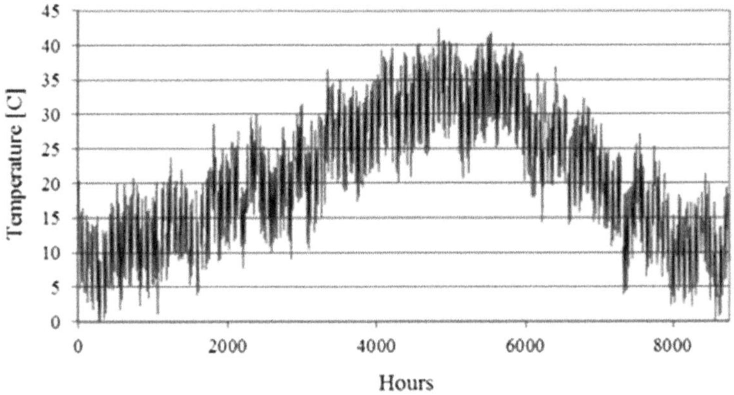

Figure 1 Monthly distribution of the outside temperature.

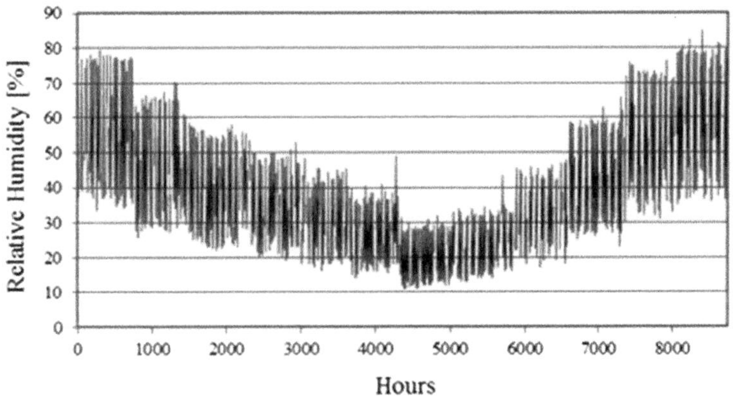

Figure 2 Monthly distribution of the relative humidity.

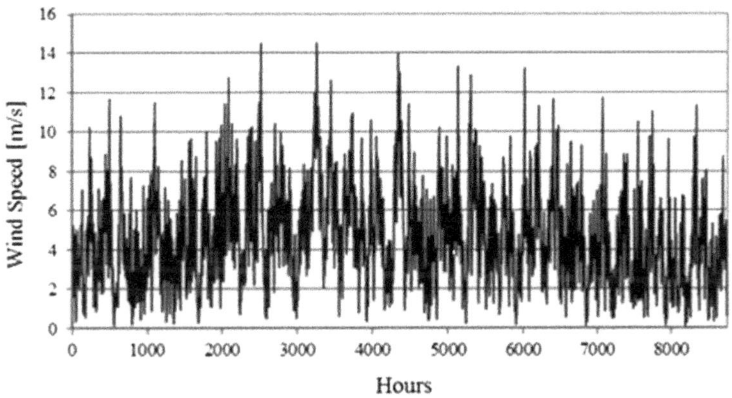

Figure 3 Monthly distribution of the wind speed.

Figure 4 Ksar of Kenadsa.

a mountainous relief, surrounded by a cliff called Barga which is composed of white rocks (silica sandstone) and fine sand in the north and the west, which offers an optimal protection against the harsh climate.

The monthly distribution of the air velocity, the temperature, and the relative humidity of the city are shown in Figures 1,2 and 3, respectively. These figures indicate that most part of the year is outside the comfort zone, except some periods of the months of September, October, March, and April. Another zone includes November to February where the building may require heating to ensure the comfort of the occupants. The third

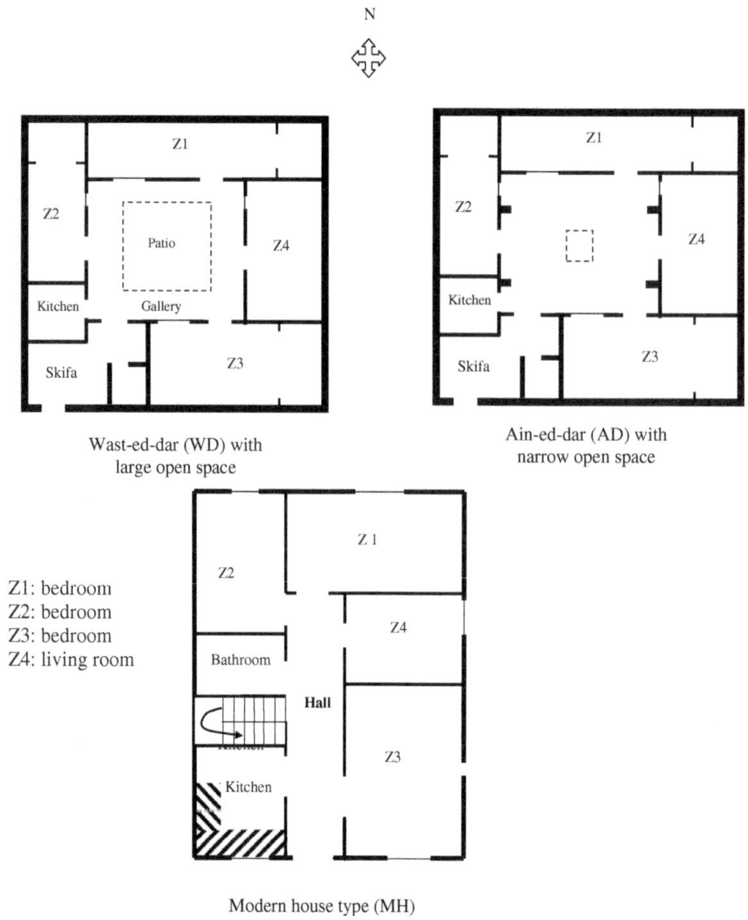

Figure 5 Model houses.

zone from May to September presents a period of hot season. The insolation exceeds 3,500 h/year, and the direct solar radiation may reach 800 W/m^2 on the horizontal surface. In summer, the temperature in the shade exceeds 40°C, and the amplitude between the day and the night is approximately 15°C, whereas the relative humidity remains low at about 27% [17].

Thermal characteristics and architectural aspects

The city is settled near the oasis to benefit from the effect of cooling by evaporation. The urban fabric appears in a form of a horizontal, dense, and compact model having a double effect of protection of the intense solar radiation in summer and the prevailing hot and cold winds (see Figure 4).

The Ksar is accessible by a route called 'derb' which is presented in the form of a labyrinthine course, which indicates the dynamics of the course during displacement. This course which is a long narrow street is shaded during the day, except when the sun is high in the sky. These streets are protected by the projection of rooms built in cantilever above the road.

The house, known under the terminology of 'dar,' is the house of the family where several families having their parental relations cohabit. It consists of four elements: bab (entry door), skifa (chicane), central space, and bit (room).

The general orientation of the house is towards the east. The questions of the orientation illustrate the convergence of the climatic, cultural, and religious considerations.

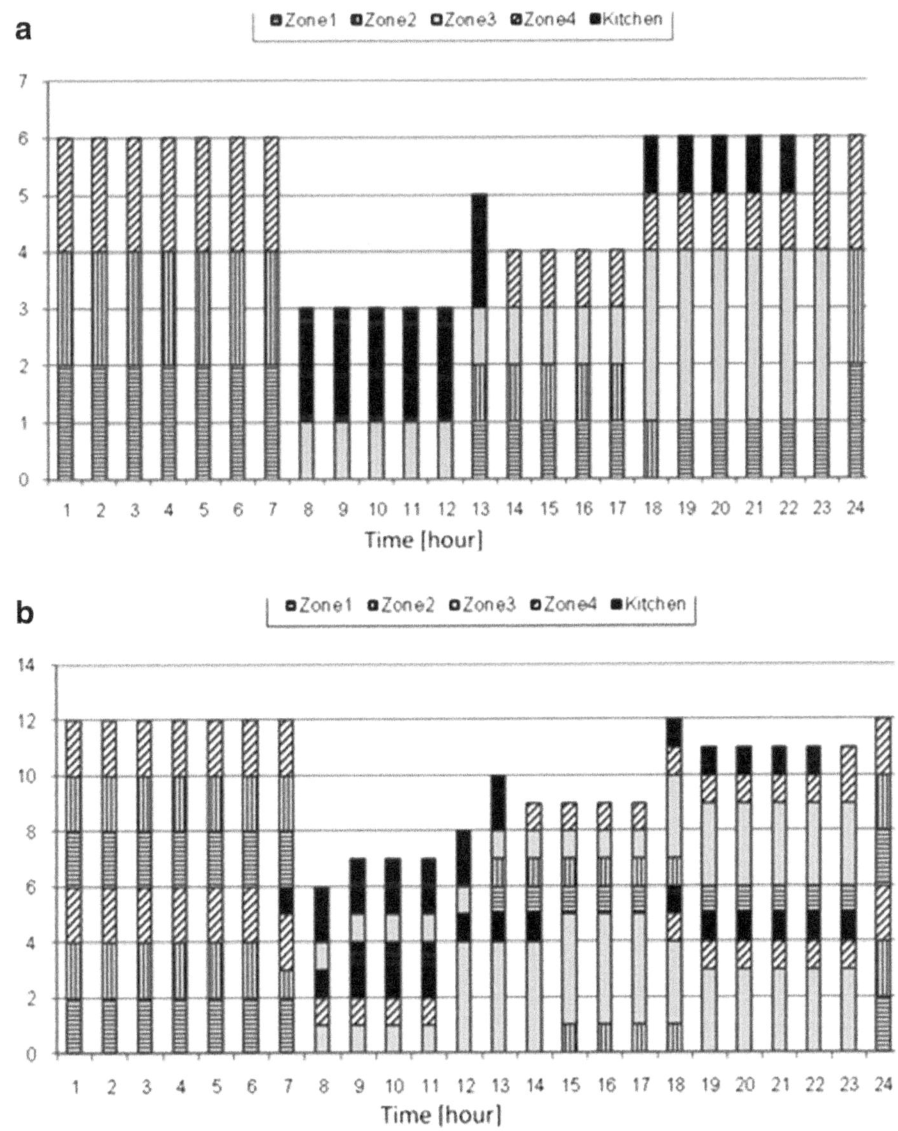

Figure 6 Occupation scenarios of the different zones ((a) weekday occupation and (b) weekend occupation).

Moslem directs their prayers and thoughts through the direction of qibla (means the direction towards Mecca). We have to distinguish two different shapes of central space. The models studied are shown in Figure 5 for the traditional housing, wast-ed-dar (WD) and ain-ad-dar (AD), and the modern housing (MH). WD has a large central space open to the sky. On the other hand, AD presents a narrow central open space (1 m × 1 m) which also opens to the sky.

The houses were completely turning inward. The walls that overlook the outside, particularly on the street (derb), are blind. All openings of the rooms are towards the inner courtyard, and the windows are long and narrow. This shape is very effective to obscure the sunrays. A very thick wall made a significant shadow over the opening. There are generally small openings for ventilation situated on the wall opposite to the courtyard near the ceiling and which remain always open during the summer.

Simulation

The aim of this paper is to examine the positive points as well as the failures of this type of housing. Detailed simulation have been carried out using the computer program TRNSYS. TRNSYS is a simulation environment and an open modular structure for the transient simulation of system used to validate new energy concepts. A TRNSYS project is typically set up by connecting components graphically in the simulation studio [18].

In the first step, we have introduced the geometrical description and the thermo-physical characteristics of the construction materials currently used in this region. The models studied are shown in Figure 5 for the traditional housing, WD and AD, and the MH. For the traditional housing, the exterior walls are made with stone of thickness of 40 cm, the interior partition walls are composed of 30-cm adobe, and the roof is in adobe of 17-cm thickness with a wooden false ceiling of 4 cm. The window glazing is clear glass of 4-mm thick. For the modern construction type, the exterior walls are made of double layer of 20-cm hollow perpend, the partition walls are in hollow perpend of 15 cm, and the roof is in hollow blocks of 17 cm. For each type of constructions, the floor is on full ground in a concrete slab with tiling of 2 cm. Since our aim is to assess

Table 1 Percentage of hours in year obtained by classification of the temperatures

House type	Zone	Tmax Tmin (°C)	34°C < T (%)	32°C < T < 34°C (%)	30°C < T < 32°C (%)	18°C < T < 30°C (%)	15°C < T < 18°C (%)	T < 15°C (%)
WD	Zone 1 (north)	35.71 7.63	1.08	5.81	7.12	42.35	13.34	30.30
	Zone 2 (west)	36.20 7.10	1.71	5.97	7.21	42.81	11.95	30.36
	Zone 3 (south)	36.72 7.72	3.35	6.14	6.49	44.06	14.18	25.91
	Zone 4 (east)	36.25 7.73	1.98	6.21	7.05	43.73	12.32	28.60
AD	Zone 1 (north)	33.54 3.07	0	2.94	10.39	44.50	11.73	31.76
	Zone 2 (west)	34.23 3.32	0.12	4.58	10.20	42.5	11.80	30.89
	Zone 3 (south)	35.36 3.49	1	8.16	7.7	42.96	12	27.08
	Zone 4 (east)	34.32 3.42	0.5	6	9.26	43	11.50	29.95
MH	Zone 1 (north)	38.69 6.7	16.59	5.78	5.17	36.64	9.58	26.23
	Zone 2 (west)	38.86 6.78	16.67	5.91	5.41	41.8	9.73	25.85
	Zone 3 (south)	39.63 8.19	17.44	5.44	5.94	39.01	11.56	20.52
	Zone 4 (east)	39.50 8.45	19.81	4.97	6.25	35.99	11.15	21.82

AD, ain-ad-dar; MH, modern housing; WD, wast-ed-dar.

and compare the thermal comfort of three different housing, therefore the same condition of occupant is set for the three dwellings. We have chosen the most realistic scenario of occupation considering six people as shown in Figure 6.

The metabolic energy of the inhabitants has been taken as 1.5 met during 8:00 am to 11:00 pm in summer and 8:00 am to 9:00 pm in winter and 1 met at night. For summer and winter, 0.5 and 1.5 clo have been considered as thermal resistance of their clothing, respectively.

The relative air speed is about 0.1 m/s. Infiltration is defined as an input according to the model of ASHRAE; the number of change per hour (ACH) is given by [19] in the following:

$$ACH = K_1 + K_2(T_{zone} - T_{amb}) + K_3 V_w \quad (1)$$

where T_{zone}, T_{amb}, and V_w are the zone temperature, ambient temperature, and wind speed, respectively. K_1, K_2, and K_3 are the empirical constants that depend on the airtightness for different construction qualities. Regarding the windows, in summer, the shutters and glazing are closed at daytime and open at night. In winter, the shutters are open during day and closed at night; the glazing being always closed. The free internal gains are due to the equipments (refrigerator, cooker, TV, and PC), and the lighting power are 20 W/m^2 for each room. TRNSYS was run hourly using a typical metrological year data recorded on the site of Bechar. We should mention here that the simulation has been done during a whole year and the ASHRAE model was used when the windows are closed. The solar radiation model available in the TRNSYS library (type 109) has been used assuming the surroundings to be black. The corridors have been modeled as thermal zones; therefore, no air movement is considered there.

Results and discussion

The objective of the simulation is to assess the level of climate adaptation of two traditional habitats. In the previous research, it has been found that the typical modern house seems to be inappropriate for a desert climate, and in the summer, the two traditional houses remain most effective to the heat problem for the two types of building materials

Table 2 Percentage of hours in year obtained by classification of the temperatures

House type	Zone	Tmax Tmin (°C)	34°C < T (%)	32°C < T < 34°C (%)	30°C < T < 32°C (%)	18°C < T < 30°C (%)	15°C < T < 18°C (%)	T < 15°C (%)
WD	Zone 1 (north)	36.76 5.72	2.85	6.08	6.49	42.57	13.03	28.97
	Zone 2 (west)	37.45 5.30	3.78	5.92	6.62	42.20	11.15	30.33
	Zone 3 (south)	37.67 5.43	5.43	5.33	6.35	42.69	12.89	27.30
	Zone 4 (east)	37.5 5.72	4.25	5.81	6.67	42.71	12.08	28.48
AD	Zone 1 (north)	34.23 2.63	0.09	3.27	8.91	44.18	11.65	31.89
	Zone 2 (west)	34.76 2.74	0.43	4.21	8.59	43.78	11.72	31.27
	Zone 3 (south)	36.09 2.80	1.82	6.18	7.10	43.93	11.95	29.01
	Zone 4 (east)	34.98 2.84	0.66	5.31	8.24	43.84	11.52	30.42
MH	Zone 1 (north)	37.17 9.02	15.50	7.37	2.64	39.76	10.87	23.91
	Zone 2 (west)	37.14 9.23	15.98	6.94	3.07	39.88	10.71	23.42
	Zone 3 (south)	37.69 11.99	17.37	6	6.11	44.58	16.58	9.34
	Zone 4 (east)	38.72 11.70	22.84	2.89	9.17	38.61	12.56	13.91

Traditional houses made of hollow perpend and modern house of adobe and stone. AD, ain-ad-dar; MH, modern housing; WD, wast-ed-dar.

tested (stone and adobe) (see Tables 1 and 2) [16]. However, the compactness of urban fabric is one of the main characteristics of the ksourienne fabrics in south Algeria.

To understand the effect of the urban compactness on the thermal behavior of the housing, a simulation of the three houses for more compact urban fabric was performed. Different scenarios of compactness have been investigated in this research. However, one case which consists in one house having two similar houses on both sides of the house is presented in this paper. The simulation results are illustrated in Table 3.

As it is shown by Table 3, the two traditional constructions present a small increase in the interval of hours of comfort (18°C < T < 30°C), and the maximal temperature is less than that obtained for the other cases. On the other hand, the minimal temperature is higher compared to the first case. For WD house, the average interval of the hours when T > 34°C has decreased from 2.03% to 0.98%, and there is a slight increase of percentage of hours when 30°C < T < 32°C from 6.96% to 7.02%. The average percentage of the hours when T < 15°C has not changed. However, we notice an increase in number of the hours when temperature is 15°C < T < 30°C from 12.94% to 14.25%.

Good results are obtained for AD house. There is an absence of temperatures higher than 34°C except for the south room. The percentage of hours of comfort (18°C < T < 30°C) presents an average of 48.48% which is much higher than the first case (43.24%). The compactness of the urban fabric gives a low percentage of hours with 32°C < T < 34°C equal to 1.49% with a decrease of 3% compared to the first case. However, for the modern house, we notice a great increase of hours when T > 34°C from 17.62% to 20.24% and a decrease in the interval of comfort from 38.36% to 34.65%. Therefore, the compactness of the urban fabric does not have a significant effect on the improvement of thermal comfort for the modern house made with hollow perpends.

We changed the construction material, and we carried out the simulation for the three houses keeping the same compactness but using the hollow perpend for traditional

Table 3 Percentage of hours in year obtained by classification of the temperatures for more compact urban fabric

House type	Zone	*Tmax Tmin* (°C)	34°C < T (%)	32°C < T < 34°C (%)	30°C < T < 32°C (%)	18°C < T < 30°C (%)	15°C < T < 18°C (%)	T < 15°C (%)
WD	Zone 1 (north)	34.97 9.64	0.33	4.34	7.39	43.65	16.07	28.13
	Zone 2 (west)	35.43 8.43	0.73	4.62	7.39	43.66	13.36	30.21
	Zone 3 (south)	35.95 8.11	2.13	5.97	6.18	43.71	14.24	27.76
	Zone 4 (east)	35.47 8.51	0.75	5.18	7.15	44.10	13.35	29.45
AD	Zone 1 (north)	33.30 3.41	0	0.33	5.60	50.28	12.34	31.44
	Zone 2 (west)	33.86 3.46	0	0.60	6.89	49.28	12.20	31.02
	Zone 3 (south)	34.80 3.43	0.22	4.13	9.26	46.03	11.88	29.07
	Zone 4 (east)	33.81 3.51	0	0.91	8.43	48.11	11.96	30.58
MH	Zone 1 (north)	38.49 7.44	18.20	5.29	5.04	35.85	10.34	25.28
	Zone 2 (west)	38.81 8.37	21.24	3.11	7	34.16	11.73	22.73
	Zone 3 (south)	39.14 8.61	20.90	3.65	6.63	34.47	11.50	22.85
	Zone 4 (east)	38.76 8.49	21.36	3.10	7.14	34.12	11.72	22.55

AD, ain-ad-dar; MH, modern housing; WD, wast-ed-dar.

constructions and adobe and stone for modern one. The results are presented in Table 4. As it can be seen, the material used affects slightly the overall thermal performance of WD housing in the comfort zone (30°C < T < 18°C). However, better performances have been obtained for AD housing for the same comfort zone.

Traditional houses made of hollow perpend and modern house of adobe and stone. AD, ain-ad-dar; MH, modern housing; WD, wast-ed-dar.

For the modern construction house with stone and adobe, we notice a slight decrease in the number of hours in thermal comfort zone. On the other side, this type of construction presents very high percentage of hours with T > 34°C equal to 36.32%. The heat stored during the day cannot be restored overnight, and the thermal inertia of the construction materials and the free internal gains contribute in the increase of the discomfort. However, this has a significant role on the winter conditions with the percentage of the hours of 0.02% when T < 15°C. This type of construction for a compact urban fabric does not ensure a good air renewal and passive refreshments. For a compact urban fabric, the air velocity is lower, and the outside temperature is higher.

Conclusion

The simulation results show that the modern typical house seems to be inappropriate for the desert climate. Indeed, except the air-conditioning in summer, there is no other solution which can ensure thermal comfort. For the two other traditional houses, in summer, they remain more effective to face the heat problem for the two types of building materials tested (stone adobe and hollow perpend).

The WD house has very important natural ventilation, but it offers less comfort than the AD type. It remains to optimize its components: patio dimension, width of the galleries, use of vegetation, and water (fountain and pond). The phenomenon of nomadism allows a comfortable use of inhabited space according to the days and seasons. It consists in shading the maximum of the outside surfaces. As a future work, we will investigate the effect of the urban fabric compactness on the cooling and heating consumptions.

Table 4 Percentage of hours in year obtained by classification of the temperatures for more compact urban fabric

House type	Zone	Tmax Tmin (°C)	34°C < T (%)	32°C < T < 34°C (%)	30°C < T < 32°C (%)	18°C < T < 30°C (%)	15°C < T < 18°C (%)	T < 15°C (%)
WD	Zone 1 (north)	36.33 7.41	2.89	5.93	6.65	43.74	14.88	26.98
	Zone 2 (west)	36.94 6.45	2.76	5.84	6.71	42.71	11.89	30.04
	Zone 3 (south)	37.13 5.96	4.52	5.51	6.43	41.99	12.45	29.08
	Zone 4 (east)	36.82 6.77	2.97	6.07	6.75	42.95	11.83	29.47
AD	Zone 1 (north)	33.38 2.98	0	0.88	7.64	47.59	12.06	31.81
	Zone 2 (west)	33.87 3.01	0	1.32	8.04	47.13	12.03	31.47
	Zone 3 (south)	34.74 2.91	0.25	4.29	7.51	46.24	11.67	30.03
	Zone 4 (east)	33.71 3.04	0	1.51	8.91	46.65	11.83	31.08
MH	Zone 1 (north)	41.41 15	37.51	7.5	7.21	40.54	7.24	0
	Zone 2 (west)	41.08 15	36.50	6.96	7.77	40.87	7.89	0
	Zone 3 (south)	40.75 14.91	34.41	5.71	8.77	42.40	8.60	0.09
	Zone 4 (east)	41.18 15	36.89	6.88	7.69	36.20	7.22	0

Competing interests
The authors declare that they have no competing interests.

Acknowledgments
The authors acknowledge the University of Bechar for providing support to finalize this paper.

Author details
[1]Civil and Architectural Engineering Department, College of Engineering, University of Sultan Qaboos, P.O. Box 33, Al-Khoudh, Muscat 123, Oman. [2]Department of Sciences, University of Bechar, P.O. Box 417, Bechar 08000, Algeria.

Authors' contributions
MK and NF carried out the calculation and drafted the manuscript. Both authors read and approved the final manuscript.

References

1. Supic, P: Vernacular architecture: a lesson of the past for the future. Energ. Build. **5**, 43–54 (1982)
2. Sharples, S, Malama, A: Thermal performance of traditional housing in the cool season in Zambia. Renew. Energ. **8**, 190–193 (1996)
3. Li, J: The bioclimatic features of vernacular architecture in China. Renew. Energ. **8**, 629–636 (1996)
4. Malama, A, Sharples, S: Thermal performance of traditional and contemporary housing in the cool season of Zambia. Build. Environ. **32**, 69–78 (1997)
5. Vyas, D: Traditional Indian architecture - the future solar buildings. In: International Conference on Passive and Low Energy Cooling for the Built Environment, pp. 19–21., Santorini, Greece (2005)
6. Krishnan, A, Baker, N, Yannas, S, Szokolay, S: Climate Responsive Architecture: A Design Handbook for Energy Efficient Buildings. Tata Mcgraw-Hill Publishing Co., Ltd., New Delhi (2001)
7. Sharma, A, Dhote, KK: Thermal comfort and heritage buildings, Institution of Engineers (India). Archit. Eng. **84**, 1–3 (2003)
8. Fanger, PO: Calculation of thermal comfort: introduction of a basic comfort equation. ASHRAE Transaction **73**(2), 4.0–4.20 (1967)
9. Achyuthan, A, Balagopal, TSP: Green architecture - traditional and modern, Institution of Engineers (India). Archit. Eng. **88**, 2–5 (2007)
10. Edward, S, Kurian, D: Thermal performance of traditional buildings in Kerala. J. Indian Instit. Archit., 7–8 (2008)
11. Hegde, LS, Ananthakrishna, KS: Indoor temperature in vernacular, conventional and alternative technology construction - a comparative investigation. J. Indian Instit. Archit., 16–18 (2008)
12. IS. SP 7: National Building Code of India. Bureau of Indian Standards, New Delhi (2005)
13. Ould-henia, A: Climatic choices and construction Arid and semi-arid regions. PhD Thesis, Ecole Polytechnique Fédérale de Lausanne., (2003)
14. Nezza, S, Gourdache, M: Thermal performance study of bioclimatic design in arid region. World Energy Council, (2003)
15. Ghrab-Marcos, N: Analysis and thermal behaviour simulation of architectural structure regarding solar gain. PhD Thesis, Faculté des sciences de Tunis., (1991)
16. Fezzioui, N, Khoukhi, M, Draoui, B, Aït-Mokhtar, A, Larbi, S: Bioclimatic architectural design of Ksar de Kenadsa: south-west area of Algeria hot and dry Climate. Archit. Sci. Rev. **52.3**, 183–190 (2009)
17. Fezzioui, N, Draoui, B, Aït-Mokhtar, A, Larbi, S: Strategy of bioclimatic architecture in Ksar of Kenadsa. In: Third Conference on Architecture and Sustainability, pp. 8–10., Biskra-Algeria (2008)
18. TRNSYS 16: Solar Energy Laboratory. University of Wisconsin-Madison, USA (2004)
19. American Society of Heating, Refrigerating and Air-Conditioning Engineers: Handbook of Fundamentals. American Society of Heating, Refrigerating and Air-Conditioning Engineers, Atlanta (1997)

3

Enhancement of energy generation from two layer solar panels

Pragya Sharma[1*] and Tirumalachetty Harinarayana[1,2]

Abstract

The enhancement of energy using solar photovoltaic in a limited space is important in urban areas due to increased land cost in the recent years. Although there exist different procedures and methodologies to focus the sunlight on solar panels, we have suggested a new approach to enhance the energy generation from the photovoltaic panels, i.e., by keeping the two layers of photovoltaic panels as collectors of energy one above the other with the same size and orientation. Our results of two layer solar panels have shown about 75% increase in efficiency as compared to a single layer solar panel. This study can also be extended to n number of photovoltaic layers piled up one above the other, if the cost economics are justified with respect to the land cost.

Keywords: Simulation, PVSYST software program, Efficiency, Shade analysis, Land cost

Background

Among all possible alternative energy options, for example, wave energy, geothermal energy, solar energy, wind energy, and hydro energy, solar energy is becoming more popular in India. This is mainly due to (1) the availability of plenty of sunlight in all the seasons and also at all the locations of India and (2) the recent initiation of solar mission by the government of India with attractive incentives to the developers [1]. If we look at the world total renewable energy generation, which is around 5×10^{20} J per year, solar thermal contributes to 0.5%, wind 0.3%, geothermal 0.2%, biofuel 0.2%, and solar photovoltaic (PV) is only about 0.04% as per statistical review of world energy during 2007 [2]. In recent years, the technology upgradation has not only made solar photovoltaic technology price competitive but also as a viable technology. It is projected that by the year 2030, the solar PV electricity will also dominate compared to other sources of energy [3]. From the study growth of photovoltaic, an average about 45% annual increase is noticed during the years 2000 to 2009 [3,4]. From the study of cost economics of a solar photovoltaic power plant, PV modules cost about 45% and the other 55% is due to components, like transformers, cables, inverters, and civil works [5]. Additionally, cost of the power plant also depends on the land value. If the solar power plant is close to the substation near the populated area, the transmission of energy losses will be minimum, but the cost of the land will be high [6]. If the power plant is at a remote location, the cost of the land is low but the energy losses will be high. On the other hand, with less population and in a remote location, the use of energy is limited to the local community. Ideally, the solar power plant needs to be located at a place where the energy generation from the plant can be connected directly to the power grid at an optimum distance from the plant. Apart from the government of India's national solar mission program, the recent initiation by the government of Gujarat to establish the solar photovoltaic plants is commendable. While Gujarat alone crosses 600 MW power through solar, the rest of the country is far behind with only about 200 MW.

The Gandhinagar Photovoltaic Rooftop Programme for solar energy generation using PV modules has set an example by government of Gujarat to save the land cost (see http://www.gpclindia.com/gpcl_rsg/index.html). Another way to save the land cost is to adopt a new methodology to get maximum output from the solar power plant in a limited area. In India, the cost of the land has grown up five to ten times for the last 10 years. This is true in all the urban and semi urban regions of India. In view of the above problem, an attempt has been made to

* Correspondence: pragyasees@gmail.com
[1]GERMI RIIC (Research and Innovation Centre), Gandhinagar, Gujarat 382007, India
Full list of author information is available at the end of the article

study different configuration of solar panels to enhance the energy generation from a solar power plant. For this purpose, the PVSYST modeling software [7] has been used, and a design with a new concept for the solar PV module is suggested, and its advantages over conventional design are discussed.

The rationale behind the present work is to enhance the energy generation for the limited space availability. In recent years several methods have been suggested. For example, in concentrating solar power technology [8], the lens or mirror for concentration of sunlight is used by refracting the rays and focusing them in a small area. In another recent study, a 3D type of solar panels is also reported [9]. Here we present another way of enhanced solar energy with two layers of solar panels as discussed in more detail below. Accordingly, the present study aimed to investigate the advantages of two layer solar panels with the same dimension and orientation lying one above the other. Additionally, the cost benefit analysis is also described to highlight the advantages of considering the suggested solar panel configuration from the present study.

Methods

The new design suggested in the present study is the result of several different design attempts using PVSYST software program. Before presenting the methodology, brief details of the software is presented in the following.

The software program

Among the various software programs, PVSYST simulation software is the most popular to analyze the detailed performance of the plant in field conditions. It can be used for many ways, for example, to investigate different loads on the system, estimate the size of the system, determine the optimal size of the panel, and assess the energy production in the system Various other capabilities and options available in the PVSYST software simulation can be seen in [7]. PVSYST, a personal computer-based software package, can also be used to study the sizing and data analysis of complete PV system. It is used for different designs and sizes of the systems. It can evaluate monthly production and performance. It also performs economic evaluation of the PV system at the design stage itself. Its application performs a detailed simulation and also shading analysis according to several dozens of variables. PVSYST also considers the shading of a diffuse radiation [7,10-12]. The limitation of the software is that it can compute only a single layer of PV module. This

Table 1 Radiation measurement details

Name of site	Ahmedabad
Latitude (degrees)	23.067° N
Longitude (degrees)	72.633° E
Altitude (meter)	55 m
Radiation model	Default (hour)
Temperature model	Default (hour)
Radiation	1981-2000
Measured parameters (WMO nr: 426470)[a]	Gh, H_Dh, H_Gk, H_Dk, H_Bn, Ta
Azimuth	0°
Inclination	25°

[a]World Meteorological Organisation - station code for Ahmedabad, India.

Table 2 Monthly data from Meteonorm

Month	H_Gh	H_Dh	H_Gk	H_Dk	H_Bn	Ta
	(kWh/m^2)	(kWh/m^2)	(kWh/m^2)	(kWh/m^2)	(kWh/m^2)	(°C)
January	147	32	201	42	211	19.6
February	157	36	195	44	193	22.4
March	203	50	227	57	225	27.9
April	214	64	215	67	205	31.6
May	225	78	208	77	202	33.0
June	184	93	165	88	125	31.7
July	139	97	128	92	56	29.1
August	137	92	131	89	63	28.1
September	163	71	171	74	131	28.9
October	171	57	201	65	180	27.9
November	144	39	188	48	181	24.1
December	137	34	188	43	188	20.3
Total	2022	742	2218	786	1961	27.1[a]

[a]Annual average temperature; H_Gh, irradiation of global radiation, horizontal; H_Dh, irradiation of diffuse radiation, horizontal; H_Gk, irradiation of global radiation, tilted plane; H_Dk, irradiation of diffuse radiation, tilted plane; H_Bn, irradiation of beam; Ta, air temperature.

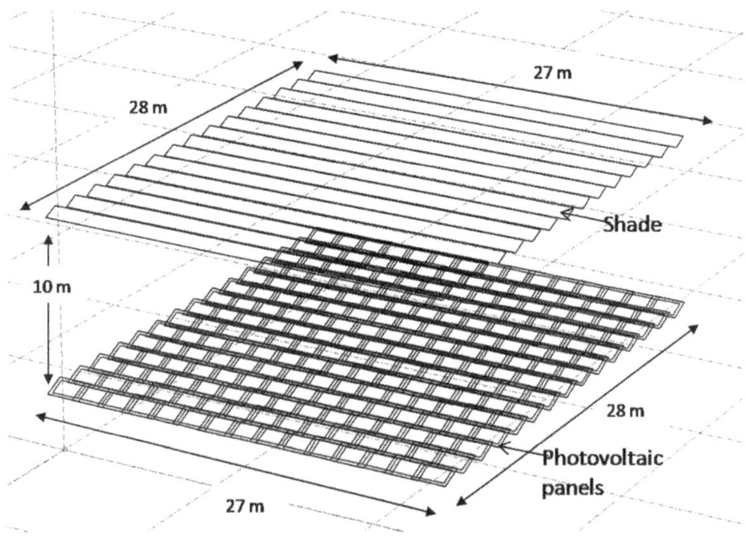

Figure 1 Schematic diagram showing the solar panel and above shade with 10 m height.

means that if there are two layers of PV modules, one above the other, the software has no provision or option to compute the solar energy. Apart from the PVSYST, there are about twelve other software tools currently in use for the simulation e.g., PV f-Chart, SOLCEL-II, PVSYSY, PVSIM, PVFORM, TRNSYS, ENERGY-10 PV, PVNet, PVSS, RETSCREEN, Renew, and SimPhoSys [10,13-24].

The data

For the grid-connected system, the basic input and model parameters required for modeling are the following - PV component database, grid inverter database, geographical site information, and monthly meteorological data for horizontal global irradiance and temperature [7]. In the present study, the meteorological data is acquired from Meteonorm version 6.1.0.23 (see Table 1), a comprehensive climatological database for solar energy applications [7,25-27].

In Table 1, the basic details of radiation measurement for Ahmedabad site are shown. The data have been measured and averaged over a period of 20 years. The radiation data is taken for 20 years period i.e., during 1981 to 2000. The meteorological data considered is given in Table 2. In this table, the information on the monthly average meteorological data of solar radiation, for Ahmedabad, is provided. The values provided are related to irradiation value of global radiation in horizontal direction (H_Gh), irradiation of diffuse radiation horizontal (H_Dh), global radiation in tilted plane (H_Gk), irradiation of diffuse radiation tilted plane (H_Dk), irradiation of the beam (H_Bn) and the air temperature (Ta). These values are used in our study to analyze the shading effects on the panels.

In Figure 1, the solar panel design configuration considered for our model study is shown. It is a schematic diagram with two sets of layers, one lying above the other in such a way that the bottom layer is a solar panel and the top layer is a blank shade with a height separation of 10 m. Since the PVSYST software cannot compute solar energy using the two solar panels with one lying above the other, the top panel is a shade without solar panel but has the same dimension and same orientation of the solar panels in our present study. Later, we will compute the solar energy without shade and add the same with the solar energy with shade to get the total solar energy generated from the two panels. In our model, the DelSolar PV modules (DelSolar Co., Ltd., Miaoli County, Taiwan) have been selected. As a sample,

Table 3 Details of the schematic model shown in Figure 1 and PV module specifications

Number of modules	225
Area of the field (land)	770 m^2 (approximately)
Active area of the modules (sensitive PV area)	372 m^2 (Each module length 1.67 m and width 99 m)
Name of the manufacturer	DelSolar
Technology	Si-polycrystalline
Year	2010
P_{MPP}	230.3 W
I_{MPP}	7.72 A
V_{MPP}	29.8 V
Module and shade tilt	23° N
Separation between rows of solar panel	2 m
Height between solar panel and shade	10 m
Orientation	0° (exactly south)

Table 4 Shading factor table for no shade over panels

Azimuth	−180	−160	−140	−120	−100	−80	−60	−40	−20	0	20	40	60	80	100	120	140	160	180
height																			
90°	1.000	1.000	1.000	1.000	1.000	1.000	1.000	1.000	1.000	1.000	1.000	1.000	1.000	1.000	1.000	1.000	1.000	1.000	1.000
80°	1.000	1.000	1.000	1.000	1.000	1.000	1.000	1.000	1.000	1.000	1.000	1.000	1.000	1.000	1.000	1.000	1.000	1.000	1.000
70°	1.000	1.000	1.000	1.000	1.000	1.000	1.000	1.000	1.000	1.000	1.000	1.000	1.000	1.000	1.000	1.000	1.000	1.000	1.000
60°	1.000	1.000	1.000	1.000	1.000	1.000	1.000	1.000	1.000	1.000	1.000	1.000	1.000	1.000	1.000	1.000	1.000	1.000	1.000
50°	1.000	1.000	1.000	1.000	1.000	1.000	1.000	1.000	1.000	1.000	1.000	1.000	1.000	1.000	1.000	1.000	1.000	1.000	1.000
40°	1.000	1.000	1.000	1.000	1.000	1.000	1.000	1.000	1.000	1.000	1.000	1.000	1.000	1.000	1.000	1.000	1.000	1.000	1.000
30°	1.000	1.000	1.000	1.000	1.000	1.000	1.000	1.000	1.000	1.000	1.000	1.000	1.000	1.000	1.000	1.000	1.000	1.000	1.000
20°	1.000	1.000	1.000	1.000	1.000	1.000	1.000	0.961	0.856	0.824	0.856	0.961	1.000	1.000	1.000	1.000	1.000	1.000	1.000
10°	Behind	Behind	1.000	1.000	1.000	1.000	0.82	0.613	0.51	0.479	0.51	0.613	0.821	1.000	1.000	1.000	1.000	Behind	Behind
0°	Behind	Behind	Behind	Behind	1.000	1.000	0.26	0.125	0.08	0.067	0.08	0.125	0.260	0.675	1.000	Behind	Behind	Behind	Behind

15 solar panel modules in *X* direction and a series of 15 rows of solar panels in another, say *Y* direction, are considered. Such a design is arbitrary and helps to compute parameters quickly. This configuration approximately provides about 50 kW of power output from the PV power plant. However, the same model can be extended to any length as required.

In Table 3, the information and details for the solar panels considered are shown. Details of the solar module and technology, power rating, and related module specifications are also provided. The technology considered is Si-polycrystalline DelSolar photovoltaic module which is available in PVSYST PV module library [7]. Each module can provide a maximum power output of 230.3 W. Accordingly, the 225 number of modules used in our study can provide a power output of about 50 kW. The modules are oriented in the south direction and accordingly, the azimuth angle is assumed as 0°. Both the modules and shade panels are tilted at the same angle of 23°. This tilt is chosen as the latitude (degrees) for the Ahmedabad site is 23.067°.

Shading factor analysis

The shading factor analysis provides the energy loss from photovoltaic panels due to near shading. Near shading means partial shading that affects a part of the panel(s) [7]. The shaded part changes during the day and also over a season. The shading factor is a ratio between the energy generated from the illuminated part and the total area of the field, or inversely, the energy loss [7].

In Table 4, the information of a single module mounting during no shade over the panels is provided. The shading loss is only a function of the sun's height and azimuth for a near shading scene. The values in the table represent the shading factor defined above, and are the ratios of the illuminated part to the total area of the field as a function of height and azimuth of the sun position. The value varies as per the season and time of the day. For example, value 1.000 represents 100% illumination or available radiation over the panels during any particular time of the day and .961 represent 96.1% illumination and so on. In 'no shade' layout, the illumination over the panel is 100% most of the times except during morning and evening hours, when

Table 5 Shading factor table for shade at a height 10 m above the solar panels

Azimuth	−180	−160	−140	−120	−100	−80	−60	−40	−20	0	20	40	60	80	100	120	140	160	180
Height																			
90°	0.194	0.194	0.194	0.194	0.194	0.194	0.194	0.194	0.194	0.194	0.194	0.194	0.194	0.194	0.194	0.194	0.194	0.194	0.194
80°	0.142	0.275	0.602	1.000	0.566	0.190	0.747	0.881	0.585	0.472	0.585	0.881	0.747	0.190	0.565	1.000	0.602	0.274	0.142
70°	0.365	0.623	1.000	0.194	0.919	0.548	0.492	0.637	0.793	0.597	0.792	0.636	0.491	0.547	0.919	0.194	1.000	0.623	0.365
60°	0.303	0.718	0.826	1.000	0.957	0.885	0.748	0.400	0.783	0.514	0.783	0.400	0.748	0.885	0.957	1.000	0.826	0.717	0.303
50°	0.896	0.352	0.899	0.665	0.642	0.847	0.390	0.476	0.538	0.352	0.537	0.475	0.390	0.847	0.641	0.665	0.899	0.352	0.896
40°	0.406	1.000	1.000	0.592	0.643	0.565	0.638	0.786	0.720	0.552	0.720	0.786	0.638	0.564	0.642	0.591	1.000	1.000	0.406
30°	1.000	1.000	1.000	1.000	1.000	0.929	0.797	0.808	0.648	0.678	0.648	0.808	0.796	0.929	1.000	1.000	1.000	1.000	1.000
20°	1.594	1.000	1.000	0.980	1.000	1.000	0.986	0.884	0.778	0.740	0.778	0.883	0.986	1.000	1.000	0.980	1.000	1.000	1.594
10°	Behind	Behind	1.000	0.960	1.000	1.000	0.821	0.613	0.512	0.479	0.512	0.613	0.821	1.000	1.000	0.959	1.000	Behind	Behind
2°	Behind	Behind	Behind	Behind	1.000	0.675	0.26	0.125	0.084	0.067	0.084	0.125	0.26	0.675	1.000	Behind	Behind	Behind	Behind

Table 6 Energy supplied to the grid by single layer with shade and two layer PV panel system

Height between the panels	No shade (a)	1 m (b)	3 m (c)	5 m (d)	10 m (e)
Total annual energy generated by solar panels and shade both of equal dimensions (372 m^2) (in kWh)	77,980	28,100	40,887	45,775	55,942
Total annual energy generated by two layer panels both of equal dimensions (372 m^2) (in kWh)	-	106,080 (a + b)	118,867 (a + c)	123,755 (a + d)	133,922 (a + e)

the height of the sun is 20° or below, with respect to site location, causes maximum shade.

Table 5 presents analysis for a shade at a height of 10 m above the photovoltaic panels. In our study, it is of the same dimension as of the bottom photovoltaic panel. Due to the presence of the shade, the shading factor in Table 5 showed lower value as compared to no shading scene in Table 4. Accordingly, the energy output reduces from the panels.

Annual energy yield

In Table 6, the energy that can be supplied to the grid, for annual generation is shown. The results of the shading analysis of photovoltaic panels for the whole year is shown with no shade and shade at different heights; and the cumulative energy form the average radiation data for the years 1980 to 2000, supplied to the grid for two layer panel system, is shown. The annual total yield of energy supplied to the grid, when there is no shade over the panels, is given in column 2, and the energy generated by the single layer solar photovoltaic system with different shade heights 1, 3, 5, and 10 m is provided in column 3, 4, 5, 6, respectively. Similarly, energy generated by the two layer solar photovoltaic systems with separation values of 1, 3, 5, and 10 m is provided in column 3, 4, 5, 6 respectively. As can be seen, the amount of energy supplied to the grid varies with respect to different height separations 1, 3, 5, and 10 m between the panels. It is observed from the present study, the energy supplied to the grid is at maximum for the case of 10 m height separation.

In Figure 2, a comparative study is shown for the energy supplied to the grid in different months of the year for the radiation data averaged for the years 1980 to 2000. The vertical axis shows the energy supplied to the grid for each month, for example, column 1 in the figure shows the total energy supplied to the grid for different months of the year.

In Figure 3, a comparative study is shown for the two layer photovoltaic panels. The details of the average energy generated per day (December 20), for the radiation

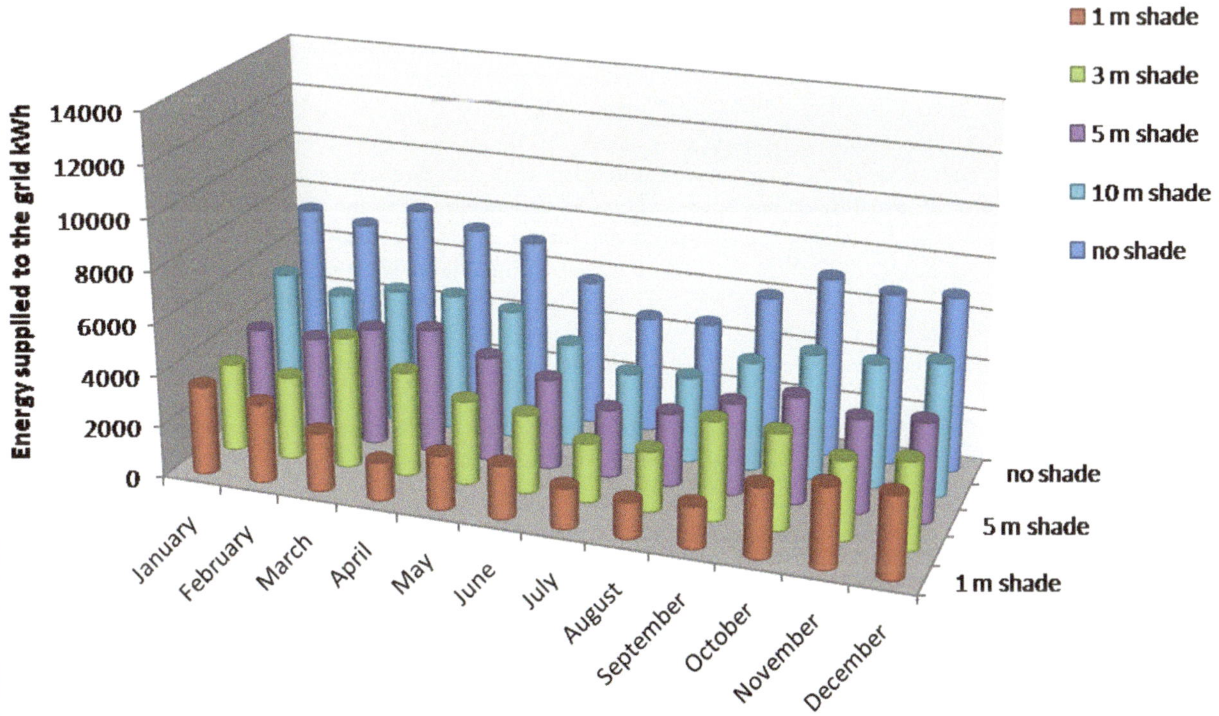

Figure 2 Monthly energy values supplied to the grid with or without the shade over the PV panels.

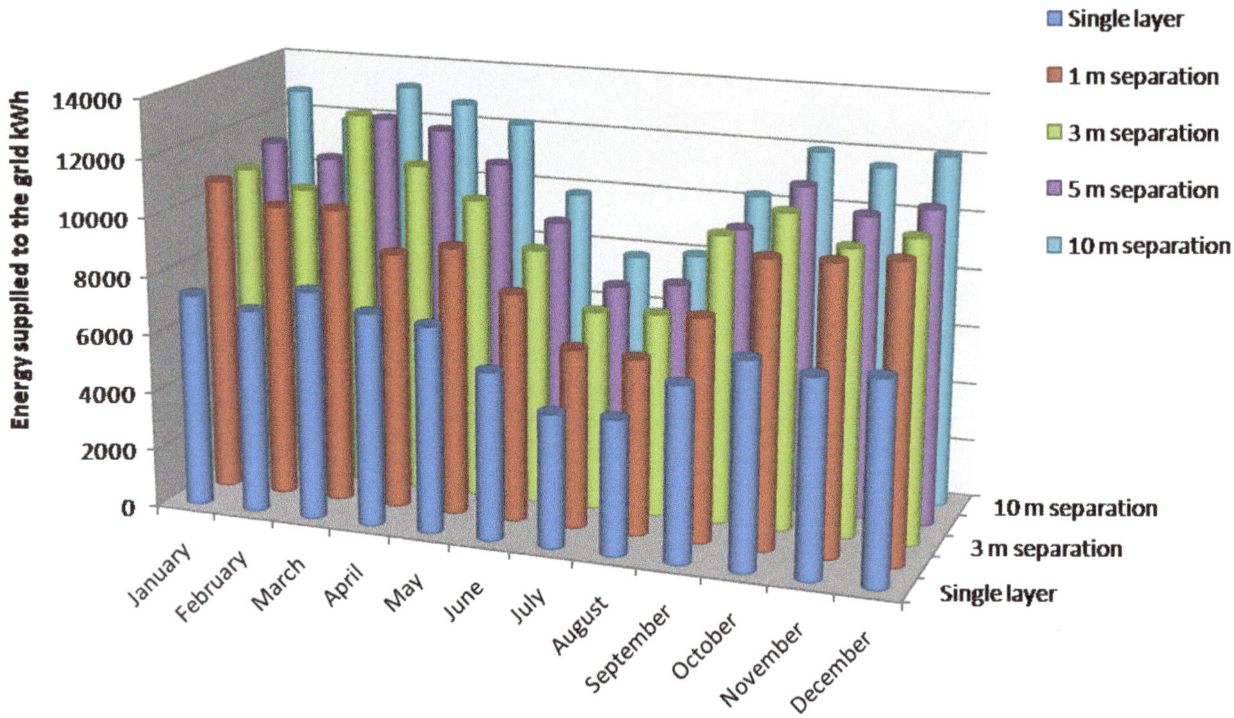

Figure 3 Monthly energy values supplied to the grid by two layer solar photovoltaic panels one above the other.

data averaged over a period of 20 years, are given in Table 7. It can be observed that the amount of energy enhanced with two layer photovoltaic panels increases with the increase in height between the panels. Obviously, the amount of energy supplied to the grid is higher for the two layer photovoltaic system as compared to single layer photovoltaic system. For example, in Figure 3 on *X*-axis, the histogram plot shows the amount of energy generated for the month of January to December. The cumulative energy yield for single and two layers system with different separations are presented as shown in the figure.

Results and discussion

In the following, the details of important results derived from our study are discussed and can be seen in Table 7 and also in Figures 4 and 5. The power output from a single layer solar photovoltaic system, with and without shade, and also for the two layer photovoltaic system separated by 1, 3, 5 and 10 m, are compiled and shown in Table 7. The results shown are for a single day i.e., December 20 and were averaged for the years 1980 to 2000. The increase in efficiency is observed for the two layer solar panels for a configuration presented in Figure 1. The result of two layer solar panels, one above the other, with different height separation between them, showed enhancement of the energy. The energy generation for no shade over the panels is about 250 kWh/day. For a single layer solar panel with shade at 10 m of height, (maximum in our study) the energy generation is about 190 kWh/day. By combining the power from the two panels, the net result increases its efficiency by

Table 7 Result of single solar panel with shade and two layer solar panels one above the other

Number	Height between the panels (meter)	Energy generated at the output of the array by single layer solar panels with shade (approximate kilowatt-hour per day)	Energy generated at the output of the array by two layer solar panels with different heights (approximate kilowatt-hour per day) (s. no.)	Increased in Efficiency for two layer solar panels with respect to single layer solar panel (approximate percentage)
1.	0[a]	252	-	-
2.	1	118	370(1 + 2)	46
3.	3	126	378(1 + 3)	50
4.	5	143	395(1 + 4)	56
5.	10	192	444(1 + 5)	76

[a]Single layer solar panel system.

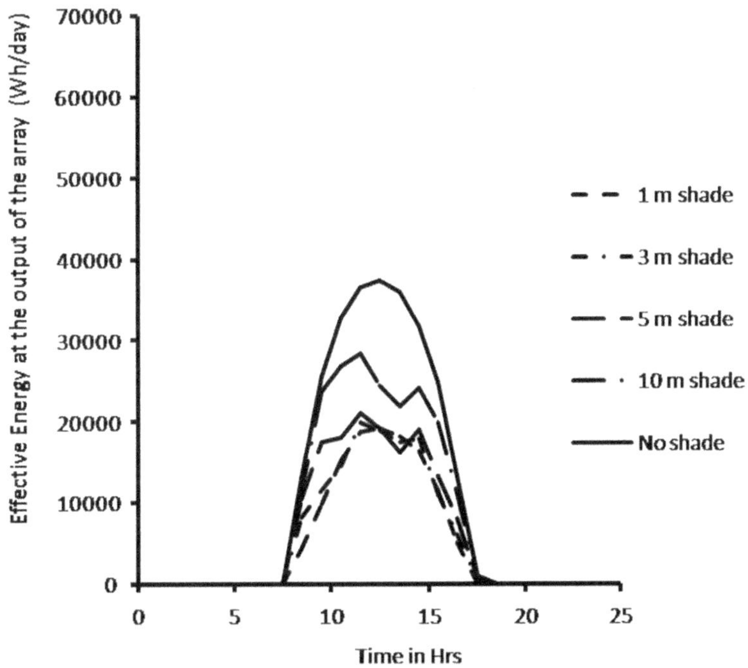

Figure 4 Effective energy at the output of array for single layer solar panel with shade at different heights.

approximately 76% as compared to the power generated by the single layer solar panel without shade. Similarly, one can see that the resultant increase in the efficiency is around 56% for 5 m height, around 50% for 3 m height, and 46% for the 1 m height between the solar panels. The reason for choosing this day (December 20) for modeling from Meteonorm radiation data [25] is due to its clear day in the month of December.

Figure 4 is a graphical plot of the effective energy at the output of array as a function of the local time of a day. In our case, we considered December 20 of Meteonorm radiation data averaged over a period of 20 years, 1981 to 2000. As can be observed from the figure, the obvious result that is closer the shade has less output. For example, for shade over the panel at a height of 1 m showed the lowest energy as compared to 3, 5, and 10 m. Figure 5 shows important result of enhancement in energy generation using two solar panels one above the other. As before, one can see that the enhancement increases with the height between the two layer solar panels. For 10 m height, considering a typical for a 50 + 50 kW photovoltaic system as an example, we observed as much as 65 kW peak around noon. In recent years, semi-transparent solar panels are also under way, and they pass on more solar energy to the bottom panels.

Although our study clearly demonstrates the enhancement of energy generation for the two layer solar panel system as compared to single layer, one should be careful about the cost economics involved for such a system. In Table 8, the economics of the two layer solar panel system have been compiled. The monetary benefit for two layer solar panel system over a single layer solar panel system is shown. For a 50 kW of system, we assumed the land cost (e.g., Ahmedabad, Gandhinagar, Rajkot in Gujarat) as 20 million Indian rupees (INR), module cost of 4 million INR, and 1 million INR for other accessories. Accordingly, single layer solar panel system provides about 10 kWh of energy per day per

Table 8 Monetary benefits

Number	Height between the panels (meter)	Land cost (in million INR)	Solar panel cost (in million INR)	Accessories (in million INR)	Energy generation per day (kWh approximate)	Output of energy per million investment (kWh/million INR/day)
1.	0[a]	20	4	1	252	10.1
2.	1	0	4	.5	370	12.5
3.	3	0	4	.6	378	12.7
4.	5	0	4	.8	395	13.2
5.	10	0	4	1	444	14.8

[a]Single layer solar panel system.

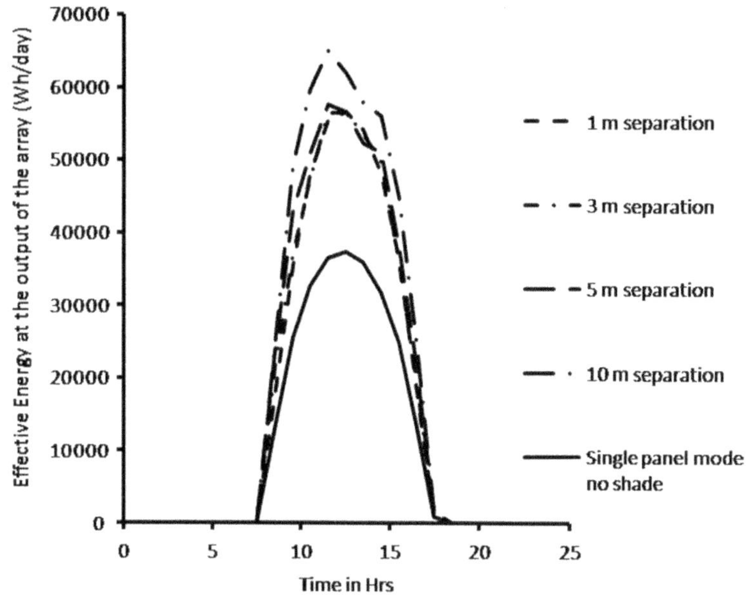

Figure 5 Effective energy at the output of array for two layer solar panel.

million INR of investment. For the two layer of solar photovoltaic system, as the area remains same the land cost are zero. But for the two layer, the added expenditure are the solar panels and other mounting accessories. Thus for 10 m separation one can have 14.8 kWh/ million (INR)/day, which is nearly 50% extra benefit for the one million INR investment.

Conclusions

An attempt has been made in our study on near shade analysis of single and two layer solar panels through modeling for a limited dimension. The energy generation from a single layer solar panel system for a day (December 20 as a sample) is 252 kWh/day for 756 sq m area. It increases up to nearly 445 kWh/day with the two layer solar panels separated by 10 m in the same area. The output varies depending on the separation between the two layers of photovoltaic panels. Due to high land cost in urban areas, the present study is significant. We have shown an increase of over 70% in the output. The present modeling results are limited to the two layer PV system with opaque modules as solar energy collectors for small dimensions as shown in Figure 1. Our result is more applicable to roof tops of the houses or small scale plants. The study, however, can easily be extended to the n layer solar PV panel system of any di mension. However, the justification of the plant cost with respect to solar panels need to be considered. Thus, one needs to have an optimal cost in designing the number of solar panel layers. It should also be based on the foundations of the site location.

Competing interests

The authors declare that they have no competing interests.

Authors' contributions

PS carried out all the computation, system designing, modeling analysis, software simulation, and drafted the manuscript. TH conceived of the study and participated in its design and coordination. Both authors read and approved the final manuscript.

Authors' information

PS, a M. Sc., M. Tech, Research Associate, Solar Research Wing, is involved in research and development program of solar energy technology in GERMI Research Innovation and Incubation Centre (GRIIC), Gujarat, India with more than three and a half year of experience in the field of solar photovoltaic. She is an M. Tech. from School of Energy, Devi Ahilya Vishwavidyalaya (DAVV), Indore, India. She has also done M. Sc. in Physics. She has worked with a dedicated unit from the Ministry of New and Renewable Energy Government of India, Solar Energy Center on 'Design and Development of 20 kWp roof top PV power plant at SEC', in Gurgaon, India. She has presented her research work in various national and International conferences.

TH, a Ph.D, Director of GRIIC, Gujarat and earlier as Head of Magnetotellurics, National Geophysical Research Institute, has done outstanding contributions in the field of deep electromagnetics both on land and also in marine environment. TH is one of the top scientist among the geomagnetism and electromagnetic scientists in the world. His pioneering works are related to oil exploration, geothermal energy assessment, deep crustal studies, tsunami studies, earthquake studies, etc. He is instrumental in introducing a new geophysical technique - marine magnetotellurics - in Gulf of Kutch for hydrocarbon exploration that has delineated 4 km thick buried sediments below the volcanic cover. This work was initiated as a part of international cooperation with Scripps Institution of Oceanography, USA. This has opened up a new scenario of search for oil in the Gulf. He has received a National Mineral Award from Ministry of Mines, Government of India at an early age of 40. He was chosen as the best scientist by Government of Andhra Pradesh. He was elected to the prestigious Russian Academy of Natural Sciences, in Moscow, Russia. He is an elected fellow of the Indian Geophysical Union and Fellow of Andhra Pradesh Academy of Sciences. He has taken up controlled source electromagnetic modeling studies for gas hydrates as a part of his research at University of Texas at Austin, USA as a visiting scientist and also carried out tsunami studies at University of Tokyo, Japan as a visiting professor.

Acknowledgements
PS is very much thankful and express sincere gratitude to Dr. Jayanta Deb Mondol, Ulster University, for providing his quick comments on the manuscript. PS and TH would like to acknowledge all the research technical and scientific staff of GERMI for their encouragement, motivation, and support.

Author details
[1]GERMI RIIC (Research and Innovation Centre), Gandhinagar, Gujarat 382007, India. [2]CSIR-NGRI (National Geophysical Research Institute), Hyderabad 500007, India.

References

1. Goswami, D: India's solar sunrise. Renewable Energy Focus **13**(2), 28–30 (2012)
2. BP Statistical Review of World Energy. (2007). www.bp.com. Accessed 22 January 2011
3. European Photovoltaic Industry association: Solar Photovoltaics competing in the Energy sector, on the road to competitiveness. EPIA (2011). http://www.epia.org/publications/epiapublications. Accessed 21 January 2012
4. European Photovoltaic Industry association: Solar generation 6 solar Photovoltaic electricity Empowering the World. EPIA (2011). http://www.epia.org/publications/epiapublications. Accessed 21 January 2012
5. Sharma, A: A comprehensive study of solar power in India and World. Renew Sustain Energy Rev **15**(4), 1767–1776 (2011)
6. Raman, P, Murali, J, Sakthivadivel, D, Vigneswaran, VS: Opportunities and challenges in setting up solar photovoltaic based micro grids for electrification in rural areas of India. Renew Sustain Energy Rev **16**(5), 3320–3325 (2012)
7. University of Geneva: User's Guide, PVsyst Contextual Help, (1994–2010). www.pvsyst.com. Accessed 8 December 2011
8. Ummadisingu, A, Soni, MS: Concentrating solar power – technology, potential and policy in India. Renew Sustain Energy Rev **15**(9), 5169–5175 (2011)
9. Bernardi, M, Ferralis, N, Wan, JH, Villalon, R, Grossman, JC: Solar energy generation in three dimensions. Energy and Environmental Science **5**, 6880–6884 (2012). doi:10.1039/C2EE21170J
10. Mondol, JD, Yohanis, YG, Norton, B: Comparison of measured and predicted long term performance of grid a connected photovoltaic system. Conversion and Management **48**(4), 1065–1080 (2007)
11. International Energy Agency: SI IC State-of-the-art of digital tools used by architects for solar esign. Task 41 – Solar Energy and Architecture T.41.B.1. http://iea-shc.org. Accessed 25 February 2012
12. Bayod-Rújula, AA, Ortego-Bielsa, A, Martínez-Gracia, A: Photovoltaics on flat roofs: energy considerations. Energy **36**(4), 1996–2010 (2011)
13. Klise, GT, Stein, JS, Sandia National Laboratories Report SAND2009-8285: Models Used to Assess the Performance of Photovoltaic Systems., (2009). http://photovoltaics.sandia.gov. Accessed 14 and 18 March 2012
14. Hoover, ER, Sandia National Laboratories Report: SOLCEL II-an improved photovoltaic system analysis program. SAND, 79–1785 (1980). http://www.sandia.gov/index.html. Accessed 12 January 2009
15. Mermoud, A: Use and validation of PVSYST, a user-friendly software for PV-system design, pp. 736–739. 13th European Photovoltaic Solar Energy Conference, Nice, France (1995)
16. King, DL, Dudley, JK, Boyson, WE: PVSIM: a simulation program for photovoltaic cells, modules, and arrays, pp. 1295–1297. 25th IEEE Photovoltaic Specialists Conference, Washington (DC, USA) (1996)
17. Menicucci, DF, Fernandez, JP, Sandia National Laboratories Report: User's manual for PVFORM: a photovoltaic system simulation program for stand-alone and grid interactive applications. SAND, 85–0376 (1988). http://prod.sandia.gov Accessed 5 February 2012
18. Mondol, JD, Yohanis, YG, Smyth, M, Norton, B: Long term validated simulation of a building integrated photovoltaic system. Solar Energy **78**(2), 163–176 (2005)
19. National Renewable Energy Laboratory Report, Balcomb, JD, Hayter, SJ, Weaver, NL: Hourly simulation of grid connected PV systems using realistic building loads. NREL/CP, 550–29638 (2001). http://www.nrel.gov. Accessed 25 February 2012
20. Bishop, JW: Computer simulation of the effects of electrical mismatches in photovoltaic cell interconnection circuits. Solar Cells **25**, 73–89 (1988)
21. Goldstein, LH, Case, GR: PVSS—a photovoltaic system simulation program. Solar Energy **21**, 37–43 (1978)
22. Bakos, GC, Soursos, M, Tsagas, NF: Technoeconomic assessment of a building-integrated PV system for electrical energy saving in residential sector. Energy Build **35**, 757–762 (2003)
23. Woolf, J: Renew: a renewable energy design tool for architects. Renew Energy **28**, 1555–1561 (2003)
24. Schmitt, W: Modeling and simulation of photovoltaic hybrid energy systems optimization of sizing and control, pp. 1656–1659. 29th IEEE photovoltaic specialists conference, New Orleans (LA) (2002)
25. Global Meteorological Database. Meteonorm, Switzerland (2012). www.metonorm.com, version 6.1.0.23. Accessed 17 Feb 2012
26. Marinopoulos, AG, Bouhouras, AS, Papaioannou, IT, Labridis, DP, Exioglou, MD, Theodoridis, AN: Siting and installation of PV systems in Greece and their contribution in the reliability of the distribution network. Electricity Market, 5th International Conference on European **28–30**, 1–6 (2008). doi:10.1109/EEM.2008.4579059
27. Hatwaambo, S, Jain, PC, Perers, B, Karlsson, B: Projected beam irradiation at low latitudes using meteonorm database. Renew Energy **34**, 1394–1398 (2009)

Analysis of free and forced convection in airflow windows using numerical simulation of heat transfer

Mohammad Ghadimi[1*], Hossein Ghadamian[1], Aliasghar A Hamidi[2], Farivar Fazelpour[3] and Mehdi A Behghadam[4]

Abstract

The present paper describes a two-dimensional finite volume numerical simulation of flow and heat transfer in airflow windows by free and forced convection techniques. The governing equations are the fully elliptic, Reynolds-averaged Navier–Stokes equations. The simple algorithm is employed to correct the pressure term. The second-order upwind scheme is used to discretize the convection terms. The (k-ϵ/RNG) turbulence model is applied for the flow simulation. The mesh used is the body-fitted, multi-plane grid system. Results on the variations of velocity and temperature profiles with geometrical parameters, at different temperature and velocity, for heat transfer by free and forced convection techniques are presented. Comparisons of the present results on temperature distribution for forced convection and for free convection with the available experimental forced convection data indicate that the airflow-influenced forced convection methods are considerably enhanced.

Keywords: Advanced envelope, Airflow window, Building energy simulation, Energy performance, Fenestration, Office building, Solar energy, Wind

Background

Improvement of building performance energy can be achieved by using airflow windows. Airflow windows in buildings are used to improve the indoor thermal comfort and/or reduce the energy use of air-conditioning systems. The airflow window is also an invention to improve the thermal comfort and reduce energy costs. It consists of a double window construction with an air cavity in between. Nowadays, the outside window is mostly a double-glazed window, and the inner window is a single glass leaf. On the bottom of the airflow cavity, exhaust air ventilation will be extracted out of the room and transported through the cavity to the exhaust duct, as can be seen in Figure 1. This results in less temperature difference between the inside glass surface and the room. This requires choosing a suitable heat transfer method to prevent missing energy in buildings. The mechanically ventilated variants include an airflow window and a supply-air window. The airflow window is a popular solution in fully conditioned office buildings.

The first studies on airflow windows were published in the 1950s in Scandinavia (Brandle and Boehm [1]; Ripatti [2]; Park et al. [3]). The issue was to improve the energy efficiency and the thermal comfort of residential fenestration. In 1957, the first patent related to airflow windows was filed in Sweden. In 1967, the EKONO Company built the first office building equipped with airflow windows in Helsinki, Finland (Brandle and Boehm [1]; Park et al. [3]).

The spirit for further development is to be found in the energy crises of 1973 and 1979. Suddenly, energy efficiency and thermal comfort were no longer an exclusive issue for Nordic climates. Experimental data for a Japanese building with an airflow window is available from Baker et al. [4], Ziller [5] and Busselen and Mattelaer [6]. Tracer gas techniques make it possible to determine the airflow rate in both naturally and mechanically ventilated active envelopes without interference with the driving forces. Busselen and Mattelaer [6], however, point out that it is difficult to illustrate the highly

* Correspondence: m.ghadimi@riau.ac.ir
[1]Department of Energy Engineering, Science and Research Branch, Islamic Azad University, P.O. Box 147789385, Tehran, Iran
Full list of author information is available at the end of the article

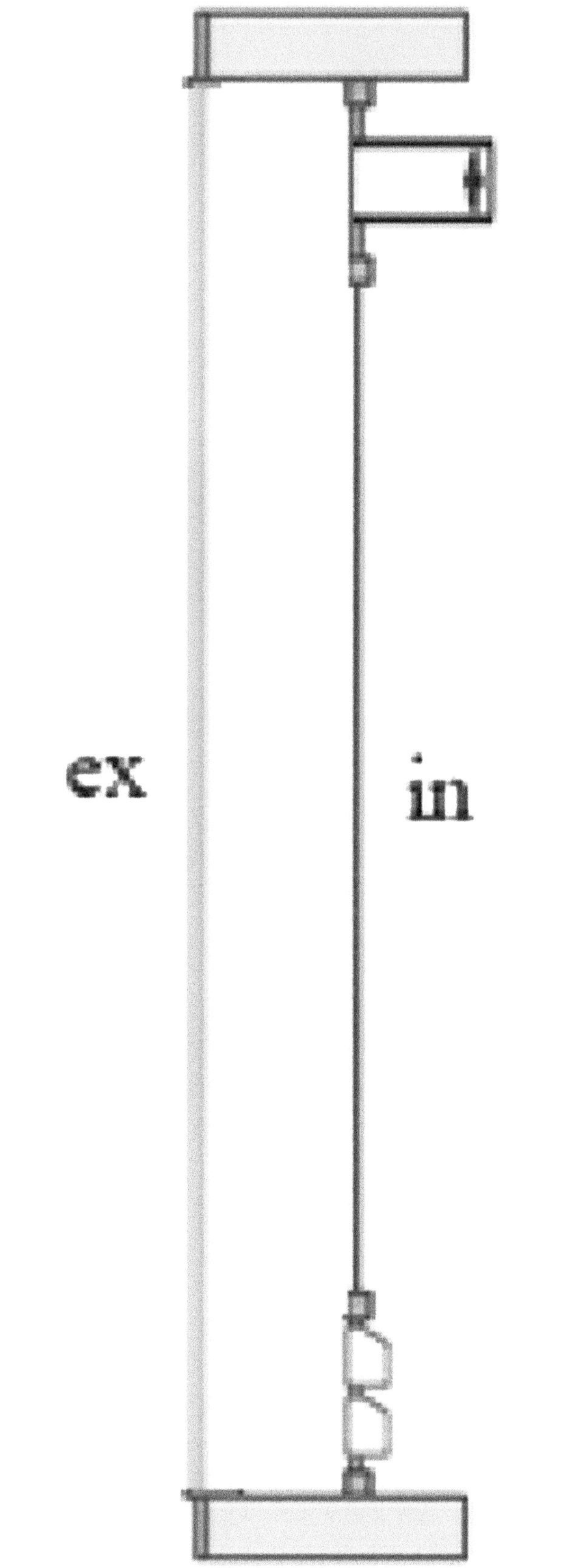

Figure 1 Experimental model of an airflow window.

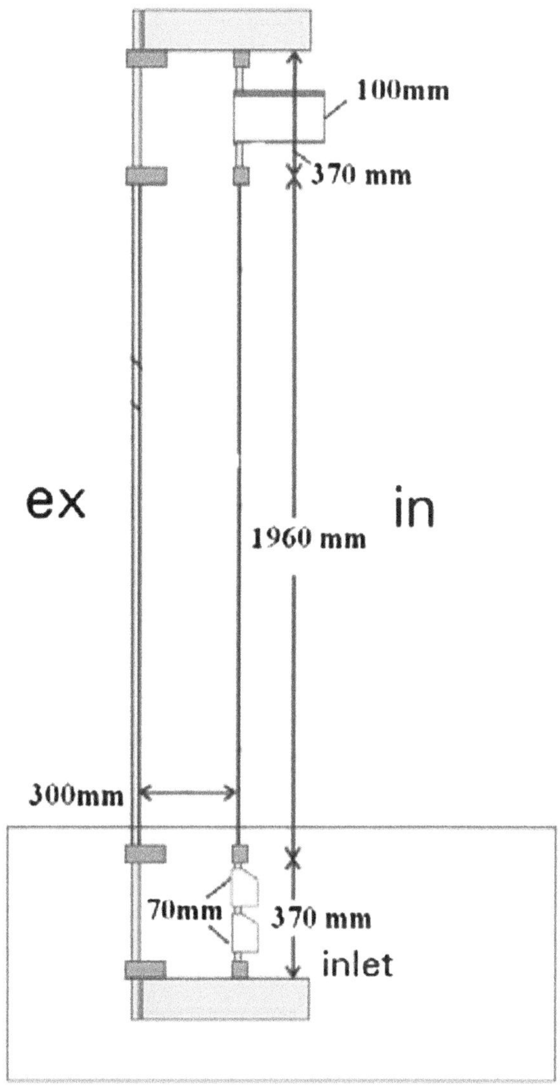

Figure 2 Detailed dimensional model.

fluctuating airflow rate accurately with the constant emission technique. The same conclusions could be drawn from the measurements at the Postcheque building.

Methods

Governing equations

The time-averaged fully elliptic, Reynolds-averaged Navier–Stokes (RANS) equations governing the motion of an incompressible, steady and constant property flow are expressed in a general form as follows [7]. The conservation of mass, momentum and energy is defined in Equations 1, 2 and 3, respectively.

$$\frac{\partial(\rho u_j)}{\partial x_j} = 0 \tag{1}$$

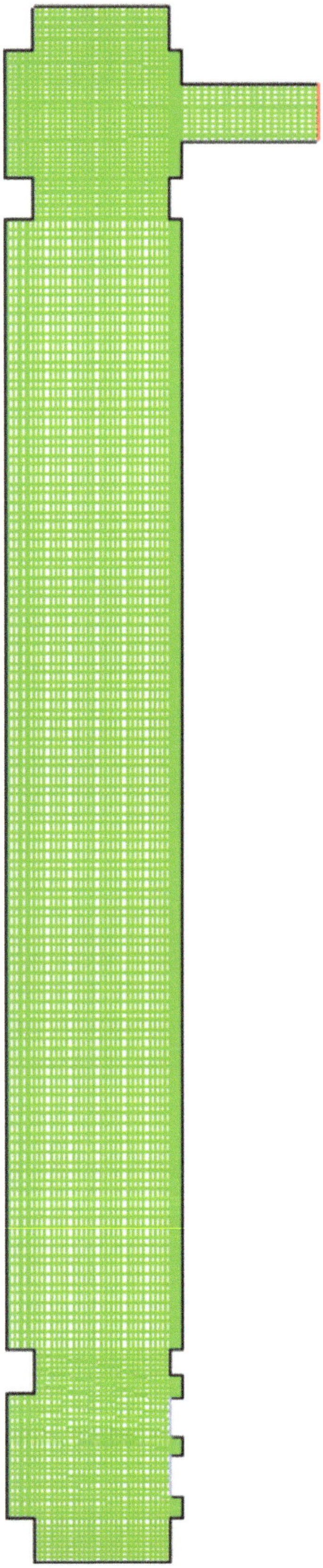

Figure 3 Structured mesh for model.

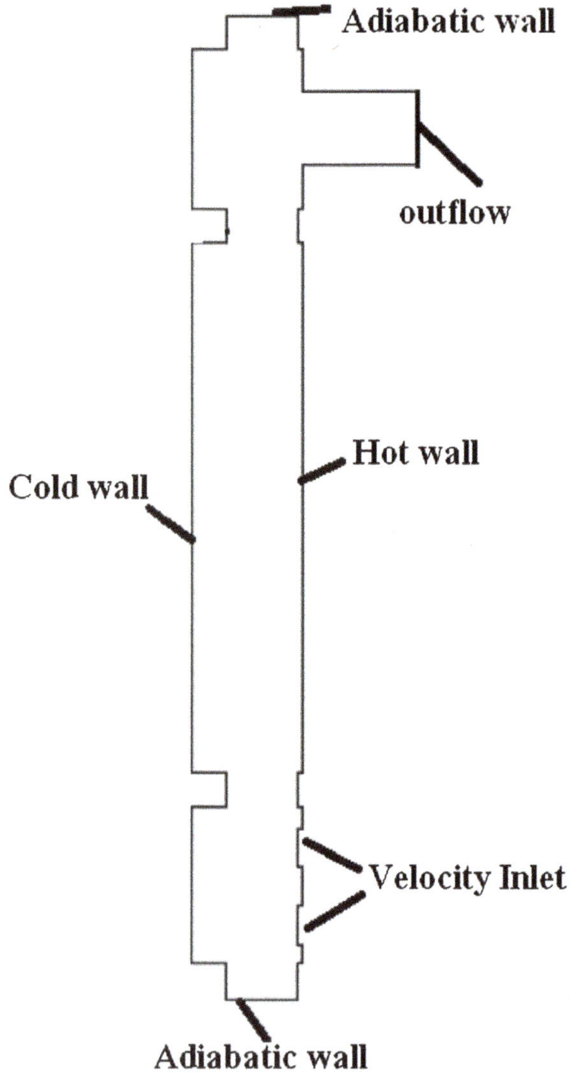

Figure 4 Geometry of the computational domain.

where ρ is density and u is velocity vector.

$$\frac{\partial(\rho u_i u_j)}{\partial x_j} = -\frac{\partial P}{\partial x_j} + \frac{\partial}{\partial x_j}\left[(\mu + \mu_t)\frac{\partial u_j}{\partial x_j}\right] \quad (2)$$

where P, μ and μ_t are static pressure, laminar viscosity and turbulent eddy viscosity, respectively.

$$\frac{\partial(\rho u_i T)}{\partial x_i} = \frac{\partial}{\partial x_i}\left[\left(\frac{\mu}{\mathrm{Pr}} + \frac{\mu_t}{\mathrm{Pr}_t}\right)\frac{\partial T}{\partial x_j}\right] \quad (3)$$

where Pr and Pr_t are the molecular and turbulent Prandtl numbers, respectively. The proper choice of a turbulence model is of great importance, and it depends on many factors such as the flow physics, the required level of accuracy and the available computational resources. The (k-ϵ/Renormalization group (RNG)) turbulence model [7] is based on the model (RNG)

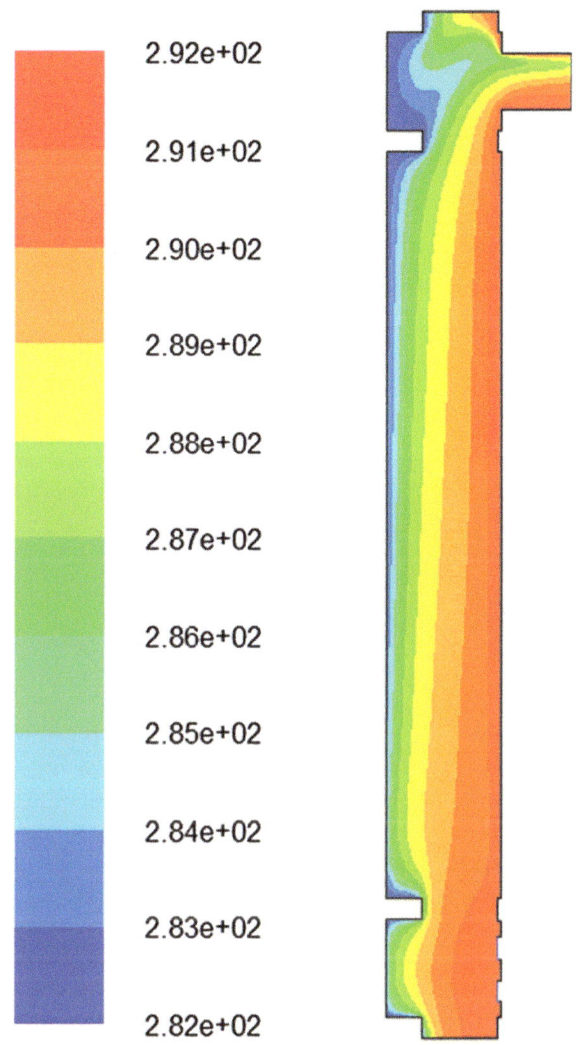

Figure 5 Temperature distribution in the airflow window (present simulation).

equations for the turbulence kinetic energy, k, and its dissipation rate, ϵ. The turbulence kinetic energy and its rate of dissipation are obtained from the following transport equations.

$$\frac{\partial}{\partial x_i}(\rho k u_i) = \frac{\partial}{\partial x_i}\left(\Gamma_k \frac{\partial k}{\partial x_j}\right) + G_k - Y_k \tag{4}$$

$$\frac{\partial}{\partial x_i}(\rho \epsilon u_i) = \frac{\partial}{\partial x_j}\left(\Gamma_\epsilon \frac{\partial \epsilon}{\partial x_j}\right) + G_\epsilon - Y_\epsilon + D_\epsilon \tag{5}$$

In these equations, G_k represents the generation of turbulence kinetic energy due to mean velocity gradients. Y_k represents the contribution of the fluctuating dilatation in compressible turbulence to the overall dissipation rate. The relevant boundary conditions, on the basis of Baker et al.'s [4] experimental work, are outlined in one of the following sections. In the RNG model, diffusion in k and ϵ equations appears as:

$$(\mu + \mu_t) \times \alpha \tag{6}$$

while in the standard k-ϵ model, it is given by

$$\mu + \mu_t/\mathrm{Pr} \tag{7}$$

For the new implementation, a UDF is needed to define a Prandtl number, Pr, as

$$\mathrm{Pr} = \mu_t/((\mu + \mu_t) \times \alpha - \mu) \tag{8}$$

in order to achieve the same implementation as the original RNG model.

Flow geometry

Figure 2 depicts a schematic view of the present airflow window model. It is similar in concept to Baker et al.'s [4] experimental model (see Figure 1). It comprises an inlet region, an outlet region and a main channel airflow. The airflow stream enters through the inlet region into the main channel airflow when it flows along the chamber, eventually flowing out through the outlet region. The main chamber is a rectangular cross section with a width equal to 300 mm and a height equal to 2,700 mm, the inlet vent has two equal widths, 70 mm, and the outlet region has width equal to 10 mm (see Figure 2). The airflow enters the channel at a rate of $G_i = 110$ m^3/h [8]. The origin of the coordinate system is positioned at the corner of the model.

Grid generation

Computational fluid dynamics (CFD) methods, based on Cartesian or cylindrical coordinate systems, have certain limitations regarding irregular geometries [9], such as the geometry of a jet cross section, which is connected to a channel. Methods based on the body-fitted grid or the non-orthogonal grid systems do not have such limitations, and hence, in the present work, the body-fitted multi-block grid is used. The GAMBIT software [10] is employed to create the present complex geometry and then to generate the appropriate mesh. Four structured mesh sizes (hexahedral elements) are used to verify the independence of present numerical solution from the mesh size. The optimum grid size for various plates is chosen as follows: inlet region (30 × 20), skin (222 × 30) and outlet region (60 × 30). Finally, the total of 9,060 mesh size is applied throughout the present work. Figure 3 represents a typical mesh generated around and within an outlet region shown in the present computational domain (see Figure 4).

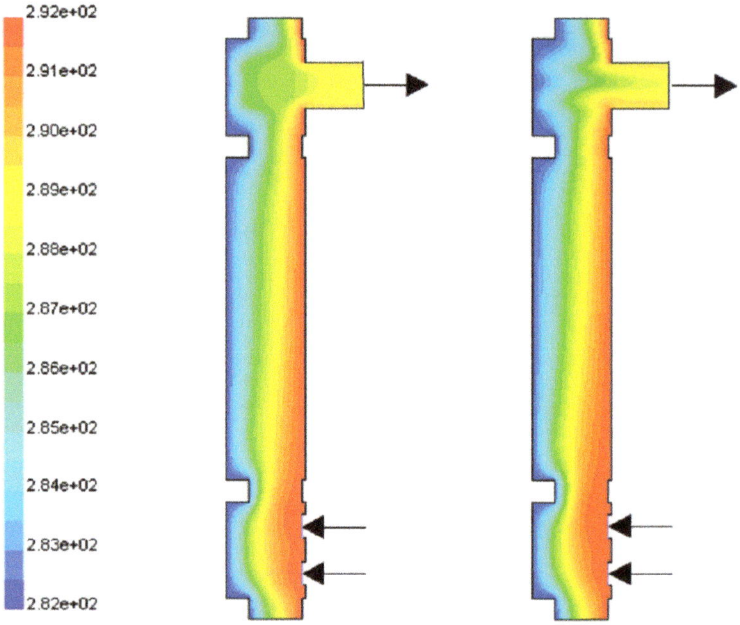

Figure 6 **Temperature distribution in the airflow window.**

Numerical details

The RANS and turbulence transport equations are solved numerically using the FLUENT code [11], which has been extensively validated against experimental data for many flow cases. The present numerical solution employs the finite volume-based finite difference approach [12,13], and it includes the following details: (1) Solution of the governing equations on the basis of three-dimensional Cartesian coordinates with adaptive, variable density grids is employed. (2) The second-order upwind scheme is used to discretize the convection terms. (3) The SIMPLE algorithm [12,13] is employed to correct the pressure term. (4) The (k-ϵ/RNG) turbulence model is used. (5) The mixing is considered as a non-reacting flow. (6) The value of 1×10^{-8} is considered as the convergence criterion for the energy equation, and 1×10^{-6} for the other equations.

Boundary conditions

The present computational domain, as illustrated in Figure 5, has five boundaries: two mass flow inlet for the airflow window and an outlet pressure plane for the outflow, two walls for the top and bottom of the airflow windows, a hot wall at the right-hand side of the airflow

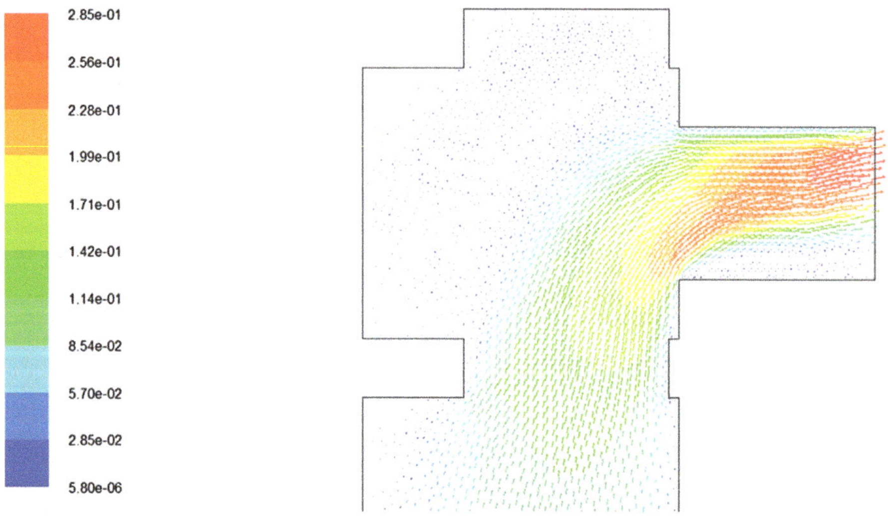

Figure 7 **Velocity vectors just passing the upper frame (present simulation).**

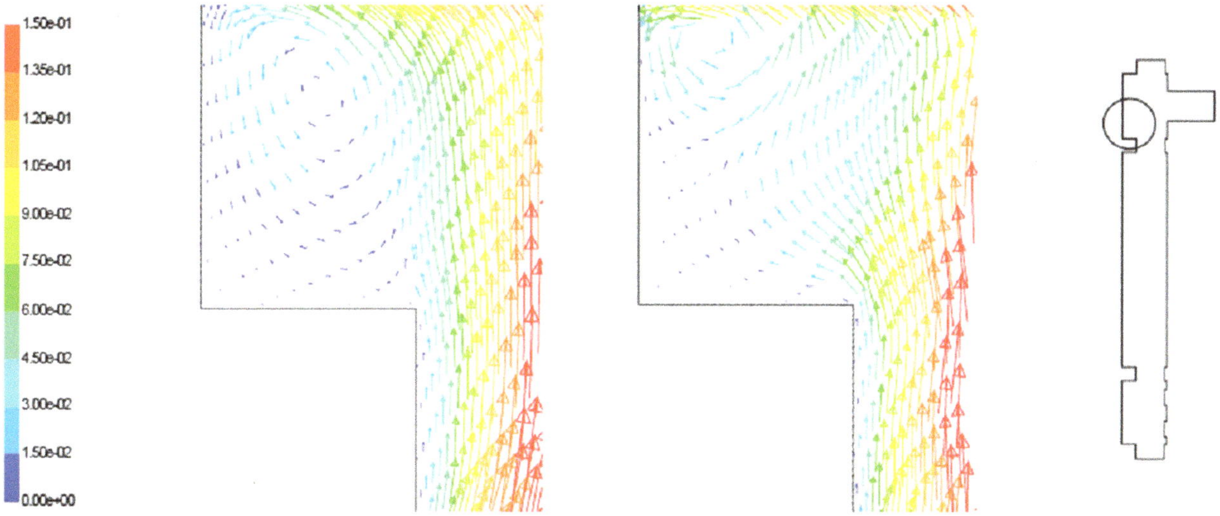

Figure 8 Velocity vectors just passing the upper frame.

window and a hot wall at the left-hand side of the airflow window. At the inlet boundaries, uniform profiles of velocity and temperature are specified from the experimental data [4], where the relevant boundary conditions are taken as follows:

Inlet vent: $\underline{G}_i = 110\ m^3/h$
Hot wall: $T_H = 292.15$ K
Cold wall: $T_C = 282.15$ K

The turbulence intensity of inlet and length measure 0.1% and 10 mm, respectively. At the outlet region, sufficiently far from the inlet vent, a pressure outlet is considered. The walls (including top and bottom walls) are all considered as adiabatic; the walls (including right and left walls) are all considered as constant temperature.

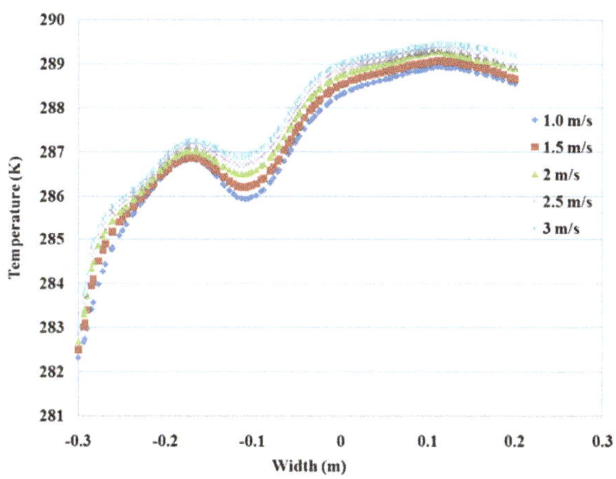

Figure 9 Temperature variations in relation to airflow window width for different velocity.

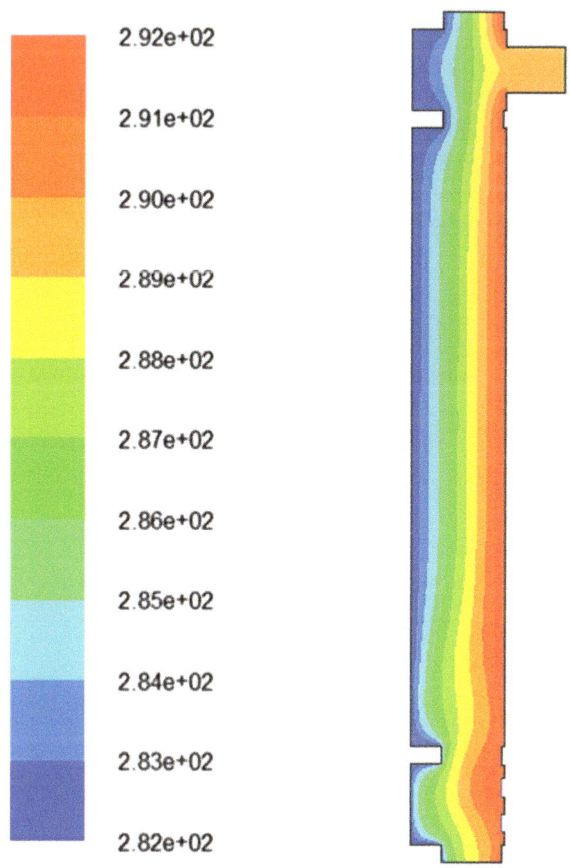

Figure 10 Temperature distribution in the airflow window for free convection.

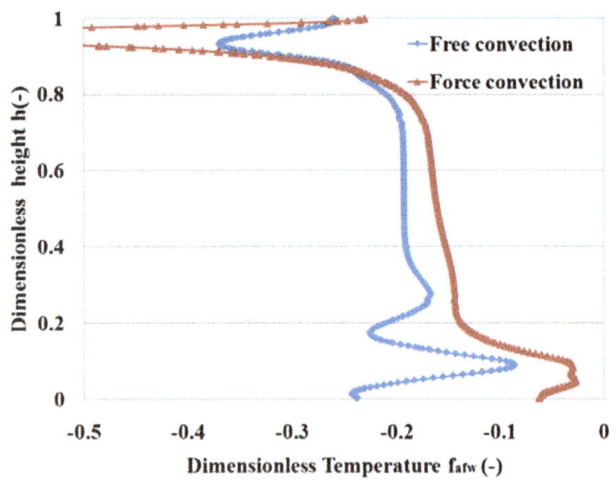

Figure 11 Cavity air vertical dimensionless temperature profile in the airflow window for free and forced convection.

Adiabatic effectiveness

The influence of the airflow rate on the vertical temperature difference in the airflow windows is illustrated, and it is defined as follows [4]:

$$f_{\text{afw}} = \frac{T_{\text{x}} - T_{\text{i}}}{T_{\text{i}} - T_{\text{e}}} \tag{9}$$

where T_x is the temperature at height x, and T_e and T_i are the exterior and interior temperature (in degree Celsius). The results on f_{afw} are presented for the middle jet hole (center-line plane) in a row, in the case of film cooling technique only.

Results and discussion

Figures 5 and 6 represent the temperature distribution in the airflow window, and a comparison is made between the present numerical results and the available experimental data of Baker et al. [4]. Figures 7 and 8 represent the velocity vectors just passing the upper frame in the airflow window, and a comparison is made between the present numerical results and the available experimental data of Baker et al. [4]. There exists a reasonably good agreement, and hence, the present numerical solution may also be used for the other analysis.

Figure 9 illustrates the typical temperature variations in relation to airflow window width for different velocity inlet adjacent outlet region. It can be seen that the temperature variations increased with velocity; also, the maximum temperature occurs near the outlet region because of the existent hot wall.

Figure 10 represents the temperature distribution in the airflow window for free convection simulation. It can be seen that the temperature is uniform near the walls, compared to forced convection simulation.

The influence of the airflow rate on the vertical temperature difference in the airflow windows for forced and free convection heat transfer is illustrated in Figure 11. It can be seen that in free convection, the airflow rate is higher compared to forced convection.

The dynamic pressure on the vertical temperature difference in the airflow windows for the forced and free convection heat transfer is illustrated in Figure 12. It can be seen that in free convection, the airflow rate is higher compared to the forced convection.

Figure 13 illustrates the influence of the airflow rate on the vertical temperature difference in the airflow

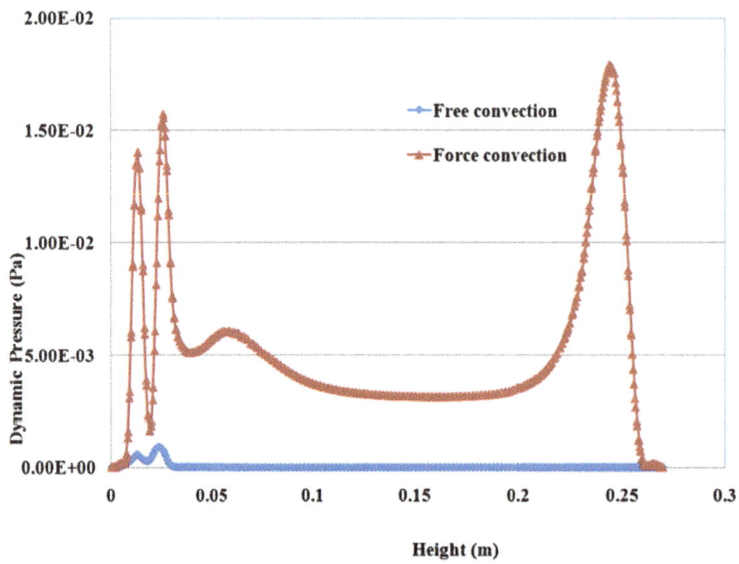

Figure 12 Cavity air dynamic pressure versus height in the airflow window for free and forced convection.

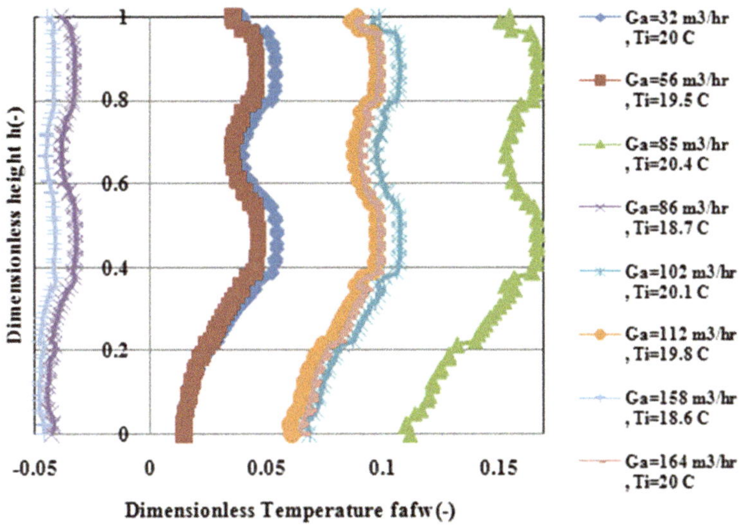

Figure 13 Cavity air vertical dimensionless temperature profile in the airflow window for different airflow rates.

windows for different airflow rates with similar solar radiation. Based on Figure 12, we can assume that the airflow rate in the exterior cavity is lower than in the interior cavity since (1) the vertical temperature gradient is higher, indicating that the air has more time to heat up while flowing through the exterior cavity, and (2) the outlet temperature trends more towards the temperature at the top of the interior cavity, which indicates that a higher amount of air from the interior cavity flows towards the outlet region.

Figure 14 illustrates the influence of the airflow rate on the vertical temperature difference in the airflow windows for different cavity widths. It can be seen that airflow influence is not proportional to width such that it has an optimum amount at 300 mm.

Figures 15 and 16 represent the streamline in the airflow window for forced and free convection simulations. It can be seen that the streamline in forced convection is uniform streamline compared to free convection. However, free convection has more rotation stream line due to high bouncy force.

Conclusions

Overall forced convective heat transfer in an airflow window was calculated using a CFD code and

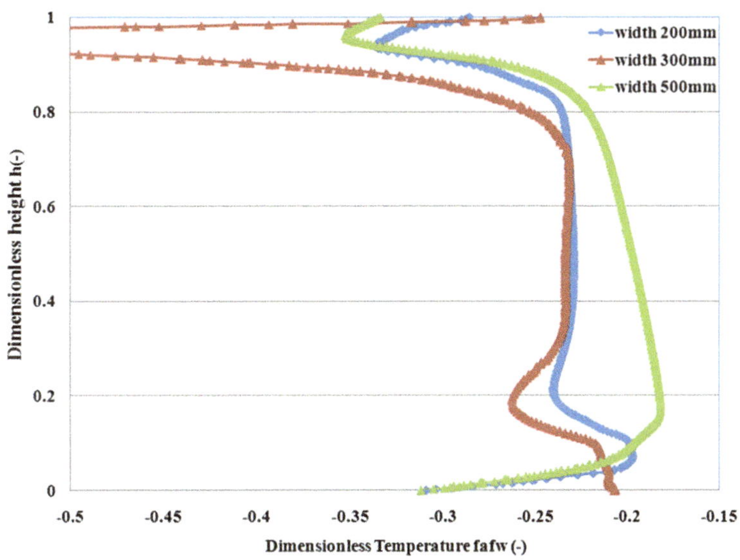

Figure 14 Cavity air vertical dimensionless temperature profile in the airflow window for different cavity widths.

Figure 15 **Variations of streamline in forced convection.**

compared with correlations based mainly on experimental results. The main conclusions may be drawn as follows:

1. A comparison between the present numerical results with the available experimental data is made. There exists a reasonably good agreement, and hence, the present numerical solution may also be used for the energy technique analyses.
2. A comparison between forced convection and free convection results illustrates that airflow influence increases in airflow windows.
3. Comparatively, it can be said that the effect of airflow in airflow window is proportional to inlet temperate and flow rate, but the effect of temperate is higher than the effect of flow rate.
4. Airflow influence in airflow window is not proportional to airflow window width and has an optimum amount at 300 mm.
5. By comparing streamline in forced and free convections, it can be said that streamline in forced convection behaves uniformly and regularly.
6. Effect of turbulence modeling is important, and comparing numerical and experimental results provides an acceptable and reasonably good agreement.

Abbreviations

f: dimensionless temperature (–); *H*: cavity height (m); *h*: convective heat transfer coefficient (W/m^2 K); G_a: airflow rate (m^3/s); *V*: velocity (m/s); *w*: cavity width (m); α: thermal diffusivity (W/m K); T_i: inlet temperature (K); T_C: cold wall temperature (K); T_H: hot wall temperature (K).

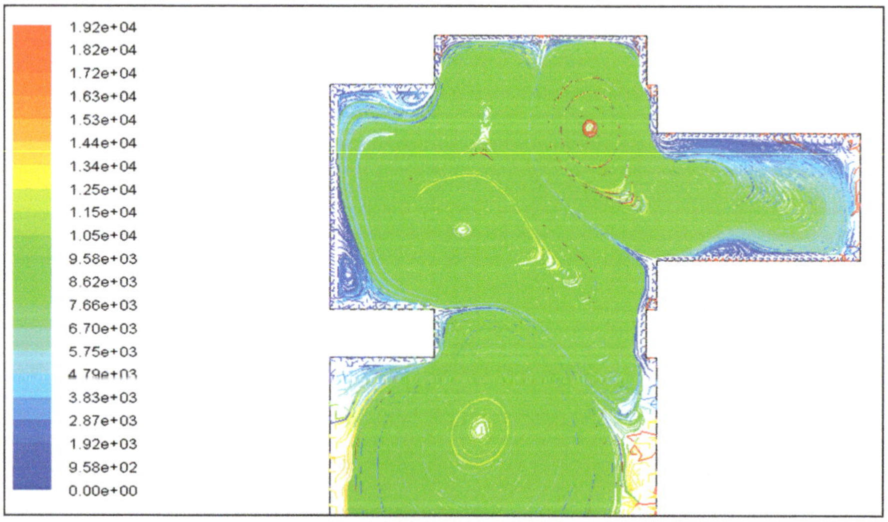

Figure 16 **Variations of streamline in free convection.**

Competing interests
The authors declare that they have no competing interests.

Authors' contributions
MG collected the airflow window data. Modeling and product mesh were created by MG and MAB. MG carried out computations and analysis. MAB and MG drafted the manuscript. All authors read, edited and eventually approved the final manuscript to be submitted.

Acknowledgments
The support from the Department of Energy Engineering, Science and Research Branch, Islamic Azad University, Tehran, Iran, is gratefully acknowledged.

Author details
[1]Department of Energy Engineering, Science and Research Branch, Islamic Azad University, P.O. Box 147789385, Tehran, Iran. [2]Department of Chemical Engineering, Tehran University, P.O. Box 147789385, Tehran, Iran. [3]Department of Energy System Engineering, Islamic Azad University of South Tehran Branch, P.O. Box 189, Tehran, Iran. [4]Department of Mechanical Engineering, Islamic Azad University of Roudehen branch, P.O. Box 189, Tehran, Iran.

References
1. Brandle, K, Boehm, RF: Airflow windows: performance and applications, Proceedings of the ASHRAE/DOE conference, pp. 361–379. Thermal Performance of the Exterior Envelopes of the Building II, Clearwater Beach (1982)
2. Ripatti, H: Airflow window system – Making the fenestration the solution rather than the problem in energy use. Proceedings of the ASHRAE Winter meeting, Atlanta (1984)
3. Park, SD, Suh, HS, Cho, SH: The Analysis of Thermal Performance in an Airflow Window System Model, Proceedings of the ASHRAE/DOE/BTECC/CIBSE Conference, pp. 361–375. Thermal Performances of the Exterior Envelopes of Buildings IV, Clearwater Beach (1989)
4. Baker, P, Saelens, D, Grace, M, Inoue, T: Advanced Envelopes – Methodology, Evaluation and Design Tools. Acco, Leuven (2000). Annex 32, Final report
5. Ziller, C: Modellversuche und Berechnungen zur Optimierung der natürlichen Lüftung durch Doppelfassaden (in German). Ph.D. Thesis. Shaker Verlag, Aachen (1990)
6. Busselen, B, Mattelaer, P: Experimental evaluatie van actieve gevelsystemen in het Vliet proefgebouw (in Dutch). MSc thesis, Laboratory for Building Physics. KU Leuven, (2000)
7. Wilcox, DC: Turbulence Modeling for CFD. DCV Industries, Inc, La Canada (1993)
8. Saelens, D, Hens, H: Low-energy design and airflow windows. Some considerations illustrated with a Case Study. Proceedings of the 10th International Symposium for Building Physics, International Building Physics Conference, September, pp. 27–29., Dresden (1999)
9. Versteeg, HK, Malalaskera, W: An Introduction to Computational Fluid Dynamics. The Finite Volume Method., McGraw-Hill, New York (1960)
10. GAMBIT: User's Manual, Version 1.1.2. Vol **2**., Anon, Lebanon, USA (1998)
11. FLUENT: *User's Manual, Version 1.1.2*, Vol **2**., Anon, Lebanon, USA (1998)
12. Patankar, SV: Numerical Heat transfer and Fluid Flow. MacGraw-Hill, New York (1980)
13. Cebeci, T, Shao, JP, Kafyeke, F, Laurendeau, E: Computational Fluid Dynamics for Engineers. Horizons Publishing Inc, Long Beach (2005)

Tamanu oil - an alternative fuel for variable compression ratio engine

Mohan T Raj[1*] and Murugumohan Kumar K Kandasamy[2]

Abstract

Biodiesel can be produced from vegetable oils and also from waste fats. Biodiesel is a mono-alkyl-ester of long chain fatty acids derived from renewable feedstock such as vegetable oils by transesterification process. The esterified cotton seed oil, pungam oil, rice bran oil, and tamanu oil are chosen as the alternative fuels. Among these oils, tamanu oil is considered for the first time as an alternative fuel. An experiment is conducted to obtain the operating characteristics of the variable compression ratio (VCR) engine run by chosen esterified oils, and the results are compared with esterified tamanu oil. From the comparison of results, it is inferred that the engine performance is improved with significant reduction in emissions for the chosen oils without any engine modification. The effective compression ratio can be fixed based on the experimental results obtained in the engine since the findings of the present research work infer that the biodiesel obtained from tamanu oil is a promising alternative fuel for direct-injection four-stroke VCR engine.

Keywords: Biodiesel, Esterified tamanu oil, Compression ratio, VCR engine, Transesterification

Background

Vegetable oils have good ignition quality since they are not branched and have very long molecular chains. Certain functional groups and the poor volatility could be responsible for their low cetane numbers. The heating value of vegetable oils is somewhat lower due to oxygen content, and viscosity and carbon residue are higher than diesel due to their larger molecular mass and chemical structure [1]. The flash point of these oils is much higher than that of diesel, indicating that they are much safer to store than diesel oil. They are about 10% denser than diesel [2]. Their cold point is higher, indicating problems of thickening or even freezing at low ambient temperatures. It is evident that vegetable oils are much less volatile than diesel. This makes their slow evaporation when injected into the engine. Vegetable oils have cetane numbers of about 35 to 50 depending on their composition [3]. It is seen that the value is very close to diesel. The compression process effectively starts only after the intake valve closes and also depends on the momentum of the flow into the cylinder, and thus, the actual realized amount of compression ratio is known as the effective compression ratio (ECR). ECR is a more suitable indicator of the compression process, and it also influences the engine operation. One of the very promising and interesting fields of study involves the use of alternative fuels including biodiesel and diesel fuels to provide effective solutions [4]. The development of treatment devices has also mitigated the emission problem to a large extent while allowing the combustion process to be optimized for maximum fuel efficiency [5]. Lehman et al. [6] obtained high ester conversion with a 6:1 M ratio of methanol to vegetable oil. In the process of peanut oil esterification, the 6:1 M ratio liberated significantly more glycerol than the 3:1 M ratio. These investigators also found that glycerol yields increased from 77% to 95% as the sodium hydroxide catalyst increased from 0.2% to 0.8% at the 6:1 M ratio. Fatty ester is the major product, and glycerol is the by-product. Barsic and Humke [7] have found that transesterification is one of the methods by which viscosity could be drastically reduced and the fuel could be adopted for use in diesel engine. The transesterification process involves reacting vegetable oils with alcohols such as methanol or ethanol in the presence of a catalyst (usually sodium hydroxide) at about 70°C to give the ester and the by-product glycerin. This esterified vegetable oil is popularly known as biodiesel which is commercially

* Correspondence: tmraj@mech.sastra.edu
[1]Department of Mechanical Engineering, SASTRA University, Thanjavur, Tamilnadu, India
Full list of author information is available at the end of the article

available in developed countries due to its distinct advantage over the conventional diesel. Free fatty acids (FFAs) present in the pongamia oil have got the greater influence in the process of converting it into biodiesel. This has been observed during the biodiesel preparation process in the laboratory level. However, their high viscosity and poor volatility lead to reduced thermal efficiency and increased hydrocarbon, carbon monoxide, and smoke emissions. Vellguth [8] studied the performance of a direct-injection single-cylinder diesel engine on vegetable oil. He conducted a variable load test on the engine with rapeseed oil, peanut oil, and soybean oil. He found that vegetable oils which are results of performance are comparable to diesel values with slightly reduced thermal efficiency. He concluded that vegetable oils could be directly used as fuel in diesel engines on a short-term basis.

Ramesh et al. [9] investigated the performance of a glow plug-assisted hot surface ignition engine using methyl ester of rice bran oil as fuel. Normal and mnemonic crown pistons were used for their tests. They reported an improvement in brake thermal efficiency of about 1% when the glow plug is on. The percentage improvement in brake thermal efficiency was higher in the case of normal piston compared with that in the case of mnemonic piston. Larry Wagner et al. [10] studied the effect of soybean oil esters on the performance and emissions of a four-cylinder direct-injection turbocharged diesel engine. They found that the engine performance with soybean oil esters is the same with diesel. Clark et al. [11] studied the effect of methyl and ethyl esters of soybean oil on engine performance and durability in a direct-injection John Deere four-cylinder diesel engine. They observed that the engine fuelled with soybean esters produced less power output with an increase in fuel consumption. Emissions results were found to be similar to diesel. Panwar et al. [12] conducted an experiment in single-cylinder variable compression ratio diesel engine at different loads. The engine performance for castor methyl ester was investigated. The lower blends of biodiesel increased break thermal efficiency and reduced fuel consumption. The work done by Gumus et al. [13] deals with the performance and emissions of a compression-ignition diesel engine without any modification using neat apricot seed kernel oil methyl ester and its blends. They found that a lower concentration of apricot seed kernel oil methyl ester in blends gives better improvement in engine performance and exhaust emissions. The work done by Celikten et al. [14] tells about the performance and emissions of diesel fuel from rapeseed and soybean oil methyl esters injected at different pressures (250, 300, and 350 bar), and they were compared. It has been found that the torque and power of diesel fuel engine were reduced with increasing injection pressure. Smoke level and CO emission were also reduced, while NO_x emission was increased as the injection pressure was increased. Jindal et al. [15] studied about the comparison of performance and emission characteristics for different compression ratios along with injection pressure, and the best possible combination for operating engine with Jatropha methyl ester has been found. It is found that the combined increase in compression ratio and injection pressure results in an increased brake thermal efficiency and reduced brake specific fuel consumption while emissions were lowered. The combustion performance and exhaust emission characteristics of turpentine oil fuel blended with conventional diesel fuel in a diesel engine was evaluated [16]. The combustion characteristics of crude rice bran oil methyl ester blended in a direct-injection compression-ignition engine were analyzed, and it was found that cylinder pressure was comparable, whereas delay period and the maximum rate of pressure rise were lower than that of diesel. Masjuki et al. [17] analyzed exhaust emissions and lube oil deterioration of a diesel engine fuelled with Malaysian palm oil diesel (ester of palm oil) and ordinary diesel emulsions containing 5% to 10% of water by volume and compared with 100% palm oil diesel (POD) on 200-h endurance tests. They observed that the exhaust of POD and emulsified fuels was much cleaner, containing less CO, CO_2, HC, NO_x, SO_x, and smoke level, than that of neat vegetable oil. Power output was slightly reduced when using POD emulsified fuels.

Monyem and Van Gerpen [18] conducted experiments to characterize the effect of oxidized biodiesel on engine performance and emissions. They used methyl soyate for testing a John Deere turbocharged direct-injection diesel engine. Oxidization was done by heating the oil up to 60°C and by bubbling. The diesel blends with 20% biodiesel were tested at two different loads and three injection timings. They found that the performance of neat biodiesel and their blends was similar to that of diesel fuel with the same brake thermal efficiency but higher fuel consumption. Compared with non-oxidized biodiesel, oxidized biodiesel produced lower exhaust carbon monoxide and hydrocarbons of 16%. Statically, there was no significant difference on the oxides of nitrogen. They found significant reduction in the Bosch smoke number with neat biodiesel and blends with diesel. The highest reduction was found for the oxidized biodiesel. Jacobs et al. [19] demonstrated drastic reduction in NO_x with high amounts of exhaust gas recirculation (EGR) on a light-duty diesel engine with 5% reduction in fuel

Table 1 Percentages of FFA values for biodiesel

Sample number	Biodiesel	FFA (%)
1	Cotton seed oil	1.9
2	Pungam oil	2
3	Rice bran oil	2.1
4	Tamanu oil	1.8

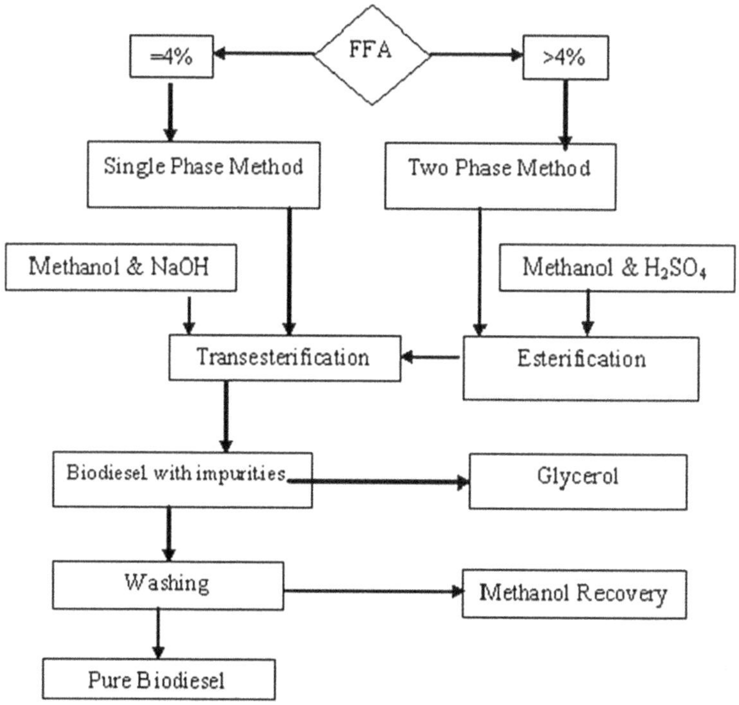

Figure 1 Schematic presentation of biodiesel production.

efficiency. Ishikawa et al. [20] performed early injection PCCI on an engine and a vehicle at low load operating conditions using high EGR rates. Using a low cetane number diesel fuel, their results showed about a 90% decrease in NO_x and an 85% reduction in particulate matter (PM) without any fuel efficiency drop. Murata et al. [21] reported a 60% reduction in both NO_x and PM with a very minute increase in fuel consumption in a single-cylinder engine. They used early fuel injection with high EGR rates and reduced ECR by intake valve closing (IVC) modulation. The model was validated against different operating points using engine data from Cummins. In addition, the model was also validated with data from a second engine of similar make at Purdue University's Herrick Laboratories. This model will be used here for the simulation study. Combustion homogenization is accomplished by early injection of fuel, allowing more time for the fuel to get thoroughly mixed with the charge inside the cylinder [22], resulting in a more uniform combustion. With early injection of fuel, combustion often occurs before it is desired, and one way to prevent this combustion is by reducing the ECR. This can be done by IVC modulation. The formation of uniform charge before the beginning of combustion enables reduction in particulate matter [23]. Also, the lower flame and post combustion temperature due to premixed combustion reduce NO_x formation. The early fuel injection with high EGR reduces the effective compression ratio by IVC modulation. It also outlined the various methods of achieving the desired burned gas fraction inside the cylinder by variable valve actuation. A very straightforward

Table 2 Comparison of biodiesel properties with diesel

Number	Properties	Esterified cotton seed oil	Esterified pungam seed oil	Esterified rice bran oil	Esterified tamanu oil	Diesel
1	Kinematic viscosity at 32°C (Cst)	11.1	17.84	18.7	9.0	6.0
2	Flash point (°C)	234	205	200	108	72
3	Density at 32°C (g/cm^3)	0.886	0.913	0.935	0.905	0.86
4	Lower calorific value (kJ/kg)	36,800	37,304	38,952	41,500	44,000
5	Pour point (°C)	15	13.2	15	11.3	12
6	Cetane number	45.5	55	53	50	45

Table 3 Variable compression ratio engine specifications

Number	Engine parts	Specification
1	Make and model	Kirloskar/PS 234
2	Number of cylinder	Single
3	Ignition system	Compression ignition
4	Bore and stroke	87.5 × 110 mm
5	Rated power	3.5 kW at 1,500 rpm
6	Cooling medium	Water-cooled
7	Combustion chamber	Open chamber (direct injection)
8	Compression ratio	12:1 to 18:1

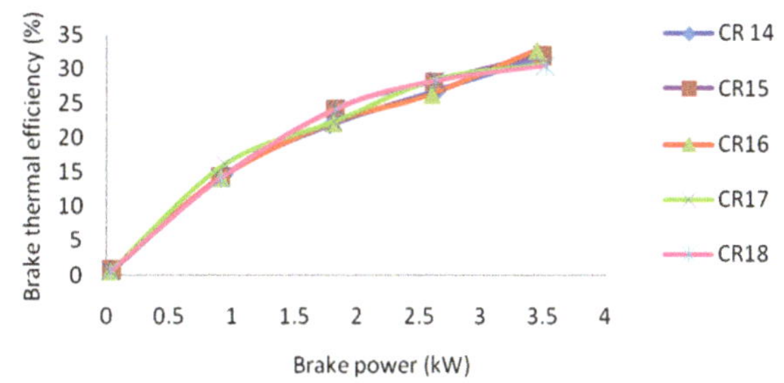

Figure 3 Variation of brake thermal efficiency with brake power.

method to compute the ECR is to take a ratio of the volume at IVC timing to the volume when the piston is at TDC. Although this method is generally good enough for initial system design, higher accuracy is expected for predicting performance and emissions [24]. Therefore, a pressure-based method is used to compute the effective compression ratio using the *P-θ* diagram of the process.

Methods

Preparation of biodiesel

Based on the percentage of FFA value, biodiesel can be prepared by either a single-phase or a two-phase method. The percentages of FFA values for biodiesel are given in Table 1.

Single-phase method

If the percentage of FFA present in the raw vegetable oil is less than 4%, the *transesterification* process by single-phase method has to be chosen. In this method, a measured amount of methanol (CH_3OH) and sodium hydroxide (NaOH) has to be mixed thoroughly with a measured amount of vegetable oil. The mixture is heated and maintained at 65°C for 2 h, and then, it undergoes natural cooling. Glycerol will deposit at the bottom of the flask, and it is separated out using a separating funnel. The remnants in the flask are the esterified vegetable oil. Another commercial name for esterified vegetable oil is biodiesel.

Two-phase method

If the percentage of FFA present in the raw vegetable oil is more than 4%, then we have to choose the two-phase method. In this method, the oil has to undergo esterification. Measured quantity of sulphuric acid and methanol has to be taken and mixed thoroughly and added with a measured amount of vegetable oil. The mixture is heated and maintained at 65°C for 2 h. The fatty ester is separated after natural cooling. At second level, the separated oil from the separating funnel has to undergo transesterification. Methoxide (methanol + sodium hydroxide) is added with the above ester and heated to 65°C. The same temperature is maintained for 2 h with continuous stirring, and then, it undergoes natural cooling for 8 h. Glycerol will deposit at the bottom of the flask, and it is separated out by a separating funnel. The remnants in the flask are the esterified vegetable oil (biodiesel).

The separated biodiesel from the above-mentioned method contains various impurities like traces of glycerol, unused methanol, soap particles, etc. Water washing is carried out to remove all impurities. *Air bubble wash* is one of the methods normally recommended in the laboratory level. In this method, the impure biodiesel is placed in a beaker initially. Water is added slowly through the side wall of the beaker (both are immiscible). It is ensured that the equal amount of water is added above the level of biodiesel. Air is made to pass through the biodiesel and the water from the bottom of

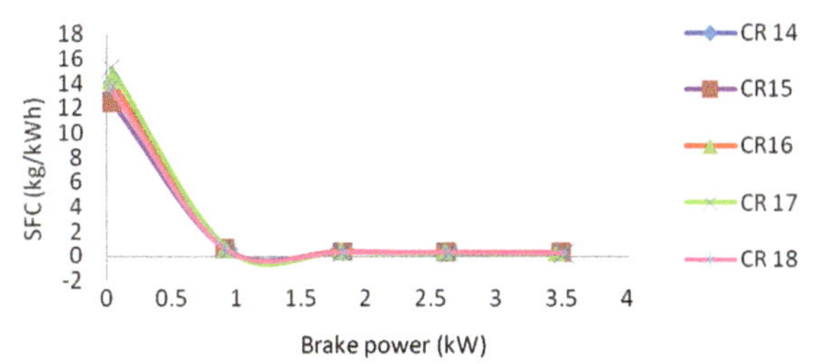

Figure 2 Variation of specific fuel consumption with brake power.

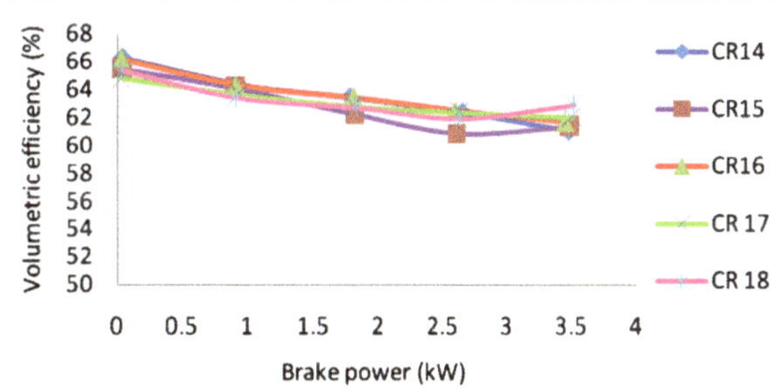

Figure 4 Variation of volumetric efficiency with brake power.

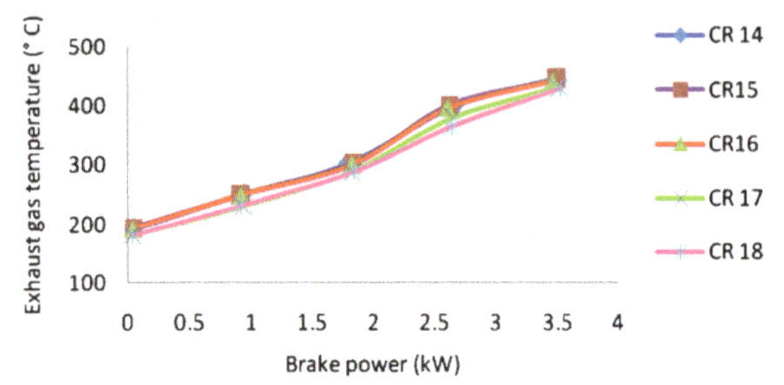

Figure 5 **Variation of exhaust gas temperature with brake power.**

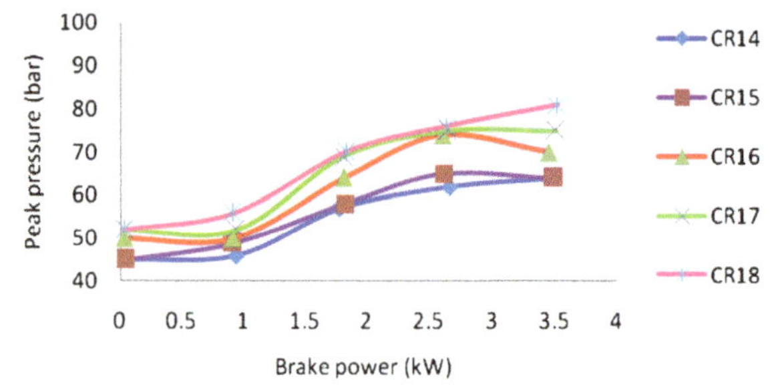

Figure 7 **Variation of peak pressure with brake power.**

the beaker with the help of a bubbler (electrically operated). The air will then take away all impurities from the biodiesel; they will move up as the bubbles move up, and they are added in the water. The unused methanol will be diluted in water. The traces of glycerol and soap particles make the water to become like soap water. Once the water becomes like soap water, the bubbler is stopped. After allowing some time for impurities to settle, the biodiesel is drained from the separating funnel, and pure biodiesel will be directly used, with or without blending, in the engine (Figure 1).

Properties of biofuel

The basic important properties of biodiesel like calorific value, flash point, viscosity, density, pour point, and cetane number are tested in the laboratory. The values are tabulated in Table 2 and compared with diesel.

Experimental setup

The specification of the engine is listed in Table 3. The engine performance analysis software package (Engine Soft) has been employed for online performance analysis. The setup consists of a single-cylinder four-stroke variable compression ratio (VCR) diesel engine coupled with an Eddy current dynamometer for loading. The compression ratio can be changed without stopping the engine and without altering the combustion chamber. A specially designed tilting cylinder block arrangement is used for varying the compression ratio. The setup is provided with necessary instruments for combustion pressure and crank-angle measurements. These signals are interfaced in a computer through an engine indicator for $P\theta$-PV diagrams. The setup has a stand-alone panel box consisting of transmitters for air and fuel flow measurements, process indicator, and engine indicator. Rotameters are provided for cooling water and calorimeter water flow measurement. The setup enables the VCR engine operating parameters such as brake power (BP), brake mean effective pressure (BMEP), indicated mean effective pressure, brake thermal efficiency, indicated thermal efficiency and mechanical efficiency, etc to be studied.

Experimental methodology

Using esterified tamanu oil in the variable compression ratio engine at a rated speed of 1,500 rpm, the performance analysis is carried out. In every test, volumetric efficiency specific fuel consumption and exhaust gas emissions such as carbon monoxide (CO), hydrocarbon (HC), nitrogen oxides (NO_x), carbon dioxide (CO_2), and oxygen (O_2) are measured. From the initial measurement, brake thermal efficiency, specific fuel consumption, BP, BMEP, mechanical efficiency, and exhaust gas temperature with respect to compression ratios 18:1 to 14:1 are calculated and recorded. At each operating conditions, the performance

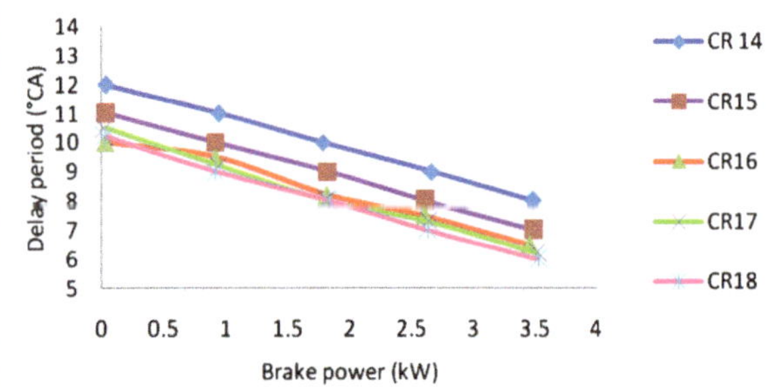

Figure 6 **Variation of delay period with brake power.**

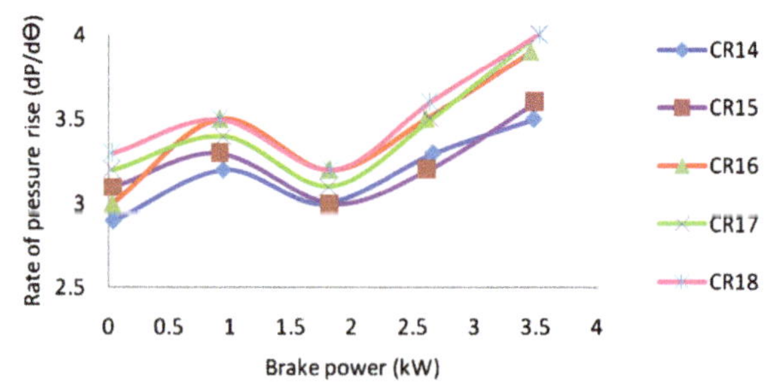

Figure 8 **Variation of rate of pressure rise with brake power.**

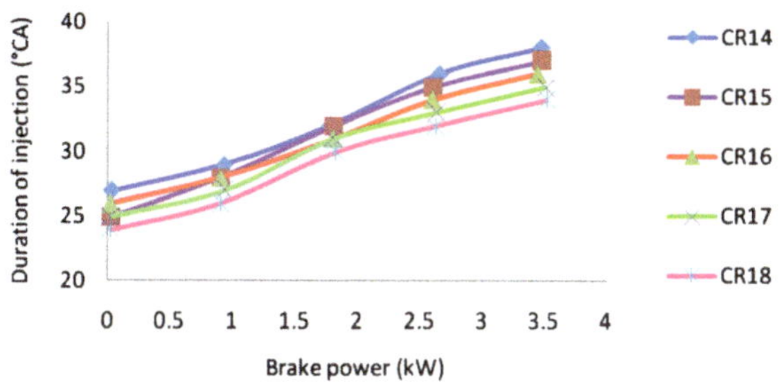

Figure 9 Variation of duration of injection with brake power.

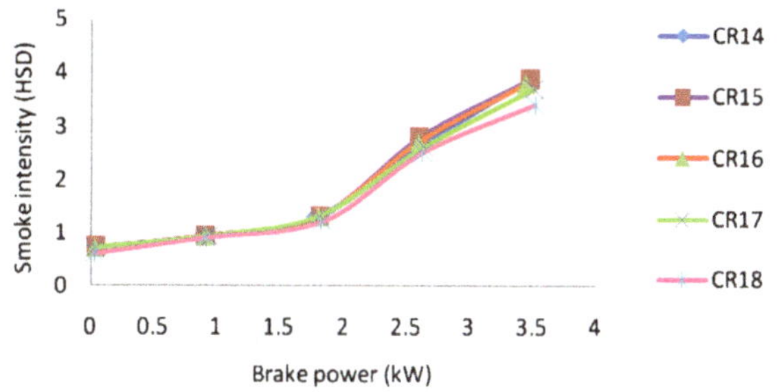

Figure 11 Variation of smoke intensity with brake power.

characteristics, combustion characteristics, and exhaust emission levels are performed and the same procedure is repeated for other loads also.

Results and discussion

The operating characteristics are analyzed by running the VCR engine with esterified tamanu oil at different compression ratios ranging from 14 to 18 at different load conditions.

Performance characteristics

The specific fuel consumption decreases with the increase in compression ratio. It is found to be lower at a compression ratio of 18. The specific fuel consumption of the esterified tamanu oil at the compression ratio of 18 is 0.24 kg/kWh, whereas at the compression ratio of 14, it is 0.29 kg/kWh for maximum load. It has been observed that the brake thermal efficiency of the biodiesel is slightly higher at the compression ratio of 16 and lower at the compression ratio of 18. The brake thermal efficiency of the esterified tamanu oil for the compression ratio of 18 is 30.4% at maximum load. By varying the compression ratio of the engine, the brake thermal efficiency also gets varied. The variation in the volumetric efficiency is comparably less for all range of compression ratios, and it is high at CR18. The reason for the increase in volumetric efficiency is due to increase in volume of incoming air to the engine. The volumetric efficiency of esterified tamanu oil for compression ratio of 18 is 62.91% at maximum load.

The result indicates that the variation in exhaust gas temperature is very minimal when the compression ratio is varied from 14 to 18. The highest temperature obtained is 446.35°C at the compression ratio of 14 and 430.5°C at the compression ratio of 18 at maximum load. These performance characteristic curves of the esterified tamanu oil have been compared by running the engine at various compression ratios, and it is shown in Figures 2, 3, 4, 5, and 6.

Combustion characteristics

The delay period is consistently low for esterified tamanu oil by increasing the compression ratio from 14 to 18. The delay period is 8° crank angle (CA) at the compression ratio of 14 and 6°CA at 18 for maximum load. The peak pressure increases linearly from 64 to 81 bar as the compression ratio increases from 14 to18 for maximum load. The rate of pressure rise is the derivative of peak pressure with respect to the crank angle degree. It varies from 3.3 to 4 bar per degree of CA for the compression ratio of 18 at maximum load. It decreases when the compression ratio is increased from 14 to 18. The duration of injection is 38°CA at the compression ratio 14 and 34°CA at 18 for maximum load. It increases with increase in compression ratios from 14 to 18. The combustion duration is 40°CA at the compression ratio of 14 and 43°CA at the

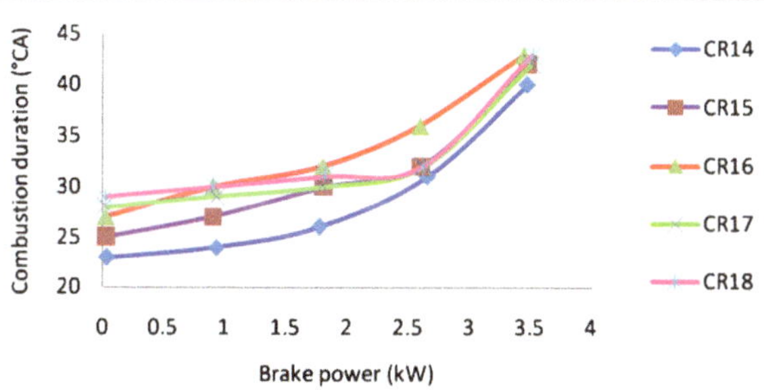

Figure 10 Variation of combustion duration with brake power.

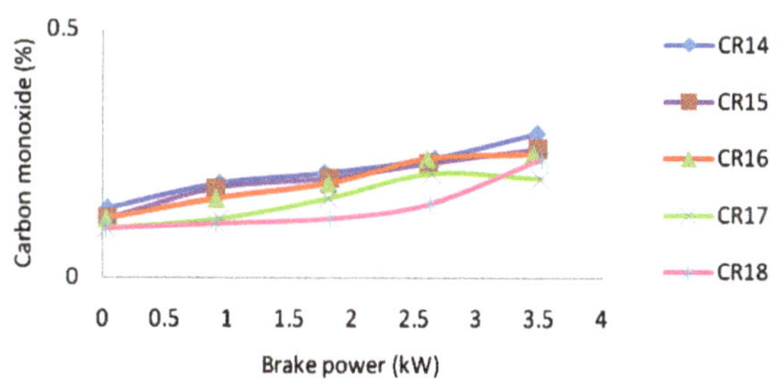

Figure 12 Variation of carbon monoxide emission with brake power.

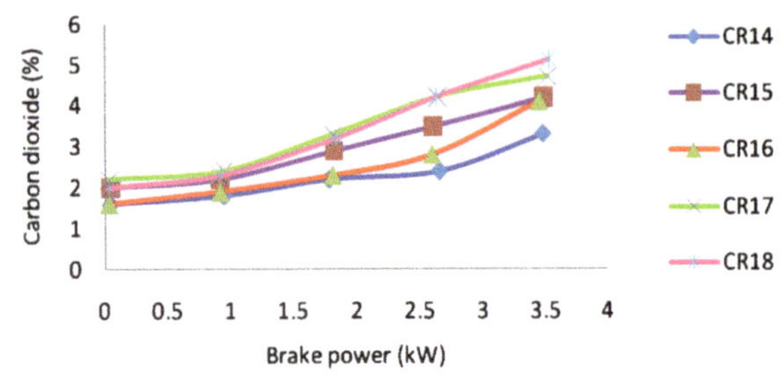

Figure 13 Variation of carbon dioxide emission with brake power.

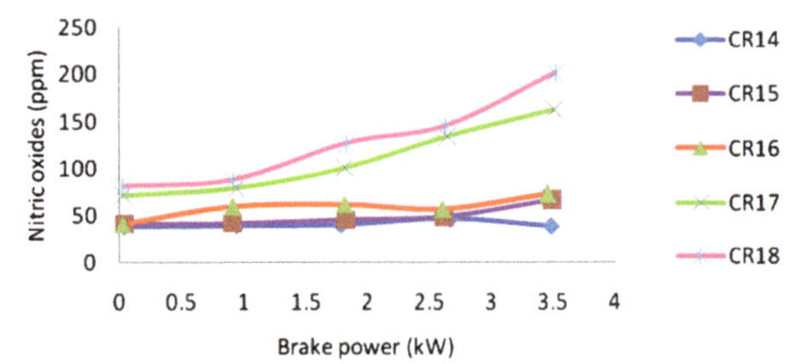

Figure 15 Variation of NO_x emission with brake power.

compression ratio of 18 for maximum load. It decreases with the increase in compression ratios from 14 to 18. The smoke intensity is 3.9 Hartridge smoke density (HSD) at the compression ratio of 14 and 3.4 HSD at the compression ratio of 18 for maximum load. These combustion characteristics are shown in Figures 7, 8, 9, 10, and 11.

Emission characteristics

These emission characteristics of the esterified tamanu oil have been compared, and they are shown in Figures 12, 13, 14, 15, and 16. The percentage of CO emission for low compression ratio increases due to the rising temperature in the combustion chamber. The CO emission of the esterified tamanu oil is found to be lower for high compression ratio of 18, and it is 0.18% at maximum load. The esterified tamanu oil emits higher percentage of CO_2 at high a compression ratio, less at lower compression ratios, and vice versa. More amount of CO_2 is an indication of complete combustion of fuel in the combustion chamber. The CO_2 emission is 3.3% at the compression ratio of 14 and 5.1% at the compression ratio of 18 for maximum load. In the case of esterified tamanu oil, the HC emission decreases with increase in compression ratio. This is due to the complete combustion of oil at a higher compression ratio. The HC emission is 152 ppm at the compression ratio of 14 and 125 ppm at the compression ratio of 18 for maximum load. From the figure, it is observed that the NO_x emission for esterified tamanu oil is higher with the increase in compression ratio. It is obvious that for higher compression ratio, NO_x emission is higher than that of low compression ratio. The reason for the higher NO_x emission for esterified tamanu oil is the higher peak temperature. The NO_x emission for esterified tamanu oil at the compression ratio of 18 is 201 ppm at maximum load. The particulate emission decreases with the increase of compression ratio from 14 to 18 for all loads. The particulate emission is 1,900 mg/m^3 at the compression ratio of 14 and 1,600 mg/m^3 at the compression ratio of 18 for maximum load.

Conclusions

The study aims to evaluate the suitability of using biodiesel as an alternative fuel in VCR engine. Experimental investigations were carried out on the operating characteristics of the engines. The following conclusions are drawn from the investigations:

- The brake thermal efficiency of the VCR engine slightly increases at higher loads when compared with that of standard engine.
- The specific fuel consumption is lower at all load conditions.
- The volumetric efficiency increases at higher loads.
- The exhaust gas temperature is the same and increases as the load increases.
- The delay period is consistently low when load increases.

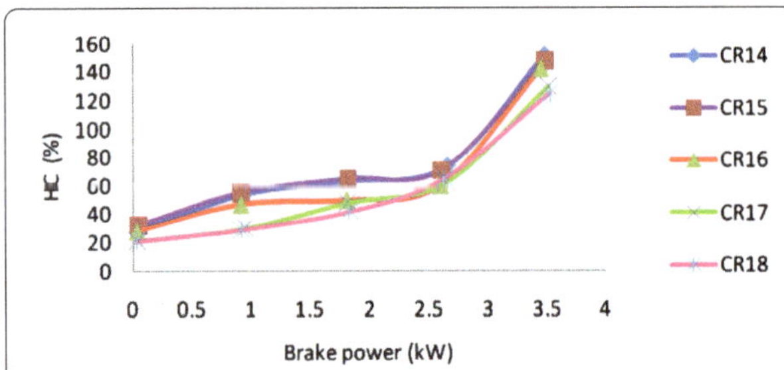

Figure 14 Variation of HC emission with brake power.

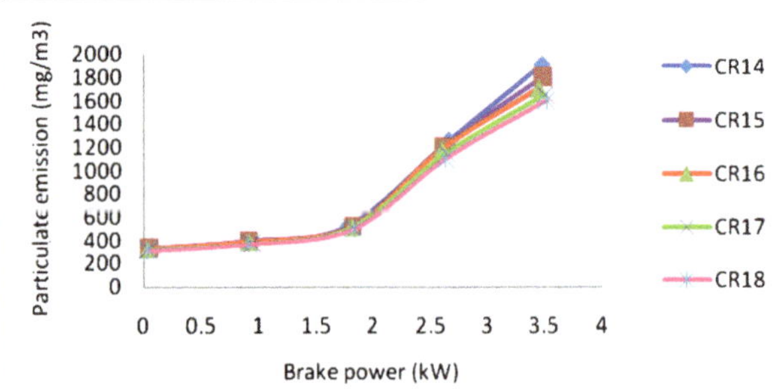

Figure 16 Variation of particulate emission with brake power.

- The peak pressure and the rate of pressure rise are higher and increase with increase in load.
- The duration of injection slightly decreases at all loads.
- Combustion duration is slightly higher at lower loads and lower at higher loads.
- There is not much variation in the HC, CO, CO_2, and NO_x emissions at all loads.
- Similarly, there is not much variation in the case of particulate matter emission and smoke intensity.

From the above conclusions, it is proved that the biodiesel could be used as an alternative fuel in VCR engine without any engine modifications.

Abbreviations

BMEP: brake mean effective pressure; CA: crank angle; CR: compression ratio; ECR: effective compression ratio; IVC: intake valve closing; PM: particulate matter; VCR: variable compression ratio.

Competing interests

The authors declare that they have no competing interests.

Authors' contributions

MTR participated in the preparation of biodiesel from cotton seed oil, pungam oil, rice bran oil, and tamanu oil and carried out the experimental investigation for these biodiesel. MKKK guided the preparation of biodiesel and helped carry out the experimental investigation. He also corrected the manuscript as per journal standards. All authors read and approved the final manuscript.

Authors' information

MTR is an assistant professor at the Department of Mechanical Engineering in SASTRA University, Thanjavur, Tamilnadu, India. He completed his MT in Energy Engineering from the National Institute of Technology, Trichy. His research interests include the preparation of biodiesel from vegetable oils and the testing of these oils in diesel engine for performance characteristics. He is currently involved in the Research and Modernization project sanctioned by SASTRA UNIVERSITY at a cost of Rs 6.2 lakhs for variable compression ratio engine (VCR engine). MKKK is the principal of Paventhar Bharathidasan College of Engineering and Technology, Trichy. He received his ME in Automobile Engineering from Anna University, MIT Campus, Chennai and his PhD in Alternate Fuel for IC Engine from Bharathidasan University, Trichy. His research interests include the biodiesel in internal combustion engine with less pollution and recent trends in automobile technology. He is an active life member of the Society of Automotive Engineering.

Acknowledgments

The authors gratefully acknowledge the Vice Chancellor of Sastra University, Thanjavur for his support through research and modernization funded project, sanctioned to variable compression ratio engine.

Author details

[1]Department of Mechanical Engineering, SASTRA University, Thanjavur, Tamilnadu, India. [2]Department of Mechanical Engineering, Pavendar Bharathidasan College of Engineering and Technology, Trichy, Tamilnadu, India.

References

1. Hawley, JG, Wijetunge, RS, Brace, CJ, Vaughan, ND: Dynamic behavior of a high speed direct injection diesel engine. Soci. Automot. Eng. **01**, 0829 (1999)
2. He, X, Durrett, R: Late intake closing as an emissions control strategy at tier 2 bin 5 engine-out NOx level. Soci. Automot. Eng. **01**, 0637 (2008)
3. Ramadhas, AS, Jayaraj, S, Muraleedharan, C: Use of vegetable oils as IC engine fuels-a review. Renew. Energy **29**, 727–742 (2004)
4. EPA: A comprehensive analysis of biodiesel impacts on exhausts emissions. Draft technical report. United States Environmental Protection Agency, (2002)
5. Szybist, J, Song, J, Alam, M, Boeham, A: Biodiesel combustion, emissions and emission control. Fuel Processing Technology. **88**, 679–691 (2007)
6. Lehman, LW, Gauglitz, EJ: The preparation of alkyl esters from highly unsaturated triglycerides. J. Am. Oil Chem. Soc. **40**, 197–198 (1963)
7. Barsic, NJ, Humke, AL: Performance and emission characteristics of a naturally aspirated diesel engine with vegetable oils. Soci. Automot. Eng, 1173–1187 (1981). paper no. 810262
8. Vellguth, G: Performance of vegetable oils and their monoesters as fuel for diesel engines. Soci. Automot. Eng, – (1981). paper no. 831358
9. Ramesh, A, Nagalingam, B, Gopalakrishnan, KV: Performance of glow plug surface ignition engine with methyl ester of rice bran oil as fuel. In: Proceedings of XI National Conference on IC Engines and Combustion. Indian Institute of Technology, pp. 11–15., Madras, India (1989)
10. Wagner, LE, Clark, SJ, Schrock, MD: Effect of soybean oil esters on the performance-lubrication oil and water of diesel engines. Soci. Automot. Eng, – (1984). paper no. 841385
11. Clark, SJ, Wagner, L, Schorck, MD, Piennar, PG: Methyl and ethyl soybean esters as renewable fuels for diesel engines. J. Am. Oil. Chem. Soc. **61**(10), 1632–1638 (1984)
12. Panwar, NL, Hemant, Y, Shrirame Rathore, NS, Sudhakar, J: Performance evaluation of a diesel engine fueled with methyl ester of castor seed oil. Appl. Therm. Eng. **30**, 245–249 (2010)
13. Gumus, M, Kasifoglu, S: Performance and emission evaluation of a compression ignition engine using a biodiesel (apricot seed kernel oil methyl ester) and its blends with diesel fuel. Biomass Bioenergy. **34**, 134–139 (2010)
14. Celikten, I, Koca, A, Arslan, MA: Comparison of performance and emissions of diesel fuel, rapeseed and soybean oil methyl esters injected at different pressures. Renew. Energy **35**, 814–820 (2010)
15. Jindal, S, Nandwana, BP, Rathore, N.S, Vashisth, V: Experimental investigation of the effect of compression ratio and injection pressure in a direct injection diesel engine running on jatropha methyl ester. Applied Thermal Engineering. **30**, 442–448 (2010)
16. Saravanan, S, Nagarajan, G, Lakshmi, NRG, Sampath, S: Combustion characteristics of a stationary diesel engine fuelled with a blend of crude rice bran oil methyl ester and diesel. Energy **35**, 94–100 (2010)
17. Masjuki, H, Abdulmuin, MZ, Sii, HS: Investigation on preheated palm oil methyl esters in the diesel engine. Proc. Inst. Mech. Eng. **210**, 131–138 (1996)
18. Monyem, A, Gerpen Van, JH: The effect of biodiesel oxidation on engine performance and emissions. Int. J. Biomass and Bio Energy **20**, 317–325 (2001)
19. Jacobs, T, Bohac, S, Szymkowicz, P, Assanis, D: Lean and rich premixed compression ignition combustion in a light-duty diesel engine. Soci. Automot. Eng. **01**, 0166 (2005)
20. Ishikawa, N, Nakada, T: Study of premixed compression ignition on multi cylinder diesel engine. JSAE. **27**, 531–538 (2006)
21. Murata, Y, Kusaka, J, Odaka, M, Daisho, Y, Kawano, D, Suzuki, H: Emissions suppression mechanism of premixed diesel combustion with variable valve timing. Int J Engine Res **8**(5), 415–428 (2007)
22. Klingbeil, A: Particulate and NO_x reduction in a heavy-duty diesel engine using high levels of exhaust gas recirculation and very early or very late start of injection. Master's thesis, University of Wisconsin-Madison, (2002)
23. Khan, I, Greeves, G, Wang, C: Factors affecting smoke and gaseous emissions from direct injection engines and a method of calculation. Soci. Automot. Eng, – (1973). paper no. 730169
24. Zheng, M, Reader, T, Hawley, J: Diesel engine exhaust gas recirculation a review on advanced and novel concepts. Energy Conserv Manage. **45**, 883–900 (2004)

Calcium oxide from *Pomacea* sp. shell as a catalyst for biodiesel production

Yosephine Yulia Margaretha, Henry Sanaga Prastyo, Aning Ayucitra* and Suryadi Ismadji*

Abstract

Heterogeneous catalysts are promising catalyst for the transesterification reaction of vegetable oils to produce biodiesel. In this study, the *Pomacea* sp. shell was used as the raw material for CaO catalyst. The calcination of *Pomacea* sp. shell was conducted at 900°C for 2 h. The raw material and the resulting CaO catalyst were characterized using X-ray diffraction, X-ray fluorescense spectrometry, and Fourier transform infrared spectroscopy analysis. From the experimental results, it was found that the maximum yield of fatty acid methyl esters was 95.61% at the following reaction conditions: reaction temperature of 60°C, a reaction period of 4 h, a ratio of methanol-oil at 7:1, and amount of catalyst at 4% *w/w*. The physical and chemical properties of biodiesel were determined based on ASTM standard, and the values obtained were confirmed with the Indonesian National Standard (SNI-04-7182-2006) and (B100)-ASTM D6751-07b.

Keywords: Biodiesel, Transesterification, Solid catalyst, CaO, Palm oil

Background

The extensive use of fossil fuel in human activities during the last several decades causes the depletion of fossil fuel source; therefore, the search of other alternative energies is crucially important [1,2]. One of the alternative fuels which have been developed over a century ago is biodiesel. Biodiesel has been chosen as an alternative fuel because it is renewable, produces lower emission, and possesses high flash point, better lubrication, and high cetane number [3].

Studies about the transesterification of vegetable oils or animal fats into biodiesel have been conducted by various researchers. At the time of reaction, the triglycerides will gradually be converted into diglycerides, monoglycerides, and glycerol. At each change of conversion, one mole of ester formed [4]. Different kinds of catalysts have been in use to produce biodiesel from a different kind of oil. The catalysts commonly used for biodiesel production are categorized into several types: homogeneous catalysts (sodium hydroxide, potassium hydroxide, sulfuric acid, etc.) [5], heterogeneous catalysts (cation-exchange resin, hydrotalcites, etc.), and enzymes (*Chromobacterium viscosum, Candida rugosa*, and *Porcine pancreas*) [6].

Homogeneous catalyst has been used extensively for biodiesel production, but the use of this kind of catalyst began to decrease because it has several drawbacks: it cannot be recovered [7] or regenerated after the reaction and also it produced toxic wastewater. To replace the homogenous catalyst, enzyme and heterogeneous catalysts have been developed and studied. The use of enzyme as a catalyst is less desirable because it is much more expensive than the homogeneous catalyst. Heterogeneous catalysts such as KOH/Al_2O_3 [8] and alumina/silica [9] have a high level of alkalinity and reliable performance for biodiesel production. Nevertheless, these catalysts have some drawbacks since they are easily dissolved in methanol, relatively difficult to synthesize, and susceptible to humidity. For acid catalyst, higher molar ratio of methanol/oil should be used, and it requires longer reaction time [8,10,11]. Therefore, a new heterogeneous catalyst should be developed, which is reactive at low temperature and pressure, has short time reaction for various raw materials, easy to get, has a high conversion, and low biodiesel production cost [12].

Golden snails (*Pomacea* sp.) are mostly found in muddy areas such as in a rice field. It is considered as pests; therefore, its presence in the rice field is not desired. However, the meat of *Pomacea* sp. has high protein content and can be utilized as animal feed or as

* Correspondence: aayucitra@yahoo.com; suryadiismadji@yahoo.com
Department of Chemical Engineering Widya Mandala Surabaya Catholic University, Kalijudan 37, Surabaya 60114, Indonesia

an alternative food for human consumption. In Indonesia, this mollusk has been cultivated in small industrial scale. Its meat has been utilized as raw materials for various kinds of food, while the shell is discarded as waste. The utilization of *Pomacea* sp. shell as an alternative raw material in the production of solid catalyst for biodiesel is expected to raise its economic value and helps to overcome the waste problems of *Pomacea* sp.

Pomacea sp. shell can be utilized as a solid catalyst for biodiesel production because it contains $CaCO_3$ which can be converted into CaO during calcination. CaO is a heterogeneous catalyst which is most widely studied because of its high alkalinity, low solubility, cheaper price than KOH/NaOH, and also the ease of handling [13,14].

This research was focused on developing a new type of catalyst for biodiesel production from *Pomacea* sp. shell. The capability of the catalyst derived from the snail shell was tested on the transesterification of palm oil into biodiesel. The process parameters for biodiesel production were also studied.

Methods

Materials and catalyst preparation

The *Pomacea* sp. shell was obtained from a small town called Pati, located in Central Java, Indonesia. Prior to use, the shell was repeatedly washed using tap water to remove dirt and other impurity material and then dried overnight at 100°C. The shell was then crushed and calcined at 900°C for 2 h in a tubular furnace to convert $CaCO_3$ into CaO. After the reaction is complete, nitrogen with a flow rate of 3 l/min was introduced to furnace to cool down the system and to prevent the air entering the system. Subsequently, the catalyst was removed from the furnace and kept in a desiccator to prevent contact with air.

Methanol used in this study was purchased as analytical grade from MERCK, Germany. Palm oil was purchased from the local supermarket, and the composition of fatty acid in the oil was analyzed using gas chromatography. The results are shown in Table 1. The gas chromatography analysis of fatty acid was performed using a Shimadzu GC-2014 equipped with FID and a DB-wax capillary column. The operating condition for fatty acid determination as follows: helium was used as a gas carrier with a flow rate of 40 m/s. The injector temperature was set at 200°C, while the detector was operated at 300°C. The initial oven temperature was 100°C and maintained in this isothermal condition for 2 min. Then, the oven temperature was increased to 220°C at a heating rate of 10°C/min and held for 33 min.

Other important parameters such as water content and acid value content in the oil are also given in Table 1. The acid value was determined according to ASTM D664, and the water content was determined according to ASTM D6304-07.

Table 1 Composition of fatty acid in palm oil and other characteristics

	Composition
Fatty acid (%)	
Lauric acid (C12:0)	0.81
Myristic acid (C14:0)	1.22
Palmitic acid (C16:0)	42.42
Palmitoleic acid (C16:1)	0.21
Stearic acid (C18:0)	4.68
Oleic acid (C18:1)	40.78
Linoleic acid (C18:2)	9.45
Linolenic acid (C18:3)	0.11
Arachidonic acid (C20:0)	0.32
Water content (%)	0.05
Acid number (mg KOH/g oil)	0.85

Transesterification reaction procedure

The transesterification of palm oil into biodiesel using CaO derived from *Pomacea* sp. shell was carried out in a 500-ml three-neck round-bottom flask equipped with a reflux condenser, thermometer, and magnetic heating mantle. The transesterification experiments were carried out at a temperature of 60°C and stirring speed at 700 rpm. The initial mass of the palm oil used in this study was 200 g. The amount of catalyst varied from 1% to 5% *w*/*w*, and oil to methanol ratios were 1:5 to 1:11. Reaction time used in this study was 4 h. After the reaction completed, the catalyst was removed from the mixture using vacuum filtration. The excess methanol was then recovered from the biodiesel mixture using a rotary vacuum evaporator. After the separation of methanol, the glycerol was removed from the biodiesel in a funnel separator. The biodiesel was obtained in the upper layer, while the glycerol as the byproduct was in the lower layer. For the comparison, biodiesel preparation using NaOH as a catalyst was also conducted under the same operating condition with those using CaO as catalyst.

Catalyst characterization

The raw material (*Pomacea* sp. shell) and the catalyst were characterized using several methods such as X-ray diffraction, X-ray fluorescense (XRF) spectrometry, Fourier transform infrared spectroscopy (FTIR) analysis, and nitrogen sorption analysis. XRD analysis was employed to investigate the crystalline structure of the raw material and catalyst. The diffraction spectra of both materials were obtained on a Philips X'Pert diffractometer (Almelo, The Netherlands) using CuKα as the radiation source. The measurement was conducted in 2θ angle between 8° to 72°. XRF spectrometry was used to identify and determine the bulk composition of a catalyst, and for this purpose, Rigaku ZSX100e X-ray fluorescence (Tokyo, Japan) was employed. FTIR measurement was carried out under FTIR spectrophotometer (Shimadzu 8400S, Nakagyo-ku, Kyoto, Japan) using the KBr technique with a wavenumber range of 4,000 to 700 cm^{-1}. Surface analysis of the catalyst was conducted by nitrogen sorption experiment at its boiling point (−196°C) using a Quadrasorb SI (Quantachrome Instruments, FL, USA). Degassing of catalyst before the nitrogen sorption measurement was conducted under vacuum condition for 24 h at a temperature of 150°C. The standard BET equation was employed to calculate the BET surface area of the catalyst. The BET calculation was performed in the relative pressure (p/p_0) range of 0.06 to 0.3.

Biodiesel characterization

Selected physical and chemical properties of biodiesel including viscosity, cetane index, fatty acid composition, flash point, water and sediment content, and heat of combustion were determined. The values were then compared with the Indonesian National Standard (SNI-04-7182-2006) and Biodiesel (B100)-ASTM D6751-07b. The cetane number was evaluated by ASTM D613 which is a standard method for cetane number of diesel fuel oil. The density was determined by ASTM D1298, while the kinematic viscosity using ASTM D445-10. Standard test method for water and sediment in middle distillate fuels using centrifuge (ASTM D2709) was employed to determine the water and sediment content in biodiesel. For flash point determination, ASTM D93 (standard test method for flash point by Pensky-Martens closed cup tester) was employed. The average values of physical and chemical characteristics of biodiesel obtained in this study are summarized in Table 2.

The fatty acid methyl ester (FAME) content in the biodiesel was determined using gas chromatography (GC) method. For this purpose, the GC (Shimadzu GC-2014) was equipped with a capillary column DB-wax (length 30 m × diameter 0.25 mm × film thickness 0.1 μm) and a flame ionization detector operated at 300°C. Helium was used as the carrier gas with a linear velocity of 40 cm/s. The injector temperature was 250°C at splitless condition. The initial oven temperature was 50°C with an equilibration time of 3 min. After isothermal period, the oven temperature was increased to 250°C at a heating rate of 10°C/min and held for 8 min. Peaks of methyl esters were identified by comparison with the reference standard. The biodiesel yield was calculated by the following expression: Yield = (weight of biodiesel × % FAME)/(weight of oil) × 100%. %FAME is the concentration of FAME obtained by GC analysis.

Results and discussion

Characterization of catalyst

The X-ray diffraction patterns of the raw material (Figure 1) and CaO catalyst (Figure 2) show the presence of $CaCO_3$ and CaO. The presence of $CaCO_3$ is indicated by 2θ at 26.2°, 33.1°, 37.8°, 45.8°, and 52.4°, while for CaO at 2θ, it is around 32.2°, 37.3°, and 53.8°. The presence of $Ca(OH)_2$ is also observed in Figure 2 as indicated by diffraction peaks at 18.0°, 28.6°, 34.1°, 47.0°, and 50.8°. Calcium oxide has strong basic property due to the presence of oxygen anion generated on its surface [15,16], and the exposure of this compound to ambient air results to the formation of $Ca(OH)_2$ due to the reaction of H_2O which is present in the air with its strong basic surface.

The bulk composition of a catalyst was determined by XRF, and the result is depicted in Figure 3. The composition of the catalyst produced by calcination of *Pomacea* sp. shell mainly consists of CaO (96.83%) with small amount of other metal as impurities as indicated in Figure 3. The FTIR spectra of both *Pomacea* sp. shell and catalyst are given in Figure 4. For both *Pomacea* sp. shell spectra, the characteristic peaks of C-O stretching and bending modes of $CaCO_3$ are observed at 2,513, 1,420, 867, and 3,117 cm^{-1}. During the calcination process, $CaCO_3$ decomposed into CaO and CO_2. This is in line with the FTIR spectra of the catalyst which indicate that the intensity of characteristic peaks belonging

Table 2 Comparison of the properties of biodiesel produced from palm oil

Property	Biodiesel produced in this study	SNI	(B100)-ASTM D6751-07b
Density at 15°C (g/mL)	0.87 ± 0.09	0.85 to 0.89	-
Kinematic viscosity at 40°C (cSt)	3.7 ± 0.2	2.3 to 6.0	1.9 to 6.0
Flash point (°C)	163 ± 2.2	100 min	93 min
Cetane number	57 ± 2.2	51 min	47 min
Acid number (mg KOH/g)	0.40 ± 0.07	0.8 max	0.5 max
Water and sediment (vol.%)	0.03 ± 0.006	0.05 max	0.05 max

With catalyst from *Pomacea* sp. shell with the Indonesian National Standard (SNI-04-7182-2006) and (B100)-ASTM D6751-07b.

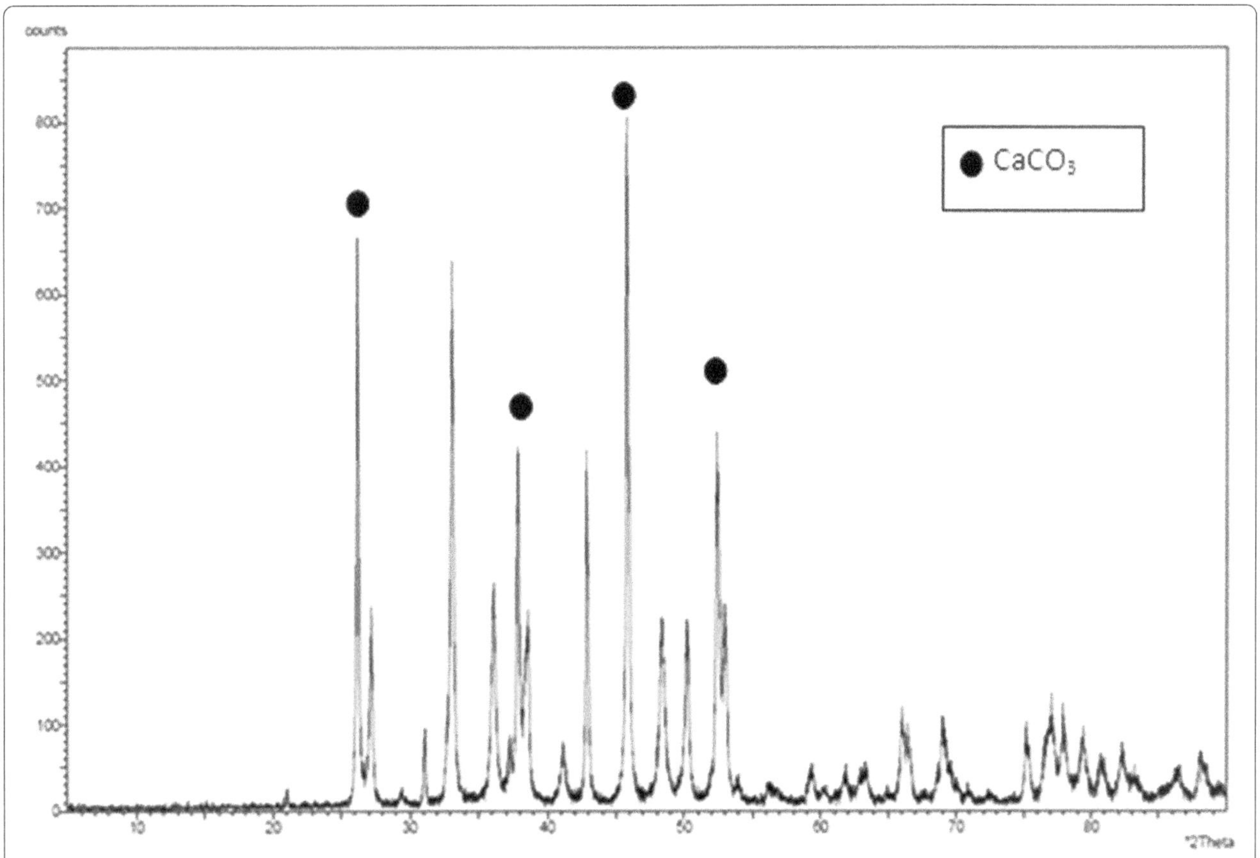

Figure 1 X-ray structure of *Pomaceae* sp. shell.

to $CaCO_3$ decreases. After the calcination process, a new peak appears at 3,620 cm^{-1}; this evidence indicates the formation of basic OH groups which attached to the calcium atoms [17]. The physical and chemical characteristics of CaO catalyst are summarized in Table 3.

Transesterification of palm oil

The transesterification of palm oil into biodiesel was conducted at different process parameters. Table 4 summarizes the yield of biodiesel obtained at different operating conditions (amount of catalyst 1% to 5% *w/w*, reaction time 4 h, and methanol to oil ratio 5:1 to 11:1). The maximum yield of biodiesel obtained in this study was 95.61 ± 2.43 at operation condition: catalyst amount of 4% *w/w* (catalyst to oil ratio) and methanol to oil ratio of 7:1. This result is comparable to that obtained by Chen et al. [18] in other heterogeneous catalyzed transesterification process and higher than the result obtained by Buasri et al. [19]. The maximum yield obtained from the process with heterogeneous catalyst was slightly lower than the process using NaOH as catalyst (97.24%). Even the heterogeneous catalyst produces lower yields of biodiesel, however, this process offers more advantages compared with homogeneous one, such as environmentally friendly, the catalyst is re-useable, etc.

By increasing the amount of catalyst, the yield of biodiesel also increased, as indicated in Table 4. In transesterification process, the CaO basic sites catalytically transform the methanol into nucleophile molecule which is much more reactive, and subsequently, this nucleophile attacks the carbonyl carbon structure of glyceride molecules [16,20]. As the amount of catalyst increases, the available catalytically basic sites also increase leading to the increase of biodiesel yield.

Stoichiometry of a transesterification reaction shows that in order to produce biodiesel, 3 mol of methanol/1 mol of triglyceride was needed [21]. Since the transesterification reaction is a reversible reaction, the adding of excess methanol drives the reaction towards the formation of biodiesel. In general, the conversion of triglyceride into fatty acid methyl esters (biodiesel) increased when the methanol to oil ratio was increased from 5:1 to 7:1. However, further increased of methanol to oil ratio caused the yield of biodiesel decreased, as indicated in Table 4. With the excessive amount of methanol, the formation of fatty acid methyl esters and glycerol becomes faster, and after the appreciable amount of glycerol was produced, the reaction between catalytically active sites of CaO with glycerol occurred and produced calcium glyceroxide [16,22]. This calcium glyceroxide is less active

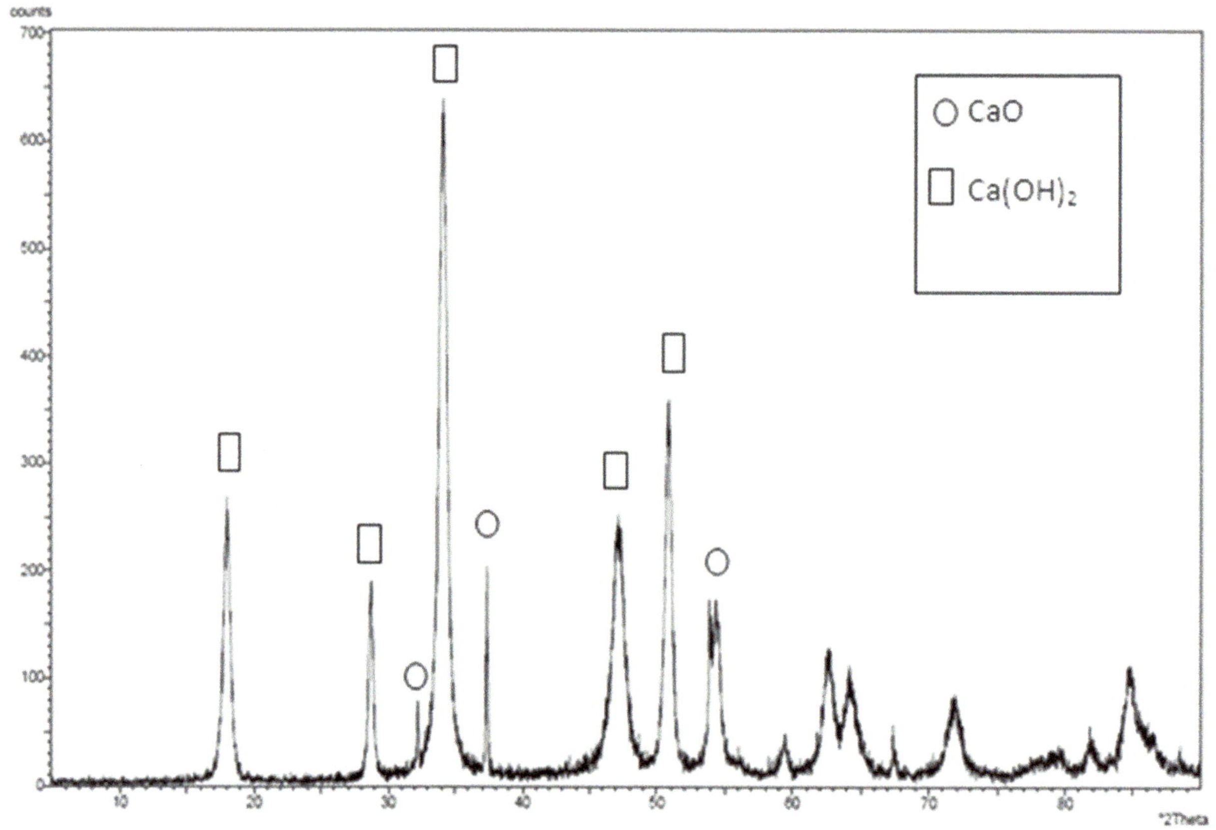

Figure 2 X-ray structure of the catalyst.

than CaO for transesterification of palm oil; therefore, less biodiesel was obtained. More detailed information about the reaction mechanism of the formation of calcium glyceroxide and its effect on the transesterification reaction may be found in the paper of Kouzu and Hidaka [16].

One of the important considerations in selecting a heterogeneous catalyst for transesterification of vegetable oils in the industrial scale production is its stability. The catalyst should be stable in a period time of operation without losing its catalytic capability or activity. In order to test the stability of CaO catalyst derived from *Pomacea* sp. shell, three reaction cycles were carried out using the following reaction condition: the amount of catalyst was 4% *w*/*w*, methanol to oil ratio was 7:1, and reaction time was 4 h. Furthermore,

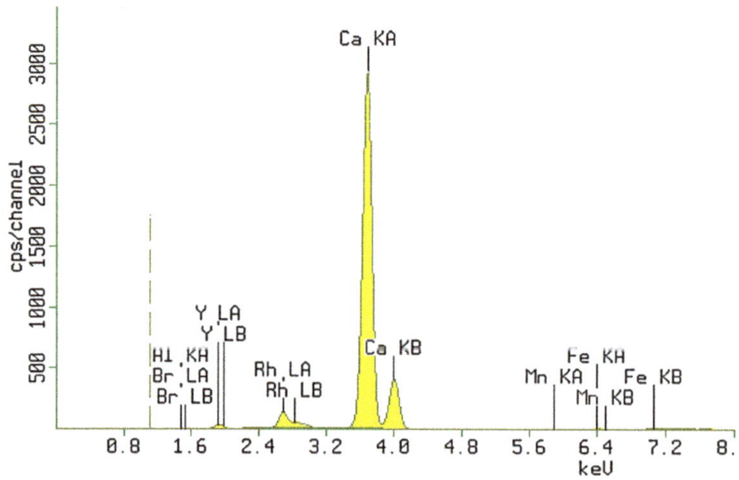

Figure 3 XRF spectra of the catalyst.

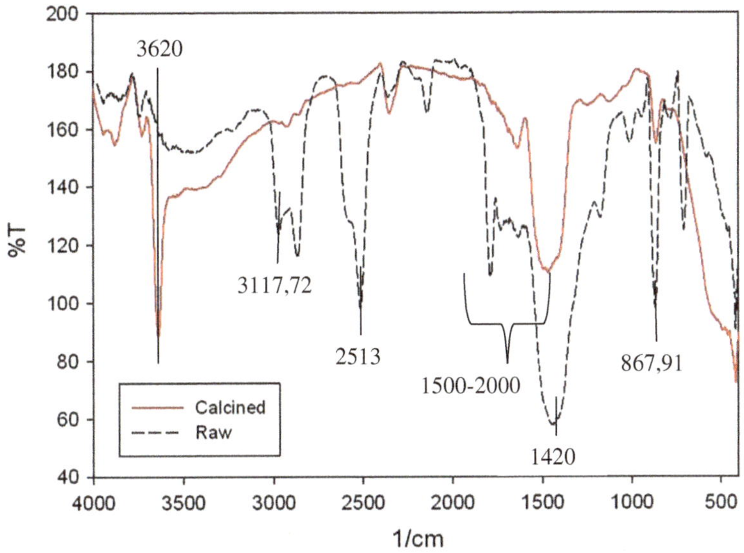

Figure 4 FTIR spectra of the raw material and CaO catalyst.

three different procedures for spent catalyst treatment were employed. The first procedure is the cheapest and easiest procedure; the spent or used catalyst was directly used without any further treatment for subsequent transesterification process. The second procedure involved the washing of spent catalyst with methanol to remove biodiesel and unconverted palm oil from the surface of the catalyst. As the last procedure, re-calcination of spent catalyst was conducted to determine the stability of the catalyst.

Figure 5 depicts the yield of biodiesel at different spent catalyst treatments under cycle reaction conditions. Figure 5 clearly shows that the treatment of spent catalyst has great influence on the yield of biodiesel. For untreated spent catalyst, the deactivation of the CaO catalyst already occurred in the second cycle, thus the yield of biodiesel significantly decreased from around 95% to 38% (Figure 5a). Further use of this spent catalyst made the conversion of palm oil into fatty acid methyl esters (biodiesel) did not occur. During the separation of spent catalyst from the liquid mixture, the adsorption of H_2O from the ambient atmosphere by the basic sites of CaO catalyst could not be avoided which caused the de-activation of catalyst. The reaction between the basic sites of CaO and the ambient H_2O was fast, and

Table 3 Physical and chemical characteristics of the CaO catalyst

Property	Value
BET surface area (m^2/g)	17
Pore volume (cm^3/g)	0.04
Mean pore size (nm)	3.2
Particle size (mesh)	80/100
CaO content (%)	96.83
SiO_2 (%)	0.60
FeO (%)	0.42
Others	2.15

Table 4 Effect of different process variables on the yield of biodiesel

Amount of catalyst (%)	Reaction time (h)	Methanol to oil ratio	Yield of biodiesel (%)
1	4	5:1	70.86 ± 3.46
		7:1	86.63 ± 4.72
		9:1	78.98 ± 3.14
		11:1	83.50 ± 2.87
2	4	5:1	81.91 ± 2.44
		7:1	86.33 ± 3.48
		9:1	83.51 ± 4.22
		11:1	84.07 ± 2.73
3	4	5:1	86.93 ± 3.03
		7:1	87.60 ± 1.98
		9:1	91.26 ± 2.51
		11:1	89.22 ± 4.02
4	4	5:1	91.91 ± 3.91
		7:1	95.61 ± 2.43
		9:1	90.16 ± 2.78
		11:1	92.55 ± 3.82
5	4	5:1	93.48 ± 3.07
		7:1	85.81 ± 4.32
		9:1	85.28 ± 2.17
		11:1	91.16 ± 3.44

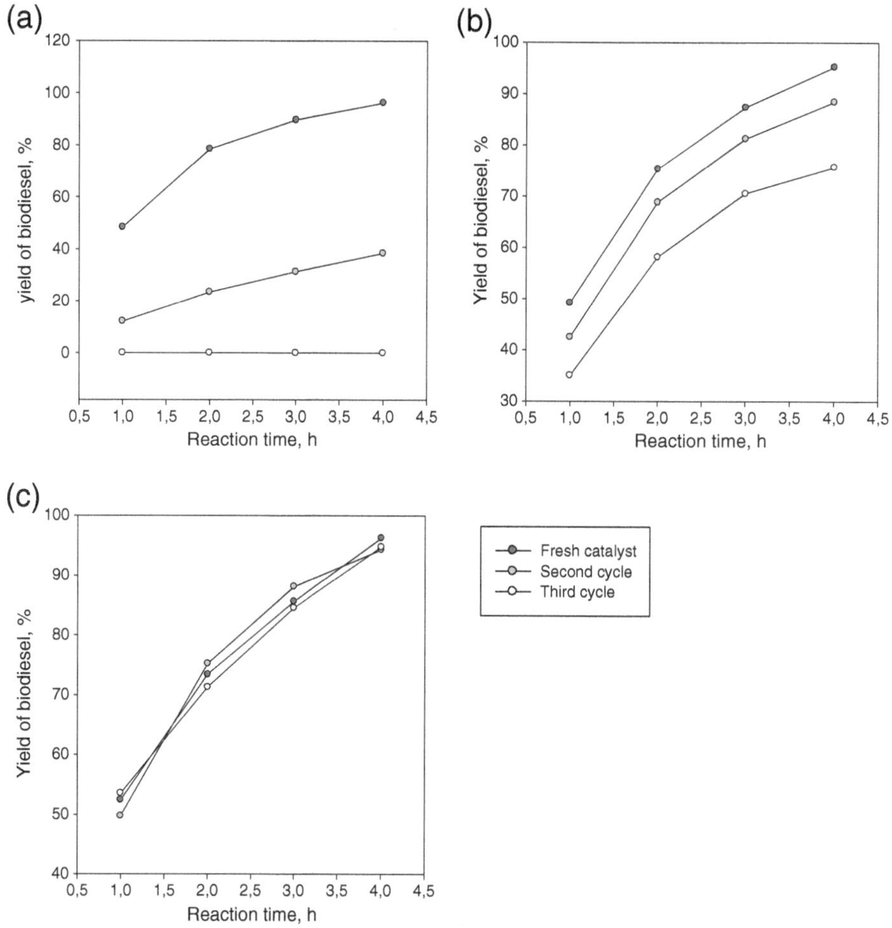

Figure 5 The yield of biodiesel at different spent catalyst treatments under cycle reaction conditions. (**a**) Untreated spent catalyst, (**b**) washing of spent catalyst with organic solvent, and (**c**) re-calcination of spent catalyst.

according to Kouzu et al. [22], it took only 3 min to appreciably reduce the catalyst activity.

The treatment of spent catalyst with methanol gave better yields of biodiesel than non-treated spent catalyst. Here, the conversion of palm oil into FAME for the second and third cycles was higher than the transesterification process using non-treated spent catalyst, as indicated in Figure 5b. As mentioned in the previous paragraph, the contact between the basic site and ambient H_2O may cause the deactivation of catalyst due to the formation of hydroxide groups on the surface of the catalyst, and during washing of this spent catalyst with methanol, the hydroxide groups on the surface of catalyst were isolated; these isolated OH groups with the help of the basic site of CaO still could act as an active site for transesterification reaction [16,23].

Figure 5c shows the yield of biodiesel synthesized using re-calcined spent catalyst. After transesterification process, the surface of spent catalyst was covered by hydrate, carbonate, and other compounds such as unreacted oil, biodiesel, glycerol, etc.; therefore, the catalytic activity of this spent catalyst drastically reduced. In the re-calcination process, all of those impurities were perfectly removed from the surface of the catalyst. By removing those impurities, the available basic sites in CaO surface also significantly increase, and the catalytic activity is similar to fresh catalyst. However, one of the drawbacks of using re-calcination method is the high energy requirement.

Biodiesel characterization

Density is one of the most important characteristics of biodiesel. In a combustion system, pumps and injectors require a precise amount of fuel to provide proper combustion [24], and fuel density is the main characteristic that determines the amount or mass of fuel injected into the combustion chamber. The density of the biodiesel produced in this study is given in Table 2, and the value is in accordance with the Indonesian National Standard (SNI-04-7182-2006).

Viscosity of biodiesel is another important characteristic of biodiesel; it affects the atomization of fuel upon injection into the combustion chamber. Fuel with high viscosity has the tendency to form deposit in the engine, but fuel with low viscosity may not provide sufficient lubrication in the combustion system. The Indonesian National Standard has set the standard viscosity for biodiesel which is in the range of 2.3 to 6.0 cSt, while the ASTM viscosity standard for biodiesel is 1.9 to 6.0 cSt. The average viscosity values of biodiesel produced from palm oil using CaO catalyst derived from *Pomacea* sp. shell is 3.7 ± 0.2 cSt. This average viscosity is in the range of the standard value of viscosity of biodiesel from SNI and ASTM.

Flash point of biodiesel relates to the ignition of the fuel. In order to handle the fuel safely, moderate value of flash point is required. The biodiesel produced in this study possesses higher flash point value than SNI and ASTM standards (Table 2); therefore, this fuel is safe for handling and storage for a period of time. The combustion quality of diesel fuel during compression ignition is indicated by an index called as cetane number. This index has great influence on the performance of the engine especially during combustion and exhaust emission. The low value of cetane index creates the problem to the engine during start up and produces noise. Also, during the combustion, smoke exhaust gas will be produced [25]. The average cetane number of the biodiesel obtained in this study is higher than the standard value of SNI and ASTM (Table 2). The cetane number of biodiesel depends on the distribution of fatty acids in the raw material (oil); the longer fatty acid carbon chains, the higher the cetane number. In this study, palm oil was employed as raw material for biodiesel production, where methyl palmitate and methyl oleate are the major compounds in palm oil. These fatty acids have cetane indexes of 59.3 and 85.9, respectively [25]. Therefore, the minimum standard of cetane index will be easily fulfilled if palm oil is used as the raw material for biodiesel production. Furthermore, the other physical characteristic of biodiesel is also affected by the composition and length of fatty acid as mentioned by Ramos et al. [26].

The control of acid number on biodiesel production is also important since this parameter indicates the quantity of free fatty acids and mineral acids present in the biodiesel. Biodiesel with high value of acidity will cause corrosion and engine deposits, especially in the fuel injectors. The increase of acid number during storage is a strong indication that the degradation of biodiesel takes place. Water contamination is also another important parameter in biodiesel quality. Several problems associated with the presence of excess water in biodiesel are low calorific value, corrosion, growth of microorganisms, oxidation, and degradation of biodiesel during long-term storage. The standard values of these parameters were fulfilled by the biodiesel produced in this study as indicated in Table 2.

Conclusion

A catalyst for biodiesel production was prepared from the *Pomacea* sp. shell. The transesterification reaction of palm oil into biodiesel using this heterogeneous catalyst was studied under various operating conditions, and the maximum yield of biodiesel (95.61 ± 2.43) was obtained on the condition as follows: catalyst 4% and methanol to oil ratio 7:1. The reusability and stability are very important criteria for large-scale production of biodiesel using CaO catalyst prepared from *Pomacea* sp. shell. The activity of spent catalyst could be restored by re-calcination process. The physical and chemical properties of biodiesel produced conform to the available standards.

Competing interests

The authors declare that they have no competing interest.

Authors' contributions

YYM and HSP collected the experimental data, AA drafted the manuscript, and SI developed the idea and corrected the manuscript. All authors read and approved the final manuscript.

Acknowledgments

The authors would like to express their sincere gratitude to the editor (Professor Farivar Fazelpour) and anonymous reviewers for their invaluable comments and suggestions, which substantially improved the overall quality of the manuscript. The funding from the Department of Chemical Engineering, Widya Mandala Surabaya Catholic University is also greatly appreciated.

References

1. Eddine, BT, Salah, MM: Solid waste as renewable source of energy: current and future possibility in Algeria. Int. J. Energy Env. Eng. **3**, 17 (2012)
2. Marasabessy, A, Kootstra, AMJ, Sanders, JPM, Weusthuis, RA: Dilute H_2SO_4-catalyzed hydrothermal pretreatment to enhance enzymatic digestibility of jatropha curcas fruit hull for ethanol fermentation. J. Energy Env. Eng. **3**, 15 (2012)
3. Kouzu, M, Kasuno, T, Tajika, M, Sugimoto, Y, Yamanaka, S, Hidaka, J: Calcium oxide as a solid base catalyst for transesterification of soybean oil and its application to biodiesel production. Fuel **87**, 2798–2806 (2007)
4. Dias, JM, Ferraz, M, Almeida, MF: Comparison of the performance of different homogeneous alkali catalysts during transesterification of waste and virgin oils and evaluation of biodiesel quality. Fuel **87**, 3572–3578 (2008)
5. Pugnet, V, Maury, S, Coupard, V, Dandeu, A, Quoineaud, AA, Bonneau, JL, Tichit, D: Stability, activity and selectivity study of a zinc aluminate heterogeneous catalyst for the transesterification of vegetable oil in batch reactor. Appl. Catal. A. **374**, 71–78 (2010)
6. Demirbas, A: Progress and recent trends in biodiesel fuels. Energy Conv. Manag. **50**, 14–34 (2009)
7. Liu, X, He, H, Wang, Y, Zhu, S: Transesterification of soybean oil to biodiesel using SrO as a solid base catalyst. Catal. Comm. **8**, 1107–1111 (2007)
8. Noiroj, K, Intarapong, P, Luengnaruemitchai, A, Jai-In, S: A comparative study of KOH/Al_2O_3 and KOH/NaY catalysts for biodiesel production via transesterification from palm oil. Renewable Energy **34**, 1145–1150 (2009)
9. Lukić, I, Krstić, J, Jovanović, D, Skala, D: Alumina/silica supported K2CO3 as a catalyst for biodiesel synthesis from sunflower oil. Biores. Technol. **100**, 4690–4696 (2009)
10. Gao, L, Teng, G, Lv, J, Xiao, G: Biodiesel synthesis catalyzed by the KF/Ca-Mg-Al hydrotalcite base catalyst. Energy Fuel **24**, 646–651 (2010)
11. Ngamcharussrivichai, C, Totarat, P, Bunyakiat, K: Ca and Zn mixed oxide as a heterogeneous base catalyst for transesterification of palm kernel oil. Appl. Catal. A. **341**, 77–85 (2008)
12. Serio, MD, Cozzolino, M, Tesser, R, Patrono, P, Pinzari, F, Bonelli, B, Santacesaria, E: Vanadyl phosphate catalysts in biodiesel production. Appl. Catal. A. **320**, 1–7 (2007)

13. Yoosuk, B, Udomsap, P, Puttasawat, B, Krasae, P: Improving transesterification activity of CaO with hydration technique. Biores. Technol. **101**, 3784–3786 (2010)
14. Viriya-Empikul, N, Krasae, P, Puttasawat, B, Yoosuk, B, Chollacoop, N, Faungnawakij, K: Waste shells of mollusk and egg as biodiesel production cata. Biores. Technol. **101**, 3765–3767 (2010)
15. Iizuka, T, Hattori, H, Ohno, Y, Sohma, J, Tanabe, K: Basic sites and reducing sites of calcium oxide and their catalytic activities. J. Catal. **22**, 130–139 (1971)
16. Kouzu, M, Hidaka, J: Transesterification of vegetable oil into biodiesel catalyzed by CaO: a review. Fuel **93**, 1–12 (2012)
17. Garcia, JLT, Alvarez, M, Aguilar, DH, Quintana, PJ: Spectroscopic, structural and textural properties of CaO and CaO-SiO2 materials synthesized by sol–gel with different acid catalysts. Non-Crystal Solids **354**, 729 (2008)
18. Chen, YH, Huang, YH, Lin, RH, Shang, NC, Chang, CY, Chang, CC, Chiang, PC, Hu, CY: Biodiesel production in a rotating packed bed using K/γ-Al_2O_3 solid catalyst. J. Taiwan Inst. Chem. Eng. **42**, 937–944 (2011)
19. Buasri, A, Chaiyut, N, Loryuenyong, V, Rodklum, C, Chaikwan, T, Kumphan, N: Continuous process for biodiesel production in packed bed reactor from waste frying oil using potassium hydroxide supported on *Jatropha curcas* fruit shell as solid catalyst. Appl. Sci. **2**, 641–653 (2012)
20. Hattori, H, Shima, M, Kabashima, H: Alcoholysis of ester and epoxide catalyzed by solid base. Stud. Surf. Sci. Catal. **130**, 3507–3510 (2000)
21. Singh, SP, Singh, D: Biodiesel production through the use of different sources and characterization of oils and their esters as the substitute of diesel: a review. Renew Sustain Energy Rev. **14**, 200–216 (2009)
22. Kouzu, M, Kasuno, T, Tajika, M, Yamanaka, S, Hidaka, J: Active phase of calcium oxide used as solid base catalyst for transesterification of soybean oil with refluxing methanol. Appl. Catal. A. Gen. **334**, 357–365 (2008)
23. Liu, X, He, H, Wang, Y, Zhu, S, Piao, X: Transesterification of soybean oil to biodiesel using CaO as a solid base catalyst. Fuel **87**, 216–221 (2008)
24. Dzida, M, Prusakiewicz, P: The effect of temperature and pressure on the physicochemical properties of petroleum diesel oil and biodiesel fuel. Fuel **87**, 1941–1947 (2008)
25. Rashid, U, Anwar, F, Knothe, G: Biodiesel from Milo (*Thespesia populnea L.*) seed oil. Biomass Bioenergy **35**, 4034–4039 (2011)
26. Ramos, MJ, Fernandez, CM, Casas, A, Rodriguez, L, Perez, A: Influece of fatty acid composition of raw materials on the biodiesel properties. Biores. Technol. **100**, 261–268 (2009)

7

Performance of a diesel engine with blends of biodiesel (from a mixture of oils) and high-speed diesel

Hifjur Raheman*, Prakash C Jena and Snehal S Jadav

Abstract

A 10.3-kW single-cylinder water-cooled direct-injection diesel engine was evaluated using blends of biodiesel (B10 and B20) obtained from a mixture of mahua and simarouba oils (50:50) with high-speed diesel (HSD) in terms of brake specific fuel consumption, brake thermal efficiency, and exhaust gas temperature and emissions such as CO, HC, and NO_x. Based on performance and emissions, blend B10 was selected for long-term use. Experiments were also conducted to assess soot deposits on engine components, such as cylinder head, piston crown, and fuel injector tip, and addition of wear metal in the lubricating oil of diesel engine when operated with the biodiesel blend (B10) for 100 h. The amount of soot deposits on the engine components was found to be, on average, 21% lesser for B10-fueled engine as compared with HSD-fueled engine due to better combustion. The addition of wear metals such as copper, zinc, iron, nickel, lead, magnesium, and aluminum, except for manganese, in the lubricating oil of B10-fueled engine after 100 h of engine operation was found to be 11% to 50% lesser than those of the HSD-fueled engine due to additional lubricity.

Keywords: Biodiesel blend, Performance, Emissions, Soot deposit, Wear metal addition

Background

Higher soot or carbon deposits on in-cylinder engine components and lubricating oil contamination are the main causes for engine wear. Wear processes due to oil contamination lead to diminished fuel efficiency, shorter useful oil service life, reduced component life, and loss of engine performance. Hence, in addition to engine performance and emissions of diesel engine, soot depositions on engine components and wear metal additions in lubricating oil of the engine are required for the selection of a fuel that would replace conventional diesel fuel. With increase in demand for using biodiesel in place of high-speed diesel (HSD), a detailed study on these aspects is very much required for the selection of a suitable blend to replace HSD for long-term use in diesel engines.

Very few studies have been conducted on soot depositions and wear metal additions in the lubricating oil of engine that are operated with biodiesel obtained from single-vegetable oil and its blends with HSD. It was reported that metal addition in the lubricating oil of engines operated with different biodiesel (soybean, palm kernel oil, linseed oil) blends was lower or similar as compared to when the engine was operated with HSD alone [1-5]. Ramaprabhu et al. [6] reported that iron wear was almost similar, whereas copper wear was higher in the lubricating oil of diesel engine operated with Jatropha and Karanja biodiesel blend as compared with when the lubricating oil of diesel engine was operated with HSD. Lesser soot deposition on engine components was reported for biodiesel blend-operated diesel engines [6,7]. With increasing demand on the use of biodiesel, more and more oils and mixtures of oils are explored for biodiesel production [8]. Mahua (*Madhuca indica*) oil (MO) with high free fatty acids (FFA) and simarouba (*Simarouba glauca*) oil (SRO) are few such potential oils suitable for biodiesel production. Higher FFA present in oil requires higher methanol; hence, an attempt was made to produce biodiesel from a mixture (50:50) of these two oils to reduce methanol requirement in biodiesel production [9]. Performance evaluation,

* Correspondence: hifjur@agfe.iitkgp.ernet.in
Agricultural & Food Engineering Department, Indian Institute of Technology, Kharagpur 721302, India

emission examination along with soot deposition on in-cylinder engine components, and wear metal addition in the lubricating oil of the engine are required for biodiesel obtained from such mixture of oils. Hence, a study was undertaken at Indian Institute of Technology, Kharagpur, India.

Methods

Biodiesel production

MO and SRO with FFA levels of 13% and 1.43%, respectively, were mixed at 50:50 *v*/*v* proportions (MSO) to reduce methanol consumption in biodiesel production. This oil mixture (MSO) had an FFA content of 7.19%. Hence, pretreatment esterification with an acid catalyst to bring down the FFA level of the oil mixture to around 1% was required.

A two-step 'acid-base' process, an acid pretreatment followed by the main base-transesterification reaction, using methanol as reagent and H_2SO_4 and KOH as catalysts for acid and base reactions, respectively, was followed to produce biodiesel from MSO. The various fuel properties of MO, SRO, MSO, and biodiesel (B100)

Table 1 Technical specification of engine and hydraulic dynamometer

	Particulars	Details
Engine	Model	DM14
	Maximum power (kW)	10.3
	Type	Water-cooled, four stroke
	Rated speed (rpm)	1,500
	Number of cylinders	1
	Compression ratio	15.5:1
	Bore × stroke (mm)	114.3 × 116
	Brake mean effective pressure at 1,500 rpm (kg/cm^2)	7.054
	Combustion	Direct injection (DI) and naturally aspirated
	Injection timing	24° before TDC
Hydraulic dynamometer	Model	AWM15
	Type	Hydraulic
	Water pressure at inlet (kg/cm^2)	1.5
	Power (hp)	
	Range	1 to 100
	Accuracy	±2%
	Data resolution	0.02%
	Maximum speed (rpm)	
	Range	5,650 to 8,000
	Accuracy	±2%
	Data resolution	0.03
Exhaust gas analyzer	Model	PEA205
	CO (%)	
	Range	0 to 15
	Accuracy	±0.06
	Data resolution	0.001
	HC (ppm)	
	Range	0 to 30,000
	Accuracy	±4%
	Data resolution	1
	NO_x (ppm)	
	Range	0 to 5,000
	Accuracy	±2%
	Data resolution	1

obtained from MSO and its blends with HSD (B10, B20) were determined as per the American Society for Testing and Materials (ASTM) standards.

Experimental setup

Performance and exhaust emissions were studied for a 10.3-kW single-cylinder four-stroke direct-injection water-cooled diesel engine using biodiesel blends with HSD (B10 and B20) by varying the engine load from no-load to 100% load in steps of 20%. A hydraulic dynamometer (SAJ Model AWM 15, SAJ International Pvt. Ltd, Pune, India; 100 hp at 5,650 to 8,000 rpm) equipped with a strain gauge-based load cell and digital readout for measuring engine torque and speed was used for loading the diesel engine. Details of the technical specification for the engine, dynamometer, and exhaust gas analyzer used are given in Table 1.

Methodology

Performance tests and emission measurement of diesel engine with biodiesel blends

Performance tests were conducted in a diesel engine using HSD and blends of biodiesel (B10 and B20) following the Indian standard (IS) 10000: part 8 [10]. The rated engine load measured was found to be 57.8 Nm at 1,500 ± 10 rpm, and the corresponding rated power developed by the engine was 9.1 kW. At rated power, the load going to the engine was taken as 100%. Accordingly, intermediate loads were calculated as 11.56, 23.12, 34.68, and 46.24 Nm (corresponding to 20%, 40%, 60%, and 80% of torque obtained at rated power). The tests were conducted for 3 h and 30 min for each test fuel, and load was applied at an interval of 30 min. The readings were recorded at an interval of 10 min for a particular engine load. During this test, performance parameters such as brake specific fuel consumption (BSFC), brake thermal efficiency (BTE), and exhaust gas temperature (EGT) were determined by measuring fuel consumption, engine torque, and EGT. During this performance test, exhaust emissions such as CO, hydrocarbon (HC), and oxides of nitrogen (NO_x) were also measured using an online exhaust gas analyzer. A suitable blend for long-term use was selected by comparing the performance and emissions of the two blends with those of HSD.

Estimation of soot deposits on in-cylinder engine components

The engine was run for 100 h (16 test cycles each of 6.25-h continuous running) with each of the two fuels (HSD and selected biodiesel blend). In 100 h, the engine was subjected to different loadings, and the lubricating oil samples were collected at an interval of 25 h of engine run. The duration for each loading was decided as per IS 10000: part 9 [11]. In each test cycle, loads were applied in a random manner for predetermined durations, i.e., 100%, 50%, 100%, no load, 100%, and 50% load for 93.75 (including 11.72 min of warm up), 93.75, 23.45, 11.72, 70.31, and 82 min, respectively. After 100 h of engine operation with each of these fuels, the engine was dismantled, and the cylinder head, piston, and fuel injector were removed carefully and kept on a clean surface. Photographs of these components were taken to visually compare the soot deposits when operated with different fuels. Later, the soot deposits were gently scraped from these components using a wooden scraper. Weights of these scraped deposits for each component were taken separately for comparison.

Detection of wear metals by atomic absorption spectroscopy analysis

The lubricating oil samples collected after each 25 h of engine run were subjected to atomic absorption spectroscopy (AAS) analysis to determine the addition of metals (such as copper, zinc, iron, manganese, nickel, lead, magnesium, and aluminum) in lubricating oil of engine. Dry ashing technique was used to prepare the sample of lubricating oil for metal analysis by AAS. Each lubricating oil sample was taken in a 250-ml conical flask and thoroughly mixed in a water bath at 50°C for 1 h at constant speed. Approximately 10 g of mixed lubricating oil sample was taken in a previously washed and dried silica crucible. The crucible was then kept on a hot plate at a temperature of 120°C until the lubricating oil gets completely dried up. Thereafter, the crucible was kept in a muffle furnace for 4 h at a temperature of 450°C and then for 2 h at 650°C. The ash that remained in crucibles was dissolved in 1.5 ml of HCl solution. The solution was then diluted with 100 ml of deionized water and stored in plastic bottles in a refrigerator at a temperature of 10°C to 15°C. This method was followed to prepare all samples for metal analysis by flame AAS. Variation of these elements was found for different engine operating hours.

Results and discussion

Fuel properties

Various fuel properties of MO, SRO, MSO, biodiesel obtained from this mixture of oils, and its blends with HSD (B10 and B20) were determined as per the ASTM standards and are summarized in Table 2. It can be seen from this table that the fuel properties of biodiesel are comparable to those of HSD and are well within the latest American (D 6751–02) and European (EN 14214) standards for biodiesel. The MO, SRO, and MSO, however, were found to have much higher values of fuel properties way above any of these standard limits, thus restricting its direct use as a fuel for diesel engines.

Table 2 Fuel properties of biodiesel obtained from MO, SRO, mixture of MSO and their biodiesels

Fuel type	Acid value (mg KOH/g)	Density (kg/m^3)	Kinematic viscosity (cSt)	Calorific value (MJ/kg)	Flash point (°C)	Pour point (°C)	Carbon residue (%)	Ash content (%)	Water content (ppm)
HSD	-	812	2.85	42.5	52	−20.0	0.15	0.01	90
MO	26	913	40.92	36.5	233	14.0	1.6	1.13	1,200
MB100	0.46	861	5.38	37.05	168	3.0	0.22	0.01	450
SRO	2.86	912	44.95	36.6	245	15.0	1.50	1.17	1,400
SRB100	0.39	862	5.58	37.0	146	4.0	0.21	0.01	445
MSO[a]	14.38	912	42.94	36.55	238	14.0	1.60	1.14	1,300
B10	0.27	817	3.02	42.0	63	−18.0	0.15	0.01	126
B20	0.27	821	3.19	41.4	74	−15.0	0.16	0.01	159
B100	0.33	857	4.56	37.02	164	3.5	0.21	0.013	450
ASTM D6751	<0.80	-	1.9 to 6.0	-	>130	-	-	<0.02	<500
EN14214	<0.50	860 to 900	3.5 to 5.0	-	>120	-	<0.30	<0.02	<500
BIS15607	<0.50	860 to 900	2.5 to 6.0	-	>120	-	-	-	<500

MB100, mahua biodiesel; SRB100, simarouba biodiesel; B100, biodiesel obtained from MSO; B10, 10% MSO with HSD by volume basis; B20, 20% MSO with HSD by volume basis. [a]MSO, mixture of MO and SRO at 50:50 *v/v*.

Engine performance

The performance parameters such as BSFC, BTE, and EGT obtained with B10, B20, and HSD are found to be affected by fuel blend and engine loading and are discussed in the following sections.

Brake specific fuel consumption

The variations of BSFC with engine load for B10, B20, and HSD are shown in Figure 1. BSFC, in general, was found to increase with an increase in proportion of biodiesel in the fuel blends with HSD. Mean BSFC values with B10 and B20 were found to be 287.29 and 298.67 g/kW h, respectively, and were 2.44% and 5.63% higher than those with HSD. Among the fuel blends tested, 10% blending gave the minimum BSFC, and it increased with the increase in biodiesel percentage in the blends.

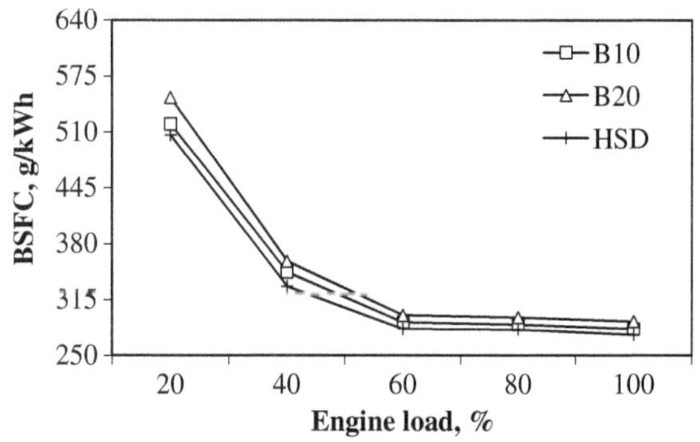

Figure 1 Variations of BSFC with engine load for B10, B20, and HSD.

BSFC was also observed to decrease sharply with the increase in engine loading for all the fuels tested due to relatively less amount of heat losses at higher loads. At full load, the mean BSFCs for HSD, B10, and B20 were found to be 274.29, 281.05, and 290.00 g/kW h, respectively, as compared with 506.76, 519.52, and 550.37 g/kW h at 20% engine loading. This decrease in BSFC with the increase in engine load might be due to the fact that percentage increase in fuel required to operate the engine was less than the percentage increase in brake power as relatively less portion of the heat losses occurred at higher engine loads.

Brake thermal efficiency

BTE of diesel engine when operated with HSD, B10, and B20 at different engine loads has been plotted in Figure 2. At 20% engine loading, BTE values were found to be 16.72%, 16.50%, and 15.80% with HSD, B10, and B20, respectively, which were increased to 30.89%, 30.50%, and 29.99% at full load condition. The BTE improved with the engine load for the main reason that a relatively less portion of the power was lost with the increase in engine load. The mean BTE with B10 and B20 was found to be 21.87% and 21.46%, respectively, as compared to 22.27% with HSD. There was a reduction in BTE with the increase in biodiesel percentage in the fuel blends due to the decrease in calorific value of fuel blend.

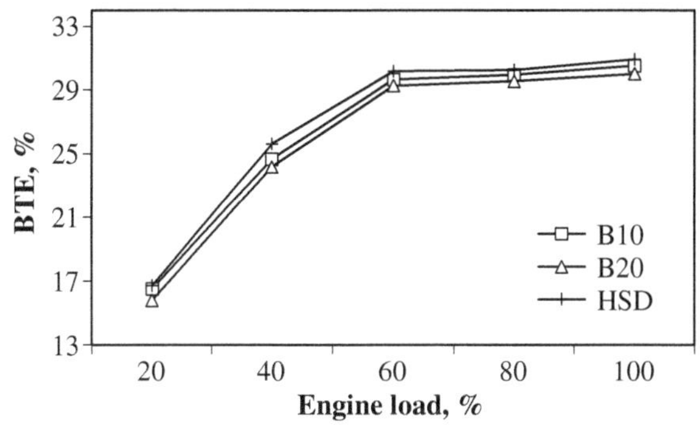

Figure 2 Variations of BTE with engine load for B10, B20, and HSD.

Exhaust gas temperature

The variations of EGT of a diesel engine when operated with HSD, B10, and B20 at different engine loads are shown in Figure 3. EGT was found to increase with the increase in both concentrations of biodiesel in the blends and engine load. The mean EGT values with B10 and B20 were found to be 308°C and 310°C, respectively, which were 4.44% and 5.12% higher than that with HSD (296°C). The maximum EGT was obtained at full load conditions with all the fuels tested and was 430°C, 438°C, and 443°C with HSD, B10, and B20, respectively, whereas it was 213°C, 224°C, and 225°C at 20% engine loading. The increase in EGT with engine load is obvious from the simple fact that a higher amount of fuel was required in the engine to generate that extra power needed to take up the additional loading.

The variations of BSFC, BTE, and EGT for blends of biodiesel (B10 and B20) obtained from the mixture of oils and HSD at different engine loadings were found to be similar to the trend as reported, while testing biodiesel blends obtained from individual oils (sunflower, linseed, Karanja, rubber seed, rapeseed, soybean, polanga, waste palm, castor, soybean, mahua, and tamanu oils) and HSD in different diesel engines [12-24].

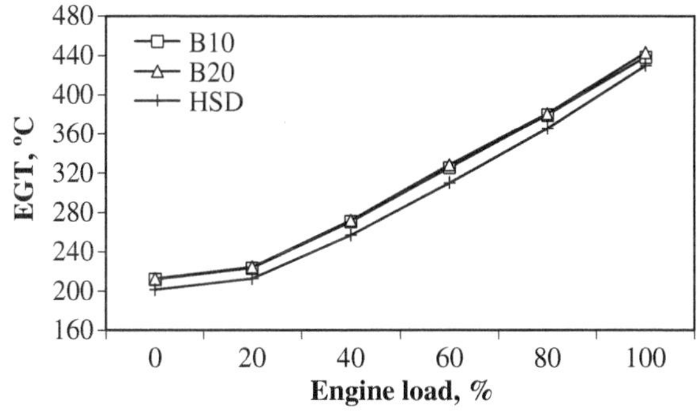

Figure 3 Variations of EGT with engine load for B10, B20, and HSD.

Engine emissions

The average exhaust emissions such as CO, HC, and NO_x from diesel engine with different fuels tested were found to be affected by biodiesel blend and engine load and are discussed in the following sections.

Carbon monoxide

The variations of CO with engine load when operated with B10, B20, and HSD are presented in Figure 4. CO emission was found to decrease with the increase in proportion of biodiesel in the fuel blends with HSD. The mean values of CO emissions with B10 and B20 were found to be 10.97% to 21.16% lower than those with HSD. Among the two fuel blends tested, B10 blending gave the maximum CO emission, and it decreased with the increase in biodiesel percentage in the blends.

The mean values of CO emissions for B10 decreased from 0.031% at no-load conditions down to 0.014% at 60% engine load and then increased up to 0.071% at 100% engine load. A similar trend was also observed for B20 and HSD. Initially, at no-load condition, cylinder temperature might be too low, and then, it increased with loading due to more fuel injected inside the cylinder. At an elevated temperature, performance of the engine improved with relatively better burning of the fuel, resulting in decreased CO. However, on further loading, the excess fuel required led to the formation of more smoke, which might have prevented the oxidation of CO into CO_2, consequently increasing the CO emissions sharply.

Hydrocarbons

The HC emissions from the diesel engine when operated with B10, B20, and HSD at different engine loadings are plotted in Figure 5. The mean values of HC emissions

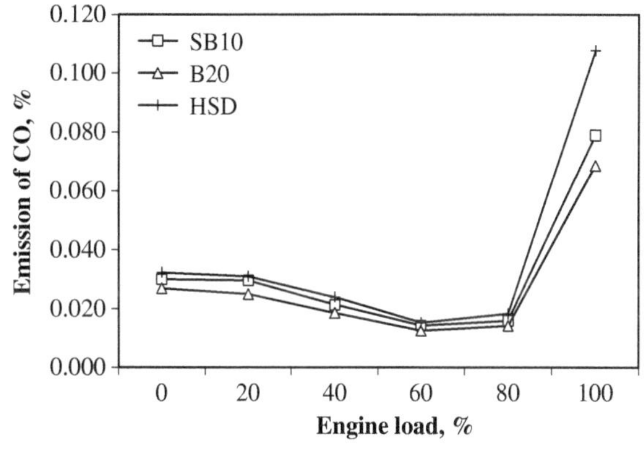

Figure 4 Variations of CO emissions with engine load for B10, B20, and HSD.

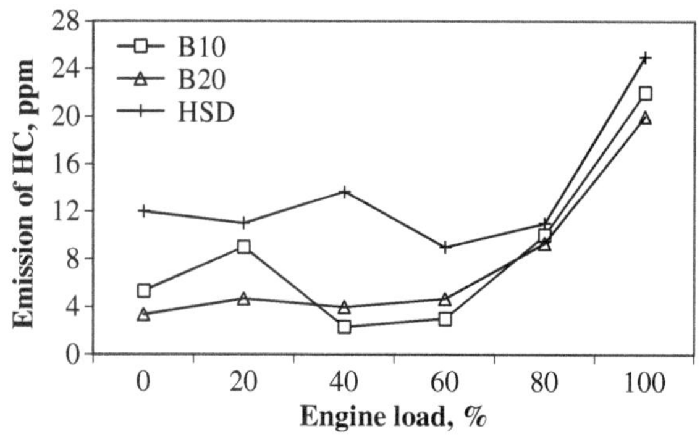

Figure 5 Variations of HC emissions with engine load for B10, B20, and HSD.

with all the fuels tested initially decreased from no load to 60% engine load, and then it increased with further increase in engine load up to 100%. It followed the same trend as that of CO emissions. The mean HC emissions with B10 and B20 were found to be 38.76% and 47.60% lesser than those with HSD. Among the two fuel blends tested, B10 gave the minimum reduction in HC emission as compared with HSD.

Oxides of nitrogen

The NO_x concentration in exhaust gas from a diesel engine when operated with HSD, B10, and B20 at different engine loadings is shown in Figure 6. It increased with the increase in both engine load and biodiesel concentration in the blends. The maximum NO_x values were obtained at full load conditions and were 1,173, 1,211, and 1,230 ppm, respectively, with HSD, B10, and B20, whereas it was 265, 296, and 344 ppm at 20% engine loading. The mean NO_x values with B10 and B20 were found to be 667.5 and 693.17 ppm, respectively, and were 5.57% and 11.45% higher than those with HSD. This higher NO_x production with biodiesel blends could be attributed to the higher EGT and the fact that

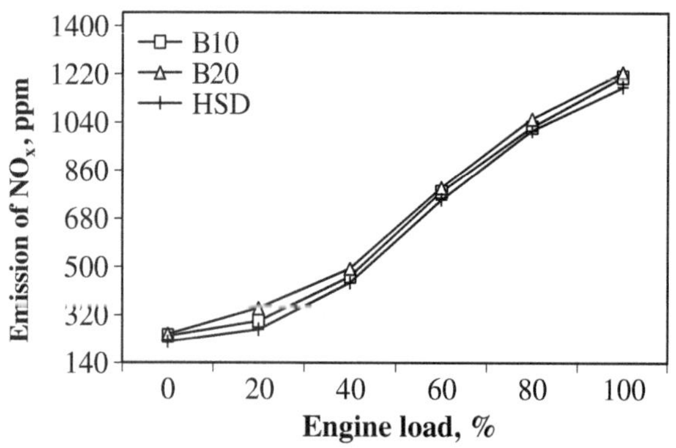

Figure 6 Variations of NO_x emissions with engine load for B10, B20, and HSD.

biodiesel had some oxygen content in it, which facilitated NO_x formation. As the engine load increased, the overall fuel-to-air ratio supplied to the engine increased, resulting in an increase in average gas temperature in the combustion chamber; hence, NO_x formation, which is sensitive to temperature, increased.

Similar findings on exhaust gas emissions were observed while operating the diesel engines with different biodiesel obtained from individual oils [23].

Selection of suitable biodiesel blend

Based on the minimum BSFC, maximum BTE, and minimum EGT, the biodiesel blend, B10, was found suitable for the running diesel engine without compromising engine performance when operated with HSD. However, from the emission point of view, this blend (B10) was found to produce higher CO and HC and lower NO_x emissions as compared with B20. However, both fuel blends tested exhibited lower CO and HC and higher NO_x emissions as compared with HSD. Considering both performance and emissions, B10 was considered as the suitable blend for replacing HSD in diesel engine and also in the long-term test of the engine that was conducted.

Soot deposits on in-cylinder engine components

The amount of soot deposits on the in-cylinder engine components when operated with different fuels were measured as per the procedure outlined in the 'Estimation of soot deposits on in-cylinder engine components' section and are summarized in Table 3.

The photographs of soot deposits formed on the cylinder head, piston crown, and fuel injector of diesel engine when operated with HSD and B10 along with a clean cylinder head, piston crown, and fuel injector are shown in Figure 7a,b,c, respectively. From these figures, it can be clearly seen that the soot deposits on the cylinder head, piston crown, and fuel injector of the engine were substantially lower when operated with B10 as compared with HSD. The amount of soot deposits on the cylinder head, piston crown, and fuel injector in the case of the B10-fueled engine was found to be 39.5%, 7.7%, and 25.0% lesser, respectively, as compared with those of the HSD-fueled engine.

A lesser amount of soot deposits on the different engine components when operated with B10 was because of the complete combustion of fuel due to the availability of extra oxygen in biodiesel molecules. The above results are in line with the findings reported by Agarwal et al. [25] when they conducted the 512-h endurance test with 20% blend of linseed oil methyl ester with HSD. They found about a 40% reduction in carbon deposits on the engine components when operated with B20 as compared with HSD.

Table 3 Soot deposits on cylinder head, piston crown and fuel injector of an engine

Engine parts	Operated with HSD (g)	Operated with B10 (g)	Variation as compared to HSD (%)
Soot deposits on cylinder head	0.38	0.23	−39.47
Soot deposits on piston crown	0.52	0.48	−7.69
Soot deposits on fuel injector	0.04	0.03	−25.00

Addition of wear metals in lubricating oil

Addition of wear metals in lubricating oil after 100 h of engine operation each with HSD and B10 was determined by following the procedure outlined in the 'Estimation of soot deposits on in-cylinder engine components' section and is summarized in Table 4. From this table, it can be seen that the concentrations of heavy metals such as Cu, Zn, Fe, Pb, Mg, and Al, except for Mn and Ni, were lower in the lubricating oil of engine when fueled with B10 as compared with when it was fueled with HSD. This could be due to the lesser friction of engine components because of additional lubricating property and presence of fatty acid compounds such as fatty acid methyl esters, FFA, monoglycerides, etc. in B10 fuel. According to some researchers, the presence of long-chain molecules, degree of unsaturation, and oxygenated moieties in the biodiesel play an important role in improving its lubricity [2,26,27]. Another reason for this could be the higher viscosity of B10 blend (5.96% higher) as compared with HSD.

Similar findings on the addition of metals on lubricating oil were also reported while conducting 100-and 512-h

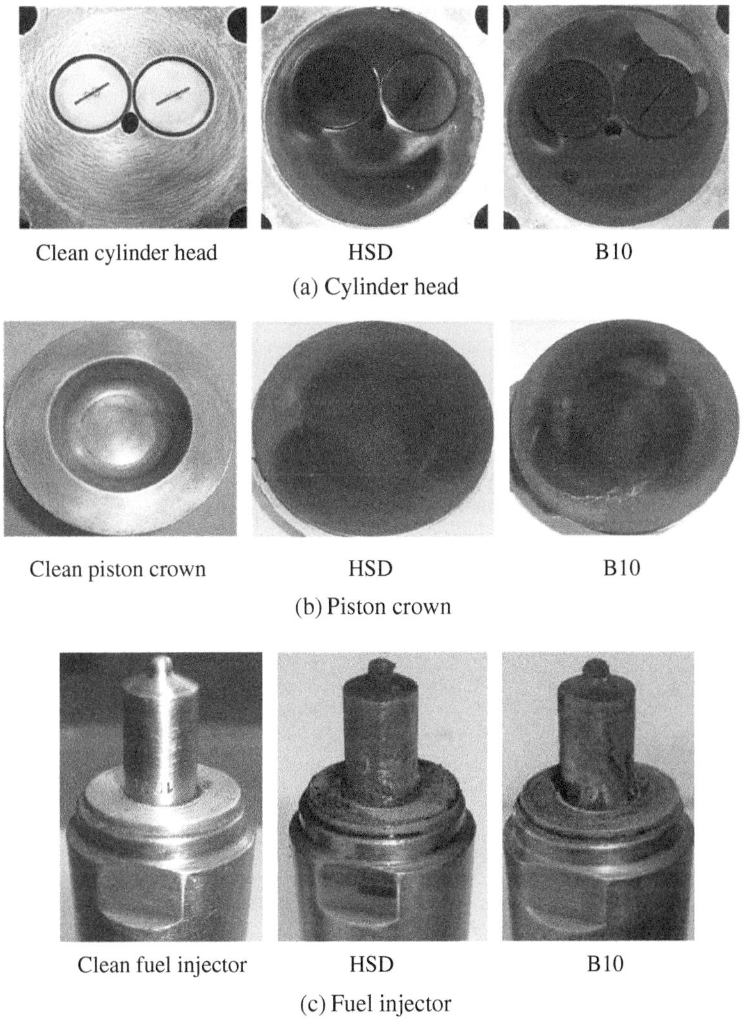

Figure 7 Soot deposits on (a) cylinder head, (b) piston crown, and (c) fuel injector of diesel engine.

Table 4 Wear metal addition in lubricating oil of engine when operated with HSD and biodiesel blends

Metals	Concentration at 0 h (ppm)	Concentration after 25 h (ppm)		Variation (%)	Concentration after 50 h (ppm)		Variation (%)	Concentration after 75 h (ppm)		Variation (%)	Concentration after 100 h (ppm)		Variation (%)
		HSD	B10		HSD	B10		HSD	B10		HSD	B10	
Cu	0.8	2.8	2.4	−14.3	4.4	3.0	−31.1	5.5	3.9	−29.1	7.02	4.7	−33.0
Zn	915.1	951.1	945.0	−0.64	1,027.8	1,012.0	−1.5	1,063.4	1,042.2	−2.0	1,260.0	1,120.8	−11.0
Fe	17.8	33.7	28.5	−15.4	43.6	37.7	−13.5	55.3	44.4	−19.7	60.9	48.6	−20.2
Mn	1.1	1.4	1.5	7.1	1.6	1.7	6.3	1.7	1.9	11.8	1.9	2.3	21.1
Ni	0.3	0.4	0.4	0	0.4	0.5	25	0.5	0.5	0	1.1	0.9	−18.2
Pb	1.3	2.2	1.4	−36.3	2.8	1.6	−42.8	3.4	1.9	−47.8	4.5	2.8	−37.8
Mg	50.6	59.8	54.2	−9.4	77.2	55.3	−28.4	89.2	57.2	−35.9	115.8	57.9	−50.0
Al	36.6	38.7	37.8	−2.5	58.5	49.3	−15.7	59.6	53.8	−9.7	66.3	57.5	−13.3

tests in diesel engines using palm kernel oil biodiesel blends (B7.5 and B15) and linseed oil methyl ester blend (B20), respectively [2,25].

Conclusions

The following conclusions have been made in this study:

1. The fuel properties of biodiesel obtained from the mixture, MSO, were found to be within the limits specified by the biodiesel standards ASTM D 6751–03, DIN EN 14214, and BIS 15607. The fuel properties of biodiesel blends approached those of HSD with a decrease in concentration of biodiesel in the blends.
2. BSFC and EGT of the 10.3-kW diesel engine when operated with biodiesel blends as compared with HSD at different engine loads were found to increase by 2.49% to 5.62% and 4.44% to 5.2%, respectively, whereas BTE was found to decrease by 1.48% to 3.22% with an increase in biodiesel concentration in the fuel blends. Among the two fuel blends tested, B10 had lower mean BSFC (287.29 g/kW h) as well as EGT (308°C) and a higher mean BTE (21.87%).
3. With an increase in engine load, the BSFC decreased, whereas both EGT and BTE increased for all the fuels tested. However, BTE for all biodiesel blends as compared with HSD was reduced on average by 2.09% at full load, and it was further reduced to 3.41% at 20% engine loading due to higher losses.
4. The CO and HC emissions of the diesel engine when operated with biodiesel blends as compared with HSD were reduced by 10.97% to 21.16% and 38.76% to 47.6%, whereas NO_x emissions increased by 5.57% to 11.45%.
5. Based on the performance (minimum increase in BSFC and EGT and with lesser reduction in BTE) and emissions (minimum reduction in CO and HC and minimum increase in NO_x), biodiesel blend B10 was selected for long-term use in diesel engine.
6. As compared with HSD-fueled engine, lesser carbon deposits on the in-cylinder parts (such as cylinder head, piston crown, and fuel injector) were observed for the B10-fueled engine due to better combustion of biodiesel blend.
7. Lower concentrations of all heavy metals (such as Cu, Zn, Fe, Pb, Mg, and Al, except for Mn and Ni) in the lubricating oil of diesel engine were found inB10-fueled engine as compared with those in HSD-fueled engine. This could be due to the lesser friction of engine components because of additional lubricity and higher viscosity of the B10 blend as compared with HSD.

Findings of this study would encourage the use of biodiesel obtained from the mixture of oils and would help in reducing the dependency on a particular oil for biodiesel production.

Competing interests

The authors declare that they have no competing interests.

Authors' contributions

All the authors - HR, PCJ, and SSJ - collectively carried out experiments, collected data, and analyzed them. HR wrote the manuscript. All authors read and approved the final manuscript.

Authors' information

HR is a professor at Indian Institute of Technology. PCJ is a scientist at Central Institute of Agricultural Engineering. SSJ is a systems engineer at John Deere India Pvt. Ltd.

Acknowledgments

To carry out this research work, the financial help provided by Petroleum Conservation Research Association, Government of India, New Delhi is sincerely acknowledged.

References

1. Schumacher, LG, Borgelt, SC, Hires, WG: Fueling a diesel engine with methyl ester soybean oil. Trans. ASAE **11**(1), 37–40 (1995)
2. Kalam, MA, Masjuki, HH: Biodiesel from palm oil- an analysis of its properties and potential. Biomass Bioenergy **23**(6), 471–479 (2002)

3. Agarwal, AK, Bijwe, JL, Das, M: Wear assessment in a biodiesel fueled compression ignition engine. J. Eng. Gas. Turbines Power **25**, 820–826 (2003)
4. Schumacher, LG, Soylu, S, Gerpen, JV, Wetherell, W: Fueling direct injection diesel engines with 2% biodiesel blends. Appl. Eng. Agric. Trans. ASAE. **21** (2), 149–152 (2005)
5. Schumacher, LG, Peterson, CL, Gerpen, JV: Engine oil analysis of biodiesel fueled engines Appl. Eng. Agric. Trans. ASAE. **21**(2), 153–158 (2005)
6. Ramaprabhu, RO, Bhardwaj, M, Abraham: Comparative study of engine oil characteristics in utility vehicles powered by turbocharged direct injection diesel engines fuelled with biodiesel blend and diesel fuel. SAE, 2008280117 (2008)
7. Jadhav, SS: Development of a biomass based hybrid power generating system M.Tech. thesis, Indian Institute of Technology, Kharagpur (2009)
8. Hugo, HV, Patricio, GVD, Armando, G, Héctor, P: The development impact of biodiesel: a review of biodiesel production in Mexico. Int. J. Energ. Environ. Eng. **2**, 91–99 (2011)
9. Jena, PC, Raheman, H, Prsanna Kumar, GV, Machavaram, R: Biodiesel production from mixture of mahua and simarouba oils with high free fatty acids. Biomass Bioenergy **34**(2), 1108–1116 (2010)
10. Bureau of Indian Standards: Methods of tests for internal combustion engines: part 8 performance tests. BIS, New Delhi (1980)
11. Bureau of Indian Standards: Methods of tests for internal combustion engines: part 9 endurance test. BIS, New Delhi (1980)
12. Moreno, F, Munoz, M, Morea-Roy, J: Sunflower methyl ester as a fuel for automobile diesel engines. Trans. ASAE. **42**(5), 1181–1185 (1999)
13. Agrawal, AK, Das, LM: Biodiesel development and characterization for use as a fuel in compression ignition engines. Trans. ASME. **123**(2), 440–447 (2001)
14. Raheman, H, Phadatare, AG: Diesel engine emissions and performance from blends of karanja methyl ester and diesel. Biomass Bioenergy **27**(4), 393–397 (2004)
15. Puhan, S, Vedaraman, N, Sankaranarayanan, G, Bharat Ram, BV: Performance emission study of mahua oil (*Madhuca indica* oil) ethyl ester in a 4-stroke natural aspirated direct ignition diesel engine. Renew. Energ. **30**, 1269–1278 (2005)
16. Ramadhas, AS, Muraleedharan, C, Jayaraj, S: Performance and emission evaluation of a diesel engine fueled with methyl esters of rubber seed oil. Renew. Energ. **30**(12), 1789–1800 (2005)
17. Labeckas, G, Slavinskas, S: The effect of rapeseed oil methyl ester on direct injection diesel engine performance and exhaust emissions. Energ. Convers. Manage. **47**(13–14), 1954–1967 (2006)
18. Canakci, M: Combustion characteristics of a turbocharged DI compression ignition engine fueled with petroleum diesel fuels and biodiesel. Bioresour. Technol. **98**(6), 1167–1175 (2007)
19. Raheman, H, Ghadge, SV: Performance of diesel engine with biodiesel at varying compression ratio and ignition timing. Fuel **87**(12), 2659–2666 (2008)
20. Sahoo, PK, Das, LM, Babu, MKG, Naik, SN: Biodiesel development from high acid value polanga seed oil and performance evaluation in a CI engine. Fuel **86**, 448–454 (2007)
21. Ozsezen, AN, Canakci, M, Turkcan, A, Sain, C: Performance and combustion characteristics of a DI diesel engine fueled with waste palm oil and canola oil methyl esters. Fuel **88**(4), 629–636 (2009)
22. Nabi, MN, Akhter, MS, Shahadat, MMZ: Improvement of engine emissions with conventional diesel fuel and diesel-biodiesel blends. Bioresour. Technol. **97**, 372–378 (2006)
23. Gumus, M, Kasifoglu, S: Performance and emission evaluation of a compression ignition engine using a biodiesel (apricot seed kernel oil methyl ester) and its blends with diesel fuel. Biomass Bioenergy **34**, 134–139 (2010)
24. Raj, MT, Kandasamy, MKK: Tamanu oil - an alternative fuel for variable compression ratio engine. Int. J. Energ. Environ. Eng. **3**, 1–8 (2012) doi:10.1186/2251-6832-3-18
25. Agarwal, AK, Bijwe, J, Das, LM: Effect of biodiesel utilization of wear of vital parts in compression ignition engine. Trans. ASME **125**, 604–611 (2003)
26. Geller, DP, Goodrum, JW: Effects of specific fatty acid methyl esters on diesel fuel lubricity. Fuel **83**, 2351–2356 (2004)
27. Knothe, G, Steidley, KR: Lubricity of components of biodiesel and petrodiesel: the origin of biodiesel lubricity. Energy Fuel **19**, 1192–1200 (2005)

8

Sugarcane energy use: accounting of feedstock energy considering current agro-industrial trends and their feasibility

Walfrido Alonso Pippo* and Carlos A Luengo

Abstract

The sugarcane agro-industry is seen as a great opportunity for economic and industrial development in many sugarcane-producing countries. Important changes happening in recent decades have converted the sugar mill from being just a food producer into a diversified production factory. The sugar mill has become a multipurpose factory since it produces food, energy, and biofuels at present. The key cause of this change is the use of sugarcane agro-industrial residues as feedstock for energy cogeneration and biofuel production.

The main aim of this paper is to carry out an assessment on sugarcane feedstock availability and its energy use in the sugar mill. The trade-off on sugarcane bagasse energy use (electricity cogeneration vs. bioethanol production), considering the agro-industrial development level, is analyzed in this work, too. The better options in each case are highlighted. The main environmental and techno-economic aspects concerning the sugarcane agro-industry were taken into account during the assessment process. The most promising trends of the sugarcane agro-industry and the barriers to overcome in its implementation are pointed out.

Keywords: Biofuels, Cogeneration, Sugarcane agro-industry, Residue energy use

Review

Introduction

The main solid residues from sugar and ethanol production are sugarcane bagasse and sugarcane trash, also named as sugarcane agriculture residues (SCAR)[a] [1]. Sugarcane bagasse is the fibrous waste that remains after the recovery of sugar juice via crushing and extraction. It also has been the principal fuel used around the world in the sugarcane agro-industry because of its well-known energy properties [2,3]. The introduction of mechanized sugarcane harvesting combined with technological improvements has empowered the cogeneration in sugar and ethanol factories. This development has taken place during the last two decades of the twentieth century, and it has changed radically the viewpoint on the use of residuals in the sugarcane agro-industry [4-6].

Sugarcane (*Saccharum officinarum*)[b] is the world's largest cash crop. There are more than 90 sugar-producing countries around the world. Especially for underdeveloped countries, sugarcane residue disposal has a first-order priority. The world's sugarcane production has experimented a dramatic growth during the last decade (Figure 1).

The world sugarcane agro-industry has processed more than 1,685 × 10^6 tons in 2010. The mentioned amount of sugarcane generated 23.6 × 10^6 tons of bagasse (dry basis) and a similar amount of SCAR. The energy content of both residues could mean about 85 × 10^6 tons of oil equivalent. In other words, the sugarcane agro-industry produces around 660 kg of solid residues for each milled ton of cane (wet basis).

Among the sugarcane-producing countries, Brazil, today the world's leader in the use of renewable energies[c] in general, in bioenergy in particular, plays a special role in this field. The Brazilian sugarcane agribusiness development is a successful reference which could be replicable partially or modified by many sugarcane-producing countries [7-12]. Brazil has the most developed and integrated biofuel program in the world. It is an indisputable fact that Brazilian liquid biofuels are most well known across the world. The bioethanol that comes from sugarcane is especially important [13-18].

* Correspondence: pippo177@yahoo.com
UNICAMP/IFGW/DFA/Grupo Combustíveis Alternativos, Cidade Universitária 'Zeferino Vaz', Campinas, Sao Paulo CEP 13083-859, Brazil

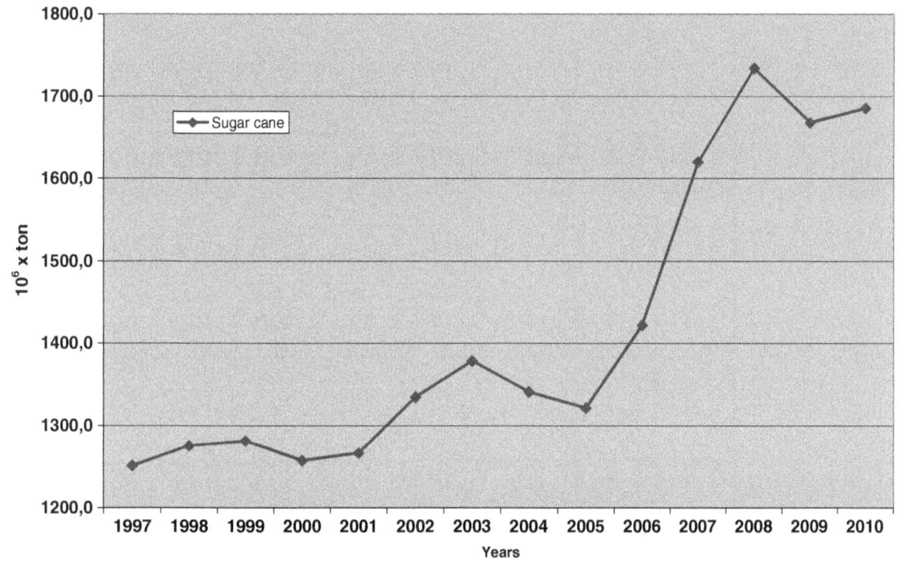

Figure 1 World's sugarcane production 1997 to 2010 [7].

From a technical and economical point of view, it is cheaper and easier to produce ethanol from sugarcane than from corn. The ethanol from sugarcane produced in Brazil is three to four times cheaper than the bioethanol from corn produced in the USA.[d] In order to produce ethanol from corn, an additional step is necessary in the production process. The starch of corn should be converted to sugar before the fermentation, while in the case of sugarcane, the sugar for fermentation is contained directly in the sugarcane juice. This advantage of ethanol from sugarcane is really an opportunity for underdeveloped sugarcane-producing countries to diversify the sugarcane industry through the introduction of new sugarcane by-products, such as fuel bioethanol and other sugarcane derivates. However, the production of fuel bioethanol from sugarcane juice competes with the production of sugar, other food products, and medicines. This fact plus other elements, such as issues concerning greenhouse gas (GHG) emissions, the rise of oil price in the international market, and other geopolitical factors associated with traditional oil supply instability, are encouraging the introduction of the second-generation biofuels as a new opportunity obtained from biomass residues and lignocellulosic biomass.

This new trend should be considered during the modernization of the sugar industry or for energy planning by companies and governments. The main reason to take into account the production of second-generation bioethanol is that, as it was already mentioned, 66% of the residues of sugarcane are lignocellulosic.

In the light of second-generation bioethanol production surplus sugarcane bagasse could be used to produce bioethanol or to produce surplus electricity. The use of bagasse for second-generation bioethanol production, instead of its traditional use as fuel for cogeneration in the sugar factory, has raised some new concerns on its possible substitute. Although SCAR energy use is still minuscule in the Brazilian sugar factories, it seems to be that SCAR should be the substitute of the bagasse for cogeneration in the sugar factory.

The main aim of this paper is to carry out an assessment on the bagasse and SCAR availability and their energy use in the sugarcane agro-industry. The trade-off on sugarcane bagasse energy use (electricity cogeneration vs. bioethanol production), considering the agro-industrial development level, is analyzed, too. The better options in each case are highlighted. The main environmental and techno-economic aspects concerning the sugarcane agro-industry were taken into account during the assessment process. The most promising trends of the sugarcane agro-industry and the barriers that should be overcome in its implementation are pointed out.

Agro-industrial features of sugarcane agribusiness

Agricultural aspects

The sugarcane is a prodigious gramineous plant which belongs to the so-called C_4 plants (because the first product is 4-carbon sugar). These kinds of plants have the biggest potential sunlight conversion capacity into biomass (until 6.7%). In fact, the average figures reported on sunlight conversion are about 1.5% to 3%. Most of the world's sugarcane is grown between latitudes 22°N and 22°S and some up to 33°N and 33°S. Because of this reason, sugarcane plantations in commercial scale can be found from Bangladesh to South Africa, from Louisiana, USA to Argentina, and from China to Australia. Several indicators and general parameters are valid for all sugarcane-producing countries, regardless of their geographical region. Considering this fact and in order to make easier the present assessment, those very specific indicators of sugarcane that do not have expressive importance for comparison were disregarded. In Table 1 are shown the most common indicators of sugarcane agriculture, their parameters, features, performances, and the particularities that deserve to be considered.

Industrial aspects

Unlike the agricultural aspects, the development of the sugarcane industry has the particular footprint of each region and country. Each country's footprint includes geographical, demographical, cultural, socioeconomic, geopolitical, anthropologic, techno-economic, and environmental factors. Because of the huge amount of regional particularities, it is impossible to have a 'one size fits all' approach to assess the sugarcane industry across the world.

The degree of technological development is not uniform in the sugar industry, even in the same country. It is possible to find these contrasting situations in African, Latin-American, and Indian sugarcane industries. For instance, all sugar mills existing in the state of Orissa, India (seven sugar mills) present a low and very low technological development. In contrast, in the Indian state of Maharashtra (24 sugar mills), there is a combination of well technologically developed sugar mills (three sugar mills) and low technologically developed sugar mills [19].

Most of the sugar mills in Louisiana, USA have good technological development. However, in this state is located the Lasuca sugar mill (daily milling capacity about 4,000 tons of cane per day) [20], which uses natural gas for sugar manufacturing. The technological development of the Lasuca sugar mill can also be considered very low since it burns fossil fuel to meet the sugar production process energy requirement, which is retrograde today, considering the readiness level of the sugar industry technology and the environmental issues.

Although there is a general trend towards modernization, sugar mills with a very low technological development and with state-of-the-art technology are often both found in sugar-producing countries.[e] For this reason, in this work, the technological development of sugar factory was conventionally classified in four possible levels:

1. Very low technological development (VLTD),
2. Low technological development (LTD),
3. State-of-the-art technology (SOTAT),
4. Future step of technology development (FSOTD).

In Table 2 are shown the main features belonging to each of the considered levels of technological development.

BPST, back-pressure steam turbine; CEST, condensation-extraction steam turbogenerator; BIG/GTCC, biomass-integrated gasifier/gas turbine combined cycle system; VHP very high pol. [a]It is known by a range of names: *gur* in India and Bangladesh, *desi* in Pakistan, *jaggery* in Africa, and *panela* in South America. This type of sugar, as a rule, is produced using relatively low-cost technology but can also be produced in medium and large factories. [b]Raw sugar is a type of sugar with sucrose content not less than 96° pol. [c]VHP not less than 99.3° pol. [d]As a rule, 50% of cane juice is used for sugar production. Half of cane juice production goes to ethanol production. The average production yield of raw centrifugal per ton of milled cane is 0.106 to 0.12 ton. The production yield of VHP sugar

Table 1 Sugarcane parameters, indicator, and features adopted in the present work

Number	Parameters, indicator, and features	Unit	Value	Observation
1	Sugarcane agriculture yield	$ton{\cdot}ha^{-1}$	≥70	
2	Sugarcane pol[a] content	%	≥13	
3	Sugarcane fiber content	%	13	
4	Nonsugar insoluble solid content	%	≤2 to 3	
5	Sugarcane plantation age	months	12 to 18	
6	Total rainfall or irrigation	mm	1,100 to 1,500	
7	Milling season duration	day	180 to 210	In dry season
8	Number of harvests	-	≤5	
9	Harvesting method yield			
	Manual[b]	$ton\ (man{\cdot}day)^{-1}$	4 to 8	Burnt cane
	Mechanical[c]	$ton\ (machine{\cdot}day)^{-1}$	500 to 800	Green cane
	Combined harvesting method[d]			

[a]Pol is a specific unit that characterizes the sucrose content of sugarcane juice. A pol (polarization) is a measure of the sucrose content of sugar. Sugar with 98 pol (or 98° pol) contains about 98% sucrose. The pol content is used to determine the price of sugarcane delivered to the sugar mill in many countries. Pol is measured using a polarimeter and is determined by the International Commission for Uniform Methods of Sugar Analysis; [b]hand cutting with mechanical grab loading; [c]stalk and top cutting with simultaneous cleaning and loading; [d]combined harvesting method is a method where both manual and mechanized harvesting is used.

Table 2 Operational parameters, main features, and products of the different levels of sugar factory technological development

Products, operational parameters, and data		VLTD	LTD	SOTAT	FSOTD
Operational parameters and features	Milling capacity (ton of cane per day)	≤3000	3,000 to 7,000	7,000	7,000 to 1,200
	Thermodynamic cycle	Rankine	Rankine	Rankine	Rankine and Brayton
	BPST (Mw, K, MPa)	≤7, ≤593, 1.9	-	-	-
	CEST (Mw, K, MPa)	-	38, 593, 4.2	30, 793, 6.3	
	BIG/GTCC (Mw, K, MPa)	-	-	-	52, 793, 8
	Steam pressure process (MPa)	0.25	0.25	0.25	0.25
	Steam pressure molecular sieve (MPa)	-	-	0.6	0.6
	Steam per ton of milled cane (%)	52	45	38	28
	Electrical and mechanical consumption per ton of cane (Kw h_e)	25	25	25	30
	Energy self-sufficiency	No[g]	Yes	Yes	Yes
Products	Non-centrifugal sugar[a] (ton/ton of cane)	Not available	Not available	-	-
	Raw centrifugal sugar 96[b] (ton/ton of cane)	0.1	0.117	0.12	0.12
	VHP sugar[c] (ton per ton of cane)[d]	-	0.117	0.12	0.12
	Hydrated ethanol[e] (L/per ton of cane)	-	70 to 75	73 to 80	80 to ?
	Anhydrous ethanol[f] (L/ton of cane)	-	70	70 to 75	75 to ?
	Electricity cogenerated per ton of cane (Kw h_e)	22.79	50	110	120 to 177
	Surplus electricity per ton of cane (Kw h_e)	0	25	85	90 to 148

and refined sugar per ton of milled cane are 0.117 and 0.083 tons, respectively. [e]Hydrated ethanol alcohol content is 95, 56 ± 0.43 degrees Gay-Lussac (° GL). [f]Anhydrous ethanol minimum alcohol content is 99, 58° GL. It is used for gasoline blend. [g]Most of the sugar mills import electricity to meet process energy demand.

This classification is necessary from a methodological viewpoint because the level of technological development determines the real possibility of sugarcane energy use. The higher is the sugar mill technological development, the bigger will be its possibility of sugarcane energy use.

It deserves to remark that this classification is not conclusive. In other words, this classification may not reflect exactly the reality of some sugar mills across the world. The classification makes easier the analysis of the main goals of the present work and contributes to make the analysis as wide as possible.

Very low technological development Commonly, this kind of sugar mills is a small facility that produces centrifugal and noncentrifugal sugar. Product diversification in these mills is practically inexistent. Ethanol production is semi-craft and very small scale (when it takes place). Bagasse management and disposal practices employed by these mills have remained, in most cases, the same as those used back in the early nineteenth century when the factories were designed without any relation to energy efficiency (most operate at a low 16% to 20% efficiency). At that time, factories consumed all the available bagasse just to meet mill energy demands, resulting in no accumulation of excess bagasse.[f] Regrettably, by that method, the possibility to recover and use an important quantity of energy from bagasse was lost. This mentioned waste of a potential energy source is especially unfortunate in underdeveloped sugarcane-producing countries where the demand for electric power exceeds the local capacity. Figure 2 shows the typical cogeneration scheme of a VLTD sugar factory.

Low technological development There are many possible variants of LTD sugar factories. Figure 3 shows a scheme with most common elements of these kinds of mills. Commonly, these facilities do not use sugarcane juice directly to produce ethanol. An insignificant amount of ethanol is produced from molasses obtained during sugar production in an LTD sugar mill. However, nowadays, several sugar facilities have begun their technological improvement in Latin America, Asia, and Africa. Considering the mentioned situation, the simultaneous sugar-ethanol production from sugarcane juice was calculated, assuming that half of the sugarcane juice is used for ethanol production (Figure 3). Mass flow

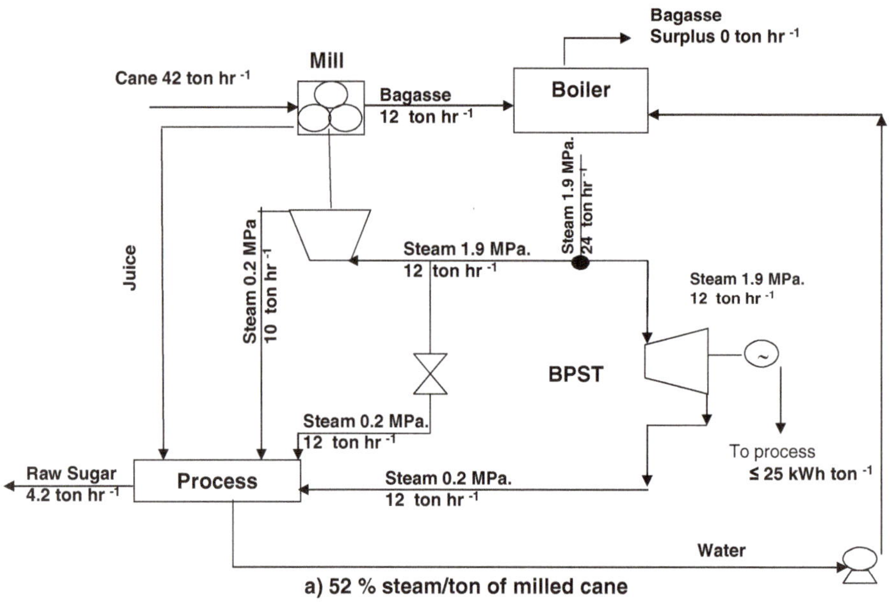

Figure 2 VLTD sugar mill typical cogeneration system [21].

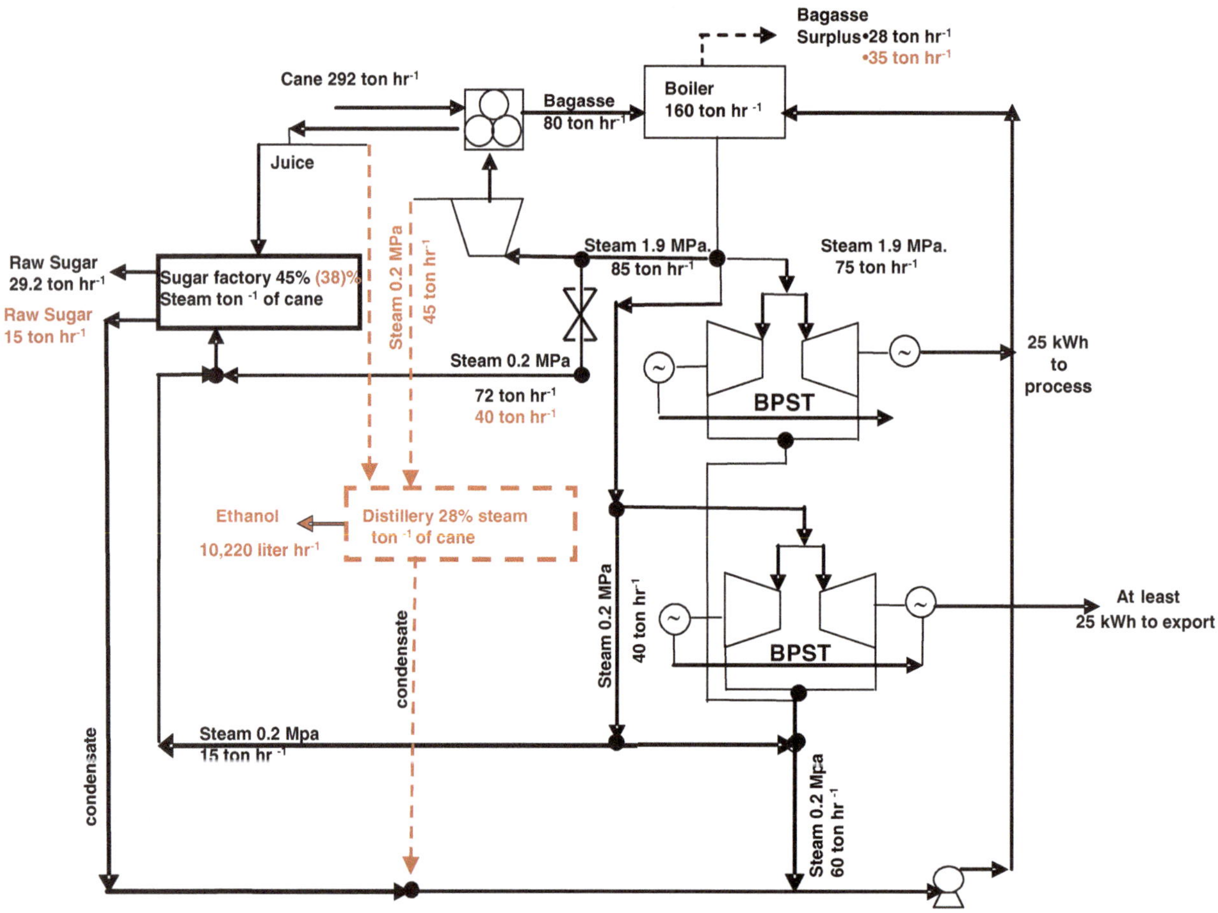

Figure 3 LTD sugar mill scheme (simultaneous sugar-ethanol production option was estimated) [22].

parameters and production rates of the simultaneous sugar-ethanol production layout are highlighted in red. It is easy to notice that when juice is not used for ethanol production, which is a common situation, the sugar production rate is twice bigger than the production when ethanol is produced.

Even in an LTD scheme sugar mill, there is a possibility of exporting some quantities of electricity if its process of mass energy balance is properly adjusted. However, because of the high-cost investment of transmission lines, the sugar mill rarely takes this opportunity. Thus, the surplus bagasse could be sold or used to produce board.

State-of-the-art technology The SOTAT level is characterized by the wider variety of products, among them are raw sugar, VHP sugar, anhydrous ethanol, hydrated ethanol, and power electricity. Many Brazilian sugar mills have reached this stage. The classical scheme of a SOTAT sugar facility is shown in Figure 4. The simultaneous sugar-ethanol production is characterized in the SOTAT stage. Several improvements were introduced in the sugar and ethanol processes in the last years. These improvements made possible a considerable decrease of steam consumption in both sugar production process and ethanol distillation. Steam-saving makes possible to increase the sugar mill cogeneration capacity.

Future step of technology development There are two principal schemes potentially foreseen for FSOTD of sugar factories until today:

1. *FSOTD: Maximization of electricity power cogeneration.* FSOTD maximization of electricity power variant hereafter is referred also as FSOTD$_{electricity}$. The sugar factory FSOTD$_{electricity}$ is a convenient variant when the goal is to increase electrification in poor regions without other alternative renewable energy supplies. The main disadvantages of this pathway are the seasonality characteristic of the sugarcane agro-industry, the relative small scale of sugar facilities,[g] and the high initial capital investment. Most well-known scheme of FSOTD$_{electricity}$ is the biomass integrated gasifier/gas turbine combined cycle (BIG/GTCC) system [24]. The BIG/GTCC system for a sugar mill is shown in Figure 5.
2. *Large-scale production of lignocellulosic ethanol including both bagasse and SCAR energy.* FSOTD second-generation ethanol maximization variant hereafter is referred also as FSOTD$_{ethanol}$. The second pathway of sugar factory FSOTD$_{ethanol}$ will be possible by the large-scale production of lignocellulosic ethanol in sugar facilities. Lignocellulosic ethanol is produced initially from

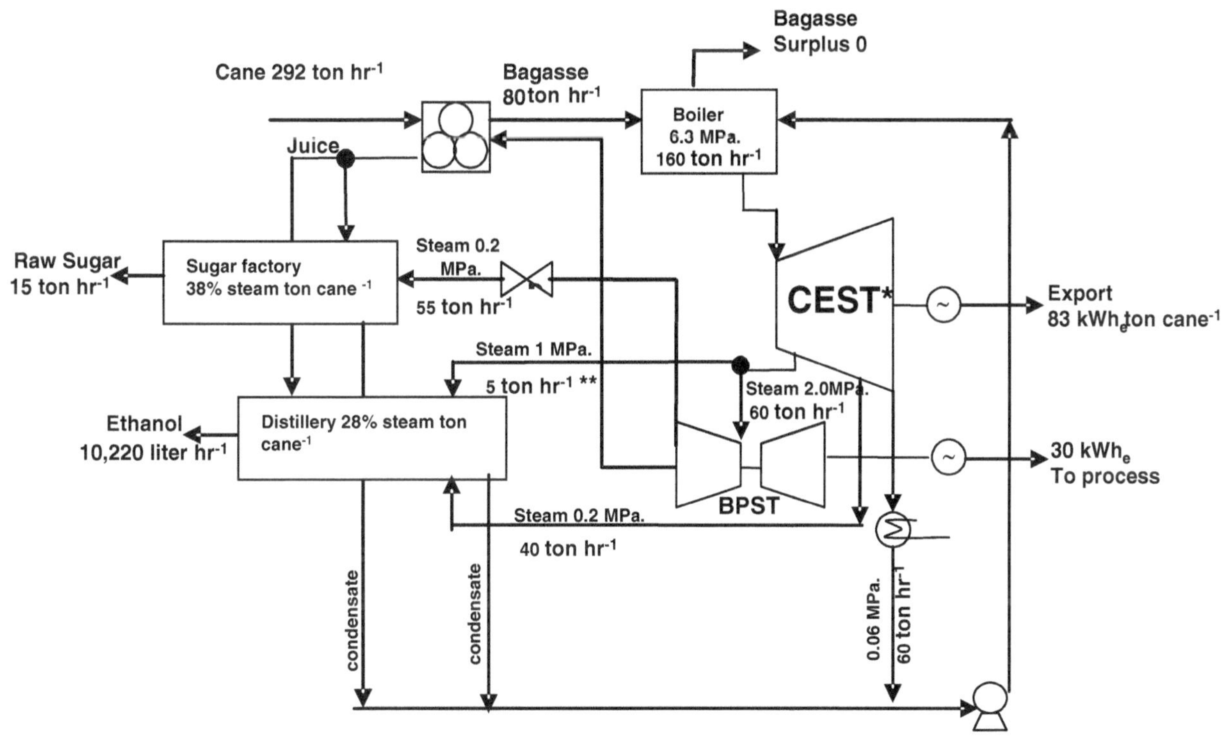

Figure 4 SOTAT sugar mill scheme and cogeneration scheme of condensation extraction steam turbogenerator (CEST). Simultaneous sugar-ethanol production: two extractions (single asterisk) and steam for molecular sieve dehydration (double asterisks) [23].

sugarcane solid residues (bagasse and SCAR). Lignocellulosic ethanol is also called second-generation ethanol because its production does not compete with sugar produced from sugarcane juice such as in the case of traditional ethanol. In this case, unlike the first pathway, it maximizes ethanol production. The advantage of this scheme is that residue energy can be stored and used in the off-season period or anytime independently from its production date. The disadvantage of lignocellulosic ethanol is that ethanol production requires large quantities of water and the disposal of vinasse.[h] Figure 6 shows the scheme of an $FSOTD_{ethanol}$ factory.

The feedstock and second-generation ethanol mass flows are highlighted in orange. There are two possible approaches to carry out the assessment of second-generation ethanol production efficiency in the FSOTD sugar facility. The first approach is to exclude pentose (hemicellulose) fermentation and just regard hexose (cellulose) use for ethanol production. The second approach is to consider both celluloses and hemicelluloses. The inclusion of hemicelluloses in the production process depends on their hydrolysis efficiency. The hydrolysis of hemicelluloses until today is an unsolved question [25]. Table 3 [26] shows the chemical composition of sugarcane bagasse and SCAR.

The hemicellulose represents 27% of bagasse and SCAR weight. From Table 3, it is easy to notice that the hemicellulose exclusion means a significant decrease of the ethanol yield.

Accounting of feedstock energy

The accounting of feedstock energy can be a good tool for a quick analysis on the pre-feasibility of a manufacturing facility during project management. The key indicator of this method is the utilization rate. The assessment of unused capacity is expressed usually through the utilization rate figures. In this work, this performance indicator was adapted to assess what share of sugarcane energy is not used for each of the production schemes analyzed.

The feasibility and sustainability of sugarcane energy use have been treated by different authors [7,9,13]. Most of published works are based on environmental and techno-economical premises. In the present work, the main criteria used were sugarcane energy content and production diversification capacity installed in sugar mills. The goal is to elaborate an additional simple and complementary tool for the preliminary assessment of feasibility of a sugarcane facility. Since the used criteria for assessment, sugarcane energy content and sugar mill technological development, are simpler to evaluate than others, based on the accounting of GHG, the proposed method is easy. It is important to underline that accounting of feedstock energy does not substitute the mass-energy balance assessment of the manufacturing process. The utilization rates cannot be considered during thermodynamic calculations. Although the efficiency is also a rate indicator, its nature is completely different

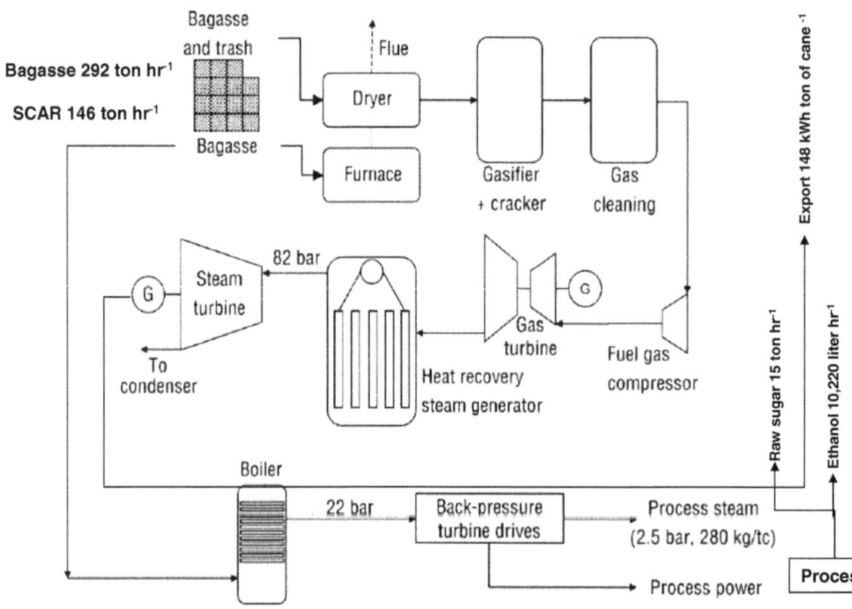

Figure 5 Schematic diagram of the BIG/GTCC system. All process steam demand is provided from the exhaust steam of the back pressure steam turbine drives. The electrical output of the gas turbine/steam turbine combined cycle augments the mechanical and electrical power process provided by the backpressure drives. Remaining electricity is available for export to the grid. Adapted from [24].

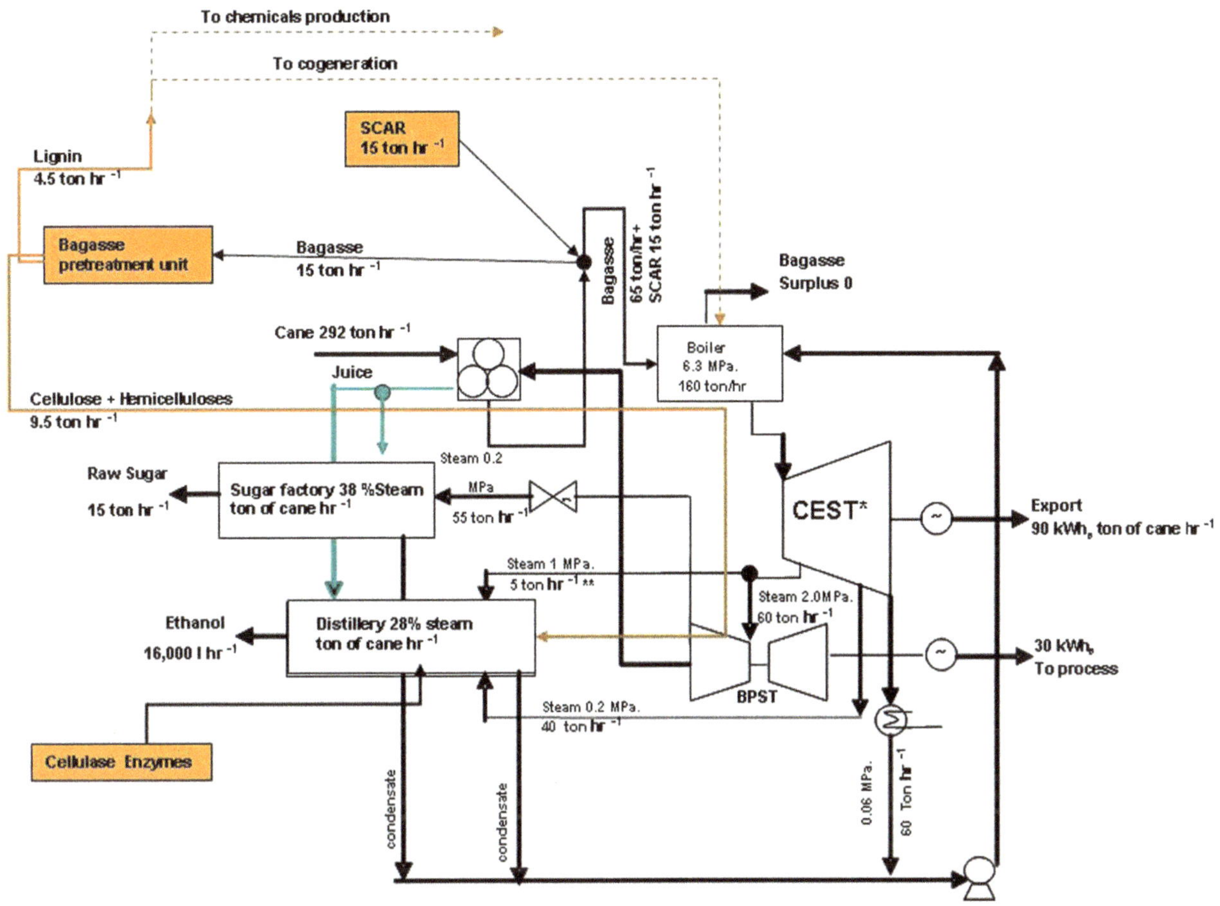

Figure 6 Sugar mill cogeneration scheme (7,000 tons of cane per day). Condensing extraction steam turbine (CEST) with SCAR use and simultaneous sugar-bioethanol production (second-generation bioethanol) [4].

from the utilization rate used during accounting of feedstock energy. The efficiency takes into account enthalpy, phase changes, and thermodynamic laws; its value should be less than one. The utilization rate calculated in this work does not consider any of the aforementioned factors.

Sugarcane energy content

The total cane energy content, excluding ash (around 2% to 3%), can be divided into three main parts (see Table 4 rows 1 to 3). Traditionally, sugarcane juice (H_2O + sacarose + molasses + insolubles) is used for sugar and ethanol production, while bagasse is used to meet the energy demand required during sugar and ethanol manufacturing processes. The SCAR energy content is similar to the bagasse energy content [1,27]. However, the modern sugarcane agro-industry does not utilize usually the energy potential of SCAR. In many places, sugarcane is burnt just before the harvest in order to facilitate easier harvesting of cane stalks. In that way, SCAR energy use is made impossible.

Table 3 Sugarcane bagasse and SCAR chemical composition

Component	Chemical composition (wt.%)	
	Bagasse	SCAR
Cellulose (hexoses)	46	48
Hemicelluloses (pentoses)	27	27
Lignin	23	13
Pectin	0	4
Ash	4	8

Adapted from [14].

Feedstock utilization rate calculation

The calculation of utilization rates was carried out step by step for each of the selected schemes. The results are shown in Table 4. Columns 4, 5, and 6 of Table 4 show the main final products obtained currently at the sugar mill and their corresponding energy content. Columns 7 to 21 of Table 4 show the three parameters that characterize each of the possible stages of technological development which are production yield, products energy content, and sugarcane utilization rate. The

Table 4 Utilization rate of sugarcane feedstock component parts

Row	Feedstock/sugarcane parts (average figures for currently commercial sugarcane varieties) [22]	Mass (kg) [11] (1 ton of cane; dry basis)	Energy content [11] (MJ)	Final products	LHV(MJ/kg), (MJ/kW h)	Unit	Rate of feedstock energy use (1 ton of sugarcane)				
							VLTD (Figure 2)			LTD (Figure 3)	
							Yield	Product energy content (MJ)	Sugarcane utilization rate (%)[a]	Yield	Energy (MJ)
	(1)	(2)	(3)	(4)	(5)	(6)	(7)	(8)	(9)	(10)	(11)
1	Juice (sucrose + molasses + others)	142	2,257	Sugar	17	kg	100	1700	0.26	58	986
2	Fiber residues (bagasse)	140	2,184	Ethanol	22	L	0	0		35	770
3	SCAR	140	2,184	Electricity	3.6	kW h	0	0		25	90
4	Total sugarcane	422	6,625	Total				1,700			1,846

Rate of feedstock energy use (1 ton of sugarcane)									
LTD (Figure 3)	SOTAT (Figure 4)			FSOTD (Figure 5) cogeneration maximization [9]			FSOTD (Figure 6) second-generation ethanol		
Sugarcane utilization rate (%)[a]	Yield	Product energy content (MJ)	Sugarcane utilization rate (%)[a]	Yield	Product energy content (MJ)	Sugarcane utilization rate (%)[a]	Yield	Product energy content (MJ)	Sugarcane utilization rate (%)[a]
(12)	(13)	(14)	(15)	(16)	(17)	(18)	(19)	(20)	(21)
0.28	58	58 986	0.31	58	986	0.35	58	986	0.33/0.34
	35	770		35	770		42/45	924/990	
	83			148	532,8		90	324	
		2,041			2,289			2,234/2,300	

LHV, Low heating value. VLTD, very low technological development; LTD, low technological development. [a]Sugarcane utilization rate = Σ Product energy (MJ)/6,625 MJ (figures shown in row 4, columns 8, 11, 14, 17, and 20 of each scheme are the numerators of the rate).

sugarcane feedstock utilization rate was estimated using the following equation:

$$\text{Utilization rate}_{\text{sugarcane feedstock}} = \frac{\sum \text{Product}_{\text{energy content}}}{\text{Total sugarcane}_{\text{energy content}}}, \quad (1)$$

where $\text{product}_{\text{energy content}}$ is the sum of the LHV of each sugarcane feedstock component (MJ/kg); $\text{total sugarcane}_{\text{energy content}}$ is the energy content of 1 ton of sugarcane (dry basis) (row 4, column 3 of Table 4 (6,625 MJ)); the calculated figures of $\text{utilization rate}_{\text{sugarcane feedstock}}$ for each level of technological development are shown in Table 4 (columns 9, 12, 15, 18, and 21).

The relevant explanations about each of the showed figures were given as footnotes and included in the last row of Table 4. A graphical explanation on the utilization rate indicator is shown in Figure 7.

VLTD is the stage where the main and unique product is the sugar. The rest of the schemes considered simultaneous sugar and ethanol production. All figures of sugar and ethanol production were calculated assuming that half of the sugarcane juice is used for each product.

There are no sugar mills running full in FSOTD level. SCAR energy use is not considered for FSOTD variants because the use of this residue in actual practice is still minuscule. SCAR use is the subject of several research across the world currently. All data used for FSOTD figure calculations were assumed, considering the main trends of the sugarcane agro-industry [28-31].

Economic impact of sugarcane energy use

The result obtained from economic impact calculation is shown in Table 5. In this table, the utilization rates of sugarcane for each of the studied levels of technological development are also shown. In Table 5, 1 ton of oil equivalent and its relevant energy content were included in order to facilitate the understanding of the analysis (row 2, columns 2 and 3). The current prices of Brent and West Texas Intermediate oils were also included (row 3). The relevant explanations about each of the shown figures were given as footnotes and included in the last row of Table 5.

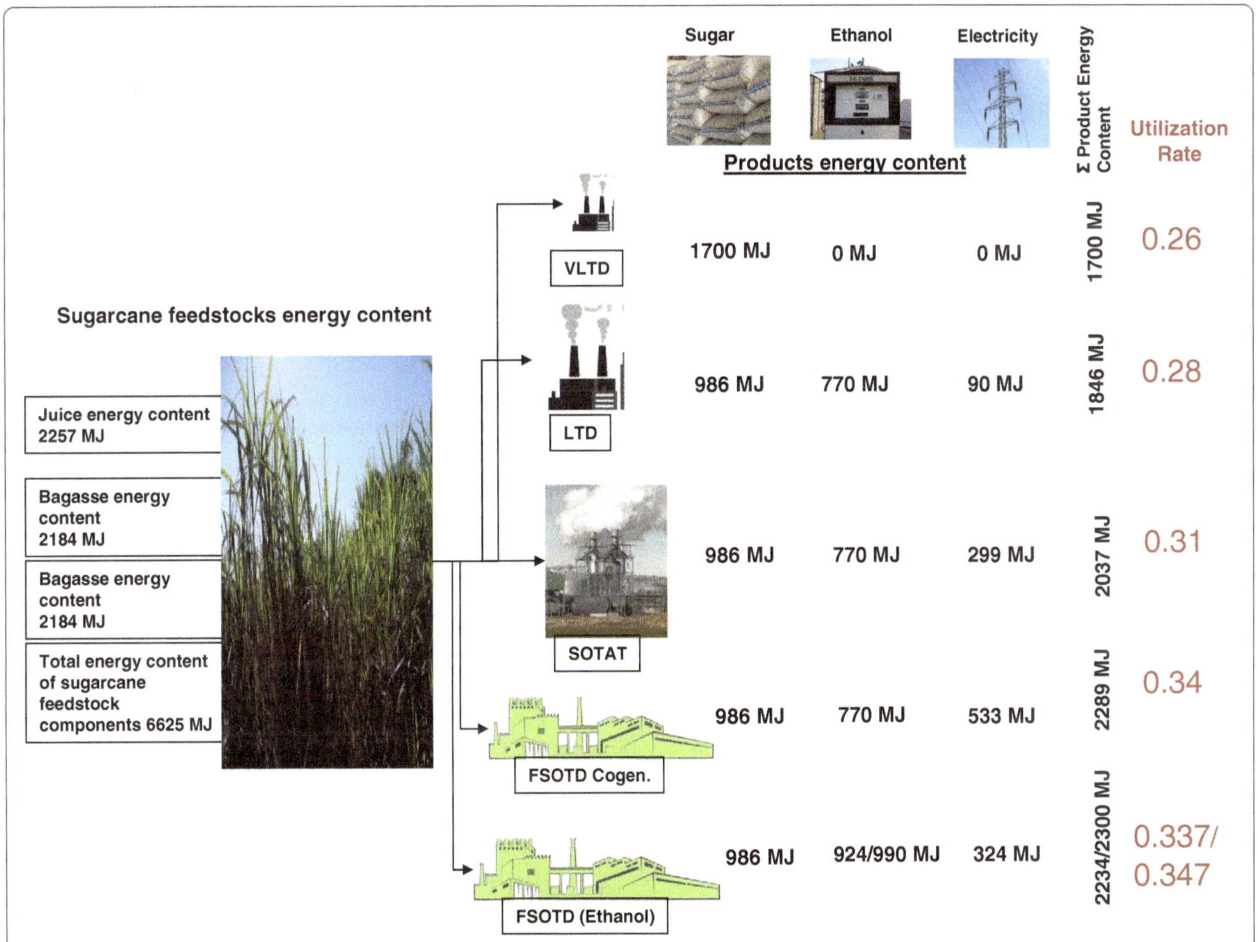

Figure 7 Graphical explanation of the utilization $\text{rate}_{\text{sugarcane feedstock}}$ calculation.

Table 5 Economic impact of sugarcane energy use

Row	1 ton (dry basis)	Mass (kg)	Energy content (MJ)	Parameter	VLTD	LTD	SOTAT	FSOTD (electricity)	FSOTD (ethanol)
	(1)	(2)	(3)	(4)	(5)	(6)	(7)	(8)	(9)
1	Total sugarcane	423	6,626	Total utilization rate (%) [a]	0.26	0.28	0.31	0.35	0.33/0.34
2	Oil equivalent	1,000	43,120	Sugarcane energy (i) (ton of oil equivalent)[b]	0.04	0.043	0.048	0.054	0.050/0.052
3	Crude oil price (US$) [32]								
	Brent			105.26	4.2	4.51	5.04	5.67	5.25/5.46
	WTI			89.90	3.59	3.86	4.31	4.85	4.49/4.67

WTI, West Texas Intermediate. [a]Utilization rate figures from Table 4. [b]Sugarcane energy (i) = 6,626 × column (i) / 43,120, where i = columns 5, 6, 7, 8, and 9.

The utilization rate of sugarcane energy does not exceed 31% nowadays. Even though SOTAT is presently the best stage of development from the viewpoint of sugarcane energy use, the total return of energy is lower than US$0.10/ton of milled cane. It is important to underline that this value is possible if sugar factories produce sugar and ethanol simultaneously. However, about 30% of Brazilian sugar factories are autonomous distilleries[i] (most of them were built recently), considering the forecasting of sugar price in the international market and other disadvantages such as less nimbleness of industrial process to face the failures and the oscillation of price at the sugar international market. It seems that the predominant tendency in the near future will still be simultaneous sugar-ethanol production and product diversification in the sugar factory.

Assuming that all the processed sugarcane in the world in 2010 ($1,685 \times 10^{6}$ tons; see Figure 1) was milled in sugar factories running on VLTD scheme, their relevant profitability (expressed in oil price units) would be US $6,232 \times 10^6$. On the other hand, considering that all aforementioned factories were running on LTD and SOTAT, the increase of profitability from the same amount of milled cane would be US$$6,696 \times 10^6$ and US $7,480 \times 10^6$, respectively. It is to say that there was an increase of US$$464 \times 10^6$ and US$$1,248 \times 10^6$, respectively.

FSOTD requires a very high cost of investment [24,29], but the increase of profit is about 5% for both $FSOTD_{electricity}$ and $FSOTD_{ethanol}$. In other works [24,28], capital investments for $FSOTD_{electricity}$ and $FSOTD_{ethanol}$ were estimated to be US$$367 \times 10^6$ (in 2004) and US$$311 \times 10^6$ (in 2011), respectively. In this way, considering the average money depreciation rate of 0.06/year, the capital investment cost is similar for both schemes. It is not possible to establish a clear economical advantage in any of the analyzed FSOTD variants. The selection of $FSOTD_{electricity}$ or $FSOTD_{ethanol}$ scheme depends on the domestic context and international economic growth tendencies.

Main barriers to overcome to further increase sugarcane energy utilization rate

The further increase of sugarcane energy use depends on several factors. Among them are physical, engineering, chemical, economical, social, biological, environmental, and political factors. For this reason, the agro-industrial development of sugarcane agribusiness is a multifactorial subject that requires a multidisciplinary approach. Regarding their influence on the issue, the following deserve special mention:

- The mechanized harvesting method should be improved to make a better SCAR collection. Currently, existing methods were designed to leave SCAR in the field. Without an efficient use of SCAR energy, a complete upgrading of the sugarcane agro-industry is unthinkable.
- The relative low capacity of known yeast to live in a substrate with high alcohol concentration.
- The low thermoresistance of known yeasts used in the fermentation process.
- The biomass cell wall resistances to microbial and enzymatic attack (also known as biomass recalcitrance) [32].
- The very high cost of cellulase enzyme production.
- The relatively low efficiency of known hydrolysis pretreatments.
- The seasonality characteristic of sugarcane agro-industry.
- The relatively small-scale sugar facilities (low agricultural yield).
- The prohibitive cost of capital investment for the modernization of sugar factory.

It is necessary that a new generation of materials with high mechanical, thermal, and electrical properties is made available in order to develop a new generation of machinery that is more efficient and cheaper than the existing machinery.

Conclusions

It can be identified nowadays that the main trend of the sugarcane agro-industry is production diversification in the sugar mill, that is, the simultaneous production of food, biofuel, and electricity. Although this tendency is growing in many places, the use of sugarcane as feedstock still being retrograde, disregarding the urgent necessity of the sustainable development. Most of the sugarcane-producing countries are poor, and many of them have no fossil fuel reserves. Sugarcane energy use is really a big chance for these countries' growth.

The utilization rate of sugarcane feedstock as performance indicator of the sugarcane agribusiness is a simple method that calculates, preliminarily, the feasibility of new investment or the assessment of any agro-industrial upgrade in an existing facility.

This method does not substitute any of the classical methods of techno-economical assessment, which include GHG calculations, and also does not substitute the mass-energy balance calculation of the production process in the sugar mill. This work is a modest contribution to make the process better and simpler.

Endnotes

[a]SCAR is composed of sugarcane leaves (green and dry) and cane tops. For a detailed explanation, see in [3].

[b]Most widespread sugarcane specie.

[c]In Brazil, 43% of the composition of the energy matrix is renewable, while the share of renewable energies, in the world energy matrix, is only 15%.

[d]The energy needed for sugar and bioethanol production (steam and electricity) is obtained from sugarcane bagasse. A well-balanced production process (mass and energy) in the sugar mill can meet the energy necessities of the production processes using only 40% of energy contained in the sugarcane bagasse. The bagasse surplus can be used to produce surplus power and export electricity to public grid.

The bioethanol from corn (dry grind production, USA), as a rule, in order to produce ethanol from corn, it is necessary to use fossil fuel (mainly gas). For this reason, in California (USA), the corn-based ethanol does not qualify as fuel that sufficiently reduces GHG emissions. The agricultural yield of corn (USA) and sugarcane (Brazil) were 9.59 and 79.19 tons ha^{-1}, respectively, in 2010. The bioethanol production yield from corn (USA) and sugarcane (Brazil) are 371 and 85 L ton^{-1}, respectively. In the USA, more than 80% of bioethanol is produced using the dry grind production process.

[e]There are small sugar mills with very low technological development in Latin America (Chumbagua, Honduras) [20], for instance, and in contrast, it is possible to find a sugar mill with state-of-the-art-technology in the same region (Santa Ana, Guatemala) [20]. The same situation is possible in Asia. In Indonesia, India, and Bangladesh, it is common to find sugar mills with very low technological development. In contrast, in Malaysia there are sugar mills with state-of-the-art technology. Regarding available data, there are more than 100 sugar mills in the VLTD situation [20].

[f]The bagasse is burned inefficiently just to avoid environmental pollution without any perspective on higher energy efficiency of the sugar-ethanol production process. The bagasse energy content and its potential capacity to cogenerate surplus electricity are not taken into consideration.

[g]The thermal installed capacity of BIG/GTCC systems mentioned in the open literature is, as a rule, 1.5 to 2 times bigger than the average thermal capacity of most sugar facilities.

[h]Vinasse is a liquid residue from ethanol production. It is also called mud. Approximately 10 L of vinasse is generated to produce 1 L of ethanol. In many places, like in Brazil, vinasse is used as fertilizer.

[i]In autonomous distillery, the simultaneous production of sugar and ethanol is the rule of the sugarcane industry. Most existent sugar factories have associated distilleries. The distilleries not associated with sugar factories are called autonomous distilleries.

Abbreviations

BPST: Back-Pressure Steam Turbine; BIG/GTCC: Biomass Integrated Gasifier/Gas Turbine Combined Cycle System; CEST: Condensation Extraction Steam Turbogenerator; FSOTD: Future Step of Technology Development; GHG: Greenhouse Gas; ° GL: degrees Gay-Lussac (vol.%); LTD: Low Technological Development; SOTAT: State-of-the-Art Technology; VLTD: Very Low Technological Development; WTI: West Texas Intermediate.

Competing interests

The authors declare that they have no competing interests.

Authors' contributions

WAP carried out the collection of data and relevant calculations. CAL checked the calculations as well as updated the revised manuscript. All authors read and approved the final manuscript.

Authors' information

WAP is a postdoctorate/senior researcher of the Alternative Fuels Group, Department of Applied Physics (DFA), Institute of Physic 'Gleb Wataghin' (IFGW) at the State University of Campinas (UNICAMP), Brazil. CAL is a full professor and coordinator of the Alternative Fuels Group of DEA/IFGW at the State University of Campinas (UNICAMP), Brazil.

Acknowledgments

The authors would like to thank the Brazilian National Council for Scientific and Technological Development (CNPq, Process 150604/2009-2) for the support to this work

References

1. Zanzi R, Sjöström K, Björnbom E (1996) Rapid high-temperature pyrolysis of biomass in a free fall reactor. Fuel 75(5):545–550
2. Jenkins BM, Baxter LL, Miles TR, Jr, Miles TR (1998) Combustion properties of biomass. Fuel Process Technol 54(1–3):17–46

3. Dermibas A (2004) Combustion characteristic of different biomass fuels. Prog. Energy Combust. Sci. 30:219–230
4. Alonso Pippo W, Luengo CA, Alonsoamador Morales Alberteris L, Garzone P, Cornacchia G (2011) Energy recovery from sugarcane-trash in the light of second generation biofuels. Part 1: current situation and environmental aspects. Waste Biomass Valor 2:1–16
5. Eddine BT, Salah MM (2012) Solid waste as renewable source of energy: current and future possibility in Algeria. Int. J. Energy Env. Eng. 3:17
6. Pandyaswargo AH, Onoda H, Nagata K (2012) Energy recovery potential and life cycle impact assessment of municipal solid waste management technologies in Asian countries using ELP model. Int. J. Energy Env. Eng 3:28
7. Goldemberg J (2007) Ethanol for a sustainable energy future. Science 315:808–810
8. Moreira J, Goldemberg J (1999) The alcohol program. Energy Policy 27(4):229–245
9. Lèbre La Rovere E, Pereira AS, Simoes AF (2011) Biofuels and sustainable energy development in Brazil. World Dev. 39(6):1026–1036
10. Macedo IC (2005) Sugarcane's energy: twelve studies on Brazilian sugarcane agribusiness and its sustainability (Original in Portuguese), 1st edition. UNICA, Sao Paulo
11. Ferraz-Dias-de-Moraes MA () Considerations on ethanol industry of Brazil. Biofuels in Brazil: realities and perspectives. (Original in Portuguese). Biocombustíveis no Brasil Realidades e Perspectivas. [http://www.itamaraty.gov.br/imagens-e-textos/temas-brasileiros-1/portugues/biocombustiveis-no-brasil]. (2006). Accessed Nov 21 2011
12. Borges JMM (1986) Proàlcool: Encontro Nacional dos Produtores de Álcool, VI Econálco. (Lecture: original in Portuguese). Proceeding Sopral, São Paulo, Brasília, pp 18–20. 1986
13. Larson ED, Jin H, Celik FE (2005) Gasification-based fuels and electricity production from biomass. Without and with carbon capture and storage. Biofuels, Princeton
14. COPERSUCAR (1989) Proálcool: Fundamentals and Perspectives. (Original in Portuguese) [PROÀLCOOL FUNDAMENTOS E PERSPECTIVAS]. COPERSUCAR, Maio
15. National Company of Supply (CONAB) (2008) Profile of productions units by states and profile of sugar and ethanol sector in Brazil: situation observed from November 2007 to April 2008, pp. 13–16 (original in Portuguese). National Company of Supply, Brasilia
16. França R, Nogueira LAH (2008) Seventy questions to understand ethanol. (Original in Portuguese) [Setenta questões para entender o etanol]. Revista Veja 2052:104–114
17. Suleiman José H, Manoel Regis Lima Verde L, de Carvalho Macedo I (2005) Biomass power generation: sugar cane bagasse and trash. PNUD-CTC, Piracicaba, Brazil
18. Merola SS, Tornatore C, Marchitto L, Valentino G, Corcione FE (2012) Experimental investigations of butanol-gasoline blends effects on the combustion process in a SI engine. Int. J. Energy Env Eng 3:6
19. ISMA () Atlas of sugar mills in India. [http://www.indiansugar.com/SugarMap.aspx] (2010). Accessed 09 Nov 2012
20. Sugar Engineers () Sugar factories of the world. [http://www.sugartech.co.za/factories/index.php] (2012). Accessed 09 Nov 2012
21. Alonso PW, Garzone P, Cornacchia G (2007) Agro-industry sugarcane residues disposal: the trends of their conversion into energy carriers in Cuba. Waste Manage. 27:869–885
22. Alonso-Pippo W, Luengo CA, Koehlinger J, Garzone P, Cornacchia G (2008) Sugarcane energy use: the Cuban case. Energy Policy 36(6):21632181
23. Alonso Pippo W, Luengo CA, Alonsoamador Morales Alberteris L, Garzone P, Cornacchia G (2011) Energy recovery from sugarcane-trash in the light of 2nd generation biofuels. Part 2: socio-economic aspects and techno-economic analysis. Waste Biomass Valor 2:257–266
24. Larson ED, Williams RH, Leal LV, Regis M (2001) A review of biomass integrated-gasifier/gas turbine combined cycle technology and its application in sugarcane industries with an analysis for Cuba. Energy for Sustainable Development 5(1):54–76
25. Sánchez TO, Cardona ACA (2007) Fuel ethanol production (Original in Spanish). In: Tizan (ed) [Producción de alcohol carburante Una alternativa para el desarrollo agro-industrial]., Manizales, Caldas, Colombia, p 380
26. Fernández RJ, Pérez JA, Pérez SO, Alonso Pippo W (2004) Characterization of industrial and agricultural residues of sugarcane for obtaining Biooil. [Conferencia Internacional de Derivados de la caña de azúcar. Diversificación 2004.] International Conference of Sugarcane Derivates. Diversification 2004. ICIDCA, La Habana
27. Hugot E (1986) Handbook of cane sugar engineering, 3rd edition. Elsevier Science, New York
28. Stefano M, Johan M, Guido Z (2012) Techno-economic evaluation of 2nd generation bioethanol production from sugar cane bagasse and leaves integrated with sugar-based ethanol process. Biotechnol. Biofuels 5:22
29. Dias MOS, Junqueira TL, Cavalet O, Cunha MP, Jesus CDF, Rosell CEV, Filho RM, Bonomi A (2012) Integrated versus stand-alone second generation ethanol production from sugarcane bagasse and trash. Bioresource Technol. 103:152–161
30. Walter A, Dolzan P, Quilodrán O, de Oliveira JG, da Silva C, Piacente F, Segerstedt A (2011) Sustainability assessment of bio-ethanol production in Brazil considering land use change. GHG emissions and socio-economic aspects. Energy Policy 39:5703–5716
31. FAOSTAT () CROPS: Statistics division. [http://faostat.fao.org/site/567/DesktopDefault.aspx?PageID = 567#ancor] (2010). Accessed July 12 2012
32. Oil-price.net . [http://www.oil-price.net/?gclid = Clai0Ym_ubECFQgEnQodjTgAwA] (2012). Accessed 27 July 2012

9

Solar energy generation potential along national highways

Pragya Sharma[1*] and Tirumalachetty Harinarayana[1,2]

Abstract

Energy generation using solar photovoltaic requires large area. As cost of the land is growing day by day, there is a strong requirement to use the available land as efficiently as possible. Here, we explored the potential of energy generation using the land above national road highways by constructing a roof structure. This space can contribute to the energy generation without extra cost for the land. It also results in energy efficiency, for example, improved vehicle movement and minimum energy for air conditioning of vehicles. Additionally, it also helps in minimum road repairs and longer vehicle tire life due to the effect of sun shade. Thus, the expenditure for wear and tear for road repairs is reduced considerably. From our modelling study, it is observed that the Ahmedabad-Rajkot highway can generate 104 MW of electricity (163 GWh of annual energy generation) and the Ahmedabad-Vadodara highway space can generate 61 MW of electricity (96 GWh of annual energy generation) for single-layer solar panels. If there are two layers of solar panels one over the other, the annual energy generation of the same highways, Ahmedabad-Rajkot and Ahmedabad-Vadodara, can be increased to 229 GWh and 140 GWh, respectively. If our concept is implemented throughout India, it not only increases the power generation to more than a few gigawatts of electricity but also has other various fringe benefits including longer road life, employment generation, reduced CO_2 emission in environment, etc.

Keywords: National highways, Modelling, PVsyst software program, Efficiency, Shade analysis, Land cost, Employment generation

Background

Many island countries, for example, Fiji islands, Papua New Guinea, etc., are still suffering from the use of any kind of electricity system. Due to poor resources, they all depend on fossil fuel, mainly diesel, for electricity generation. In fact, 85% of their total electricity production is being generated mainly using diesel generators [1]. In recent years, energy conservation measures are on the rise. Every nation is looking at the various resources available and plans to use them more optimally to get the best out of them. For example, in a recent detailed study in Nigeria, several measures are suggested on the energy use in ventilating equipment, lighting, electrically operated industrial machines and engines, design for energy-efficient buildings, etc. for energy sustainability [2]. Our study also aimed at the same direction but aimed to effectively use the existing national highways for solar energy generation.

Solar photovoltaic (PV) is already a proven technology for energy generation [3]. Worldwide, the grid-connected solar PV continued to be the fastest growing power generation technology [4,5], with a 55% increase in 1 year with a cumulative installed capacity of 3.1 GW in 2005, up from 2.0 GW in 2004 [6]. In a similar way, the Indian solar energy sector has been growing rapidly, for the past few years, mainly due to the government's initiative such as tax exemption and subsidies. The other factors are the country's geographical location and huge gap between energy demand and generation. All these factors are opportunities for India to become one of the most rapidly emerging solar energy markets in the world. The government of India is ambitious about its national solar mission that targets to achieve 10,000 MW by the year 2017 and 20,000 MW by the year 2022 [7]. To reach this target, one requires many innovative ways including an attractive national policy for solar industries. Looking at the grid-

* Correspondence: pragyasees@gmail.com
[1]GERMI-RIIC, Gandhinagar, Gujarat 382007, India
Full list of author information is available at the end of the article

interactive power generation, the total installed capacity of grid-connected solar power as of June 2011, which is the cumulative capacity based entirely on PV technology, is 39.6 MW and 979.4 MW in April 2012, and off-grid capacity is 85 MW [8]. This shows that the total cumulative solar capacity has increased 25 times in less than a year due to thrust on solar power. From the study growth of photovoltaic in India, on an average, about 45% annual increase is noticed during the years 2000 to 2009. If we look at the state-wise grid-connected solar power generation, Gujarat alone contributed to a total of 654.8 MW. This shows that the government of Gujarat is contributing more than 65% grid-interactive solar power generation [9,10]. Fortunately, the state of Gujarat is located in the sunny belt (>2,000 kWh/m^2) with a clear bright weather for 250 to 300 days per year and receives maximum annual global radiation [11]. More importantly, Gujarat has attracted many industries with its proactive policies and firm commitment.

From the study of cost economics of a solar photovoltaic power plant, the PV module cost is about 45% and that of the other accessories like transformers, cables, inverters, civil works, etc. comes to about 55% [12]. Additionally, the cost of the power plant also depends on the land value. As the cost of solar photovoltaic is continuously decreasing, the major challenge now lies on the land cost. Land is becoming a scarce resource in India in recent years, and per capita land availability is low [6]. Land is often considered as the topmost challenge for deploying solar energy technology [13]. In view of the above, our study is based on using the available land in an effective way. Towards this direction, national highways are studied for energy generation using solar photovoltaic technology. To demonstrate our approach on the use of land and its solar potential, two areas, namely Ahmedabad-Rajkot national highway road and Ahmedabad-Vadodara national expressway, are taken for modelling. Our concept in the form of a schematic diagram is shown in Figure 1. The total span of the Ahmedabad-Rajkot national highway is 205 km, and that of the Ahmedabad-Vadodara national express highway is 93 km [14,15]. For the purpose of modelling, we have used two software programs - Google Earth Pro and PVsyst simulation softwares [16,17].

There are other approaches and methods of using solar photovoltaic technology on highways, for example, using solar-powered LED roadway lighting [18], security lighting, highway changeable message [19], etc. Although these are very important applications, in terms of power generation or power saving, they are almost negligible in comparison to the approach that we have suggested in our paper. Also, one can see the cost-oriented modelling and design optimization for alignment [20] to determine operation and engineering cost of the highway.

The rationale behind the proposed study is to explore the effectiveness of using the national highway and to enlarge all the factors to generate energy. The additional advantage of using the space above the road highways for installation of solar panels is the shading on the roads. This results in improved vehicle efficiency by reducing energy losses due to heat inside the vehicle [21] and also improves the life of the tires of all vehicle wheels due to the shade derived from the road. Combining, the two will result in both improved economy by

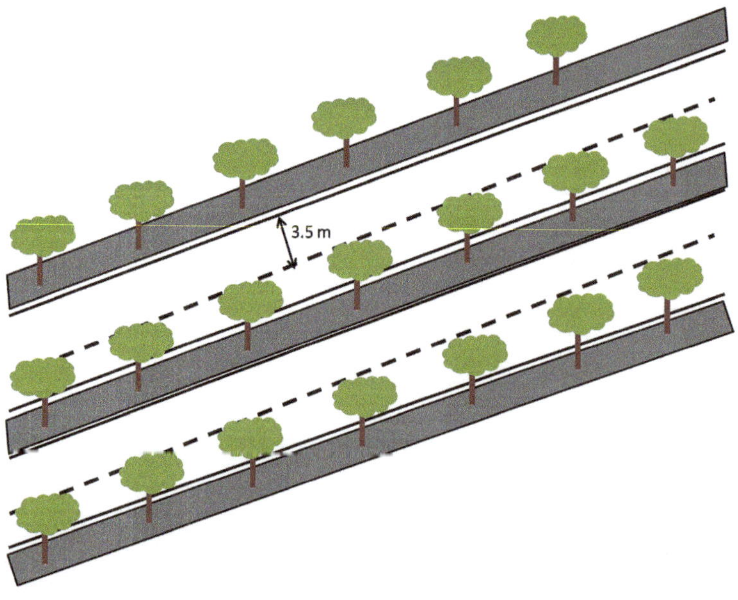

Figure 1 A schematic layout of the national highway.

energy generation and improved efficiency of the vehicle at the same time. One more advantage of the shade over the road is the life of the road and reduction of the cost due to road maintenance. It is a usual phenomenon on many roads that soon after the rains, pot holes usually develop. This will become a history if all roads are provided with solar panels above them.

Our study thus proposes a new idea on the effective use of roadways and also an opportunity to meet the major challenge for the industry, i.e., land availability for the grid-interactive solar power plants. Although our study shows many advantages, one needs to take a few extra measures for proper implementation. For example, one needs to take care of fixing the solar panels to the base firmly onto the structure. Fixing the solar panels at an elevated location, say about 9 to 10m above the ground level, is vulnerable to heavy wind during the storm or rainy days. It may pose a problem for the stability of the solar panels as compared to the ones close to the ground. With extra care of fixing, the panels may overcome this problem. Another problem may be from the dust and smoke particles due to movement of the vehicles. However, as many national roadways are wide and generally much cleaner than other roads, for example, village roads, narrow city roads, or state roads, in any case, one needs to make special arrangement of cleaning the panels on a daily basis as compared to normal panels on the ground away from the road traffic or remotely located panels, where cleaning of panels is done once in a week or 10 days. If all or most of the vehicles follow the Euro standards as regards to smoke emissions, then there will be less pollution problems related to vehicular traffic on the panels. Thus, although there are a few issues to be considered if one follows the ideas of our present study, the disadvantages can be tackled with proper planning - mainly on firm fixing of the panels and regular cleaning of the solar panels.

Methods

The modelling of our present study is done through PVsyst and Google Earth Pro software programs. Before presenting the methodology, brief details of the software are presented in the following.

The software programs

Among various software programs, PVsyst simulation software is popular to analyze the detailed performance of the solar plant in field conditions. It can be used in many ways, for example, to investigate the different loads on the system, to estimate the size of the system, to determine the optimal size of the panel, to assess the energy production in the system, etc. The other various capabilities and options available in the PVsyst software simulation can be seen in [17]. PVsyst, a PC-based software package, can also be used to study the sizing and data analysis of complete PV system. It is used for different designs and sizes of the system. It can evaluate monthly production and performance. It also performs economic evaluation of the PV system at the design stage itself. Its application performs detailed simulation and also shading analysis according to several dozens of variables.

Google Earth is a popular software to view satellite imagery, maps, terrain, 3D building, etc. and for research studies [22-24]. It also provides tools to analyze and explore places with realistic views of an area and navigation through directions. The different features and high quality of satellite maps [25] available through Google Path allow navigating through the Earth, finding addresses and location, marking location, etc. It is also helpful in obtaining the coordinates of different locations. The other various capabilities and options available can be seen in [16]. This can be used by linking the data with GIS software. Another better way of studying is to use the more advanced version of Google Earth - namely 'Google Earth Pro.' This is used in our study for more accurate results. With high-resolution imagery, it offers a comprehensive geospatial database for roads and historical imagery. One can visually see the surface of the Earth with all the field information. This is a very useful feature as one can get the field information without actually visiting the location. Since our present study is more of a regional nature, Google Earth Pro is found to be suitable in our study. Google Earth Pro, with the integrated measurement and drawing tools, is also used for radius and area measurement. Thus, this software helps for identification of thick vegetation, bridges, and unusable road locations with its dimensional parameters.

Figures 1 and 2 show a schematic diagram prepared to show an example of a typical national highway. Each lane width is 3.5 m as specified by the National Highway Authority of India [14] and ground-checked by the authors. The roadside plantations with trees are also shown symbolically in the figure. One can design different configurations based on the area available on the road and the size and technology of the solar panel as detailed below.

In the following, details of the two national highways, namely Ahmedabad-Rajkot and Ahmedabad-Vadodara, are presented.

Details of the sites

Figure 3 shows the Ahmedabad-Rajkot national highway road map with small towns on its way. For modelling, the potential of solar energy generation using solar photovoltaics, an area of about 1,355 m^2 is chosen for the four-lane road. More details are given in Table 1.

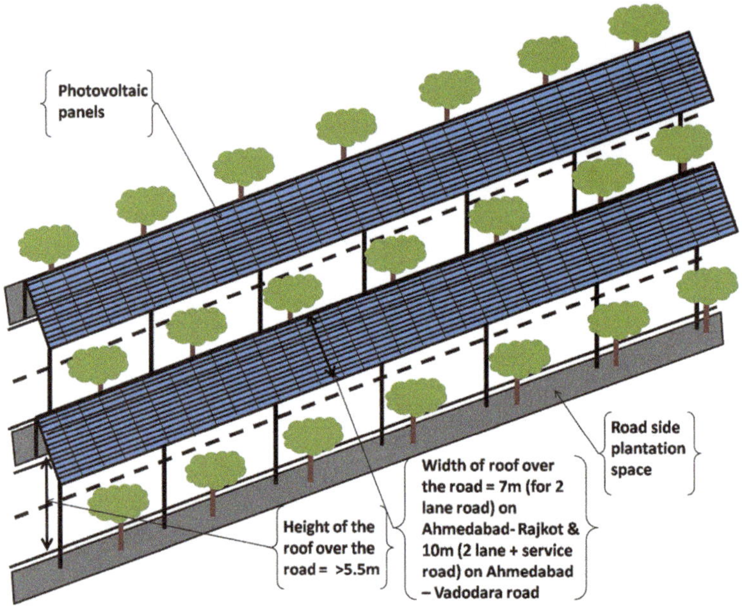

Figure 2 A schematic layout of the national highway with solar photovoltaic panels on its roof.

Figure 4 shows the Google map for the Ahmedabad-Vadodara national express highway. The same area as before has been used for computation to find out the potential of solar energy generation using solar photovoltaics.

For the modelling part, we have chosen the four-lane Ahmedabad-Rajkot national highway and six-lane Ahmedabad-Vadodara expressway (four-lane highway with two service roads) for finding the potential of solar energy generation. The total span of the Ahmedabad-Rajkot national highway length is 205 km. Each lane on the national highway has a width of 3.5 m [26], and thus for two lanes, the total width is 7m. For both sides (to and fro), the total width available for solar panels above the road is 14m. The width of the road has been also measured by the authors with measurement tapes from various places along the national highway. This also helps to measure the accuracy of Google Earth Pro software for modelling purposes. Similarly, for the Ahmedabad-Vadodara express highway, the span of the road is 93 km in length [14,15]. The width of the express highway is 20 m, out of which 14 m is for the two-lane national express highway. Additional 3 m is the extra road available on both sides of the national express highway or service road. This is again verified and ground-checked by the authors while travelling all along the national highways and measuring road width using measurement tapes. Additional information on the national highways is also provided from virtual observation of the road using Google Earth Pro software, for example, the lengths of the road unavailable due to certain environmental reasons like bridges, crossing, unusable road, thick vegetation, etc. These parts have been identified with high resolution with high accuracy in measurements [16] using Google Earth Pro software modelling without actually making the ground observation (but verified on various different places by travelling all along the national highway). More information on these aspects is shown separately in Annexure I in Additional file 1 for the Ahmedabad-Rajkot national highway and Annexure II in Additional file 1 for the Ahmedabad-Vadodara national express highway

The data

For the grid-connected system, the basic input parameters required for modelling are PV component database, grid inverter database, geographical site information, and monthly meteorological data for horizontal global irradiance and temperature. In the present study, a silicon-polycrystalline module Titan 24–100 of the Titan Energy systems (Secunderabad, Andhra Pradesh, India) has been used. Each solar module has a maximum power output of 100 W. For the grid inverter, Sputnik Solarmax 50 C (Länggasse, Bienne, Switzerland) has been used that works on a 50-Hz frequency, a voltage range of 430 to 800 V, and a power rating of 50kW as one unit for our calculation. The unit considered here is extended to the entire length of the highway. The meteorological data are acquired from METEONORM version 6.1.0.23 (Fabrikstrasse, Bern, Switzerland) [12,27].

In Table 1, the information with more details for the solar panels such as solar module technology, power rating, and related module specifications for a 50-kW photovoltaic system is provided. This is considered as

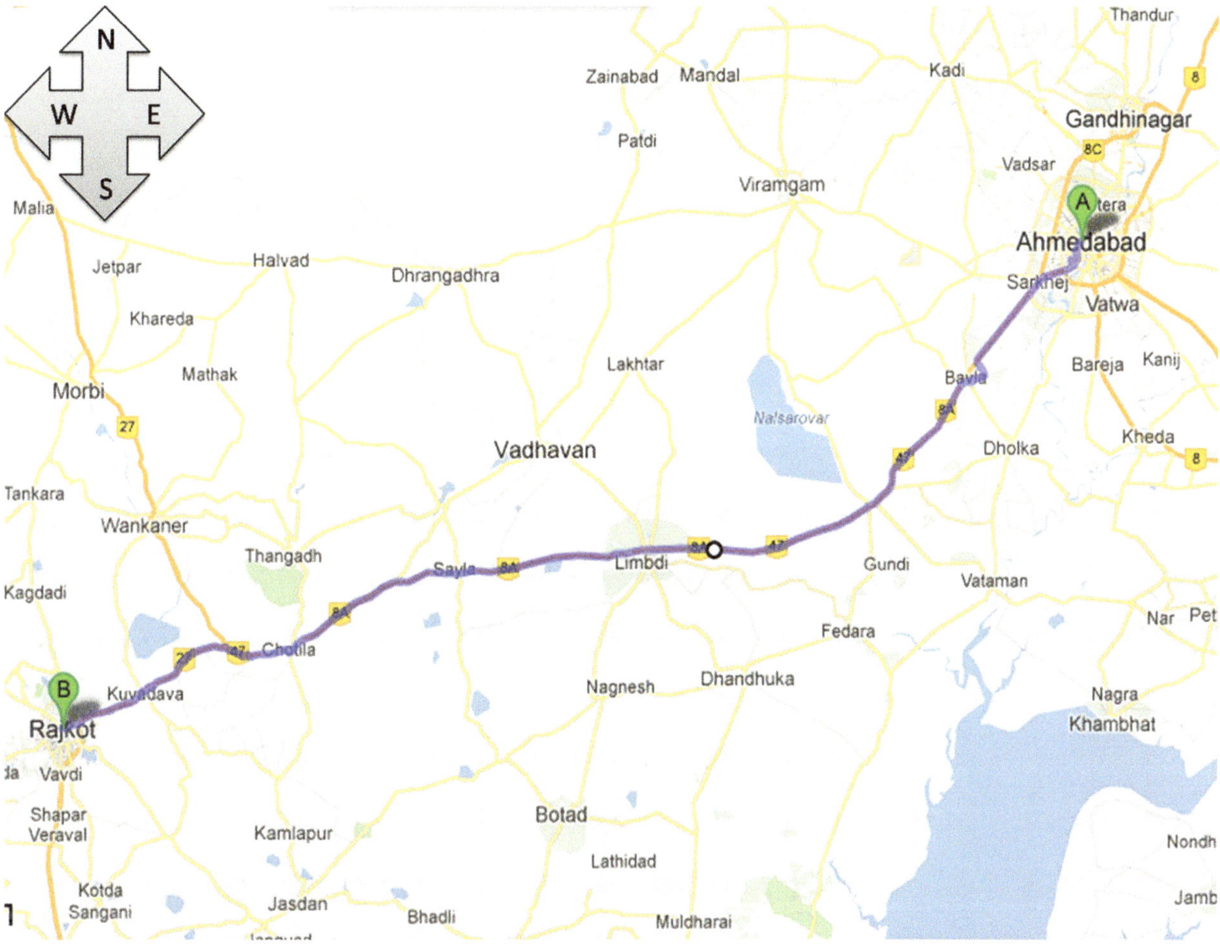

Figure 3 Ahmedabad-Rajkot national highway shown using Google maps.

Table 1 Details of 50-kW solar photovoltaic system for PVsyst software, a unit considered for national highway modelling

Parameters	Specifics
Number of modules	504
Area of the field(land)	1,355 m^2 (approximate)
Dimension of the solar panel (length) above the road	89 m (for Ahmedabad-Rajkot highway)
	70 m (for Ahmedabad-Vadodara highway)
Dimension of the solar panel (width) above the road	17 m (for Ahmedabad-Rajkot highway)
	23 m (for Ahmedabad-Vadodara highway)
Active area of the modules (sensitive PV area)	423 m^2 (each module length 1.29 m and width 0.65 m)
Maximum power generation capacity (at peak solar radiation, i.e., 1,000 Wp standard)	50 kW
Energy supplied to the grid (kWh)	78,074
Module tilt	23° N
Separation between rows of solar panel	2.5 m
Orientation of the panels	Exactly south
Solar module technology	100-W silicon polycrystalline of Titan Energy systems

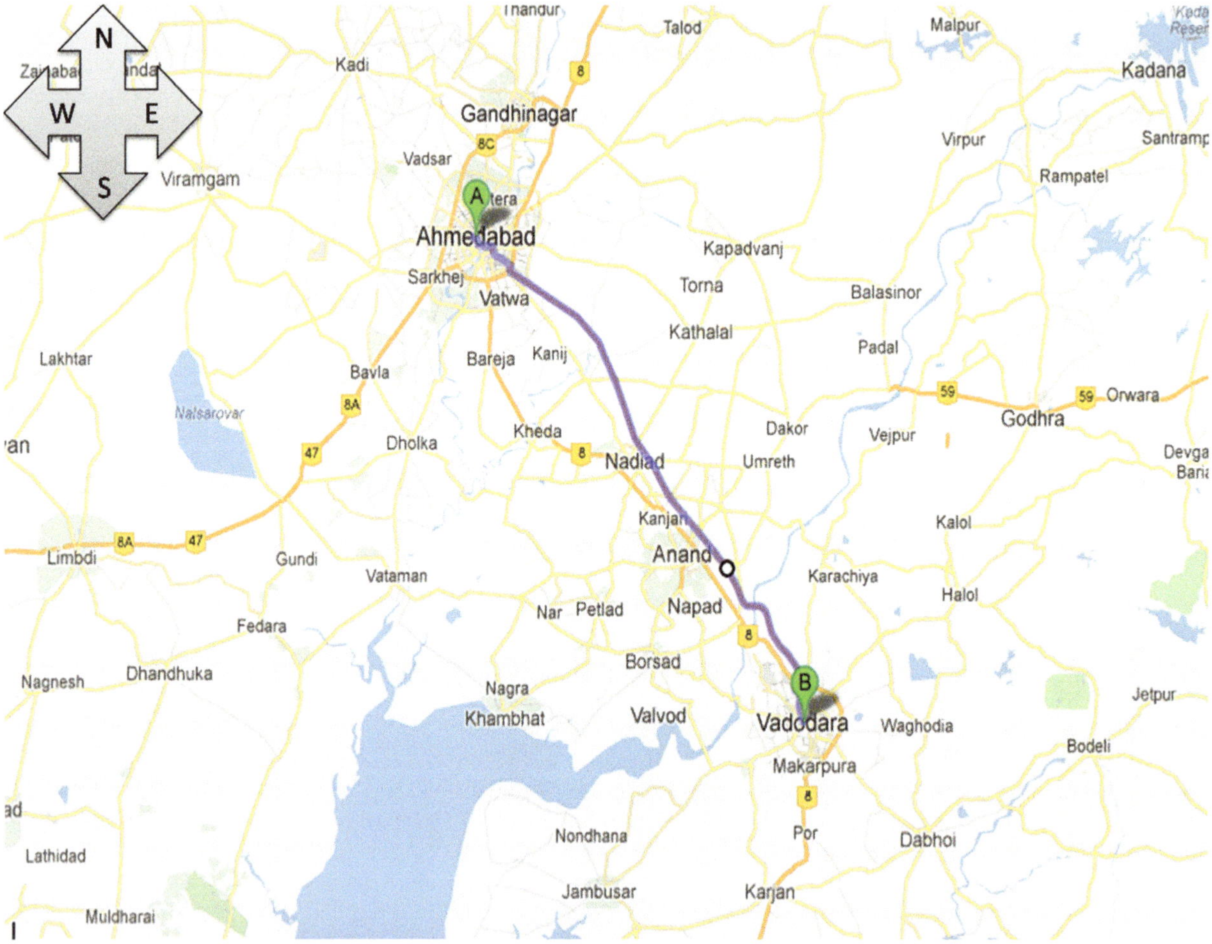

Figure 4 **Ahmedabad-Vadodara national express highway shown using Google maps.**

one unit for the purpose of calculation, which will be extended for the entire length of the highway. Accordingly, 504 numbers of modules are used for both Ahmedabad-Rajkot national highway (four lanes) and Ahmedabad-Vadodara highway (four lanes wo service roads). The panels are tilted at an angle of 23°. This tilt is chosen equal to the latitude for the Ahmedabad site (23.067°).

Solar panel orientation

It is well known that the highest amount of energy generation can be produced through solar photovoltaic panels oriented exactly to the south in the northern hemisphere [28,29]. Also, if the orientation of the solar panel is other than south, then the amount of energy generation is less. The orders of decrease in energy generation for different directions are the East, the West, and the North, based on the compilation of a 50-kW system using PVsyst software [17]. We suggest that the panel orientation should be taken extra care while designing the mounting of panel; this helps to utilize energy available from solar panel as maximum as possible by keeping the orientation towards the south direction. As described before, due to the limitation of the PVsyst software, the simulation is considered only for 50 kW, a small area covering the national highway roads.

In Figure 5, the solar panel design configuration considered for our study of the Ahmedabad-Rajkot highway is shown as a schematic diagram, for two lanes. As described earlier, the width of the lane is 3.5 m for a multi-lane national highway [26] and the same is considered in our model study. Considering the width of the two lanes of a national highway, the schematic diagram shows a 7-m width of solar panels for each of the four lanes with a separation gap between the roads for tree plantation, etc. of 3 m, as shown in Figure 5. This configuration is computed using PVsyst simulation software in order to get the total solar energy generated from the panels on the four lanes. As a sample, 84 solar modules in each single row of, say, *X* direction (lengthwise) and a series of six rows of solar panels in another, say, *Y* direction (widthwise) are considered. Based on the dimensions of the Ahmedabad-Rajkot highway, these possibilities of

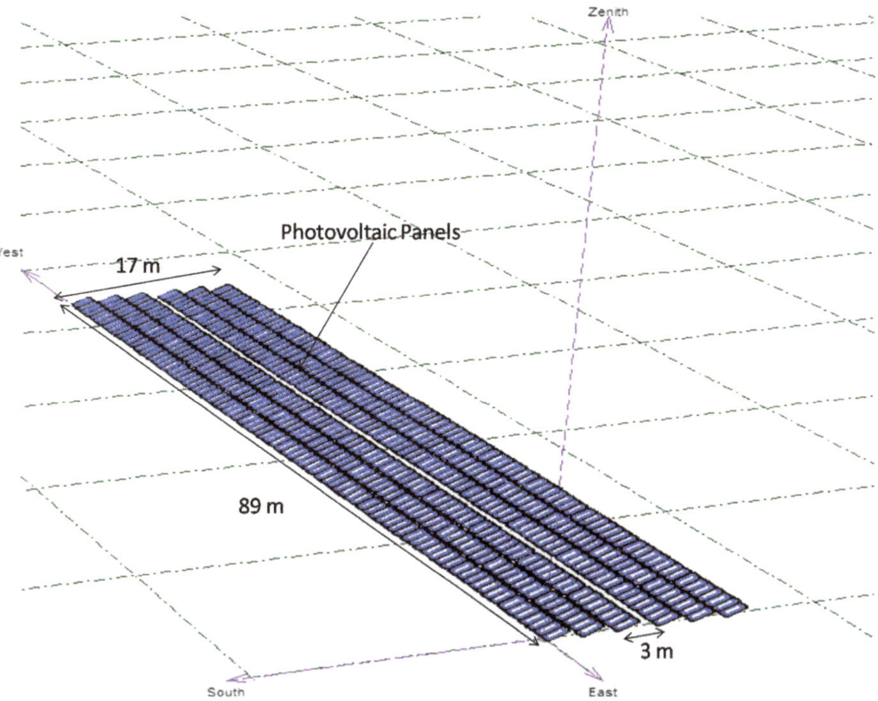

Figure 5 A design configuration for Ahmedabad-Rajkot national highway used for simulation in PVsyst simulation software.

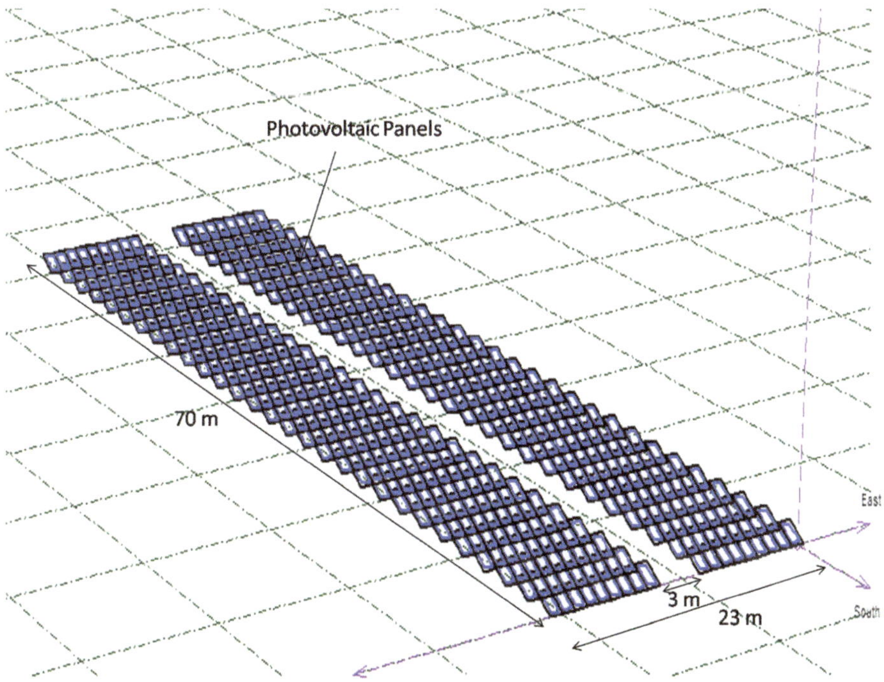

Figure 6 A design configuration for Ahmedabad-Vadodara national express highway used for simulation using PVsyst simulation software.

keeping solar modules are chosen. Such a design is arbitrary but helps to compute the parameters quickly. This configuration approximately provides a maximum of 50 kW of power output from PV modules. However, the results from this model can be extended to any length of a national highway as required.

This design configuration suggested in our study requires a rigid structure for the photovoltaic panel mounting. The type of structure is based on the number of solar panels for installation in the available area. For example, for the four-lane Ahmedabad-Rajkot national highway, 84 modules of 100 W each can be configured in one row, and thus, six such rows in 14-m width of the road require 504 modules for 89-m length along the road. Such a design is arbitrary and can be varied depending on the size and power rating of a particular photovoltaic technology.

In the same fashion, Figure 6 shows a schematic layout for the photovoltaic installation over the Ahmedabad-Vadodara national express highway. Similar to the Ahmedabad-Rajkot highway, we have chosen (9 + 9) modules widthwise and 56 such rows (28 + 28) with a separation of 2.5 m between rows of the modules from the module-to-module shading calculation. It is observed that the height of the module is 1.69 m; a total of 504 such modules of the same size with 100-W capacities can be configured along the Ahmedabad-Vadodara express highway that has a width of 23 m (10 + 10 + 3 m). Such a design is arbitrary and can vary depending on the size and power rating of a particular photovoltaic technology being used.

The different schematic configurations shown for the Ahmedabad-Rajkot national highway and the Ahmedabad-Vadodara express highway are necessary due to different orientation directions of the road. It may be noted that most parts of the Amedabad-Rajkot route is oriented towards near the east-west direction and for the Ahmedabad-Vadodara is near south-north direction. Accordingly, the unit considered for modelling using PVsyst software is the east-west direction for one model and the north-south direction for another model.

As shown in a schematic diagram, one needs to construct a structure above the highway road using firm RCC beams or stainless steel structure materials. These structures as a roof probably with stainless steel cladding provide a rigid structural support to the solar panel installations. Depending on the site location, one can construct a structure avoiding the trees, thick vegetation, level crossing, or other obstructing structures en route.

As an example, the effective energy at the output of an array as a function of local time of a day for December 20 using meteonorm radiation data for Ahmedabad averaged over a period of 20 years, 1981 to 2000, can be seen [12]. The amount of effective energy supplied to the grid per day as a function of time of the day is also shown with the figure [12].

Modelling the national highways

In Table 2, we describe the various parameters and dimensions considered to estimate the solar potential along the Ahmedabad-Rajkot national highway, i.e., the length of the highway identified suitable or unsuitable for the solar photovoltaic system installation as a roof of the road for energy generation. As described earlier, it has a length of 205 km with a two-way road (four lanes) with a width of 14 m. Thus, the total area (Area = Length ×Width) available for solar panel installation is 2.87 km^2. Although the length of the highway is 205 km, all along the highway, the road is not usable due to certain parameters, for example, environmental issues like thick vegetation, bridges on ponds, or overhead bridges for road crossings with other roads. We have also not considered the portions of the road either under construction or not clearly visible with the Google Earth Pro software with high-quality resolution and clarity [16]. Accordingly, Table 2 shows that out of 2.87-km^2 total area available on the Ahmedabad-Rajkot national highway, 0.2 km^2 is found to be unavailable or, say, inappropriate for installation of the photovoltaic system due to certain factors as stated above. Hence, the effective area available on Ahemdabad-Rajkot national highway is 2.67 km^2.

Table 2 Environmental constraints for Ahmedabad-Rajkot national highway

Serial number	Issues/ constraints for installation	Ahmedabad-Rajkot national highway		Ahmedabad-Vadodara national express highway	
		Length (approximate km)	Area (approximate km^2)	Length (approximate km)	Area (approximate km^2)
1	Unusable road	8.8	0.12	3.0	0.06
2	Bridges	0.9	0.01	3.1	0.06
3	Thick vegetation	7.0	0.01	1.0	0.02
4	Road crossing	2.8	0.04	-	-
Total (approximately)		19.5	0.18	7.1	0.14

Similar to the Ahmedabad-Rajkot highway, we have considered the Ahmedabad-Vaododara expressway. As described before, it has four lanes each with 3.5-m width and additionally has a service road of 3-m width on both sides of the road. The length of the expressway is 93 km with a total available road width of 20m (4 × 3.5 + 2 × 3). Accordingly, the total area available for solar panel installation is 1.86 km^2. Out of the available area, a total of 0.14 km^2 is unavailable due to the same environmental issues which have been discussed earlier for the Ahmedabad-Rajkot highway. Thus, similar to the Ahmedabad-Rajkot highway, the approximate effective area available on the Ahmedabad-Vaododara express highway is 1.72 km^2.

Apart from the computation of solar potential for the national highway using a series of solar panels mounted on the road roof area, we have also considered two-layer solar panel installations. In an earlier study [12], it was demonstrated that two-layer solar panels one above the other showed about 70% increase in solar potential. In this context, it is of interest to observe the variation of the output for a two-layer solar panel in two perpendicular directions - one solar panel oriented towards south and lying nine (in rows) along the east-west direction and the other solar panel oriented towards south and lying nine (in rows) along the north-south direction. Interestingly, a large variation is observed between the two-layer solar panel configurations. The results are presented in Table 3.

Table 3 shows a comparative study with detailed information of single- and two-layer solar panels for enhancement of energy generation based on the concepts discussed in great detail in our previous paper [12]. The results shown here are for all the solar panels oriented towards the south direction. However, the distribution of the rows are more towards the north-south direction in one case and more towards the east-west distribution in the other case as shown in the schematic diagrams (Figures 5 and 6). Such configuration in modelling helps to understand the results for the Ahmedabad-Rajkot national highway (near east-west direction) and the Ahmedabad-Vadodara (near north-south direction) express highway. The analysis and the results presented are for different months and for the annual energy generation. It can be seen that the two-layer solar panel energy generation is more viable for the north-south-oriented highways as compared to east-west-oriented roadways. This is mainly due to the position of the sun in the sky which moves from the east to the west. Such moment casts more shadow on the east-west -distributed rows of the solar panels as compared to north-south-distributed solar panels.

Results and discussion

Based on the modelling results along the two national highways, a few interesting observations can be made as compiled in Table 4. Although the Ahmedabad-Rajkot and Ahmedabad-Vadodara national highway lengths are

Table 3 Comparative study of two-layer solar panel system for different plans of horizon (south- and east-oriented highways)

Month	Annual energy generation by 50-kW solar panels spreading along east-west and north-south directions (south-oriented solar panel) (kWh)							
	Single-layer solar panel		Two-layer solar panel with separation between panels (m)					
			1 m		2 m		3 m	
	East-west	North-south	East-west	North-south	East-west	North-south	East-west	North-south
January	7,308	7,308	8,917	9,010	9,875	9,881	10,196	10,257
February	6,984	6,984	8,658	8,921	9,862	10,072	9,765	10,045
March	7,874	7,874	9,227	9,642	11,092	11,677	11,861	12,425
April	7,328	7,328	7,930	8,317	9,287	9,972	10,343	11,098
May	7,089	7,089	7,605	7,954	8,422	9,082	9,110	99,12
June	5,730	5,730	6,350	6,582	7,069	7,494	7,536	8,063
July	4,531	4,531	5,120	5,274	5,783	6,035	6,179	6,464
August	4,587	4,587	5,171	5,359	5,933	6,227	6,422	6,736
September	5,985	5,985	6,862	7,148	8,208	8,652	8,925	9,367
October	7,051	7,051	8,608	8,936	10,008	10,348	10,239	10,608
November	6,724	6,724	8,264	8,422	9,238	9,301	9,334	9,435
December	6,883	6,883	8,297	8,291	9,144	9,041	9,808	9,702
Year (total)	78,074	78,074	91,011	93,856	103,921	107,782	109,718	114,112

The values shown are the annual energy supplied to the grid.

Table 4 Megawatt power generation capacities at national highway routes chosen - Ahmedabad-Rajkot national highway and Ahmedabad-Vadodara express highway

National highway	Constraint parameters considered during modeling	Length of road (km) (round off)	Installation capacity possible (MW)	Annual energy generation (GWh)			
				Single-layer solar panel	Two-layer solar panel with separation		
					1 m	2 m	3 m
Ahmedabad-Rajkot national highway	Useful road = TotalAhmedabad-Rajkot road length[a]	205	115	180	210	239	253
	Useful road = Total road – Unusable road[a]	196	110	172	201	229	242
	Useful road = Total road – Unusable road – Portion of bridges[a]	195	110	171	200	228	241
	Useful road = Total road – Unusable road – Portion of bridges – Portion of crossing[a]	192	108	169	197	225	237
	Useful road = Total road – Unusable road – Portion of bridges – Portion of crossing – Portion of thick vegetation[a]	185	104	163	190	216	229
Ahmedabad-Vadodara national express highway	Useful road = TotalAhmedabad-Rajkot road length[b]	93	66	104	125	143	152
	Useful road = Total road – Unusable road[b]	90	64	100	121	139	147
	Useful road = Total road – Unusable road – Portion of bridges[b]	87	62	97	117	134	142
	Useful road = Total road – Unusable road – Portion of bridges – Portion of crossing[b]	86	61	96	115	132	140

[a]See Annexure I in Additional file 1;[b]see Annexure II in Additional file 1.

205 and 93 km, respectively, the effective length of the highways for erecting the road rooftop structures for placing the solar panels are 185.4 and 85.8 km, respectively. This way of selection of the highways is based on various factors as described earlier and also shown in Table 2. From our modelling results, a 115-MW power generation along the Ahmedabad-Rajkot national highway is possible. However, due to the environmental and other restrictions, only a 104-MW electricity generation is feasible. However, if we use innovative concepts of two-layer solar panels one above the other, the energy generation can be enhanced. If we compute the energy generation for the whole year, the cumulative energy that can be generated is 163 GWh. This computation is based on the results shown in Table 4. For example, the cumulative annual energy generation for a 50-kW system is 78,074 kWh (Table 3). Accordingly, for a 1-MW system, the energy generation is 1,561,480 kWh (78,074 kWh × 1,000 kW/50 kW) which is equivalent to 1.56 GWh.

For the Ahmedabad-Rajkot highway, the energy generated by a 50-kW system (Table 3) is 78,074 kWh and the dimension for solar panel layout is 89 m × 17 m. This means that a highway road with a width of 17 m requires 89 m in length for 50-kW photovoltaic system installations; hence, for 1 MW (i.e., 50 kW × 20), the required length of the highway, i.e., 89 m × 20, is equal to 1,780 m (1.78 km). More clearly, one can say that by maintaining the width of the road as 17 m, one requires a 1.78-km length of the road to generate 1-MW power. This value can be used to compute the potential for any length of the road.

The total length of the Ahmedabad-Rajkot national highway is 205 km, out of which the total effective length available is 185.4 km (based on the computation given in Table 2; 185.4 km/1.78 km), and 104 MW is possible for the entire length of the highway. This way, for 104-MW systems, the energy generation is 163 GWh. This is for a single-layer solar panel system. If the solar panels lay one above the other with 1-m separation, the energy generation (Table 3) increases to 190 GWh; for 2-m separation, it is 216 GWh; and for 3-m separation, it is 229 GWh. One should note here that these computations are based on 50-kW systems as one unit and the direction of the national highway is oriented exactly in the east-west direction. Similar computations are shown in Table 4 for the Ahmedabad-Vadodara national expressway. Here again, the computations are based on 50-kW systems as one unit, and the solar panels are distributed in the south-north direction. Accordingly, 61-MW electricity generation is possible for the Ahmedabad-Vadodara expressway, and 96 GWh of cumulative annual generation is possible for single layer, 115 GWh for two layers of 1-m separation, 132 GWh for 2-m panel separation, and 140 GWh for 3-m separation, as presented in Table 4.

Conclusions

In this paper, we have suggested, first time ever of its kind, an approach to utilize solar energy to meet the global challenges like climate change, pollution, and energy insecurity [13] and also to address the biggest challenge for the photovoltaic technology, i.e., land cost. Indian roads are paying a very vital role in the economy through transportation, whose share has been increased before 2000 only to 60% in total sharing of 40% with railways [30]. This means that efficiency of the vehicle and life of the road are equally important. Through this paper, we are also answering some of these issues. It is well known that the availability of land is becoming more expensive and also becoming a rare commodity. In view of this, our study proposed a new concept to use the existing land in a more effective way. Towards this direction, an attempt has been made for the possible use of the national highways presently being used for transportation only. Our suggestion is that they can also be utilized for power generation. For this purpose, with one-time investment, one can construct a simple elevated structure covering the national highway roads. For this purpose, we have considered two case studies, one along the Ahmedabad-Rajkot highway and the other along the Ahmedabad-Vadodara expressway. It may be noted that the Ahmedabad-Rajkot highway is oriented near the east-west direction and the Ahmedabad-Vadodara expressway is oriented near the south-north direction. Accordingly, using PVsyst software, solar panel configurations have been constructed that have an area equivalent to generate 50 kW of electricity. This is considered as one unit for computation to estimate solar potential on a national highway to enhance the power generation. We have also incorporated two layers of solar panels that lie one above the other. Such an idea has already been demonstrated on enhanced power generation [12].

Using Google Earth Pro software, the national highways are carefully scanned and the part of the road that cannot be used for structural construction due to environmental and other factors has been deleted in the estimation, and the details are given in Annexures I and II in Additional file 1.

From the results of our study, 104 MW of electricity is feasible along the Ahmedabad-Rajkot national highway. It is observed that two-layer solar panels installed along the national highway oriented towards south lying in the south-north direction (this direction is the spread of solar panels in rows) produce more energy as compared to solar panels oriented towards south lying in the east-west direction (this direction is the spread of solar panels in rows) on the national highway. The results are summarized in Table 3.

We can suggest that the same concept can be extended for use on the 52,584-km national and state highways in India that have four lanes or more [14], for example, the four-lane Golden Quadrilateral highway which is 5,839 km. The Golden Quadrilateral [14] is a highway network connecting India's four largest metropolises: Delhi, Mumbai, Chennai, and Kolkata, thus forming a quadrilateral that can have an installation power of approximately 4,418 MW. Similarly, it is possible for the four-lane north-south to east-west (NS-EW) Corridor highway of 7,300 km. The NS-EW Corridor highway is the largest ongoing highway project in India. It consists of four/six-lane expressways connecting Srinagar, Kanyakumari, Porbandar, Silchar [14], etc. Accordingly, the NS-EW Corridor highway has 5,524-MW capacity of power generation. In a similar manner, we recommend to initiate solar panel installations on all the train trackways also. This will open up large potential of power generation in a very short period of time which is very much needed in India to have aggressive industrial growth which is presently in strong need for power.

As detailed earlier, a total of 104-MW potential of solar energy generation exists along the Ahmedabad-Rajkot national highway and a total of 61-MW potential of solar energy generation exists on the Ahmedabad-Vadodara national express highway. This way, huge potential of solar energy generation exists on the national highways with capacities based on the length. This leads to not only meeting the energy demand but also reduction in CO_2 emissions by minimizing dependency on conventional sources [31], hence promoting renewable energy resource on a very large scale, and this is the best answer for the national threats called GHG emission for any country [32]. Power generation using this concept generates huge amount of electricity without combustion of fossil fuels, as electricity generation through fossil fuels is the largest source of CO_2 emission [33], and hence, this helps to reduce the dependency on fossil fuels and transportation of fossil fuels. A more detailed study on environmental damages by CO_2 emission and transportation can be seen in [34,35].

In summary, the national highway space can contribute to the huge amount of energy generation without extra cost for the land. By this, we can also say that this idea can be a part of the key concept of solar grand plan [36]. Apart from this, there are various fringe benefits. For example, if implemented, it improves vehicle efficiency, use of minimum energy for air conditioning vehicles, more tire life due to shade, etc. Also, another fringe benefit of this road rooftop solar project is increase of the life of the road from wear and tear on the highways, and this helps to reduce the fund requirement for road repairs. Another benefit is rainwater harvesting at selected locations. This increases the groundwater level and helps to reduce the fluoride and arsenic content in the groundwater. In summary, we can say that this approach is also the answer for the various concerns of sustainable transport and helps in reaching a goal for a sustainable transport system [30]. For all the above ideas, to see the light of day, there is an urgent need for an attractive policy by the government of India to increase the solar potential if one dreams to realize it in 1-year duration.

Additional file

Additional file 1: Annexures I and II. Annexure I: List of the places in length (m) along with latitude and longitude. Length of these places is eliminated in the energy calculation for Ahmedadad-Rajkot national highway. Annexure II:List of the places in length (m) along with latitude and longitude. Length of these places is eliminated in the energy calculation for Ahmedadad-Vaoddara national highway.

Competing interests
The authors declare that they have no competing interests.

Authors' contributions
PS carried out all the computation, system designing, modelling analysis, software simulation, and field measurement and drafted the manuscript. TH conceived of the study and participated in its design and coordination. Both authors read and approved the final manuscript.

Authors' information
PS, an M.Sc., M.Tech, and research associate in the Solar Research Wing, is involved in a research and development program of solar energy technology in GERMI Research Innovation and Incubation Centre (GRIIC), Gujarat, India, with more than 5 years of research and industrial experience in R&D of solar photovoltaic energy technology. She is involved in setting up various solar research laboratories. She is an M.Tech. from the School of Energy, Devi Ahilya Vishwavidyalaya (DAVV), Indore, India. She has also done M.Sc. in Physics with solar specialization. She has worked with a dedicated unit from the Ministry of New and Renewable Energy Government of India, Solar Energy Center on 'Design and development of 20 kWp roof top PV power plant at SEC' in Gurgaon, India. She has published her work in international journals and presented her research work in various national and international conferences. TH, a Ph.D., Director of GRIIC, Gujarat, and earlier Head of Magnetotellurics, National Geophysical Research Institute, has done outstanding contributions in the field of deep electromagnetics both on land and marine environment. TH is one of the top scientists among the geomagnetism and electromagnetic scientists in the world. His pioneering works are related to oil exploration, geothermal energy assessment, deep crustal studies, tsunami studies, earthquake studies, etc. He is instrumental in introducing a new geophysical technique - marine magnetotellurics - in the Gulf of Kutch for hydrocarbon exploration that has delineated 4-km-thick buried sediments below the volcanic cover. This work was initiated as a part of international cooperation with Scripps Institution of Oceanography, USA. This has opened up a new scenario of search for oil in the Gulf. He has received a National Mineral Award from Ministry of Mines, Government of India, at an early age of 40. He was chosen as the best scientist by the Government of Andhra Pradesh. He was elected to the prestigious Russian Academy of Natural Sciences, Moscow, Russia. He is an elected fellow of the Indian Geophysical Union and a fellow of the Andhra Pradesh Academy of Sciences. He has taken up controlled source electromagnetic modelling studies for gas hydrates as a part of his research at the University of Texas, Austin, USA, as a visiting scientist and also carried out tsunami studies at the University of Tokyo, Japan, as a visiting professor.

Acknowledgements
PS and TH would like to acknowledge VCMT Sri Tapan Ray for all his encouragement and also all the research, technical, and scientific staff of GERMI for the support.

Author details
[1]GERMI-RIIC, Gandhinagar, Gujarat 382007, India. [2]CSRI-NGRI, Hyderabad 500007, India.

References
1. Lal, S, Raturi, A: Techno-economic analysis of a hybrid mini-grid system for Fiji islands. Int. J. of Energy and Environ. Engineering **3**, 10 (2012)
2. Oyedepo, S: Efficient energy utilization as a tool for sustainable development in Nigeria. Int. J. of Energy and Environ. Engineering **3**, 11 (2012)
3. Zweibel, K: Should solar photovoltaics be deployed sooner because of long operating life at low, predictable cost? Energy Policy **38**(11), 7519–7530 (2010)
4. Dinçer, F: Overview of the photovoltaic technology status and perspective in Turkey. Renew. Sustain. Energy Rev. **15**(8), 3768–3779 (2011)
5. Bhutto, AW, Bazmi, AA, Zahedi, G: Greener energy: issues and challenges for Pakistan—solar energy prospective. Renew. Sustain. Energy Rev. **16**(5), 2762–2780 (2012)
6. Renewable Energy Policy Network: Renewables global status report. http://www.ren21.net/Portals/0/documents/activities/gsr/RE_GSR_2006_Update.pdf (2006) Accessed 21 June 2012
7. Ministry of New and Renewable Energy, Government of India:Jawaharlal Nehru National Solar Mission: Towards Building SOLAR INDIA. http://www.mnre.gov.in/file-manager/UserFiles/mission_document_JNNSM.pdf (2009). Accessed 20 Feb 2011
8. Ministry of New and Renewable Energy, Government of India: Achievements. http://www.mnre.gov.in/mission-and-vision-2/achievements/ (2013). Accessed 19 Sept 2012
9. Press Information Bureau, Government of India: Progress under Jawaharlal Nehru National Solar Mission. http://pib.nic.in/newsite/erelease.aspx?relid=83632 (2012). Accessed 19 Sept 2012
10. Gujarat Energy Development Authority: Gujarat - Solar Power Plants as of May-9-2012. http://geda.gujarat.gov.in/media/Solar_Power_PLants_at_a_Glance.pdf (2012). Accessed 19 Sept 2012
11. GeoModel Solar: Solar GIS India. http://solargis.info/doc/_pics/freemaps/1000px/ghi/SolarGIS-Solar-map-India-en.png (2011). Accessed 9 April 2012
12. Sharma, P, Harinarayana, T: Enhancement of energy from the two layer solar panels. International Journal of Energy and Environmental Engineering **3**, 12 (2012). doi:10.1186/2251-6832-3-12
13. Jacobson, MZ, Delucchi, MA: Providing all global energy with wind, water, and solar power, Part I: technologies, energy resources, quantities and areas of infrastructure, and materials. Energy Policy **39**(3), 1154–1169 (2011)
14. NHAI (National HighwaysAuthority of India): http://www.nhai.org/ (2013). Accessed 20 April 2012
15. Google: Google Maps. http://maps.google.co.in (2013). Accessed 17 July 2012
16. Google: Google Earth Pro software. http://support.Google.com/earth/?hl = en (2013). Accessed 11 April 2012
17. PVsyst, SA: User's Guide: PVsyst Contextual Help. www.pvsyst.com (2012). Accessed 8 Dec 2011
18. Wu, MS, Huang, HH, Huang, BJ, Tang, CW, Cheng, CW: Economic feasibility of solar-powered led roadway lighting. Renew. Energy **34**(8), 1934–1938 (2009)
19. Young Jr, WR: Photovoltaic application for disaster relief. FSEC- CR-849-95. Florida Solar Energy Center/University of Central Florida. http://www.fsec.ucf.edu/en/publications/pdf/FSEC-CR-849-95.pdf (1995). Accessed 8 Oct 2012
20. Jian-xin, C, Qing, L: Research on cost-oriented modeling and optimization for highway alignment. Procedia Engineering **15**, 3931–3935 (2011)
21. Centre for Urban Forest Research:Why shade streets?: The unexpected benefit. http://www.fs.fed.us/psw/programs/uesd/uep/products/cufr_673_WhyShadeStreets_10-06.pdf (2007). Accessed 17 July 2012
22. Sundvall, E, Nystrom, M, Forss, M, Chen, R, Petersson, H, Ahlfeldt, H: Graphical overview and navigation of electronic health records in a prototyping environment using Google Earth and open EHR archetypes. Stud. Health Technol. Inform. **129**(Pt 2), 1043–1047 (2007)
23. Conroy, GC, Anemone, RL, Van Regenmorter, J, Addison, A: Google Earth, GIS, and the great divide: a new and simple method for sharing paleontological data. J. Hum. Evol. **55**(4), 751–755 (2008)
24. Galway, LP, Bell, N, Al, SSAE, Hagopian, A, Burnham, G, Flaxman, A, Weiss, WM, Rajaratnam, J, Takaro, TK: A two-stage cluster sampling method using gridded population data, a GIS, and Google EarthTM imagery in a

population-based mortality survey in Iraq. Int. J. Health Geographics **11**, 12 (2012)

25. Chang, AY, Parrales, ME, Jimenez, J, Sobieszczyk, ME, Hammer, SM, Copenhaver, DJ, Kulkarni, RP: Combining Google Earth and GIS mapping technologies in a dengue surveillance system for developing countries. Int. J. Health Geographics **8**, 49 (2009)
26. Facts-About-India.com: http://www.facts-about-india.com/indian-roads.php (2013). Accessed 17 July 2012
27. Global Meteorological Database: Meteonorm Switzerland. Version 6.1.0.23. http://meteonorm.com/ (2007). Accessed 17 Feb 2012
28. Mehleri, ED, Zervas, PL, Sarimveis, H, Palyvos, JA, Markatos, NC: Determination of the optimal tilt angle and orientation for solar photovoltaic arrays. Renew. Energy **35**(11), 2468–2475 (2010)
29. Chang, TP: Performance analysis of tracked panel according to predicted global radiation. Energy Conversion and Management **50**(50), 2029–2034 (2009)
30. Reddy, AKN, Anand, YP, D'Sa, A: Energy for a sustainable road/rail transport system in India. Energy Sustain. Dev. **4**(1), 29–44 (2000)
31. Hoffert, MI, Caldeira, K, Benford, G, Criswell, DR, Green, C, Herzog, H, Jain, AK, Kheshgi, HS, Lackner, KS, Lewis, JS, Lightfoot, HD, Manheimer, W, Mankins, JC, Mauel, ME, Perkins, LJ, Schlesinger, ME, Volk, T, Wigley, TML: Advanced technology paths to global climate stability: energy for a greenhouse planet. Science **298**, 981 (2002). doi:10.1126/science.1072357
32. Yan, X, Crookes, RJ: Energy demand and emissions from road transportation vehicles in China. Progress in Energy and Combustion Science **36**(6), 651–676 (2010)
33. EPA(United States Environmental Protection Agency): Greenhousegas emissions. http://www.epa.gov/climatechange/ghgemissions/gases/co2.html (2012). Accessed 2 Jan 2013
34. Guerrero-Lemus, R, Martínez-Duart, JM: Renewable Energies and CO2: Cost Analysis, Environmental Impacts and Technological Trends. In: (ed.) Lecture Notes of Energy, vol. 3. Springer, Heidelberg (2012)
35. Krupnick, AJ, Rowe, RD, Lang, CM: Transportation and Air Pollution: The Environmental Damages. Springer, Berlin (1997)
36. Zweibel, K, Mason, J, Fthenakis, V: A solar grand plan. http://www.nature.com/scientificamerican/journal/v298/n1/full/scientificamerican0108-64.html Scientific American **298**, 64–73 (2008). doi:10.1038/scientificamerican0108-64

10

Simulation of total dust emission from palm oil mills in Malaysia using biomass fuel composition and dust collector efficiency models

Yau Chwan Kun* and Ahmad Makmom Haji Abdullah

Abstract

Palm oil mills are generally self-sufficient in terms of energy. However, burning of biomass fuel in a boiler has generated serious air pollution problem due to inappropriate boiler operation and inefficient dust collection system. This study aims to simulate the total dust emission from palm oil mills in Malaysia by modelling biomass fuel composition and dust collector efficiency. It has been found that before removal, the simulated average dust load after biomass boiler was 51.67 ± 5.39 g/s at a confidence level of 95%, and total dust load was 389,961 tonnes/year. Simulation values of total dust emission at different percentages of dust collector efficiency at 60%, 70%, 80%, 90% and 95% were 155,984, 116,988, 77,992, 38,996 and 19,498 tonnes/year, respectively. The simulation has also shown that dust concentration at a boiler outlet was 4.5 g/Nm^3. From the study, dust removal efficiency should be above 91% to ensure the compliance of dust concentration at 0.4 g/Nm^3 as prescribed by the Environment Quality (Clean Air) Regulations of 1978. Whereas in 2009, Sabah state has processed the most quantities of fresh fruit bunches; it has posed a potential impact on the environment and public in terms of total dust emission. Besides stack sampling that is used to monitor dust emission status, simulation can be a more simple way and an alternative to estimate the emission.

Keywords: Boiler, Biomass, Dust collector efficiency, Palm oil mill

Background

Malaysia is one of the largest palm oil producers in the world. However, the process of producing palm oil has generated million tonnes of empty fruit bunches, fibre and palm shell every year as wastes. Fibre and shell are mixed at certain ratios and used as fuel to generate energy to run the palm oil mill. Utilization of this palm waste as boiler fuel is creating a serious emission problem in the industry. The emissions are not only posing threat on human health, but also affecting agricultural crops, forest species and ecosystem [1-4]. Under the enabling legislation provided by the 1974 Environmental Quality Act, the Malaysian Government has issued the Environment Quality (Clean Air) Regulations, which came into force on 1 October 1978 and which limited air emissions from stationary fuel burning sources. For the palm oil industry, the principal concern of the regulation is toward the control of dark smoke and dust emission from the burning waste, which are mixtures of fibre and shell. Under Regulations 21 and 25, the maximum permitted level of dust concentrations in gas emitted will be 0.5 g/Nm^3 before 1 October 1978 and a limit of 0.4 g/Nm^3 for plant erected after the date, expressed in terms of 12% CO_2.

A study conducted by Rozainee et al. [5] shows that the mean particulate concentration emitted from water-tube boilers equipped with and without particulate control equipment was 1.11 ± 0.58 g/Nm^3 and 1.93 ± 1.40 g/Nm^3, respectively, with only one of the boilers studied met the Department of Environment limits of 0.40 g/Nm^3 whilst others were still violating the standards. From another study by Rashid and Rozainee [6], it was found that the boilers, using palm waste as fuel at palm oil mills, are producing very much higher levels of dust emission of up to 11.6 g/Nm^3. According to De Kock and Yap [7], many dust collectors are installed without proper design and quality control procedures and many were simply copied from original suppliers. Dust collectors are often installed without verifying the performance during commissioning

* Correspondence: ckyau23@yahoo.com
Faculty of Environmental Studies, Universiti Putra Malaysia, Serdang, Selangor 43400, Malaysia

with only stack emission measurements carried out. By gross estimation, dust generated from typical 25 tonnes/h boiler combustion is around 4.5 g/Nm^3. Nowadays, most of the palm oil mill boiler operations are equipped with multi-cyclones to filter out the dust generated. For 60% of dust collection efficiency on average, it can be forecasted that the emission would be around 1.8 g/Nm^3 after the multi-cyclone, which create an estimated dust load of 450 tonnes/year.

In 2009, there were around 417 oil palm processing plants in the country [8] As compared with the number of mills in 1999, there are additionally more than 80 new mills that were put into operation as of today (over a span of 10 years). It is not difficult to foresee the air pollution from palm oil mill boilers that are yet to comply with the standard of 0.4 g/Nm^3, which is still continually creating a great burden to the local atmosphere. The palm oil industry has made a tremendous effort to increase its productivity over the past decades, but at the same time, the palm waste has also increased. Many studies focus on handling palm oil mill effluent or to maximize the utilization of biomass as renewable energy, yet there is lack of data and information about dust emission monitoring and follow-up as well as methods and technologies to lower emissions from biomass boiler.

Simulation of total dust emission can be used to estimate the trend of dust emission in the future and help in setting its reduction target for palm oil mills in Malaysia. It can also be used to facilitate decision-making processes that are related to air pollution control as part of emission inventory estimations. Therefore, the objective of this study is to estimate the total dust emission from palm oil mills in Malaysia from the perspectives of fuel compositions and efficiency of dust collection system to elucidate the air pollution problem that might have been neglected in the past. Simulation was carried out using several models: dust load and dust concentration model as well as dust collector efficiency. Dust load and dust concentration models are based on the composition of fibre and shell (Tables 1 and 2 and Figure 1), and the dust loading rate is calculated based on monthly average operation hours of a palm oil mill (Table 3). The selected efficiency range of dust collector was used to show the total dust emission rate, based on simulated dust load and dust concentration.

Methods

Source of emission

Baseline information collected from Malaysian Oil Palm Statistics 2009 [9] such as number of mills, fruit yield, oil palm planted area and mill operation hours (Table 3) in every district of Malaysia in 2009 was used as input data for the simulation. Biomass boilers are often used to supply steam for palm fruit sterilization processes and turbines to generate power for plant operation. Due to the availability of adequate quantities of fibre and shell materials, which are used as solid fuel in the steam boiler, palm oil mills are generally self-sufficient in terms of energy. However, burning of biomass waste to generate energy has caused serious air pollution problem.

Table 1 Chemical composition on dry basis of fibre and shell

Element	Fibre	Shell	Fuel mixture of fibre and shell (simulated)
H (%)	6.0	6.3	6.09
C (%)	47.2	52.4	48.76
S (%)	0.3	0.2	0.27
N (%)	1.4	0.6	1.16
O (%)	36.7	37.3	36.88
Ash (%)	8.4	3.2	6.84

Generally, as suggested by USEPA, air pollutant emission can be estimated by using the following equation [10]:

$$E = A \times \mathrm{EF} \times (1 - \mathrm{ER}/100),$$

where, E = emissions, A = activity rate, EF = emission factor and ER = overall emission reduction efficiency as percentages.

Dust load model

Dust load can be defined as the concentration of solids in the inlet gas stream [11]. This study used the unit of mass flow, in gramme per second (g/s) for the dust load to show the loading rate to dust collector. The emission factor and the activity rate, which in this study refer to the fuel feeding rate to the boiler, were calculated in order to estimate the dust load.

The emission factor was determined by the amount of ash in gas flow and the ash content of palm oil biomass (fuel mixture of fibre and shell) in kilogramme of dust emitted per kilogramme of biomass burned. From Figure 1, it is found that the percentage of fibre and shell in fresh fruit bunches (FFB) are 13.5% and 5.5%, respectively [12] Therefore, the ratio of fibre and shell mixture in 1 kg of bulk fuel can be calculated by the following equations:

$$\text{Fibre mass} = \frac{\text{Percentage of fibre in FFB}}{(\text{Percentage of fibre in FFB} + \text{Percentage of shell in FFB})},$$

$$\text{Shell} = 1 - \left[\frac{\text{Percentage of fibre in FFB}}{(\text{Percentage of fibre in FFB} + \text{Percentage of shell in FFB})}\right].$$

Table 2 Calculation steps in determining dust concentration after biomass boiler

Step 1: From ultimate analysis, amount of oxygen required and air/fuel ratio were calculated[a]							Step 2: Calculated wet and dry composition of flue gas condition[b]								Step 3: Air volume was calculated based on ideal gas law at normal condition, and dust concentration was corrected to 12% carbon dioxide[c]		
Ultimate analysis on a dry basis	Mass (mass%)	Molar mass (kg/mol)	$\frac{\text{Mass (\%)}}{\text{Molar mass}}$ (mol/100 kg)	O_2 required	Product	mol/100 kg	From combustion/100 kg fuel (dry)				Wet composition		Dry composition		Per kilogramme of fuel	Dry	Wet
							Flue gas calculation	kmol	Molar mass	Mass	(mass%)	(vol.%)	(mass%)	(vol.%)			
Sulphur	0.27	32	0.0084	0.0084	SO_2	0.0084	H_2O	4.35	18	78	7.46	11.95			Ash (kg)	0.068	0.068
Hydrogen	6.09	2	3.045	1.5225	H_2O	1.5225	CO_2	4.06	44	179	17.05	11.16	18.42	12.68	Ash in flue gas (%)	50	50
Carbon	48.76	12	4.06	4.063	CO_2	4.06	SO_2	0.01	64	0.54	0.05	0.02	0.06	0.03	Fly ash (kg)	0.034	0.034
Oxygen	36.88	32	1.15	−1.153			N_2	25.96	28	727	69.31	71.34	74.90	81.02	Air volume at 0°C, 101.3 kPa (Nm^3)	7.2	8.2
Nitrogen	1.16	28	0.041				O_2	2.01	32	64.3	6.13	5.52	6.63	6.27	Dust concentration (g/Nm^3)	4.76	4.19
Ash	6.84						Total	36.39		1049					Regular standard carbon dioxide (vol.%)	12.0	12.0

[a]Amount of oxygen needed = 4.442; Theoretical air/fuel ratio (Af_{theo}) (kg air/kg fuel) = 6.125; Amount of excess air (%) = 50; Air/fuel ratio (kg air/kg fuel) = 9.19. [b]Excess water: from air = 18.38 kg/100 kg fuel, 1.021 mol/100 kg dry fuel; from fuel = 32.50 kg/100 kg fuel, 1.806 mol/100 kg dry fuel. [c]Corrected to carbon dioxide (both for dry and wet) = 4.51. (Adapted from "Fundamentals of Classical Thermodynamics, Third Edition," by Van Wylen, G.J., Sonntag, R.E., 1985. *Chapter 12: Chemical Reaction*, p.473. Wiley, Singapore).

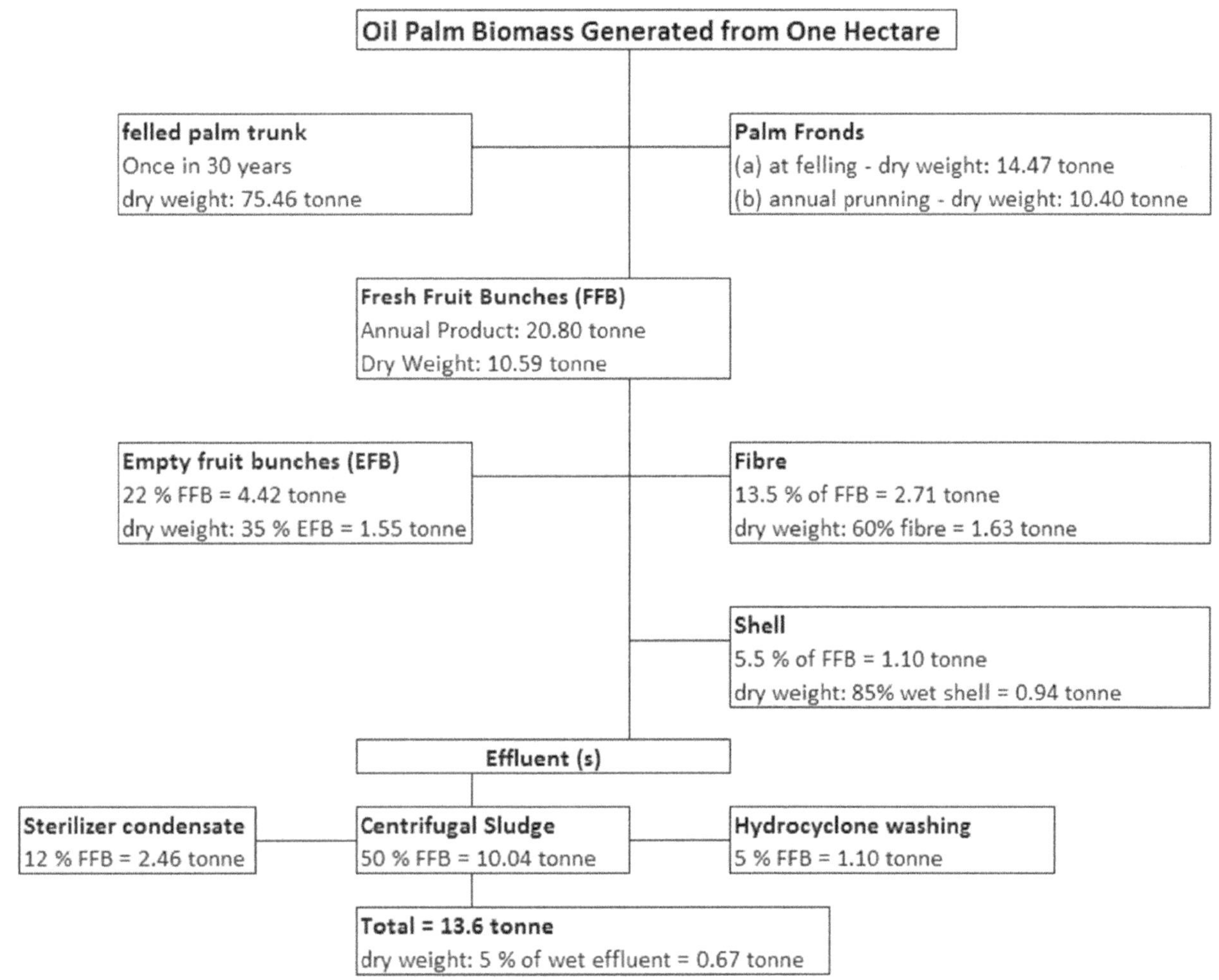

Figure 1 Oil palm biomass generated from 1 ha [12].

The following are the sample calculations:

$$\text{Fibre} = \frac{0.135}{(0.135 + 0.055)} = 0.7\ \text{kg},$$
$$\text{Shell} = 1\text{–}0.7 = 0.3\ \text{kg}.$$

Note that the ratio of fibre and shell (0.7 and 0.3 kg, respectively) may differ since shell is a valuable by-product that may be used in lower percentages as fuel.

From the chemical compositions on the dry basis of palm oil waste as shown in Table 1, the ash content for fibre and shell are 8.4% and 3.2%, respectively [3]. Hence, the ash content for the fuel mixture is estimated as follows:

$$\text{Ash percentage} = (\text{Fibre ash percentage} \times \text{Fibre mass}) + (\text{Shell ash percentage} \times \text{Shell mass}).$$

Here is its sample calculation:

$$\text{Ash percentage} = (8.4 \times 0.7) + (3.2 \times 0.3) = 6.84\%.$$

As most of the boiler firing systems like the suspension firing system and the fluidized bed combustion methods have very high dust entrainment (as high as 80%), older boiler design which uses spreader stoker has about 20% to 30% of the fuel being burned in suspension and thus has much less dust entrainment in the gas stream [13]. However, the data and statistics for the above boiler information are not available; it is assumed that the ash percentage in gas stream is 50%. Therefore, the emission factor can be calculated as follows:

$$\begin{aligned}\text{Emission factor} &= 6.84\%\text{ash content} \times 50\%\text{in gas stream}\\ &= 0.0342\ \text{kg dust/kg biomass}.\end{aligned}$$

Dust generation rate from biomass fuel combustion in boiler can be estimated by taking into account the mill

Table 3 Monthly average operating hours of palm oil mill in 2009

State	January	February	March	April	May	June	July	August	September	October	November	December	Average
Johor	394.67	356.24	370.12	366.33	420.10	440.56	488.55	431.30	422.23	551.10	443.22	414.74	424.93
Kedah	397.85	420.37	433.33	395.53	423.66	476.69	577.16	494.81	487.94	524.84	409.82	382.10	452.01
Kelantan	267.27	229.19	276.91	297.33	270.71	274.28	306.10	336.82	331.53	445.99	303.30	319.32	304.90
Melaka	468.15	455.96	441.31	476.76	484.34	502.47	495.88	458.75	487.49	597.81	409.19	434.45	476.05
N. Sembilan	451.32	452.73	456.48	467.37	459.63	474.88	516.05	475.70	481.04	546.72	419.78	373.17	464.57
Pahang	360.60	334.92	387.06	362.92	402.57	394.19	429.87	433.29	428.28	534.00	411.32	371.16	404.18
Perak	367.52	368.36	402.74	404.09	451.18	464.65	516.41	447.19	425.13	505.99	402.46	411.48	430.60
P. Pinang	429.71	477.58	476.61	483.85	519.53	528.93	625.42	578.27	505.76	590.22	519.68	516.65	521.02
Selangor	379.15	386.27	406.49	414.12	404.52	447.80	472.55	391.97	391.25	457.88	393.20	371.47	409.72
Terengganu	333.48	273.77	348.10	332.15	344.54	327.55	421.18	413.03	421.53	588.35	437.59	424.55	388.82
P. Malaysia	384.97	375.54	399.92	400.05	418.08	433.20	484.92	446.11	438.22	534.29	414.96	401.91	427.68
Sabah	407.79	326.58	357.50	343.76	353.87	357.00	347.79	366.74	410.19	520.56	465.88	467.63	393.77
Sarawak	358.14	306.75	326.55	314.27	347.12	325.94	333.28	374.15	408.73	486.30	407.19	386.86	364.61
Sabah/Sarawak	382.97	316.67	342.03	329.02	350.50	341.47	340.54	370.45	409.46	503.43	436.54	427.25	379.19
Malaysia	384.64	365.73	390.27	388.21	406.81	417.91	460.85	433.50	433.43	529.15	418.55	406.13	419.60

Adapted from [8].

capacity as total fresh fruit bunches processed over a certain milling operation hours. By multiplying the mill capacity with the percentage of biomass generated from processed FFB, the biomass fuel feeding rate to the boiler can be estimated as follows:

$$\text{Fuel feed rate} = \frac{\text{Total FFB}}{\text{Mill operation hours}} \times [(\text{Fibre percentage} \times \text{Dry weight percentage}) + (\text{Shell percentage} \times \text{Dry weight percentage})].$$

This implies that the dust load equation can be deduced as follows:

$$\text{Dust load} = \text{Emission factor} \times \text{Fuel feed rate}.$$

Statistical analysis

The average of total dust load is calculated to typify the overall situation of dust load per mill before the dust collector. In the future, any upgrade of mills or additional new mills in the country might contribute to the changes of the average dust load, which can be used to indicate the level of pollution source. Descriptive statistics was used to summarize the simulated average dust load. Furthermore, the normality of the data was tested using Anderson-Darling and Lilliefors test [14-16].

Dust concentration model

Dust concentration before the dust collector was simulated using ultimate analysis and air/fuel ratio calculation (Table 1). An example of the calculation for this model is shown in Table 2. A ratio of 70% fibre and 30% shell is applied for the mass percentage estimation in the ultimate analysis.

The combustion process involves the oxidation of constituents in the fuel that are capable of being oxidized [17] and can therefore be represented by chemical equations as follows:

$$S + O_2 \rightarrow SO_2$$
$$2H_2 + O_2 \rightarrow 2H_2O$$
$$C + O_2 \rightarrow CO_2$$
$$O_{2\,(\text{needed})} = \sum O_{2\,(\text{required for combustion})} - O_{2\,(\text{from air})}$$
$$\text{Air/fuel ratio } (\text{AF}_{\text{theo}}) = \frac{O_{2\,(\text{needed})}\,(1 + 3.76) \times 28.97}{100}$$

where 3.76 is the amount of N_2/O_2 in air (as a ratio), 28.97 is the mol mass of atmospheric air and calculation is based on 100 kg of fuel.

At normal conditions, air volume is calculated at 0°C, 101.3 kPa based on the ideal gas law:

$$V = \frac{nRT}{P},$$

where V = volume of gas (Nm^3), n = moles, R = universal gas constant, 8.314 J K^{-1} mol^{-1} and T = temperature (K).

Dust concentration before the dust collector is therefore calculated by mass of fly ash divided by air volume and corrected to 12% CO_2.

Dust collector efficiency

Even though the Malaysian Standard of Performance Evaluation for Mechanical Dust Collector has been introduced since 2003, implementation of the application is not as encouraged as expected. Moreover, as reported by De Kock and Yap [7], the performance of dust collectors installed at palm oil mills are generally low in efficiency due to improper design and lack of quality control as well as poor maintenance. The efficiency of a typical multicyclone after a biomass boiler can be as low as 40% only. This study simulates the dust emissions for dust collection system efficiencies of 60%, 70%, 80%, 90% and 95%.

Results and discussion

Results

Figure 2 shows the comparison of state total dust load. It can be seen that Sabah, Johor, Pahang, Sarawak and Perak are the states with significantly higher total dust load as compared with other states. Table 4 shows the results from districts of Malaysia based on simulated total mill capacity, total fuel feeding rate, total dust loading to dust collector and average dust load. The results were sorted from largest to smallest value of total dust load. The descriptive statistics, normality test and confidence interval of average dust load are reported in Table 5. The selected simulated data were found to be normally distributed ($p > 0.05$) based on Anderson-Darling and Lilliefors test, with the average dust load of 51.67 ± 5.39 g/s at a confidence level of 95%.

Table 6 shows the results for comparison of simulated state dust emission at 60%, 70%, 80%, 90% and 95% of dust removal efficiency. It is no doubt that the highest dust collector efficiency has the lowest outlet emission. The significant difference in values shows that whereas the lowest efficiency emitted 155,984 tonnes/year, the highest efficiency emitted only 19,498 tonnes/year. The relationship between simulated dust concentration and dust removal efficiency is shown in Figure 3. To achieve dust concentration with emission limit of 0.4 g/Nm^3, the removal efficiency should be above 91%.

Discussions

Dust load

From Figure 2, it was shown that while Johor, Pahang, Perak, Sabah and Sarawak are among the states that

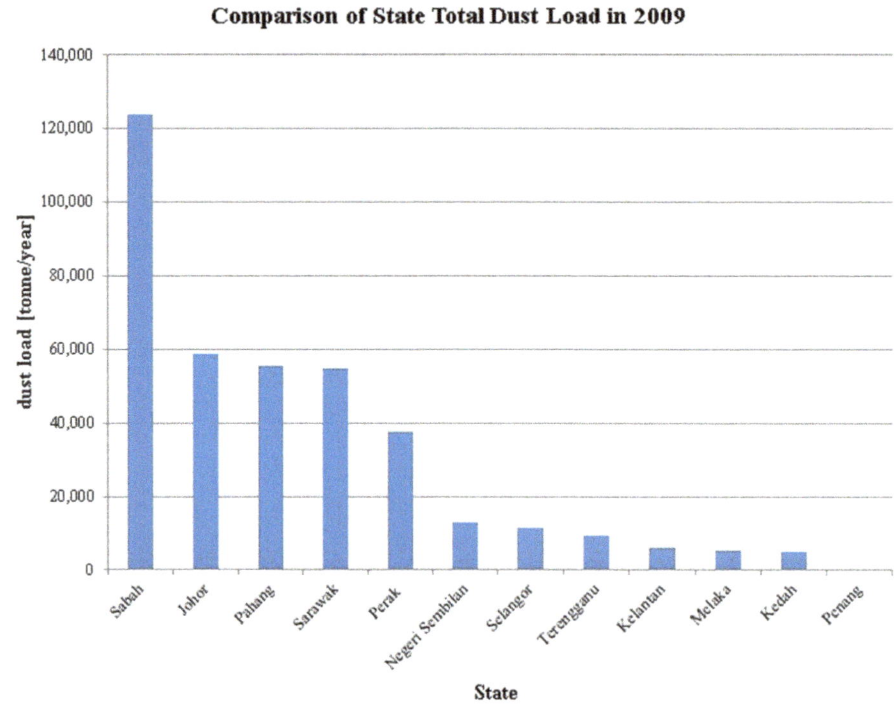

Figure 2 Comparison of state total dust load in 2009.

have the most quantities of mills, Sabah has the highest dust load among all states, which has a potential impact on the environment if not handled well. Due to the high processing rate of FFB, it is expected that more biomass fuel will be burned in boilers, thus resulting in higher dust load if compared with other states. The average dust load of 51.67 g/s from Tables 4 and 5 quantified the pollution source of a palm oil mill in 2009. Since there is no record of dust collector efficiency from palm oil mills in Malaysia, the average dust load can be used to estimate the overall emission from a range of dust collector efficiency. However, there is no limit stipulated by the Malaysia Environmental Quality Act 1974 for dust load, as a bigger mill definitely has higher dust load. Therefore, dust concentration is universally used as standard limit, although information of dust load is useful to track dust volume that has been released to the environment within a certain period of time, regardless of the size of a mill. Considering a case of two palm oil mills, 100 tonnes/h and the other is 50 tonnes/h, both emitting 0.4 g/Nm^3, they fall within the limit of the standard. However, dust load for the 100 tonnes/h mill (approximately 120 g/s) is double the dust load of 50 tonnes/h (approximately 60 g/s). This simply explains the pollution situation in a more direct way when compared to dust concentration.

From Table 6, at a different grade of dust removal efficiency, the total dust emission as simulated reduced drastically from 155,984 to 19,498 tonnes/year. This further illustrates the importance of pollution control where pollution sources have increased because of new development without proper control; the consequence can be very serious. High emission trends from palm oil mills were believed that these have not been rectified for the past decades, thus the low dust removal efficiency.

Dust concentration

The simulated dust concentration at the boiler outlet was 4.5 g/Nm^3 at 12% CO_2. This is in agreement with the results reported in the literature. For wood-fired boilers, Buonicore and Davis [18] reported a value of 3,000 mg/Nm^3, whilst De Kock and Yap [19] reported that the boiler flue gas dust loads varied from 1490 to 4300 mg/Nm^3. These results were not corrected for carbon dioxide; hence, the dust load could have been higher after 12% CO_2 correction. Furthermore, it can be seen from Figure 3 that even a small difference in dust collection efficiency can make a significant difference in dust concentration. For example, at 95% efficiency, the dust concentration is 0.225 g/Nm^3, whereas at 90% efficiency, which is lower, the dust concentration is 0.450 g/Nm^3. The latter result will not comply with the Clean Air Regulation of 1978 at 0.4 g/Nm^3.

Efficiency of dust collection system

There are a few considerations that might cause lower efficiency in the dust collection system on boilers. One

Table 4 Simulated total mill capacity, total fuel feed rate, total dust load and average dust load

		Data from [8]				Simulation results			
State	District	Mill quantity	FFB yield	Planted area	Total capacity	Total mill capacity	Total fuel feed rate	Total dust load	Average dust load
			(tonnes/ha)	(ha)	(tonnes/year)	(tonnes/h)	(tonnes/h)	(g/s)	(g/s)
Sabah	Kinabatangan	30	21.73	348,450	7,571,819	1,602.40	204.71	1,944.72	64.82
Sabah	Lahad Datu	24	20.39	275,622	5,619,933	1,189.33	151.94	1,443.40	60.14
Sarawak	Miri	15	16.71	257,832	4,308,373	984.71	125.80	1,195.07	79.67
Sabah	Tawau	14	22.62	233,809	4,610,160	975.64	124.64	1,184.06	84.58
Sabah	Labuk Sugut	19	20.7	138,493	4,108,805	869.54	111.08	1,055.29	55.54
Sabah	Sandakan	14	21.32	168,660	3,595,831	760.98	97.21	923.54	65.97
Johor	Kluang	19	19.91	138,604	3,755,106	736.42	94.08	893.73	47.04
Pahang	Rompin	23	18.92	176,792	3,344,905	689.65	88.10	836.97	36.39
Sarawak	Bintulu	9	17.23	158,869	2,737,313	625.63	79.92	759.28	84.36
Perak	Hilir Perak	15	25.91	106,793	2,767,007	535.49	68.41	649.89	43.33
Johor	Kota Tinggi	15	17.17	145,050	2,490,509	488.42	62.40	592.75	39.52
Pahang	Bera	9	21.93	94,442	2,071,113	427.02	54.55	518.24	57.58
Johor	Segamat	9	19.70	102,048	2,010,346	394.25	50.37	478.47	53.16
Sarawak	Mukah	5	13.1	130,990	1,715,969	392.20	50.10	475.98	95.20
Pahang	Kuantan	9	18.54	89,916	1,667,043	343.71	43.91	417.13	46.35
Pahang	Pekan	6	20.47	78,377	1,604,377	330.79	42.26	401.45	66.91
Johor	Muar	5	22.17	68,802	1,525,340	299.14	38.21	363.04	72.61
Perak	Sri Manjong	3	27.18	54,088	1,470,112	284.51	36.35	345.29	115.10
Kelantan	Gua Musang	7	12.32	83,081	1,023,558	279.76	35.74	339.52	48.50
Johor	Batu Pahat	6	19.32	65,007	1,275,255	250.09	31.95	303.52	50.59
Terengganu	Kemaman	6	15.96	69,600	1,110,816	238.08	30.41	288.93	48.16
Johor	Johor Baru	5	18.42	65,256	1,202,016	235.73	30.11	286.09	57.22
Perak	Batang Padang	6	20.56	58,179	1,196,160	231.49	29.57	280.94	46.82
Pahang	Maran	7	17.88	60,888	1,088,677	224.46	28.67	272.41	38.92
Sabah	Kunak	7	21.98	47,930	1,053,501	222.95	28.48	270.58	38.65
Sarawak	Samarahan	4	13.31	64,679	860,877	196.76	25.14	238.79	59.70
Sarawak	Sibu	5	13.21	63,706	841,556	192.34	24.57	233.43	46.69
Perak	Perak Tengah	5	20.79	47,172	980,706	189.79	24.25	230.34	46.07
Pahang	Kuala Lipis	3	17.85	48,927	873,347	180.06	23.00	218.53	72.84
N. Sembilan	Jempol	5	18.21	53,498	974,199	174.75	22.32	212.08	42.42
Selangor	Kuala Selangor	5	25.05	33,614	842,031	171.26	21.88	207.85	41.57

Table 4 Simulated total mill capacity, total fuel feed rate, total dust load and average dust load *(Continued)*

Selangor	Kuala Langat	6	22.33	37,582	839,206	170.69	21.81	207.15	34.52
Pahang	Jerantut	4	19.49	38,626	752,821	155.22	19.83	188.37	47.09
Johor	Mersing	5	16.41	47,475	779,065	152.78	19.52	185.42	37.08
Sarawak	Kuching	3	15.43	41,417	639,064	146.06	18.66	177.27	59.09
Sabah	Semporna	4	23.24	28,734	667,778	141.32	18.05	171.51	42.88
Sarawak	Kapit	3	13.09	45,377	593,985	135.76	17.34	164.76	54.92
Sabah	Keningau	3	18.75	31,756	595,425	126.01	16.10	152.93	50.98
N. Sembilan	Port Dickson	2	23.8	29,399	699,696	125.51	16.03	152.32	76.16
N. Sembilan	Tampin	3	19.85	34,500	684,825	122.84	15.69	149.08	49.69
Melaka	Jasin	2	25.14	26,749	672,470	117.72	15.04	142.87	71.43
Perak	Kuala Kangsar	2	22.89	25,289	578,865	112.03	14.31	135.96	67.98
Perak	Krian	4	21.53	26,197	564,021	109.15	13.94	132.47	33.12
Kedah	Kulim	3	21.95	25,762	565,476	104.25	13.32	126.52	42.17
Terengganu	Dungun	2	12.75	36,894	470,399	100.82	12.88	122.36	61.18
Sarawak	Betong	1	14.01	29,086	407,495	93.14	11.90	113.03	113.03
Pahang	Temerloh	3	10.75	40,924	439,933	90.70	11.59	110.08	36.69
Pahang	Raub	3	17.78	24,323	432,463	89.16	11.39	108.21	36.07
Melaka	Alor Gajah	1	22.4	22,697	508,413	89.00	11.37	108.01	108.01
Pahang	Bentong	3	19.1	22,452	428,833	88.42	11.30	107.30	35.77
Selangor	Ulu Selangor	4	18.81	22,610	425,294	86.50	11.05	104.98	26.24
Johor	Pontian	1	14.44	29,206	421,735	82.71	10.57	100.37	100.37
Perak	Larut Matang	8	20.69	20,505	424,248	82.10	10.49	99.64	12.46
Perak	Kinta	1	22.59	16,075	363,134	70.28	8.98	85.29	85.29
Sabah	Beaufort	2	12.82	24,614	315,551	66.78	8.53	81.05	40.52
Selangor	Sepang	4	17.26	18,691	322,607	65.61	8.38	79.63	19.91
Sarawak	Sri Aman	1	9.58	29,606	283,625	64.82	8.28	78.67	78.67
N. Sembilan	Seremban	3	19.68	17,635	347,057	62.25	7.95	75.55	25.18
Terengganu	Ulu Terengganu	3	12.97	20,916	271,281	58.14	7.43	70.56	23.52
Kedah	Baling	1	22.29	13,112	292,266	53.88	6.88	65.39	65.39
Terengganu	Setiu	1	13.19	18,382	242,459	51.96	6.64	63.07	63.07
Kelantan	Kuala Krai	2	15.27	11,933	182,217	49.80	6.36	60.44	30.22
Kedah	Bandar Bahru	2	23.22	10,880	252,634	46.58	5.95	56.53	28.26
N. Sembilan	Jelebu	1	18.09	14,216	257,167	46.13	5.89	55.98	55.98

Table 4 Simulated total mill capacity, total fuel feed rate, total dust load and average dust load *(Continued)*

Perak	Selama	1	18.27	12,215	223,168	43.19	5.52	52.42	52.42
Sarawak	Limbang	1	14.68	12,452	182,795	41.78	5.34	50.70	50.70
Kelantan	Tanah Merah	1	13.82	10,377	143,410	39.20	5.01	47.57	47.57
Sabah	Tenom	1	20.35	5,695	115,893	24.53	3.13	29.77	29.77
Selangor	Klang	2	18.03	5,261	112,886	22.96	2.93	27.86	13.93
Selangor	Petaling	1	22.28	4,737	105,540	21.47	2.74	26.05	26.05
Sabah	Kota Marudu	2	11.71	3,293	97,111	20.55	2.63	24.94	12.47
Penang	S. Prai Selatan	2	15.35	6,863	105,347	16.85	2.15	20.45	10.22
Terengganu	K. Terengganu	1	9.24	3,901	36,045	7.73	0.99	9.38	9.38
Other districts without palm oil mills				120,497	2,114,759				
Malaysia	Total mills	417		4,685,085	89,255,221	17,726.34	2,264.54	21,513.1	51.67

Table 5 Descriptive statistics, normality test and confidence interval of average dust load

Selected data	Average dust load (g/s)
Size	73
Mean	51.6696
Standard deviation	23.5164
Range	9.38 to 115.10
Skewness	0.6285
Kurtosis	0.4650
Normality test	
Anderson-Darling	score = 0.6528
(p value)	($0.05 = < p < 0.10$)
Lilliefors D = max [D−, D+]	0.0871 = max [0.0440, 0.0871]
(p value)	($0.15 = < p < 0.20$)
Confidence interval (95%)	51.67 ± 5.39 g/s

of the reasons is the deterioration of a dust collection system. Even properly designed systems from reputable suppliers may show a decrease in performance characteristics due to wear and poor maintenance. In principle, the dust collector is like any other equipment, requiring regular maintenance, servicing, evaluation and testing to ensure its operation is efficient and optimal. Poor awareness among the millers who only focus on maximizing mill productivity and who try to minimize the cost of production, instead of investing in high-efficiency pollution control system and good maintenance practice to protect the environment, is a major concern. The reason might be that there is no return or profit from the investment; it is merely a requirement of the Clean Air Regulation of 1978. Because of this, many mills simply use low-cost dust collectors just to fulfil the basic requirement of having a solid fuel boiler. Furthermore, improper design of dust collection systems, including poor quality control of material, can shorten the lifespan or increase the deterioration rate of a dust collection system, or it might even escalate the problem. For example, putting a wet scrubber instead of dry mechanical cyclone after a biomass boiler might create further problem since wastewater needs an extra wastewater treatment system to prevent water pollution problems.

Dust particle size

Dust particle size is one of the critical factors that must be carefully considered when designing a mechanical dust collector. A properly designed filter will be able to take out finer particles from the gas stream and will result in a lower dust concentration at the stack. A poorly designed dust collector will not be able to capture fine particles, thus affecting its overall efficiency.

Boiler operation

Under normal operating conditions in a palm oil mill, the boiler plant is subjected to fluctuating loads which are accompanied by variables. These fluctuating loads are inherent in the process plant operation and are the main cause for lowered boiler efficiency, which directly influence the air pollution problem, production capacity in terms of throughput of fresh fruit bunches, products recovery efficiency and the quality of products produced [20]. This means that if the boiler efficiency is low, it might need a higher load of fuel to heat up the boiler itself in order to get the required steam demand. When fuel feeding rates increase, more ash will be generated from fuel combustion, resulting in higher dust loads to the dust collection system, hence higher dust emission to the atmosphere. Therefore, it is critical for millers to make sure the boiler operation is always at optimal conditions.

Fuel conditioning

In older boilers, inefficient combustion is due to the condition of fuel which is mostly wet and lumpy, inconsistent fuel feeding and poor fuel distribution on the furnace grate of boiler. Pre-conditioning of fuel can help in better combustion in the boiler and increase the boiler efficiency, at the same time, eliminating dark smoke. Due to the increased combustion efficiency, the carbon dioxide content will be increased, resulting in lower dust emission when expressed in terms of 12% CO_2 [21,22].

Performance evaluation and reduction target

The performance evaluation of mechanical dust collector (MS 1723: 2003) involves simultaneous isokinetic measurements at the inlet and outlet of the dust collector in order to determine the actual dust collection performance characteristic of the system installed. This performance characteristic as measured can be used for verifying the design specifications. After the performance characteristic has been determined, future measurements will show if any deterioration has taken place. These measurements should be part of the maintenance programme of the mill. This will assist not only in ensuring that the emission complies with the set of standards, but also in planning maintenance work on dust collector in advance. Evaluation results should be kept as records by state branches under the Department of Environment. Establishing a databank is not only for performance evaluation and monitoring purpose, but also for the identification of inefficient systems so that action can be taken to control the pollution. Setting a short-term or long-term reduction target can help to improve dust emission. For example, with reference to Table 6, by knowing the current status of the industry's

Table 6 Comparison of simulated state total dust emission at different dust removal efficiency

State	Before removal	District dust emission (tonnes/year)				
	(tonnes/year)	60%	70%	80%	90%	95%
Sabah	123,870	49,548	37,161	24,774	12,387	6,194
Johor	58,805	23,522	17,641	11,761	5,880	2,940
Pahang	55,502	22,201	16,651	11,100	5,550	2,775
Sarawak	54,924	21,969	16,477	10,985	5,492	2,746
Perak	37,431	14,973	11,229	7,486	3,743	1,872
Negeri Sembilan	12,945	5,178	3,884	2,589	1,295	647
Selangor	11,567	4,627	3,470	2,313	1,157	578
Terengganu	9,310	3,724	2,793	1,862	931	466
Kelantan	5,895	2,358	1,768	1,179	589	295
Melaka	5,159	2,064	1,548	1,032	516	258
Kedah	4,851	1,941	1,455	970	485	243
Penang	460	184	138	92	46	23
Malaysia	389,961	155,984	116,988	77,992	38,996	19,498

total emission of 155,984 tonnes/year at overall efficiency of 60% and considering the industry and its affordability in terms of profit, the most cost-effective solution available can be recommended and a target can be set at increasing the efficiency to 95% to reduce the emission to 19,498 tonnes/year.

Conclusions

This study has proven that simulation can be used as an alternative to estimate dust emission and recommends that the minimum efficiency for a dust collector required for a boiler besides actual measurement be done on the stack. However, regular measurement of the dust collector performance, which can be done by simultaneous isokinetic measurements at the inlet and outlet of the dust collector, is critical to accurately identify the problem so that subsequent action can be taken to overcome it.

Issues such as factors that affect dust collector efficiency, boiler operation, fuel conditioning, performance evaluation and reduction target have been discussed. Unless proper implementation and necessary improvement of the above actions have been taken, high dust emission will continue to pose serious problems to the industry and environment.

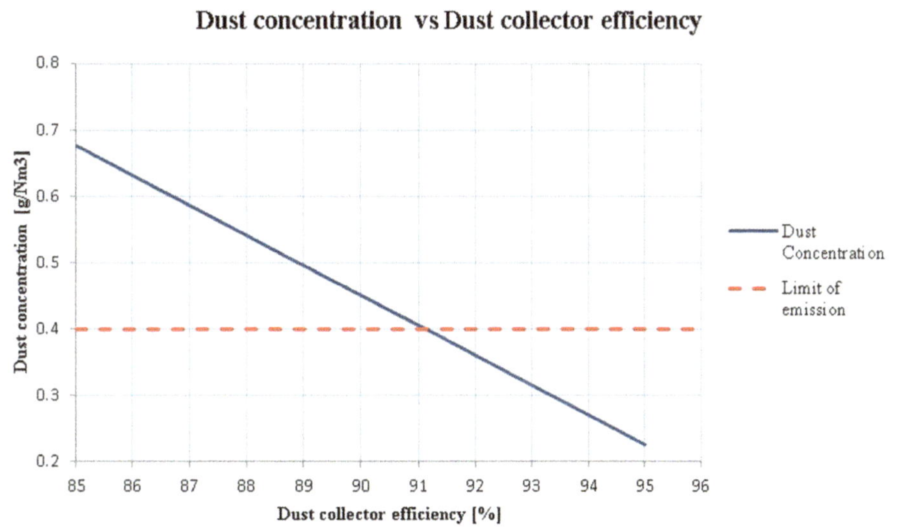

Figure 3 Correlations between simulated dust concentrations and dust removal efficiency.

Competing interests
The authors declare that they have no competing interests.

Authors' contributions
YCK carried out the data collection and formulation. AMHA participated in the design of the study and performed the statistical analysis. Both authors read and approved the final manuscript.

Acknowledgements
The authors would like to acknowledge Excel Air Engineering Sdn Bhd and Malaysia Palm Oil Board (MPOB) for providing useful information in completing this paper.

References
1. Abdullah, LC: Wong, LI, Saari, M, Salmiaton, A, Abdul Rashid. MS: Particulate matter dispersion and haze occurrence potential studies at a local palm oil mill. Int. J. Environ. Sci. Tech. **4**(2), 271–278 (2007)
2. Harimi, M, Megat Ahmad, MMH, Sapuan, SM, Idris, A: Numerical optimisation of excess air with respect to fibre-to-shell ratio during incineration process. Int. Energy J. **7**(2), 126 (2006)
3. Mahlia, TM, Abdulmuin, MZ, Alamsyab, TMI, Mukhlishien, D: An alternative energy source for palm oil wastes industry for Malaysia and Indonesia. Energy Conserv. Manage. **42**, 2109–2118 (2001)
4. Yaccob Khan, M, Rahman, A, Rashid, M, Yusuf, M: Air pollution control in palm oil mill industry in Malaysia. In: Symposium on Environmental Perspectives Towards the Year 2000 and Beyond. , Bangkok,Thailand (November 1989)
5. Rozainee, M, Ramli, M, Rashid, M: A field evaluation of particulate emissions from palm oil mills. Jurnal Teknologi. **29**, 1–6 (1998)
6. Rashid, M, Rozainee, M: Particulate emissions from a palm oil mill plant–a case study. Jurnal. Teknologi, Universiti. Teknologi. Malaysia **22**, 19–24 (1993)
7. De Kock, J, Yap, D: Performance criteria of dust collectors for palm oil mill boilers. In: Proceedings of the 2000 National Seminar on Palm Oil Milling, Refining Technology, Quality and Environment, pp. 185–194. Genting Highland (3–4 July 2000)
8. MPOB: Monthly average milling hours by mills: [hours]. Processing Performance 2009. http://econ.mpob.gov.my/economy/annual/stat2009/ie_tajuk-PROCESSING.htm (2009). Accessed on 16 October 2011
9. Malaysian Palm Oil Board: Malaysian Oil Palm Statistics 2009. Malaysian Palm Oil Board, Ministry of Primary Industries. , Kuala Lumpur (2010)
10. United States Environmental Protection Agency (USEPA): Emissions factors & AP 42, compilation of air pollutant emission factors. http://www.epa.gov/ttnchie1/ap42. Accessed 8 December 2011
11. Wang, LK, Pereira, NC, Hung, Y-T: Handbook of Environmental Engineering: Air Pollution Control Engineering, vol. 1, p. 604. , Humana, New Jersey (2004)
12. Malaysian Palm Oil Council (MPOC): (2012). www.mpoc.org.my. Accessed 22 Oct 2011
13. Boon, KY: Control of air pollution from palm oil mill boilers. In: Seminar on Developments in Palm Oil Milling Technology and Environmental Management. , Genting Highlands (16–17 May 1991)
14. Ellison, SLR, Barwick, VJ, Duguid Farrant, TJ: Distribution and types of data. In: A Practical Statistics for the Analytical Scientist: A Bench Guide, 2nd edn, pp. 16–24. RSC publishing, Cambridge (2009)
15. Henderson, AR: Testing experimental data for univariate normality. Clinica Chimica Acta **366**, 112–129 (2006)
16. Mohd Razali, N, Yap, BW: Power comparisons of Shapiro-Wilk, Kolmogorov-Smirnov, Lilliefors and Anderson-Darling tests. J. Statis. Model. Anal. **2**(1), 21–33 (2011)
17. Van Wylen, GJ, Sonntag, RE: Fundamental of Classical Thermodynamics, 3rd edn. Wiley, Singapore (1985)
18. Buonicore, AJ, Davis, WT: Air & Waste Management Association. Air Pollution Engineering Manual. New York, Van Nostrand Reinhold (1992)
19. De Kock, J, Yap, D: Air pollution problem definition and cost-effective solutions for palm oil mill boiler and incinerator flue gas exhausts. Kuala Lumpur, Palm Oil International Palm Oil Congress (PIPOC) (1999)
20. Cooper, IW: Boiler operation. In: Proceedings of Regional Workshop on Palm Oil Mill Technology and Effluent Treatment, pp. 48–58. , Kuala Lumpur (17–18 August 1982)
21. De Kock, J, Yap, D: PIPOC 2001. In: PIPOC 2001, Istana Hotel. Istana Hotel, Kuala Lumpur, Malaysia (20–22 August 2001)
22. Environmental Quality (Clean Air) Regulations: Section 25, Environmental Quality Act 1974 (Act 127) & Subsidiary Legislation (As at 5 May 2007), Malaysia (1978)

Effect of species, particle size and compacting pressure on relaxed density and compressive strength of fuel briquettes

Stephen J Mitchual[1*], Kwasi Frimpong-Mensah[2] and Nicholas A Darkwa[2]

Abstract

Densification of biomass waste materials has provided a great boost to the utilization of wood and agricultural waste for domestic and industrial fuel. However, the processes involved in the production of this fuel make it more expensive than fossil fuel. This is because densification of wood waste into fuel briquettes is not simple. This paper reports the results of research conducted to determine the effect of species, particle size and compacting pressure on relaxed density and compressive strength in cleft of briquettes produced from sawdust of tropical hardwoods. Briquettes were made using a laboratory hydraulic press. Compacting pressure was varied from 10 to 50 MPa at an interval of 10 MPa. Species used were *Triplochiton scleroxylon*, *Ceiba pentandra*, *Aningeria robusta*, *Terminalia superba*, *Celtis mildbreadii* and *Piptadenia africana*. The results indicate that species, compacting pressure and particle size of sawdust at 5% level of significance have significant effect on the relaxed density and compressive strength in cleft of briquettes produced. The multiple correlation coefficient (R) and adjusted R^2 for the regression model between relaxed density of briquettes, and species density, particle size and compacting pressure were 0.93 and 0.87, respectively. Additionally, the multiple correlation coefficient and adjusted R^2 for the regression model between compressive strength of briquettes, and species density, particle size and compacting pressure were 0.83 and 0.69, respectively. The regression models suggest that species density, particle size and compacting pressure are good predictors of relaxed density and compressive strength in cleft of briquettes produced from sawdust of tropical hardwoods.

Keywords: Briquette; Species; Compacting pressure; Particle size; Relaxed density; Compressive strength

Background

Renewable source of energy is the fastest growing source of world energy, with consumption increasing by 3% per year [1]. This is due to its environmental friendliness as against the rising concern about the environmental impact of fossil fuel use and also strong government incentives for increasing renewable penetration in most countries around the world [1]. Globally, biomass currently provides around 46 EJ of bioenergy in the form of combustible biomass and wastes, liquid biofuels, solid biomass/charcoal and gaseous fuels. This share is estimated to be over 10% of global primary energy, with over two-thirds consumed in developing countries as traditional biomass for household use [2]. Previous studies conducted to examine the economic impacts of using biomass energy clearly show that the benefits of production of briquettes for many economies clearly exist. However, there are several important factors that limit their utilization. The main reason is the high production cost resulting from high energy input in the production of various bioenergy fuel. From the past to the present, various scientists have carried out studies that have led to the improvement and better understanding of the technology for production of biomass briquettes. For example, it has been reported that pre-heating the biomass raw materials before pressing could reduce the power requirement for pressing of briquettes [3]. In a study conducted to improve biomass briquetting, it was observed that an average electrical energy saving at the heater, motor and overall system were 23.5%, 10.8% and 10.2%, respectively. This was achieved

* Correspondence: Stephen.mitchual@yahoo.com
[1]Department of Design and Technology Education, University of Education, Winneba, Kumasi Campus, Kumasi, Ghana
Full list of author information is available at the end of the article

when the biomass raw material was heated before pressing [3]. Other studies suggest that blending of two biomass raw materials could improve the durability characteristics of briquettes produced [4]. Material and process variables that could additionally influence the physical and mechanical characteristics of briquettes are compacting pressure, moisture content, particle size distribution of biomass raw material and temperature. According to [5], briquettes manufactured at lower pressures (30 to 60 MPa) fall to pieces easily. However, briquettes produced at higher pressures (150 to 250 MPa) are consistent and compact. Furthermore, the density and durability (mechanical strength) of briquettes are inversely proportional to the particle size since smaller particles have greater surface area during densification [6]. In this paper, the researchers investigated the effect of species, particle size and compacting pressure on relaxed density and compressive strength in cleft of fuel briquettes made at room temperature using low compacting pressure.

Methods

Materials and material preparation

Sawdust from the following selected tropical hardwood species were used for the study: *Triplochiton scleroxylon*, *Ceiba pentandra*, *Aningeria robusta*, *Terminalia superba*, *Celtis mildbreadii* and *Piptadenia africana*. These species were classified into lower density (*T. scleroxylon* and *C. pentandra*), medium density (*A. robusta* and *T. superba*) and high density (*C. mildbreadii* and *P. africana*). The sawdust was sun-dried at an average relative humidity and temperature of 75% and 28°C, respectively, between 5 and 7 days. Thereafter, the sawdust of each species was classified into particle size (PS): PS ≤ 1 mm, 1 mm < PS ≤ 2 mm and 2 mm < PS ≤ 3.35 mm using an automatic sieve shaker with serial number A060-01/ZG/0038 and model A060-01.

Moisture content

The moisture content, on oven-dry basis, of the sawdust was determined in accordance with [7]. A sample of 2 g of sawdust of each species was weighed and placed in a laboratory oven at a temperature of 103°C and dried until the difference in mass between two successive weighing, separated by an interval of 2 h, was 0.01 g or less. The moisture content of the specimen was then computed as follows:

$$Moisture\ content = \frac{M_1 - M_o}{M_o} \times 100, \tag{1}$$

where M_1 and M_o are mass of test samples before drying and when oven-dried, respectively (g).

Density of species

Density of the six timber species from which sawdust was collected and determined was in accordance with [8]. Fifteen clear specimens with dimensions of 20 mm × 20 mm × 30 mm were prepared for each species. The oven-dried masses of the specimens were determined. Thereafter, they were dipped one by one in paraffin wax and then kept in a desiccator. The volume displacement method which employs the use of Eureka can and a measuring cylinder was used to determine the volumes of the specimen. The density of each specimen was then computed as:

$$Density = \frac{Mass\ of\ specimen}{Volume\ of\ specimen}. \tag{2}$$

Briquetting process

A 55.3-mm ID × 52.5-cm height cylindrical mould was used to produce the briquettes. Ninety grammes of sawdust of each species and particle size was weighed and filled into the mould. The average moisture content of the sawdust was 11.46%. A manual hydraulic press with a gauge and a piston was used to compress the raw material without a binder against the other end of the mould to form the briquettes. A clearance of about 0.1 mm was provided between the piston and the inner wall of the mould to allow for air escape. The samples were pressed using the following predetermined compacting pressure levels: 10, 20, 30, 40 and 50 MPa. The dwelling time for each press was maintained at 10 s for all the pressing made. This process was repeated for all the six species and particle size.

Physical and mechanical properties of briquettes

The relaxed density and compressive strength in cleft of the briquettes were investigated using standard testing methods.

Relaxed density

Relaxed density of the briquettes was determined 30 days after removal from the press in accordance with [9]. The mass of the briquettes was determined using a laboratory electronic balance with an accuracy of 0.01 g. The diameter and length of the briquette were measured at three points with a digital vernier calliper. Relaxed density (RD) was then computed as:

$$RD\ (g/cm^3) = \frac{108000 \times M(g)}{\pi[d_1(mm) + d_2(mm) + d_3(mm)]^2 \times [l_1(mm) + l_2(mm) + l_3(mm)]} \tag{3}$$

where d_1, d_2 and d_3 were diameters of briquettes at points one, two and three, respectively, measured in millimetres; l_1, l_2 and l_3 were lengths of briquettes at points one, two and three, respectively, measured in millimetres. M is the mass of briquette in grammes.

Compressive strength

Compressive strength in cleft of briquettes was determined in accordance with [10] using an Instron Universal Strength (Norwood, MA, USA) testing machine with load cell capacity of 100 kN. The cross-head speed was 0.305 mm/min. A sample of briquette to be tested was placed horizontally in the compression test fixture and a load was applied at a constant rate of 0.305 mm/min until the briquette failed by cracking. The compressive strength in cleft was then computed as follows:

$$\text{Compressive strength in cleft } (N/mm) = \frac{3 \times \textit{The load at fracture point } (N)}{[l_1(mm) + l_2(mm) + l_3(mm)]} \quad (4)$$

where l_1, l_2 and l_3 were lengths of briquettes at points one, two and three, respectively (mm).

Results and discussion

Relaxed density

The result in Table 1 (particle size < 1 mm) indicates that the relaxed density of briquettes made from *C. pentandra*, the species with the lowest density (409 kg/m^3), ranged from 398 to 716 kg/m^3, while that of *C. mildbreadii*, species with the highest density (764 kg/m^3) ranged from 453 to 706 kg/m^3. The lowest relaxed density for all the briquettes produced from particle size < 1 mm was 366 kg/m^3 (*T. scleroxylon* at compacting pressure = 10 MPa), and the highest was 741 kg/m^3 (*P. africana* at compacting pressure = 50 MPa). The result in Table 2 for 1 mm ≤ particle size < 2 mm also shows that the relaxed density of briquettes made from *C. pentandra* ranged from 386 to 692 kg/m^3, whilst that of *C. mildbreadii* ranged from 435 to 658 kg/m^3. The result also indicates that the lowest relaxed density for 1 mm ≤ particle size < 2 mm was 354 kg/m^3 (*T. scleroxylon* at compacting pressure = 10 MPa), and the highest was 723 kg/m^3 (*P. africana* at compacting pressure = 50 MPa). Considering Table 3 which indicates the relaxed density for 2 mm ≤ particle size < 3.35 mm, it shows that briquettes made from *C. pentandra* had relaxed density ranging from 373 to 651 kg/m^3, whilst that of *C. mildbreadii* ranged from 430 to 655 kg/m^3. Additionally, the lowest relaxed density for this particle size was 324 kg/m^3 (*T. scleroxylon* at compacting pressure = 10 MPa), and the highest was 720 kg/m^3 (*P. africana* at compacting pressure = 50 MPa). Briquettes' relaxed densities obtained from this study were consistent with suggestions by [6] that briquettes made from hydraulic piston press are usually less than 1,000 kg/m^3 and are usually between 300 and 600 kg/m^3 in density [11]. Correlation analysis between relaxed density on one hand and compacting pressure and particle size on the other hand indicates that there was a weak significant negative correlation between relaxed density and particle size of briquettes produced (Pearson's $r = -0.188$, p value = 0.000; $N = 450$; one-tailed, $\alpha = 0.05$). Compacting pressure was also found to have a very strong positive significant correlation with the relaxed density of the briquettes produced (Pearson's $r = 0.901$, p value = 0.000; $N = 450$; one-tailed, $\alpha = 0.05$). These results suggest that the relaxed density of the briquettes produced increases with increasing compacting pressure level and that briquettes produced from sawdust of tropical hardwoods species with smaller particle size are likely to have higher relaxed density than those with larger particle size. This result confirms that of other researchers [2,12]. It is reported that in reality, in briquetting, when a large

Table 1 Relaxed density (kg/m^3) of briquettes made from six tropical hardwood species (PS < 1 mm)

Species	Compacting pressure				
	10 MPa	20 MPa	30 MPa	40 MPa	50 MPa
C. pentandra	398	523	622	666	716
T. scleroxylon	366	458	545	594	632
A. robusta	440	543	607	636	695
T. superba	447	557	631	675	727
P. africana	465	567	637	679	741
C. mildbreadii	453	533	611	662	706

Table 2 Relaxed density (kg/m^3) of briquettes made from tropical hardwood species (1 mm ≤ PS < 2 mm)

Species	Compacting pressure				
	10 MPa	20 MPa	30 MPa	40 MPa	50 MPa
C. pentandra	386	503	606	646	692
T. scleroxylon	354	452	535	591	627
A. robusta	362	466	538	586	658
T. superba	437	548	615	651	680
P. africana	441	534	610	664	723
C. mildbreadii	435	491	562	621	658

Table 3 Relaxed density (kg/m^3) of briquettes made from tropical hardwood species (2 mm ≤ PS < 3.35 mm)

Species	Compacting pressure				
	10 MPa	20 MPa	30 MPa	40 MPa	50 MPa
C. pentandra	373	489	598	618	651
T. scleroxylon	324	435	508	552	597
A. robusta	345	463	519	571	573
T. superba	396	514	591	649	673
P. africana	417	530	604	658	720
C. mildbreadii	430	487	543	612	655

proportion of the raw material is of smaller particles, the briquette produced will have a higher density [2,12]. The reason for this trend is that increasing the compacting pressure will lead to the particles of biomass material being closely packed due to reduction of void ratio and plastic deformation of the sawdust particles, therefore leading to increased density of the briquettes [13]. Additionally, if the raw material is finer, it gives a larger surface area for bonding which results in the production of briquette with higher density. Table 4 which shows the analysis of variance (ANOVA) of the relaxed density of briquettes produced indicates that at 5% level of significance, species, particle size, compacting pressure and their interactions have significant effects on the relaxed density of briquettes produced. The multiple coefficient of determination value and the root mean square error for the ANOVA model were 0.9907 and 11.24, respectively. Thus, it could be deduced that species, particle size and compacting pressure, and their interactions explained about 99.07% of the variability in the relaxed density of the briquettes produced.

Multiple linear regression analysis to establish the relative contribution of species density, particle size and compacting pressure in the prediction of the relaxed density of briquette and the mathematical relationship between the dependent variable relaxed density and the independent variables, species density, particle size and compacting pressure, is indicated in Table 5. The result (Table 5) shows the unstandardized (β) and standardized (Beta) regression coefficients, the multiple correlation coefficient (R), adjusted R^2, the value of t and its associated p value for each of the variables. As shown in Table 5, species density, particle size and compacting pressure collectively explained 87.1% (adjusted R^2 = 0.871) of the variance in the relaxed density of briquette produced. This suggests that the linear regression model is a good predictor of briquettes' relaxed density (R^2 = 0.872, p value = 0.000).

Table 4 ANOVA of effect of biomass material, PS and compacting pressure on relaxed density of briquettes

Source	*df*	ANOVA SS	Mean square	*F* ratio	*p* value
Species	5	452,443.680	90,488.736	715.19	< 0.0001[a]
Particle size	2	174,914.253	87,457.127	691.23	< 0.0001[a]
CP	4	4,106,254.742	1,026,563.686	8113.63	< 0.0001[a]
Species × CP	20	45,562.164	2,278.108	18.01	< 0.0001[a]
Species × Particle size	10	38,280.947	3,828.095	30.26	< 0.0001[a]
Particle size × CP	8	3,449.324	431.166	3.41	0.0009[a]
Species × Particle size × CP	40	22,406.809	560.170	4.43	< 0.0001[a]
Error	360	45,548.400	126.523		

df degree of freedom, *CP* compacting pressure. [a]Statistically significant at 0.05 level of significance.

Table 5 Regression of relaxed density of briquettes on species density, particle size compacting pressure of briquettes

Variables	β	Beta	R	Adjusted R^2	t	p value
Constant	334.651		0.934	0.871	34.238	0.000
Species density	0.125	0.160			9.446	0.000
Particle size	−23.997	−0.188			−11.109	0.000
Compacting pressure	6.639	0.901			53.234	0.000

Additionally, based upon the Beta values, it could be deduced that compacting pressure explained the bulk of the variance in the relaxed density of the briquette produced (Beta = 0.901, t = 53.234, p value = 0.000) and was the best predictor of relaxed density of the briquette. However, species density and particle size of sawdust used significantly contributed to the model (species density: Beta = 0.160, t = 9.446, p value = 0.000; particle size: Beta = −0.188, t = −11.109, p value = 0.000). Furthermore, it could be deduced from the β values that the mathematical relationship between the dependent variable Relaxed density and the independent variables, namely species density (S), PS and compacting pressure (CP) is

$$\text{Relaxed density}\ (\text{kg/m}^3) = 334.651 + 0.125\text{S}–23.997\text{PS} + 6.639\text{CP}. \quad (5)$$

Compressive strength in cleft

Briquettes' compressive strength is one of the indices used to assess its ability to be handled, packed and transported without breaking. Tables 6, 7 and 8 indicate the compressive strength in cleft of the briquette produced. The result for particle size < 1 mm (Table 6) indicates that at all compacting pressure levels, the compressive strength in cleft of *C. pentandra*, was exceptionally high compared to that of the other species. Compressive strength in cleft of *C. pentandra* ranged from 15.81 to 44.58 N/mm for compacting pressure levels 10 to 50 MPa. *C. mildbreadii* comparatively had the lowest compressive

Table 6 Compressive strength (N/mm) in cleft of briquettes made from tropical hardwood species (PS < 1 mm)

Species	Compacting pressure				
	10 MPa	20 MPa	30 MPa	40 MPa	50 MPa
C. pentandra	15.81	29.23	39.26	40.40	44.58
T. scleroxylon	3.33	9.35	14.27	20.66	30.56
A. robusta	3.97	8.23	16.28	19.48	26.63
T. superba	2.13	6.16	10.44	13.64	18.92
P. africana	1.76	5.95	10.13	17.06	21.14
C. mildbreadii	1.30	3.55	6.69	10.51	12.45

Table 7 Compressive strength (N/mm) in cleft of briquettes made from tropical hardwood species (1 mm ≤ PS < 2 mm)

Species	Compacting pressure				
	10 MPa	20 MPa	30 MPa	40 MPa	50 MPa
C. pentandra	7.28	24.34	34.74	42.62	52.60
T. scleroxylon	6.97	14.66	27.53	36.57	38.31
A. robusta	3.47	9.51	14.42	21.62	27.50
T. superba	3.26	6.61	12.50	14.89	18.03
P. africana	2.60	6.99	12.65	17.80	24.05
C. mildbreadii	2.59	5.64	9.63	12.49	16.40

strength in cleft, ranging from 1.30 to 12.45 N/mm for the same compacting pressure levels. The result for 1 mm ≤ particle size < 2 mm (Table 7) and 2 mm ≤ particle size < 3.35 mm (Table 8) reflected the trend for particle size < 1 mm. In both cases, the compressive strength in cleft of *C. pentandra* was exceptionally higher than that of the other species. Correlation analysis indicated a weak significant positive correlation between particle size and compressive strength in cleft (Pearson's $r = 0.179$, p value = 0.000; $N = 450$; one-tailed, $\alpha = 0.05$), that is, as particle size increased the compressive strength in cleft of briquettes produced also increased. This result contradicts the assertion by [2,6] and [12] that, in general, the durability (mechanical strength) of pellets are inversely proportional to particle size since smaller particles have greater surface area for moisture addition during steam conditioning therefore resulting in increased starch gelatinization and better binding. The reason for the deviation is that, in this study, the briquettes were formed at low temperature (i.e. room temperature); thus, the formation of solid bridge resulting from the natural bonding of chemicals may be absent or minimal.

Therefore, the major contributing factors to the bond formed during this densification may be the mechanical interlock of the fibres of the biomass and adhesive force between the particles. Increase in fibre length of the biomass raw material resulting from increased particle size could lead to mechanical interlocking of relatively

Table 8 Compressive strength (N/mm) in cleft of briquettes made from tropical hardwood species (2 mm ≤ PS < 3.35 mm)

Species	Compacting pressure				
	10 MPa	20 MPa	30 MPa	40 MPa	50 MPa
C. pentandra	13.80	26.71	36.73	45.40	51.45
T. scleroxylon	6.98	15.24	27.64	36.86	40.89
A. robusta	2.69	11.90	17.29	21.35	26.88
T. superba	4.22	9.58	13.05	15.80	24.67
P. africana	3.97	15.28	24.24	41.83	55.45
C. mildbreadii	3.62	5.78	10.18	12.66	19.18

Table 9 ANOVA of effect of biomass material, PS and CP on compressive strength in cleft of briquettes

Source	*df*	ANOVA SS	Mean square	*F* ratio	*p* value
Species	5	29,475.439	5,895.088	1,316.33	< 0.0001[a]
Particle size	2	2,688.496	1,344.248	300.16	< 0.0001[a]
CP	4	36,597.173	9,149.293	2,042.97	< 0.0001[a]
Species × CP	20	4,451.833	222.592	49.70	< 0.0001[a]
Species × Particle size	10	3,592.241	359.224	80.21	< 0.0001[a]
Particle size × CP	8	902.633	112.829	25.19	< 0.0001[a]
Species × Particle size × CP	40	2,186.188	54.655	12.20	< 0.0001[a]
Error	360	1,612.234	4.47843		

df degree of freedom, *CP* compacting pressure. [a]Statistically significant at 0.05 level of significance.

longer fibres [14] which, therefore, result in the formation of a stronger bond and increased compressive strength in cleft. Additionally, a strong significant positive correlation between compressive strength in cleft and compacting pressure of briquettes produced was established (Pearson's r = 0.670, p-value = 0.000; N = 450; 1-tailed, $\alpha = 0.05$). This result means that the compressive strength in cleft of the briquettes produced increased with increasing compacting pressure. Increased compacting pressure result in increased binding force between the particles [13]. The increase in binding force resulted from increased mechanical interlocking of sawdust particle as well as increased adhesion between the particles [6].

The ANOVA of compressive strength in cleft of briquettes produced (Table 9) indicates that at 5% level of significance, compacting pressure, species, particle size and their interactions had significant effects on the compressive strength in cleft of the briquettes produced (p value < 0.05). The multiple coefficient of determination value and root mean square error for the ANOVA model were 0.9802 and 2.1162, respectively. It could therefore be deduced that the species type of the biomass raw material, its particle size and compacting pressure could explain about 98.02% of the variability in the compressive strength in cleft of the briquettes produced. The summary of the result of linear multiple regression to establish the relative contribution

Table 10 Regression of compressive strength in cleft of briquettes on species density, particle size and compacting pressure

Variables	β	Beta	R	Adjusted R^2	t	p value
Constant	19.923		0.829	0.685	10.081	0.000
Species density	−0.046	−0.454			−17.147	0.000
Particle size	2.957	0.179			6.770	0.000
Compacting pressure	0.637	0.670			25.267	0.000

of species density, particle size and compacting pressure towards the prediction of compressive strength in cleft of the briquette and the mathematical relationship between the dependent variable compressive strength in cleft and independent variables, species density, particle size and compacting pressure, is as indicated in Table 10. Table 10 also indicates the β and Beta regression coefficients, R, adjusted R^2, the value of t and its associated p values for each of the variables.

The result indicates that species density, particle size and compacting pressure collectively explained 68.5% (adjusted R^2 = 0.685) of the variance in the compressive strength in cleft of the briquette produced. This suggests that the regression model is a good predictor of compressive strength in cleft of the briquettes produced (R^2 = 0.685, p value = 0.000).

The Beta value for the three predictive variables suggests that compacting pressure explained the bulk of the variance in the compressive strength in cleft of the briquette produced (Beta = 0.670, t = 25.267, p value = 0.000) and was the best predictor of compressive strength in cleft of the briquette produced. That notwithstanding, species density and particle size of sawdust used significantly contributed to the regression model (species density: Beta = −0.454, t = −17.147, p value = 0.000; particle size: Beta = 0.179, t = 6.770, p value = 0.000). The values of β also suggest that the mathematical relationship between compressive strength in cleft of briquettes produced and S, PS and CP is

$$\text{Compressive strength in cleft (N/mm)} = 19.923 - 0.046\text{S} + 2.957\text{PS} + 0.637\text{CP}. \quad (6)$$

Conclusions

This study examined the effect of species, particle size and compacting pressure on relaxed density and compressive strength of fuel briquettes. From the study, it could be concluded that (1) briquettes with adequate compressive strength in cleft could be produced from *C. pentandra* at room temperature using compacting pressure as low as 20 MPa; (2) the type of species, compacting pressure and particle size as well as their interactions have a significant effect on the relaxed density and compressive strength in cleft of briquettes produced from sawdust of *T. scleroxylon*, *C. pentandra*, *A. robusta*, *T. superba*, *C. mildbreadii* and *P. africana*; (3) the mathematical relationship between RD of briquette and species density, particle size and compacting pressure is as follows: Relaxed density = 334.651 + 0.125S − 23.997PS + 6.639CP; and (4) the mathematical relationship between compressive strength in cleft and species density, particle size and compacting pressure is as follows: Compressive strength in cleft = 19.923 − 0.046S + 2.957PS + 0.637CP. This study provides a better understanding of some of the factors that influence production of briquettes from sawdust of tropical hardwood species.

Competing interests

The authors declare that they have no competing interests.

Authors' contributions

SJM conceived the study, participated in the design of the study, drafted the manuscript and participated in the sequence alignment. KFM participated in the design of the study and sequence alignment. NAD participated in the design of the study, performed the statistical analysis and participated in the sequence alignment. All authors read and approved the final manuscript.

Acknowledgements

The authors are grateful to the Civil Engineering Department of Kwame Nkrumah University of Science and Technology and Forest Research Institution of Ghana for providing laboratory support for this study. Special thanks also go to Miss Agnes Ankomah of Crop Research Institute of Ghana for providing statistical software for this research work.

Author details

[1]Department of Design and Technology Education, University of Education, Winneba, Kumasi Campus, Kumasi, Ghana. [2]Department of Wood Science and Technology, Kwame Nkrumah University of Science and Technology, Kumasi, Ghana.

References

1. EIA: Annual energy outlook with projections to 2030. Energy Information Administration. http://www.eia.gov/oiaf/aeo (2009). Accessed
2. FAO: The briquetting of agricultural wastes for fuel: environment and energy paper, vol. 11. FAO, Rome (1990)
3. Bhattacharya, SC, Leon, MA, Rahman, MM: A study on improved biomass briquetting. Energy for sustainable development **6**(2), 106–110 (2002)
4. Yaman, S, Sahan, M, Haykiri-Acma, H, Sesen, K, Kucukbayrak, S: Fuel briquettes from biomass-lignite blends. Fuel processing technology **72**, 1–8 (2001)
5. Kers, J, Kulu, P, Aare Aruniit, A, Laurmaa, V, Križan, P, Šooš, L, Kask, Ü: Determination of physical, mechanical and burning characteristics of polymeric waste material briquettes. Estonian Journal of Engineering **16**(4), 307–316 (2010)
6. Tumuluru, JS, Wright, CT, Kenny, KL, Hess, R: A review on biomass densification technologies for energy application. U.S. Department of Energy. http://www.inl.aov (2011). Accessed
7. BSI: BS EN 13183-1: moisture content of a piece of sawn timber. Determination by oven-dry method. BSI, London (2002)
8. International, ASTM: ASTM standard D2395-2007a: standard test methods for specific gravity of wood and wood-based materials. ASTM International, West Conshohocken (2008)
9. ISO: International standard ISO 3131: standard test method for density of regular solids. ISO, Geneva (1975)
10. International, ASTM: ASTM standard D2166-85: standard test method of compressive strength of wood. ASTM International, West Conshohocken (2008)
11. Saeidy, EE: Technological fundamentals of briquetting cotton stalks as a biofuel. Humboldt Universität zu Berlin, Dissertation (2004)
12. Križan, P: Research of factors influencing the quality of wood briquettes. Acta Montanistica Slovaca Ročník **12**(3), 223–230 (2007)
13. Lindley, J, Vossoughi, M: Physical properties of biomass briquettes. Transaction of the ASAE **32**(2), 361–366 (1989)
14. Tabil, L, Adapa, P, Kashaninejad, M: Biomass feedstock pre-processing—part 2: densification. In: Dos Santos Bernardes, MA (ed.) Biofuel's Engineering Process Technology, pp. 439–464. InTech, New York (2011)

Renewable municipal solid waste pathways for energy generation and sustainable development in the Nigerian context

Olaleye M Amoo[1*] and Richard Layi Fagbenle[2]

Abstract

The continued concerns over energy prices, increase in population, and climate change issues have led towards a need for alternative and new energy sources. Municipal solid waste (MSW) is generally accepted as a renewable energy resource. This research study presents a techno-economic assessment potential to utilize the energy obtainable from MSW for the generation of electrical power. The assessment was carried out for energy generation by thermochemical (incineration or combustion) and biochemical (landfilling and anaerobic digestion) processes and based on the available data from seven selected municipalities. Due to the broad scope of this topic, life cycle impact of waste management, social acceptance, policy aspects, and emission reduction or fossil fuel offset are not considered and are not part of the assessment. Results presented in tabular form indicate, for example, that the price of steam generated by a fluidized steam generator is in the range of US\$0.018/kWh_t (Lagos) to US\$0.044/kWh (Nsukka) and compares favorably with the typical cost of steam at US\$0.015/kWh. Electrical power generation using a combined heat and power plant provides electricity at a cost that is in the range of US\$0.017/kWh_e (Lagos) to US\$0.040/kWh_e (Nsukka) and also compares well with the typical cost of electricity in Nigeria at US\$0.14/kWh_e (as of 2012).

Keywords: Municipal solid waste; Combustion; Energy; Electricity; Environment; Economics; Nigeria

Background

Nations are today faced with an overwhelming social problem of processing and disposal of municipal solid waste (MSW). With increasing population growth, rapid urbanization, rising levels of affluence, and resource scarcity, waste-to-energy (WTE) is reestablishing itself as an attractive technology option to promote low carbon growth among other renewable energy technologies. WTE is a proven process that provides electricity and steam generation in a sustainable way.

Nigeria is the most populated nation in Africa. With a 2011 population estimate of 162.4 million, growing annually at an estimated rate of 3% [1] and generating 0.55 to 0.58 kg of municipal solid waste per person per day [2], Nigeria can be said to be equally experiencing significant waste-related environmental problems. These problems can be said to have begun in the period when Nigeria gained independence, became rapidly urbanized, and thus generated more solid waste than nature can absorb or that Nigeria can efficiently dispose. MSW is produced daily all over the world, and it is a renewable energy resource with the potential to produce energy via WTE plants while also reducing the volume of waste. WTE can address the twin issues of land use and pollution from landfills and the well-established environmental perils of fossil fuels known as greenhouse gas emissions (GHG). Figure 1 shows the map of Nigeria showing Abuja, the capital.

Rapid urbanization, oil spills, and loss of arable land have created several environmental problems in Nigeria [3]. Among these environmental problems are sheet and gully erosion, coastal and marine erosion and land subsidence, flooding, drought and desertification, oil pollution, urban decay and squatter settlements, industrial pollution and waste, municipal solid waste, biodegradable petrochemical products, concrete jungles or cities, loss of fauna and flora, and climatic change [4].

MSW can be regarded as 'useless', unwanted, and discarded materials that resulted from society's normal or

* Correspondence: oamoo@stevens.edu
[1]Stevens Institute of Technology, Hoboken, NJ 07030, USA
Full list of author information is available at the end of the article

Figure 1 Map of Nigeria.

routine daily activities and, in general, may be in solid, liquid, or gaseous form [5]. It is often accompanied by commercial waste, whether solid or semisolid. It also comes in the form of food wastes, paper, cardboard, plastics, textiles, leather, wood, glass, and other household items. In most industrialized nations, from the time wastes are generated, they go through a solid waste chain, consisting of waste handling and separation; storage and processing at source; collection, separation, processing, and transformation of solid wastes; transfer and transport; energy generation; and disposal. This is however not the case in Nigeria or most developing countries. In many parts of the developing world, solid waste disposal is done through dumping and burning without control [6]. These practices pollute land, water, and air. Although these practices are illegal and their consequences recognized and felt in Nigeria, they persist in both urban and rural areas because of weak law enforcement regimen and limited public sector financial resources. A sustainable solution to this gigantic problem, which affects every populace, is through more efficient solid waste management. The generation of energy from MSW (though largely dependent on MSW composition) has the propensity to reduce greenhouse gas emissions, to generate renewable energy, and to provide energy diversity while also ensuring safe, hygienic, and reliable disposal practices.

The volume of waste generated in the world today is enormous. This volume of waste generated is to some degree due to inefficiencies (e.g., in usage, sorting, and processing) and, in principle, might be reduced [7]. As of 2011, the world generated an estimated two billion tons of MSW, and this number is expected to grow much higher [8]. MSW is a renewable form of energy that includes both commercial and residential wastes generated in municipalities. The US Environmental Pollution Agency (USEPA) goes further to classify MSW as a source of clean energy. From a sustainable development perspective, the focus is on reduction of waste, followed by recycling, both of which are advantageous in terms of reducing greenhouse gas emissions. Several analysis done using the USEPA models show that WTE avoids 36 million tons of greenhouse gases yearly [9]. However, not all wastes are recyclable, and as such, an energy recovery method becomes essential.

The interest in the practical applications of WTE dates back to several decades. WTE is hardly a new or novel idea. What is however new is the confluence of factors that have increased the attractiveness of WTE. These factors include rising oil prices, urban air pollution, energy supply security, reduction in foreign oil imports, carbon dioxide (CO_2) emissions, and climate change. These considerations are not confined to a single nation or part of the world and thus render the concept of WTE as abundantly and equitably available to humanity. Energy recovery through WTE can be defined as a waste treatment process that allows for the generation of energy in the form of electricity or heat from wastes that would have otherwise been disposed off in landfills [7-13]. It involves the use of modern combustion technologies for the recovery of energy from a mix of MSW. Today, more than 800 thermal WTE plants currently operate in nearly 40 countries around the globe [14]. WTE systems will treat at least 261 million tons of waste annually by 2022, with a total estimated output of 283 TWh of electricity and heat generation, which increased from 221 TWh in 2010. Under a more optimistic scenario, WTE will potentially treat 396 million tons of MSW a year, producing 429 TWh of power [14]. The global market for thermal and biological WTE technologies will reach about US$6.2 billion in 2012 and grow to about US$29.2 billion by 2022, according to forecasts by Cleantech market intelligence [14], while under an optimistic scenario, market value could reach US$80.6 billion by 2022. WTE facilities are integrated into broader waste management practices and policies with the goal of reducing the use of landfills. Although combustion technologies continue to lead the market, advanced thermal treatment technology deployments such as pyrolysis are expected to pick up as diminishing landfill capacity improves WTE economics. The utilization of biological technologies is also expected to increase worldwide.

It is imperative for every nation to find environmentally benign methods related to sound management of MSW for sustainable development. The energy potential of MSW is contained in materials that are either biogenic or anthropogenic in origin [15]. Biogenic materials are those that include paper, food, and yard wastes and are considered to be renewable. Anthropogenic materials, on the other hand, are those that include plastics which are derived from fossil fuels and are not sustainable. Several methods exist to analytically determine the biogenic and anthropogenic fractions of energy-generating components of MSW. Unlike some other renewable energy resources such as tidal, solar, and wind, MSW is always available. Benefits of WTE include waste volume reduction, sanitation and detoxification, stabilization, and energy recovery. The technologies that exist to produce energy from wastes are anaerobic digestion, combustion, gasification, and landfill gas to energy.

WTE has been well proven in Europe, where energy and solid waste policies have evolved from the Kyoto protocol [16] GHG reduction goals, to include renewable energy (RE) and decreased waste emissions from landfills [16]. This has been invariably good in advancing the state of technology for waste management and conversion to energy such that it is an integral component of solid waste management in Europe. About 18 European Union countries engage in thermally treating 58 million tons of wastes in WTE plants to produce energy annually [17]. As such, RE aspects of solid waste management policies work well together with waste reduction policies.

Waste management practices in Nigeria

Waste management practices then and now

The collection and disposal of solid waste in Nigeria were controllable and not a problem before the country's independence in 1960. Before the 1960s, the population was low and not concentrated in urban areas, and people lived in traditional ways [18]. Previous studies described the major cities of Nigeria, such as Ibadan, Calabar, and Port Harcourt, as clean and beautiful. However, when the country gained independence from the colonial masters and oil was subsequently discovered, migration from rural areas to urban areas increased as a result of people wanting to live more comfortably. Nigeria began importing many varieties of foods and luxury goods with consequent increase in solid wastes. Plastics and polythene produced negligible waste in 1971. Today, they make up a large part of the refuse. Empty beverage cans are found littered in every corner of the road. In fact, consumption patterns have dramatically changed since the 1960s and 1970s. For example, according to Paris-based magazine *Jeane Afrique*, Nigerians consumed 593,000 bottles of champagne in 2010, 50% more than the richer rival nation of South Africa - possibly due to the population effect with Nigeria being at least three times larger in population than South Africa.

Solid waste is now a troubling problem to both the government and the public [18]. As with many other nations, solid waste generation in Nigeria is a function of the population, its industrialization level, socioeconomic status, and dominant commercial activities [18]. Some of the factors, which influence the generation and stockpiling of solid waste, are the lack of modern waste disposal technology; the lack of facility to separate the waste at source; the effectiveness of existing solid waste management policy and its enforcement, environmental education, and awareness; and the income status. The indiscriminate dumping of secondhand electronic gadgets has also contributed to the waste disposal problem [18]. Waste storage and collection are the responsibility of state environmental agencies, private companies, or both. The wastes collected are dumped together, that is, unsegregated, indicating a lack of awareness of their nature and compatibility with one another. The populace also burns their refuse in open dump sites, e.g., documents, rags, and tires, often without sanitary or environmental control. This attracts scavengers at the dumps where they gather plastic cans and bottles, glass, and other recyclables mainly for sale to other secondhand goods markets. More than 50% of the people's wastes are highly decaying food surpluses, and more than 15% are plastic wastes. Wastes are generally recycled by the informal sector while scavengers scan for usable items, as described earlier [19].

The solid waste problem is acutely felt in Lagos State because it is the most populated city and among the most industrialized in sub-Saharan Africa. In 1995 alone, it produced approximately four million tons of solid wastes. The long history of the waste problem in Lagos goes back to the transfer of waste management from the local governments to the state. Policies kept on changing, leading to management lag. The wastes are generally not treated but moved to landfill sites where they are openly burned. The two incineration plants in Lagos are capable of treating wastes with less than 20% water, but they have never been used as the state's wastes contain 30% to 40% water. Lagos is currently undertaking efforts to generate revenue from waste management [20,21].

A brief on waste management policies in Nigeria

Intervention by the Federal Ministry of Environment has been in the form of a revised policy on environment in 1999 and the National Agenda 21 published also in 1999 [19]. These measures complemented existing guidelines and standards for environmental pollution, the waste management regulation of 1991, the Environmental Impact Assessment Decree number 86 of 1992, and an environmental edit of 1997. However, as earlier mentioned, nature's way of absorbing waste and human efforts combined have been surpassed by the enormity of the problem and its impact on the environment. Every material or product goes through a long cycle, consisting of activities, which produce greenhouse gas emissions. These are through energy consumption, methane emissions, and carbon storage. The current state of waste management in Nigeria is clearly a picture of neglect and failure in the enforcement of solid waste management policies. The effects of this neglect include unnecessarily lost resources, increased adverse impact on biological processes, disruption of infrastructure, greater risks of flooding and distribution of vermin and pests, and the deterioration of air quality.

Waste-to-energy technologies and opportunities for Nigeria

Technologies for WTE production have been rapidly evolving and changing communities and countries [20] yielding dual benefits from effective solid waste management practices. Not only will WTE deliver useful energy that is needed in many countries, but it will also aid to dispose of MSW effectively and safely. A community or country's solid waste management goals and objectives, the environmental impact, and production and consumption patterns will help determine if WTE production technology is appropriate for that community or country [20]. For example, with current government interest and the necessary investment, over the coming years, Lagos State is expected to generate revenue from better waste management, which would be beneficial to remote and rural areas of the state overall [21].

The main categories of waste to energy technologies are (1) physical technologies which process waste to produce fuel (i.e., refuse-derived fuel or solid-recovered fuel), (2) thermal technologies which can produce heat, fuel oil, or syngas from both organic and inorganic wastes, and (3) biological technologies whereby bacterial fermentation is used to digest organic wastes to produce useful fuel. These can be further elaborated as gasification (pyrolysis - thermal and plasma-arc types), anaerobic digestion, and combustion. Gasification of waste involves thermochemical conversion reactions, which will induce gases under high temperature and low concentration of pure oxygen or air. Methane is produced and can be applied to run an internal combustion engine to generate electricity. Gasification can reduce waste by 70% to 80% while preserving the land area.

Anaerobic digestion is a biochemical conversion process, which produces fuel for energy through an enclosure called a *digester*. It is currently used in both developed and developing countries to treat both wet and dry biomass resources and to convert MSW and other wastes and residues. It has been successfully used to generate electricity for rural and remote areas of developing countries.

Combustion or incineration consists of burning the whole mass of waste in an incinerator. Incineration of MSW can drastically reduce the volume of MSW by up to 80% to 90%. The process releases gaseous pollutants such that a pollution control system needs to be set up along with it. This technology operates at 800°C to 1,000°C. It is also important to note that it is however not as popular as gasification.

Since WTE would be a relatively new concept in Nigeria, the populace would need effective public awareness and sensitization programs and campaigns. The government, for its part, must establish partnership with the organized private sector entities for the needed investments [20-24]. With the exception of recent progress on waste issues in Lagos State, issues related to sound MSW management such as recycling programs, waste reduction, and waste disposal have not been adequately addressed in Nigeria. Investigation of the potential contribution of such underutilized fuels like MSW to the energy mix in a sustainable way while minimizing GHG emissions currently produced from landfilling, open dumps, and open burning of solid wastes is a worthwhile undertaking in a developing country like Nigeria.

Methods

Materials and methodology for the estimation of the potential

We begin this work by estimating the calorific value in MSW for selected municipalities in Nigeria based on available data from the literature. The calorific value is derived from the MSW weight distribution (wt.% for as-received MSW on a wet basis) that was obtained from [25-31]. The MSW was categorized as food waste or vegetables as well as putrescibles, paper, plastics, metals, textile or rubber, and inert or miscellaneous. The calorific values in the MSW collected for Lagos, Port Harcourt, Abuja, Ibadan, Kano, Makurdi, and Nsukka are estimated from their respective MSW data. We consider three ways of handling MSW which have been mentioned earlier. Solid waste is not recycled or composted and is hence disposed to the MSW collection system. It is assumed that plastics, metals, and glass will be recycled with a high degree of success and will not be introduced into a landfill, an incinerator, or an anaerobic digester. From the known moisture and ash content, the dry percentage weight (wt.%) for each category of the MSW was determined.

Calorific value by thermochemical conversion or direct combustion

The calorific value has been determined using a universal correlation for the process where MSW is incinerated. Two processes - using grate furnace and fluidized furnace steam generators - were considered to convert chemical energy to thermal energy (steam). Conventional Rankine cycle and combined heat and power (CHP) plant were considered to convert chemical energy to electrical energy. The thermal energy (steam generation) or electrical energy generated was calculated using assumed conversion efficiencies for the processes. In the first instance, calorific value was determined for the following categories: vegetables and putrescible or food waste, paper, textiles, and plastics. Glass and metals with zero or negligible free carbon content were not included in the calculation of the calorific value. It is also assumed that these materials will eventually be recycled and will not enter the MSW stream meant for recovering its calorific value through some process. The inputs that are necessary to determine the calorific value, that is, the high heating value (HHV; in MJ/kg dry MSW), are the mass of MSW collected, weight percentage on a dry and ash-free basis, ash content from a proximate analysis, and the ultimate analysis.

Two correlations were used, both of which depend on the knowledge of ash content determined by a proximate analysis and the elemental compositions of carbon, hydrogen, nitrogen, oxygen, sulfur, phosphorous, and chlorine obtained from an ultimate analysis. The unified correlations used were presented by Channiwala and Parikh [32] and has been found to estimate the calorific value (HHV) of a variety of MSW with acceptable accuracy. Another correlation by Grabosky and Bain, used for comparison, has been found to have 1.5% accuracy in the predicted calorific value for biomass [32]. The MSW collected in most cities is predominated by biomass contribution. The calculated HHV was then converted to low heating value (LHV in MJ/kg dry MSW) using Equation 1:

$$\text{LHV (MJ/kg)} = \text{HHV (MJ/kg)} - (9 \times \%\text{H} + \%\text{H}_2\text{O}) \times 2.44\ \text{MJ/kg}, \tag{1}$$

where %H and $\%H_2O$ are the weight percentages of atomic hydrogen and water, respectively. HHV and LHV per unit mass were converted to calorific value potential per month using the following equation:

$$\text{LHV or HHV (MJ/month)} = \text{LHV or HHV (MJ/kg dry mass)} \times \text{dry mass (kg/month)}. \tag{2}$$

Landfill methane generation

The second scenario that was considered is where the organic wastes (food and yard waste, wood, and textiles) [25-31] are sent to (a landfill) and undergo a biodegradation process. In order to calculate methane generation potential (L_O) from the landfill, decomposable organic carbon (%DOC) on a dry mass basis was determined for the categories of MSW mentioned earlier. Furthermore, the calculation assumes that only putrescible and other organic wastes are discarded in the landfill. The net quantity of the methane generation potential from the landfill (combusted in an internal combustion engine) may be lower due to some of the methane not collected (i.e., flared or emitted as GHG). The %DOC was calculated as follows:

$$\%\text{DOC} = (\%C_{\text{textile}} \times \%W_{\text{textile}}) + (\%C_{\text{gardenwaste}} \times \%W_{\text{gardenwaste}}) + (\%C_{\text{foodwaste}} \times \%W_{\text{foodwaste}}) + (\%C_{\text{wood}} \times \%W_{\text{wood}}), \tag{3}$$

where %C is the carbon percentage on a dry mass basis obtained from an ultimate analysis of the specific category of MSW. The calculated %DOC is converted to a fraction and is used in the equation below. The quantity %W is the weight percent of the category of MSW that is considered. Mass of methane potentially generated in a month (M_{methane} (Gg/month)) is calculated as follows [30]:

$$\text{M}_{\text{methane}}(\text{Gg/month}) = \text{MSW}_T \times \text{MSW}_F \times \text{MCF} \times \text{DOC} \times \text{DOCF} \times F \times \left(\frac{16}{12} - R\right) \times 1 - O_x, \tag{4}$$

where MSW_T is the total MSW generated and collected (Gg/month), MSW_F is the fraction of MSW that is collected and disposed in the landfill (taken to be 0.74), MCF is the methane correction factor (taken to be 0.4 for a shallow unmanaged landfill), DOC_F is the fraction of DOC converted to landfill gas by decomposition, F is the fraction of methane (by volume) in the landfill gas, set at a default value of 0.5, and R is the amount of methane recovered and is set to zero. This enables the calculation of the maximum methane generation potential, i.e., the amount that will be potentially emitted as GHG, O_x is the oxidation factor, taken as zero, which implies that the biodegradation takes place strictly anaerobically, and the ratio (16:12) is the mass ratio of 1 g mol methane for each gram mole of carbon. The value of DOC_F was determined as follows:

$$DOC_F = (0.014 \times T(^\circ C)) + 0.28 \tag{5}$$

An average temperature T of 28°C was assumed for the municipalities considered. It should be pointed out that this calculation determines the methane generation potential only. The fraction out of the generated total that is not recovered is what is lost to the atmosphere as a GHG emission. The calculated mass of methane potentially generated ($M_{methane}$) per month can be converted to methane generation potential, L_O:

$$L_o(m^3/Gg\ MSW) = \frac{(M_{methane}(kg))/0.717(kg/m^3)}{MSW_T(Gg/month)} \tag{6}$$

or reported as gravimetric ratio ($Gg_{methane}/Gg_{MSW}$).

These calculations do not determine the time dependence of methane generation potential and the effect on the methane generation potential as more MSW is added to an operating landfill over its lifetime. However, the calculated L_O has been used to determine the methane generated over the lifetime of a landfill and is presented in the next section that discusses the economics of energy generation using the landfill methane that is generated.

Economic analysis of landfill methane generation

It is assumed that the organic part of the MSW will be sorted and that 74% of MSW_F collected will be discarded in the landfill [33]. We consider a situation where the landfill already exists or existing ones will be improved and that they are shallow and poorly managed with a MCF of 0.4. The electrical energy generated by combustion of methane was calculated using a LHV of methane of 37.2 MJ/Nm^3.

The economic analysis assumes the following: (1) The landfill will be operational in 2013. (2) The landfill will collect MSW at the same level as mentioned for 30 years at the end of which it will be shut down; after the shutdown, the landfill will no longer accept any MSW. (3) However, the landfill will continue to produce methane. The landfill will be monitored, and with the inherent methane collected for the next 30 years, (4) only 50% (by volume) of the methane that is produced is captured by the recovery system. Remainder is emitted and constitutes a greenhouse gas emission, (5) the capital investment is assumed to be financed for 30 years at the rate of 9%, (6) the operating costs are not financed, (7) methane will be combusted in an internal combustion engine that is coupled to an electric generator with a thermal efficiency of 0.33 [34] and (8) the plant is operational for 7,488 h/year, and (9) tipping fees collected from the municipality or other revenue streams are not considered. In this sense, the costs calculated below for the cost of electricity is the minimum pricing at which electricity can be sold to the consumer or to the grid; (10) for the direct gas use scenario, the monetary value for greenhouse gas offset is set at US$0 per tonne of carbon equivalent. In other words, no credit is allowed for the offset of greenhouse gas emissions. Capital investment and operating costs only include investment on the landfill and gas handling. The costs associated with the internal combustion engine plant are excluded, and (11) 'direct gas' is assumed to be purified and compressed to 344,737.5 Pa (or 50 psi) and piped to a customer 8 km (or five miles) away [35] for direct use. The costs calculated are for a gas that is predominantly methane. The capital investment includes the investment made in developing/improving the landfill and in establishing the ICE power plant. The purpose of the economic analysis is to determine the specific unit price of electrical energy obtainable from methane combustion in an ICE-driven power generation plant.

LandGEM a software developed by the USEPA [36,37] was used to determine the amount of methane generated from 2013 till 2073, for a period of 60 years. The program requires the following for an input and proceeds as described in the Equation 7: (1) amount of MSW added to the landfill each year (M_i), (2) values for maximum methane generation potential L_O (in m^3/Mg MSW) were used from Table 1, (3) fraction of methane in the landfill gas 50%, with the other 50% being predominantly CO_2, (4) a MCF value of 0.4, and (5) reaction rate constant (k) of 0.05/year. This can be represented as

$$V_{CH_4}(m^3/year) = \sum_{i=1}^{n}\sum_{j=0.1}^{1} kL_o\left(\frac{M_i}{10}\right)e^{-kt_{ij}}. \tag{7}$$

V_{CH4} is the volume of methane generated (m^3/year) in the year of calculation, i is a 1-year time increment, n is year of calculation, that is, initial year of waste acceptance. The value of n changes from 0 to 30 based on the assumption that the landfill is operational for 30 years, k

Table 1 Methane generation potential in selected municipalities and the corresponding degradable organic carbon

Location	%DOC	L_O (m^3/Mg MSW)
Lagos	32.36	63.09
Port Harcourt	17.84	22.29
Abuja	26.26	48.56
Ibadan	32.24	59.62
Kano	23.75	35.62
Makurdi	25.32	41.14
Nsukka	27.32	50.54

is the reaction rate constant, j is a 0.1-year time increment, L_o is the potential methane generation capacity (m^3/Mg), M_i is the mass of waste accepted in the ith year (Mg), and t_{ij} is the age of jth section of waste mass M_i accepted in the ith year.

Anaerobic digestion of MSW

The anaerobic digestion (AD) process has been around for a long time and has found varied success [38]. There is a renewed effort to improve this process and recognize its full potential to handle the biodegradable organic matter (yard, kitchen, and agricultural waste) and combustible organic fraction (paper, wood) in a municipal solid waste. There is a variety of anaerobic digestion processes, but the general reaction block diagram of the process consists of the following:

1. Waste sorting (to separate plastics, glass, and metals) and shredding
2. Feeding system and retention in an anaerobic bioreactor in the presence of microorganisms that promote acidogenesis and methanogenesis [39]. (This reaction occurs under well-controlled temperature, pH, elemental ratio, and feed dilution with water.)
3. Biogas removal from bioreactor and purification (removing carbon dioxide and ammonia from the product methane stream)
4. Methane gas handling for the distribution or feeding system into a single- or dual-fuel engine plant

The degradation of organic matter in a municipal solid waste in an anaerobic-controlled environment of a digester serves as a responsible way to handle municipal solid waste [38]. In order to calculate the methane generation potential, the elemental compositions of C, H, N, and O that are available from an ultimate analysis of various organic fractions of an MSW were used as suggested by Parkin and Owen [40]. The authors suggest that for an organic compound with the composition $C_nH_aO_bN_c$, the amount of methane generated by anaerobic digestion, where water is used as a reactant, proceeds as per the reaction:

$$C_nH_aO_bN_c + \left[n-\frac{a}{4}-\frac{b}{2}+3\frac{c}{4}\right]H_2O\left[\frac{n}{2}-\frac{a}{8}+\frac{b}{4}+3\frac{c}{8}\right]CO_2\left[\frac{n}{2}+\frac{a}{8}-\frac{b}{4}-3\frac{c}{8}\right]CH_4 + cNH_3. \quad (8)$$

Anaerobic digestion tends to be a preferred method, over landfilling, for handling the organic content of municipal solid waste. The organic content consists of vegetative matter, food waste, and agricultural wastes. The advantage of anaerobic digestion over landfilling is its high decomposition potential and production of biogas, which is predominantly methane and compost as a final product [40]. The methane generation potential is calculated using the mass fraction (on a dry mass MSW basis) of elemental C, H, O, and N obtained from the ultimate analysis for each of the following categories indicated by index I: vegetables and putrescibles, paper, and textiles. The mass fraction is converted to gram moles of the elements n_C, n_H, n_O, and n_N from which the maximum potential for methane generation is calculated as

$$\frac{(n_{CH_4})_i}{\text{kg MSW}} = \left(\frac{n_C}{2}+\frac{n_H}{2}+\frac{n_O}{2}+\frac{n_N}{2}\right)_i. \quad (9)$$

The total number of moles of methane generated is the sum of the number of moles of methane generated from each of the categories of waste that is considered and represented as

$$n_{CH_4}(\text{g moles } CH_4/\text{kg MSW}) = \sum_i (n_{CH_4})_i. \quad (10)$$

Equation 10 is then converted to the actual mass or volume of the methane generated for a given dry mass or as-received mass of municipal solid waste.

Economic analysis of methane generation in an AD process

Among the several AD processes, the Valorga process is the one that is widely used and, as of May 2010, had a combined installed methane production capacity of 263 million cubic meters [39]. Hence, the economic analysis was based on the Valorga design. Typical biogas generation capacity for this plant is in the range of 80 to 160 Nm^3/Mg MSW and typical methane fraction in the biogas is 0.6. *While performing the economic analysis, an average methane production of 72 m^3/Mg MSW was used.* This number is much lower than the calculated potential presented in Table 2.

The economic analysis assumes the following:

- Methane constitutes about 60% (by volume) of the biogas produced.

Table 2 Calculated methane production potential (m^3/Mg MSW) via an AD process for various locations in Nigeria

Location	MSW generated (Mg/month)	VGF (wet basis) (Mg/month)	Calculated potential (m^3 CH_4/Mg MSW)
Lagos	3.71×10^5	3.04×10^5	296
Port Harcourt	1.81×10^5	9.07×10^4	168
Abuja	2.78×10^4	1.94×10^4	257
Ibadan	1.35×10^5	1.16×10^5	313
Kano	1.57×10^5	9.41×10^4	231
Makurdi	2.42×10^4	1.62×10^4	247
Nsukka	1.19×10^4	8.68×10^3	266

VGF: Vegetables, Garden and Fruit waste.

- An average biogas production of 120 Nm^3/Mg MSW and specifically an average methane production of 72 m^3/Mg MSW. Using a calorific value for methane of 37.2 MJ/Nm^3, this translates to a specific energy of 707 MJ/Mg MSW.
- Methane will be combusted in an internal combustion engine that is coupled to an electric generator with a thermal efficiency (η_e) of 0.26. This efficiency is lower than the value of 0.33 assumed for a methane landfill plant as the AD process uses some of the gas and electricity produced for its operation.
- The capital investment is assumed to be financed for 20 years at the rate of 9%.
- The operating costs are not financed.
- The plant is operational for 7,488 h/year or for 24 h/day for 26 days/month.
- Tipping fees collected from the municipality or other revenue streams are not considered. In this sense the costs calculated below for the cost of electricity or the cost of methane is the minimum pricing at which electricity or gas can be sold to the consumer.

The capital costs were then estimated using a power law based on capacity ratios as given below:

$$\text{AD plant}/C_{\text{cpl}} = C_{\text{cpl}_{\text{reference}}}\left(\frac{M_{\text{MSW}}\,(\text{tonnes/year})}{\text{Capacity}_{\text{reference}}}\right)^{0.6}, \tag{11}$$

$$\text{ICE power plant}/C_{\text{cpp}} = C_{\text{cpp}_{\text{reference}}}\left(\frac{\text{Nominal power generation capacity}}{\text{Power generation capacity}_{\text{reference}}}\right)^{0.6} \tag{12}$$

The reference ICE power plant was identified for cost comparison based on the nominal power generation capacity expected for the AD plants being considered. The capital costs and operating costs for the reference AD and the power plant are given in Table 3. The nominal generation capacity was carried out as shown in Equation 13:

$$\text{Plant capacity}\,(\text{W},\ \text{MW}_e) = \frac{120\ (\text{Nm}^3\ \text{biogas/Mg MSW}) \times M_{\text{MSW}}(\text{Mg/month}) \times Q_{\text{CH}_4}(\text{MJ/m}^3) \times \eta e \times 0.6\ \text{Nm}^3\ \text{methane/Nm}^3\ \text{biogas}}{26\ \text{days/month} \times 24\ \text{h/day} \times 3{,}600\ \text{s/h}} \tag{13}$$

where Q_{CH4} is the calorific value of methane and taken to be 37.2 MJ/m^3 and η_e is the thermal-to-electrical efficiency of methane gas-powered ICE plant and taken to be 0.26.

The typical capital cost (US\$/kW_e) ranges from US \$1,100 to US\$1,300 for a co-firing biogas power plant to US\$3,000 to US\$4,000 for a gasification and CHP plant. The reference capital costs used here (US\$1,200/kW_e) compares to the capital cost of a co-firing plant.

The unit cost of electricity is calculated as follows:

$$C_{5/\text{kWh}_e} = \frac{\left[(\text{CO}_{\text{AD}} + \text{CC}_{\text{pp}}) - \text{PMT}\left(\frac{9.9\%}{12},\ 30\ \times\ 12, \text{CC}_{\text{AD}} + \text{CC}_{\text{PP}}\right)\right]}{E(\text{kWh}_e/\text{year})}, \tag{14}$$

where CO_{AD} and CC_{pp} are respectively the annual operating cost for the AD and power plant. CC_{AD} and CC_{pp} are the respective capital cost for the AD and power plant. E is the electrical energy generated per year using the methane separated from the biogas and is calculated using Equation 15:

$$E_{\text{kWh}_e/\text{year}} = \frac{M_{\text{MSW}}(\text{Mg/year}) \times 120\ \text{m}^3\ \text{biogas/Mg MSW} \times 0.6\ \text{m}^3\ \text{CH}_4/\text{m}^3\ \text{biogas} \times 37.2\ \text{MJ/m}^3\ \text{CH}_4 \times \eta_e}{3.6\ \text{MJ/kWh}_e}, \tag{15}$$

where η_e (0.26) is the thermal-to-electrical energy efficiency. The unit cost of gas is calculated as

$$C_{5/\text{m}^3\,\text{CH}_4} = \frac{\left[\text{CO}_{\text{AD}} - \text{PMT}\left(\frac{9\%}{12},\ 20\ \times\ 12,\ \text{CC}_{\text{AD}}\right)\right]}{V(\text{m}^3\ \text{CH}_4/\text{year})} \tag{16}$$

$$V(\text{m}^3\ \text{CH}_4/\text{year}) = M_{\text{MSW}}(\text{Mg/year}) \times 120\ \text{m}^3\ \text{biogas/Mg MSW} \times 0.6\ \text{m}^3\ \text{CH}_4/\text{m}^3\ \text{biogas}. \tag{17}$$

Table 3 Plant capacity, capital, and annual operating costs for reference AD and ICE power plants

Cost		Value
Reference AD plant (Valorga at Tillburg)	Capacity capital cost ($C_{cpl_{reference}}$)	52,000 VGF tonnes/year
		US$17,500,000
	Annual operating cost	US$800,000/year (labor)
		US$ 800,000/year (supply/maintenance)
Reference ICE power plant (generating capacity 5 MW_e)	Capital cost ($C_{cpp_{reference}}$)	US$6,000,000
	Operating cost	US$657,000

Incineration and combined heat and power

The 'fuel' or waste can be incinerated in a grate furnace or a fluidized furnace. Detailed engineering analysis is required to determine which furnace design is the optimal choice. The calorific value of the fuel is converted to electrical energy or thermal energy or both in a CHP plant. The reference plant used is shown in Table 4. The electrical (E_e) or thermal energy (E_{th}) generated by each of the energy conversion mechanisms was derived from the calorific value as follows:

$$E_{th}\, or\, E_e(\mathrm{MJ/Mg\ MSW}) = \eta(\mathrm{LHV(MJ/Mg\ MSW)}), \tag{18}$$

where $\eta = E_e/\mathrm{LHV}$ for a thermal power plant and also $\eta = E_{th}/\mathrm{LHV}$ for a steam generation boiler (E_{th} is ≥ the enthalpy change between the liquid water and steam at a specified temperature and pressure).

For a CHP plant, two efficiencies may be defined - electrical efficiency and thermal efficiency. Also relevant to determining the total energy generation is the steam-to-power ratio. For the CHP process, the total energy generated ($E_{th} + E_e$) per megagram MSW is presented below. The preliminary calculations assume the use of four different options:

- Grate furnace boiler/steam generator – Thermal efficiency = 0.4.
- Fluidized bed steam generator – Thermal efficiency = 0.85.
- Rankine cycle-based thermal power plant with fluidized bed furnace – Overall efficiency = 0.35.
- The CHP plant electrical efficiency is taken as 0.27, and thermal efficiency is taken as 0.64, while for the overall efficiency, we use 0.88 with a power to heat ratio of 0.42 [41].

Economic analysis of MSW incineration

The economic analysis of MSW incineration proceeds in two different ways:

1. Given the selling price of energy (heat or electrical power), what does it cost (capital investment C_{cp} and operating costs C_o) the company to generate energy at that cost? In this case, the energy sold is revenue. This approach assumes that electrical energy is sold at US$35/MWh and thermal energy (steam) is sold at US$15/MWh.
2. Given the costs of capital investment and operating costs, what should be the price of energy to the consumer, i.e., determine the unit cost estimate?

Baseline capital costs and operating costs

Baseline costs were determined from an article by the ASME committee [42]. This article evaluates the cost of incinerating wastes with or without polymer polyvinylchloride and across incineration plants that are categorized as small (200 metric tons (MT)/day), medium (800 MT/day), and large (2250 MT/day). The small- and medium-sized plants had two incinerator lines each, while the large plant had three incinerator lines. The plants are equipped with air pollution control devices (APCD) and continuous emission monitoring systems (CEM) and use dry lime injectors, fabric filters, and/or spray drying for air pollution control. The small plants were designed to produce steam, while the medium- and large-sized plants generate electricity. The capital investment costs and operating costs for the reference plants were averaged for each size classification, and the average costs were used in the calculations. The capital cost includes the cost of the incinerator plant and cost of APCD and CEM facilities. The plant is assumed to operate for 7,488 h/year. The annual operating cost includes both direct and indirect operating costs. The amount of MSW collected per month was compared to the plant size in the baseline that was the closest. The actual fixed capital cost (C_{cp}) for the plants were estimated by applying the power factor formula to the plant capacity ratio as shown [43]:

$$C_{cp} = C_{cp\ \mathrm{reference}}\left(\frac{(M_{MSW}(\mathrm{tonnes/year}))_{\mathrm{actual}}}{(M_{MSW}(\mathrm{tonnes/year}))_{\mathrm{reference}}}\right)^{0.7}, \tag{19}$$

where M_{MSW} is the mass of MSW collected annually and $C_{cp\ \mathrm{reference}}$ is the known capital investment for the reference plant. The operating cost was assumed

Table 4 Parameters of the reference MSW incineration plant (capacity, capital investment, and operating costs)

Reference plant size (tonnes/day)	Capital investment ($Ccp_{reference}$) (US$)	Annual operating cost (reference) (US$)	Operating cost/capital investment
62,500	21.3 million	6.76 million	0.32
703,000	212.7 million	41.4 million	0.19

to be the same fraction of the capital cost for both the reference plant and the actual plant. The reference capital investment and operating costs were assumed to be the same for a plant of a given MSW handling capacity for a thermal power plant, a fluidized or grate boiler steam generator, and a CHP plant. It was assumed that the cost of capital included the yearly repayment of the borrowed capital investment and the interest on the borrowing (at 9% per annum for 30 years). The PMT function in Microsoft Excel was used to calculate the yearly payments (US$/year) on the borrowed capital. The PMT function returns a number with a negative sign implying that it is a cost for the plant operator or company. If the funds to meet the operating costs are not borrowed, then there is no cost involved in acquiring the operating capital. When the generated heat (E_{th} in MWh/year) and electric power (E_e in MWh/year) are sold at US$15/MWh and US$35/MWh, respectively, the net annual cost factor (C_F in US$/Mg MSW) may be calculated as

$$C_F = \frac{C_O\text{-PMT}\left(\frac{9\%}{12},\ 30 \times 12,\ C_{cp}\right) - \left[(E_{th} \times \text{US15/MWh}) + (E_e \times \text{US35/MWh})\right]}{\text{Mg MSW}}. \tag{20}$$

It should be noted that the tipping fee collected by the incinerator facility is not included in the revenue stream. Hence, the cost factor is positive in most instances, except for a CHP plant in Lagos. A positive cost factor implies that the expense exceeds the revenue. The positive cost factor thus indicates the additional revenue that must be generated through tipping fee and subsidies that must be secured. Alternately, one may determine the specific unit price of energy (C in US$/kWh) as shown in Equation 21:

$$C(\text{US/kWh}) = \frac{\left(C_O\text{-PMT}\left(\frac{9\%}{12},\ 30\times12,\ C_{cp}\right)\right)}{M_{MSW}(\text{Mg/year}) \times E(\text{kWh/Mg})}, \tag{21}$$

where E is the annual energy generation per metric ton MSW. The specific energy costs are in US$/kWh$_e$ for a thermal plant and in US$/kWh$_{th}$ for fluidized and grate furnace plants. The CHP plant generates both steam and electricity, with a power-to-heat ratio of 0.42, and the specific energy cost is the average of both thermal and electrical energies.

Landfill gas and ICE power plant scaling: the reference data

In order to estimate the capital cost of a landfill (C_{cpl}) and the relevant operating costs, reference cost information was established based on two USEPA reports [35,44]. The capital cost for establishing a landfill is scaled as per a power law based on the capacity of the landfill over its lifetime (known as *the amount of waste in place* (WIP)). The landfill used as a reference consisted of gas handling and flare systems and has a total WIP of one million metric tons [35]. As per USEPA classification, these landfills range from very large (>15 million metric tons (MMT)) to medium (>5 MMT) in size. The landfill that is used as a reference is located in Munster, IN, USA. The LandGEM program was executed to determine the volume of methane generated per year. Using that information, the average electrical energy produced (in MW) over the 60-year period was calculated. This provided an estimate of the nominal power generation capacity of the plant to be installed onsite. The capital investment cost for the actual plant (C_{cpl} and C_{cpp}) was then scaled based on the power generation capacity as per a power law relation as given Equations 22 and 23.

$$\text{Landfill}/C_{cpl} = C_{cpl,\text{reference}} \times \left(\frac{M_{MSW}(\text{tonnes/year}) \times 12\ \text{months/year} \times 30\,\text{years}}{\text{Capacity}_{\text{reference}}}\right)^{0.6}, \tag{22}$$

$$\text{Power plant}/C_{cpp} = C_{ccp,\text{reference}} \times \left(\frac{\text{Nominal power generation capacity}}{\text{Power generation capacity}_{\text{reference}}}\right)^{0.6}. \tag{23}$$

The annual operating cost for the landfill is scaled as a power of 0.6 based on the mass of MSW handled per year. The annual operating cost of the ICE power plant is assumed to be a fixed percentage (9.9%) of the capital cost of the ICE power plant. This assumption is justified when the estimated costs are presented below. The continuous power generation capacity is determined based on methane generated for a given year. The methane production rate is assumed to be constant and that only 50% of the methane that is generated is captured and utilized (the recovery factor of 0.5 in Equation 24). The plant is assumed to operate for 7,488 h/year.

$$\text{Power generation, year}\,i\,(W_i, \text{MW}_e) = \frac{V_{CH_4}(\text{m}^3/\text{year}) \times Q_{CH_4}(\text{MJ/m}^3) \times \eta_e \times 0.5}{7,488\ \text{h/year} \times 360\ \text{s/h}}, \tag{24}$$

$$\text{Nominal power generation}\,(\text{MW}_e) = \frac{\Sigma_{i=1}^{60} W_i}{60}. \tag{25}$$

The nominal size of the plant is the average of the power generation capacity averaged over the 60 operating years of the plant. The capital cost is determined based on this nominal plant capacity:

$$E_{avg}\ (\mathrm{GWh}_e/\mathrm{year}) = \left[\left(\frac{\frac{\Sigma_i^{60} (V_{CH_4})_i (\mathrm{m}^3/\mathrm{year}) \times 0.5 \times Q_{CH_4}\ (\mathrm{MJ}/\mathrm{m}^3) \times \eta_e}{60}}{1{,}000\ \mathrm{MJ/GJ} \times 3{,}600\ \mathrm{GJ/GWh}} \right) \right]. \tag{26}$$

The unit-specific cost of electricity generated (C_e) after suitable unit conversions is calculated from Equation 27:

$$\mathrm{Cost}\ (\mathrm{s/kWh}_e) = \frac{\left[(C_{ol} + C_{op}) - \mathrm{PMT}\left(\frac{9\%}{12},\ 30 \times 12,\ C_{cpl} + C_{cpp}\right) \right]}{E_{avg}\ (\mathrm{kWh}_e/\mathrm{year}),} \tag{27}$$

where C_{ol} is the operating cost of the landfill and C_{op} is the operating cost of the power plant. E_{avg} is the average electrical energy generated over a 60-year period. The direct use, the landfill methane, *as a fuel for combustion in household or commercial applications is considered as well.* The capital cost includes the costs associated with the development and operation of the landfill, collecting, flaring, compressing, handling, purification, and piping of the landfill gas. The ICE power plant costs are excluded as there is no power generation on site.

Table 5 gives the pertinent information for the reference landfill capital and operating costs as well as the costs for direct gas plant. The capital cost of the landfill and direct gas plant are scaled as a power of 0.6 of the ratio of the WIP amount, as specified in Equation 13. The annual operating cost for the landfill is scaled as a power of 0.6 based on the mass of MSW handled per year [42]. The landfill that is used as a reference is located in Munster, IN, USA. Equation 7 gives the annual methane production rate which is averaged over 60 years to give $\bar{V}_{CH4}$. The unit-specific cost of methane recovered (C_{gas}) for use is calculated as

$$C_{gas}\ (\mathrm{US/MJ}) = \frac{\left[\left(C_{ol} + C_{o,gas}\right) - \mathrm{PMT}\left(\frac{9\%}{12},\ 30 \times 12,\ C_{cpl} + C_{op,gas}\right) \right]}{\frac{\left[\bar{V}_{CH_4} \times 0.5\ \left(\mathrm{m}^3/\mathrm{year}\right)\right]}{Q_{CH_4}\ \left(\mathrm{MJ}/\mathrm{m}^3\right)}} \tag{28}$$

Table 6 gives the capital and operating costs for the reference landfill and ICE power plant.

Waste composition of the selected municipalities and assumptions

The calorific value of the MSW collected in each city was calculated for vegetable and yard waste, paper, and textiles only. It was assumed that plastics, bottles, and metals will be captured by a recycle system and excluded from incineration.

Lagos

The MSW collected in Lagos has significant food waste and putrescible content (68 w/o on an as-received basis). Papers make the next major category. The moisture and ash contents in the MSW are not explicitly known. Hence, typical values were used for the moisture content taken from [45]. The ash content used was obtained [28] as determined for MSW collected in Port Harcourt.

Port Harcourt

The MSW data and the proximate and ultimate analysis data for Port Harcourt are very comprehensive as reported by [28]. Published ash and moisture contents were used to arrive at the ash- and moisture-free elemental ultimate analysis. These values were used in the correlations to arrive at the calorific value for MSW. It is interesting to note that the moisture content is very high in the food waste and wood waste categories at 65.2 and 19.2 w/o. The mass distribution of food waste is low (29.2 w/o); on the other hand, the combined mass

Table 5 Capital and operating cost information for a reference landfill and direct gas plants

Cost		Value
Reference landfill	Capacity (WIP)	1 million MT
	Capital cost ($C_{cpl_{reference}}$)	US$700,000
	Mass MSW received per year	8,000 metric tonnes/year
	Annual operating cost (C_{ol})	US$196,647/year (collection)
		US$765,761/year (sort and dispose)
Reference direct gas plant	Capacity (WIP)	1 million MT
	Capital cost ($C_{cpgas_{reference}}$)	US$1,639,000
	Annual operating cost (C_{ogas})	US$136,000

Table 6 Capital and operating cost information for a reference landfill and ICE power plants

Cost		Value
Reference landfill	Capacity	1 million metric tonnes
	Capital cost ($C_{cpl_{reference}}$)	US$700,000
	Mass MSW received per year	8,000 metric tonnes/year
	Annual operating cost	US$196,647/year (collection)
		US$765,761/year (sort and dispose)
Reference ICE power plants	Generation capacity	5 MW_e
		3 MW_e
		2 MW_e
		1.5 MW_e
		1 MW_e
	Capital cost ($C_{cpp_{reference}}$)	US$6,000,000
		US$3,957,000
		US$2,517,000
		US$1,927,000
		US$1,322,000
	Operating cost	US$657,000
		US$394,000
		US$263,000
		US$197,000
		US$131,000

distribution from plastics, glass, and metal waste is significant at nearly 41 w/o.

Abuja

The MSW distribution in Abuja is dominated by food wastes that contribute 55 w/o. The ash content and ultimate analysis were obtained from a report prepared by the Center for People and Environment prepared for USEPA [46]. This report particularly considered the landfill methane generation potential in Ibadan and Abuja, Nigeria. The moisture content was assumed to be the same as in Lagos [43]. The ash content used was obtained from [28], determined for MSW collected in Port Harcourt.

Ibadan

The MSW distribution in Abuja is dominated by food wastes that contribute 64.9 w/o. The ash content and ultimate analysis were obtained from a report prepared by the Center for People and Environment prepared for USEPA [46]. The moisture content was assumed to be the same as in Lagos [45]. The ash content used was obtained from [28], determined for MSW collected in Port Harcourt. The total mass of the MSW collected per month was not available; hence, representative values published in the literature were used for Ibadan [29].

Kano

MSW collected in Kano has a low contribution from food wastes (38 w/o). Papers contributed as high as 15 w/o, and glass and plastics contributed a combined 22 w/o; these are materials that can potentially be recycled or their usage reduced. The moisture content was assumed to be the same as in Lagos [45]. The ash content used was obtained from [28], determined for MSW collected in Pt. Harcourt.

Makurdi

Makurdi has a significant amount of vegetable waste (52 w/o). The moisture content was assumed to be the same as in Lagos [45]. The ash content used was obtained from [28], determined for MSW collected in Port Harcourt.

Nsukka

Municipal waste collected in Nsukka has a significant amount of vegetable and paper waste. It also contains significant quantity of metals and plastics that can be recycled and not introduced into the MSW stream.

Results and discussion

Landfill methane generation

The methane generation potential was calculated as described earlier. These calculations determine the methane generation potential from the landfill. The results are summarized in Table 1.

The calculated methane generation potential (in m^3 methane/Mg MSW) is lower than the typical range of about 50 to 100 kg methane per tonne MSW (70 to 140 m^3 methane per Mg MSW) for a landfill methane generation estimated by the International Energy Agency (IEA) [47]. The difference could be attributed to the fact that the IEA estimate is based on landfill methane plants operated in developed countries where the degradable carbon content may be higher. Furthermore, the landfill WIP at the end of 30 years for the various cities in Nigeria ranges from approximately 133 million metric tonnes (MMT) for Lagos to 4.3 MMT for Nsukka.

Table 7 gives the summary of the monthly MSW collection and the cost (not the net cost) per tonne of MSW handled. This is the cost incurred at the landfill for collection and sorting operations and does not include the cost associated with electricity generation or direct gas use. The costs of electricity and direct landfill methane are also presented. The daily MSW collection rate ranges from 12,350 to 400 tonnes/day for the selected municipalities. These numbers are derived from the data that was provided. USEPA guidelines for full cost accounting data state that in this range of daily MSW collection rates, the net cost per ton of MSW handled ranges from less than about $36/ton (at 1,500 tons/day) to about $85/ton (at 400 tons/day). The numbers estimated reveal the same

Table 7 Specific cost of electricity and direct landfill methane for the selected municipalities

Location	Waste collection (tonnes/month)	Operating cost of landfill (US$/tonne)	Total CH_4 recovered in 60 years (m^3)	Nominal power (MW_e)	Price of electricity (US$/$KWh_e$)	Price of gas (US$/GJ CH_4)
Lagos	370,556	9.60	2.88×10^9	21.6	0.29	27.60
Port Harcourt	181,450	13.40	3.33×10^8	2.5	1.56	152.00
Abuja	27,795	22.00	1.56×10^8	1.2	1.08	108.00
Ibadan	135,391	14.40	1.86×10^9	13.9	0.25	23.40
Kano	156,676	13.50	5.53×10^8	4.1	0.88	84.60
Makurdi	24,242	28.60	1.06×10^8	0.8	1.47	144.00
Nsukka	12,000	37.90	6.99×10^7	0.5	1.47	145.00

trend but are slightly lower because the cost of capital to establish the landfill is not included. The calculations also assume that the MSW in a place like Lagos goes to a single appropriately sized large landfill. The EPA estimates consider a landfill with 1,500 tonnes/day for the highest daily capacity. The landfill rule was promulgated in the USA as a part of the Clean Air Act of 1996. According to this rule, future landfills, with the size larger than 2.5 MMT and volume greater than 2.5 million m^3, will have to install features that would either collect or flare the landfill gas. The USEPA report has determined the economics of electricity generated from direct gas use of the landfill methane [37]. For landfills that are not subject to the landfill rule and whose range in size is from 50,000 to 11,000,000 MT, the break-even gas price was calculated. For these landfills, electricity generation was not considered feasible. The break-even gas price is the price of gas at which the project results in US$0 NPV (net present value) at the end of the 15-year life of the project. For the landfill project to be viable, the WIP must be greater than or equal to the break-even WIP. The results are summarized in Table 8.

Comparing the estimated price of electricity for Nigerian towns to the USEPA projections, the price of landfill gas electricity for Nigeria is nearly a factor of 5 higher at best. The market price of natural gas is about US$2.74 per one million British thermal units (MMBtu), and the average industrial gas methane price is about US$3.42 per MMBtu. The cost of coal mine gas is also in the same range, about US$2.53 per MMBtu. For a landfill generating electricity to be viable, the break-even WIP is 2.9 MMT and the price of electricity would be US$0.04/$kWh_e$. For a landfill that produces gas for direct use, the price of gas is about US$2.74 per MMBtu (US$2.59/GJ) for a break-even landfill with a WIP of about 1 MMT. Both these projections assume US$0/TCE, i.e., no extra value for the offset of greenhouse gas emission, and do show incremental emission reductions. This renders the cost of landfill direct gas scenario to nearly 17 to 20 times higher for the municipalities considered. A significant reason for the higher estimate for Nigeria is because the calculations assume a minimally managed landfill with gas collection in place but only capturing 50% of the generated methane. This same factor contributes to the higher estimate for the price of gas. In addition, for a given landfill, the capital investments and operating costs appear to be much higher for a direct gas plant versus an ICE power plant. The extra cost comes from the cost of compression that is about 344,737.5 Pa and piping gas to the consumer from the plant. It must be emphasized that this is an estimate that does not take into account any revenue-generating stream such as tipping fee or garbage collection fee.

Table 8 Break-even landfill direct gas price and landfill WIP

Size of landfill WIP (break-even WIP)	Break-even price (US$/GJ CH_4)
50,000 MT	52.00
500,000 MT	6.38
1,000,000 MT	2.04
11,000,000 MT	1.28

Anaerobic digestion

The methane is combusted in an ICE power plant or can be sold as a direct gas at higher pressures transported through a pipe network. The volume of methane generated by anaerobic digestion for each location that was considered is given in Table 2.

It may be noted that the cost of methane generated through anaerobic digestion is much cheaper than landfill methane. The price of the low pressure gas compares very closely with the average cost of natural gas (US$2.59/GJ) as presented in Table 9. When the product gas has to be sold as direct gas to the consumer, it may have to be purified, compressed, and transported through pipes to the consumer. Thus, a factor of 15% has been added to the price of low-pressure product gas to arrive at the price of direct gas. The percentage increase in price is the same as what was estimated to deliver landfill methane for direct gas use. The price of the direct gas obtained through anaerobic digestion is very much competitive with the price of natural gas. In the scenario where methane is generated using an AD process, the methane

Table 9 Specific cost estimate for electricity and gas produced by AD of MSW for various locations in Nigeria

Location	Nominal capacity of power plant (MW_e)	Cost of electricity (US\$/kWh_e)	Cost of low-pressure gas (US\$/GJ CH_4)	Cost of direct gas (US\$/GJ CH_4)
Lagos	117	0.0209	0.89	1.27
Port Harcourt	57	0.0239	0.8	1.25
Abuja	9	0.0560	2.28	3.24
Ibadan	43	0.0317	1.37	1.94
Kano	49	0.0268	1.04	1.48
Makurdi	8	0.0584	2.35	3.35
Nsukka	4	0.0792	3.28	4.66

is subsequently combusted in an ICE power plant or is sold as direct gas at higher pressures, transported through a pipe network. The price of methane is competitive with the market price of natural gas at about US\$2.74 per MMBtu, and the average industrial methane gas price is about US\$3.42 per MMBtu for most of the selected municipalities except perhaps Nsukka. In addition and as an added benefit, the AD process produces compost that has use in agricultural practices.

Incineration or combustion

Table 10 shows the potential for energy generation (E_e in MJ_e) or thermal (E_{th} in MJ_t) by incinerating the *organic part* of the MSW. It is true that at the end of incinerating the MSW or combusting the methane gas, CO_2 will be produced. On the one hand, it is less potent than methane, and it is hoped that local environmental regulations will regulate the CO_2 emissions as well.

The World Bank has determined that for the viability of a WTE project, a minimum LHV of MSW of 6,000 MJ/Mg throughout the season and an annual average of 7,000 MJ/Mg are required [48]. As the calorific value estimates are revealed earlier in Table 3, incineration appears to be *technically* viable in the selected municipalities, except in Port Harcourt. It is clear that the cost factor for incinerating a ton of waste in a grate furnace steam generator is high. In [48], it was also noted that the net cost factor for treating a tonne (Mg) of MSW in an incinerator ranges between US\$10 and US\$100/Mg with an average cost factor of about US\$50/Mg. The cost factor for a CHP-WTE plant is at or below the average cost factor. It may be noted that a fluidized bed steam generator and a thermal power plant have similar cost factors. The specific cost of electrical power generated using CHP compares well with the price of electrical power at US\$0.14/kWh_e. The price of steam generated using a fluidized steam generator also compares favorably with the average price of steam at about US\$0.015/kWh_t as shown in Table 11. The International Energy Agency has reviewed the economic and environmental benefits of a CHP technology based on the installed and projected capacity in G8+5 countries. The report calls for increased adaptation of the CHP technology and projects about 19% reduction in CO_2 emissions in 2030 in comparison with projected emissions from fossil fuel-based power generation without a combined heat cycle [49]. The method and approach presented in this work are very much comparable to that reported in [46] and thus provide credence to the accuracies of the present results.

Implications and recommendations

The effectiveness of the proposed technologies of this study can be described as follows:

1. Propensity to reduce landfill dumping and large amounts of land area designated as dump sites. Such

Table 10 Energy generation potential by incinerating MSW in Nigeria

Location	Calorific value (LHV) (MJ/Mg)	Electrical or thermal energy generated			
		Thermal power plant (MJ_e/Mg MSW)	Fluidized bed steam generator (MJ_t/Mg MSW)	Grate incinerator boiler (MJ_t/Mg MSW)	CHP (MJ/Mg MSW)
Lagos	1.18×10^4	4,141	9,937	4,732	10,766
Port Harcourt	6.76×10^3	2,367	5,680	2,705	6,153
Abuja	1.04×10^4	3,650	8,761	4,172	9,491
Ibadan	1.27×10^4	4,436	10,647	5,070	11,534
Kano	9.49×10^3	3,320	7,968	3,794	8,632
Makurdi	1.00×10^4	3,514	8,433	4,016	9,136
Nsukka	1.08×10^4	3,787	9,088	4,328	9,845

Table 11 Cost factor and specific unit price of energy generated by various processes using MSW as fuel

Location	Thermal plant		Fluidized steam generator		Grate furnace generator		CHP	
	CF	*C*	CF	*C*	CF	*C*	CF	*C*
	US$/Mg	US$/kWh	US$/Mg	US$/kWh	US$/Mg	US$/kWh	US$/Mg	US$/kWh
Lagos	10.4	0.044	9.3	0.018	31.0	0.039	−11.9	0.017
Port Harcourt	39.8	0.095	39.1	0.040	51.5	0.084	27.0	0.037
Abuja	49.9	0.084	48.9	0.035	68.0	0.074	30.0	0.032
Ibadan	25.4	0.056	24.2	0.023	47.4	0.049	1.50	0.021
Kano	33.3	0.071	32.4	0.030	49.8	0.062	15.4	0.027
Makurdi	54.8	0.091	53.8	0.038	72.2	0.080	35.8	0.035
Nsukka	73.0	0.104	72.0	0.044	91.8	0.091	52.6	0.040

lands can be used for other purposes while incineration of waste can produce a much needed energy and reduce the volume of waste by as much as 90%.

2. Propensity to reduce dependence on fossil fuels. The advanced technologies of WTE can produce fuel that does not require mining or drilling for non-renewable fossil fuel resources.
3. Capacity to offset GHG emissions and pollution caused by burning fossil fuels, and
4. Production of clean energy since WTE technologies have significantly advanced as a function of policy measures such as the Kyoto Protocol [16].

Furthermore, the technologies analyzed in this study have been well demonstrated and employed in varying capacities [17]. Electricity generation from MSW is highly feasible in Nigeria considering the population factor and the subsequent large amounts of waste generated which are not effectively managed. Another rationale for exploring underutilized fuels like MSW is for minimizing GHG emissions currently produced by open dumping. WTE is a reasonable and sustainable MSW management strategy in terms of weight and volume reduction coupled with energy recovery, and it is recommended that this should be vigorously explored.

Regulatory and policy measures to promote WTE technologies and market penetration should be established by federal, state, and local governments. Such measures should have clear strategies for implementation and enforcement. Likewise, the weak and almost non-existent financial instruments need to be strengthened and broadened to attract the necessary private sector investments for the adoptions of MSW management via WTE technologies.

Future studies should consider (1) feasibility studies of other municipalities in Nigeria; (2) life cycle assessment of waste disposal options; (3) emissions aspects, carbon offsets, and fossil fuel displaced; and (4) policy and social aspects.

Conclusions

WTE in Nigeria can be developed as an important integrated waste management strategy while simultaneously producing energy, can displace fossil fuels and can reduce pollutant emissions. The operational reliability of WTE systems will be improved over time as more systems are commissioned and operated at a commercial level. WTE should be recognized as part of a future, balanced mix of energy technologies and policies. The location of WTE plants in order to consolidate MSW from various locations to achieve economies of scale will be important. In addition, waste composition is highly unlikely to remain stable, and changes in waste policies and population habits can contribute to waste composition changes which would also affect the performance of WTE plants. Waste having high calorific value will result in more energy recovery. As more societies in the world increasingly adopt the reduce, reuse and recycle strategies, the amount of MSW collected would reduce significantly. While the authors in [50-53] have equally estimated in one capacity or the other the potential for energy generation from other renewable energy sources such as biomass, corn cob, and animal waste for Nigeria, they however did not consider the economics of the potential. Technical and economic assessments of three waste treatment methodologies have been reported in this study. The calorific value and methane generation potential for MSW generated in various towns in Nigeria has been determined. From the estimates of the calorific value and the methane generation potential, the technical feasibility of power generation and/or direct (methane) gas availability has been determined. Within the study, it appears that incineration (particularly a CHP) or anaerobic digestion may be more economical. The costs for power produced by incineration and AD compare very well with what it is in developed countries.

This study has not addressed the emission aspects and carbon offset arguments of the three MSW management approaches considered here. A compelling argument is self-evident if we consider what happens to MSW today in developing countries that are also energy hungry. Currently, MSW is discarded in most developing nations in a landfill. These landfills are undeveloped swath of land that pollutes the environment. Methane and CO_2 are produced in the landfill, often without any means for methane capture and recovery. Given this scenario, the direct benefits of offsetting methane emission are compelling. It is true that at the end of incinerating the MSW or combusting the methane gas, CO_2 will be produced. On the one hand, it is less potent than methane, and it is hoped that local environmental regulations will regulate carbon dioxide emissions as well. Within the stated study limitations, the following were found:

- An incinerator plant based on CHP or a fluidized combustion technology is a viable option for Lagos, Ibadan, and Kano.
- The viability of landfill methane generation may not be excluded, but it requires further analysis, specifically determining the capital cost with better specificity and methane recovery benefits of a landfill with a MCF of 0.7 and a methane recovery greater than 50%. Landfills tend to take up vital land resource, and such lands cannot be put to any other use for years. This may act as a reason against adopting landfill as a MSW management strategy.
- Anaerobic digestion certainly appears to be a promising and viable MSW management strategy for Nigeria. The price of gas or electricity produced is competitive with fuel currently used and electricity generated through other fossil fuel means.
- The economic performance of the different waste management options is sensitive to their capital costs and this would depend on plant scale, suppliers and local area logistics. As such, there are some inevitable and inherent uncertainties in the estimated values.

This work serves as a preliminary assessment aimed at identifying which MSW management option would be technically and economically viable. The benefits of energy recovery from waste fuels are such that any nation's waste management policy should embrace energy recovery irrespective of the individual local strategic preferences.

Abbreviation

ICE: Internal combustion engine.

Competing interests

The author(s) declare that they have no competing interests.

Authors' contributions

OMA conceived the problem and drafted the manuscript. RLF supervised and approved the draft. All authors read and approved the final manuscript.

Acknowledgements

The authors wish to acknowledge the anonymous reviewers for their recommendations to improve the paper. We appreciate the numerous individuals who have contributed in one capacity or the other towards the successful completion of this publication.

Author details

[1]Stevens Institute of Technology, Hoboken, NJ 07030, USA. [2]Mechanical Engineering, Obafemi Awolowo University, Ile-Ife, Osun State, Nigeria.

References

1. UN Department of Economic and Social Affairs (UNDESA): Population Division. http://www.un.org/esa/population/unpop.htm (2012). Accessed 1 Dec 2012
2. Babayemi, JO, Dauda, KT: Evaluation of solid waste generation, categories and disposal options in developing countries: a case study of Nigeria. http://www.bioline.org.br/pdf?ja09042 (2009). Accessed 23 June 2013
3. Central Intelligence Agency: Nigeria. https://www.cia.gov/library/publications/the-world-factbook/geos/ni.html (2012). Accessed 25 Nov 2012
4. Adeyinka, MA, et al.: Environmental statistics: situation in Federal Republic of Nigeria. Workshop on Environmental Statistics, Dakar Senegal (2005)
5. Fagbenle, LR: The solid waste problem in Ibadan municipality: a 1980 case study. Polytechnic J. **1**, 56–81 (1982)
6. Couth, R, Toris, C: Carbon emissions reduction strategies in Africa from improved waste management: a review. Waste Manag. **30**, 2336–2346 (2010)
7. Castaldi, MJ, Themelis, NJ: The case for increasing the global capacity for waste to energy (WTE). Waste Biomass Valor. **1**, 91–105 (2010)
8. Pike Research: Global waste to energy market to reach 29.2 billion by 2022. http://www.pikeresearch.com/newsroom/global-waste-to-energy-market-to-reach-29-2-billion-by-2022 (2012). Accessed 25 Nov 2012
9. United States Environmental Protection Agency (USEPA): http://www.epa.gov/cleanenergy/energy-and-you/affect/municipal-sw.html (2012). Accessed: 25 Oct 2012
10. Ciglotti, V: Biomass and waste as sustainable resources. In: McPhail, SJ, Ciglotti, V, Moreno, A (eds.) Fuel Cells in the Waste to Energy Chain: Green Energy Technology, pp. 23–44. Springer-Verlag, London (2012)
11. Kim, MH, Song, HB, Song, Y, Jeong, IT, Kim, JK: Evaluation of food waste disposal options in terms of global warming and energy recovery: Korea. Int. J. Energy Environ. Eng. **4**, 1 (2013)
12. Pandyaswargo, A, Onoda, H, Nagata, K: Energy recovery potential and life cycle impact assessment of municipal solid waste management technologies in Asian countries using ELP model. Int. J. Energy Environ. Eng. **3**, 28 (2012)
13. Eddine, B, Salah, M: Solid waste as renewable source of energy: current and future possibility in Algeria. Int. J. Energy Environ. Eng. **3**, 17 (2012)
14. Environmental and Energy Study Institute: Reconsidering municipal solid waste as a renewable energy feedstock. http://www.seas.columbia.edu/earth/wtert/sofos/eesi_msw_issuebrief_072109.pdf (2009). Accessed 30 Oct 2012
15. Pytlar, TS: Inclusion of municipal solid waste as a renewable energy source. http://www.swananys.org/pdf/MunicipalSolidWasteEnergyJune09.pdf (2009). Accessed 25 Nov 2012
16. Steiner, CG: Kyoto protocol-compliant waste-to-renewable energy with zero air, water and solids pollution. Bull Energy Efficiency. **5**(3), 12–14 (2004)
17. Confederation of European waste-to-energy plants. http://www.cewep.eu (2012). Accessed 25 Sept 2012
18. Sridhar, MKC, Bammeke, AO, Ademola Omishakin, M: A study on the characteristics of refuse in Ibadan, Nigeria. Waste Manag Res. **3** (1985)
19. Babayemi, JO, Dauda, KT: Evaluation of solid waste generation, categories and disposal options in developing countries: a case study of Nigeria. J Appl Sci Environ Manage. **13**, 3 (2009)
20. Solomon, UU: The state of solid waste management in Nigeria: waste management. http://www.elsevier.com/locate/wasman (2009). Accessed: 5 June 2012
21. Punch News: Lagos to earn income from waste. http://www.punchng.com/business/homes-property/lagos-to-earn-income-from-waste-management/ (2012). Accessed 25 Nov 2012
22. Christian, EI: Potential impacts of climate change on solid waste management in Nigeria. http://www.earthzine.org/2010/10/04/potential-impacts-of-climate-change-on-solid-waste-management-in-nigeria/ (2010). Accessed 15 May 2012

23. Miranda, ML, Hale, B: Waste not, want not: the private and social costs of waste-to-energy production. Energ Pol. **25**(6), 587–600 (1997)
24. Mokhtar, AS: Renewable power generation opportunity from municipal solid waste: a case study of Lagos Metropolis (Nigeria). J Energy Technol Policy. **2**(2), 1–14 (2012)
25. Nagebu, AB: An analysis of municipal solid waste in Kano metropolis, Nigeria. J. Hum. Ecol. **31**(2), 111–119 (2010)
26. Imam, A, Mohammed, B, Wilson, DC, Cheeseman, CR: Solid waste management in Abuja, Nigeria. Waste Mgmt. **28**, 468–472 (2008)
27. Olanrewaju, OO, Ilemobade, AA: Waste to wealth: a case study of the Ondo State Integrated waste recycling and treatment project, Nigeria. European J Social Scie. **8**(1), 7–16 (2009)
28. Igoni, AH, Ayotamuno, MJ, Ogaji, SOT, Probert, SD: Municipal solid-waste in Port Harcourt, Nigeria. Appl Energy. **84**(6), 664–670 (2007)
29. Ogwueleka, TC: Municipal solid waste characteristics and management in Nigeria. Int J Environ Health Sci Eng. **6**(3), 173–180 (2009)
30. Sha'Ato, R, Aboho, SY, Oketunde, FO, Eneji, IS, Unazi, G, Agwa, S: Survey of solid waste generation and composition in a rapidly growing urban area in central Nigeria. Waste Mgmt. **27**, 352–358 (2007)
31. Oresanya, O: Integrated waste management shifting the paradigm. Paper presented at the 47th Annual International Conference of the Nigerian Mining and Geosciences Society (NMGS), pp. 6–11. , Minna (2011)
32. Parikh, PP, Channiwala, SA: A unified correlation for estimating HHV of solid, liquid and gaseous fuels. Fuel. **81**(8), 1051–1063 (2002)
33. Kumar, S, Gaikwad, SA, Shekdar, AV, Kshirsagar, PS, Singh, RN: Estimation method for national methane emission from solid waste landfills. Atmos Environ. **38**, 3481–3487 (2004)
34. USEPA: Landfill Methane Opportunities Program. http://EPA.gov/lmop. Accessed 15 May 2013
35. EPA: US methane emissions 1990–2020: inventories, projections, and opportunities for reductions. EPA 430-R-99-013 report. EPA, Washington, DC (1999)
36. Alexander, A, Burklin, C, Singleton, A: Landfill gas emissions model (LandGEM) version 3.02 user's guide. Office of Research and Development, USEPA, Philadelphia (2005)
37. USEPA: USEPA technology transfer network: clean air technology center. http://www.epa.gov/ttncatc1/products.html (2005). Accessed 20 May 2013
38. Verma, S: Anaerobic digestion of biodegradable organics in municipal solid wastes.. http://www.seas.columbia.edu/earth/vermathesis.pdf (2002). Accessed 18 May 2012
39. Arsova, L: http://www.seas.columbia.edu/earth/research_associates/CVs/arsova.pdf (2010). Accessed 20 May 2012
40. Owen, GF, Parkin, WP: Fundamentals of anaerobic digestion of wastewater sludges. J Environ Eng. **112**(5), 867–920 (1986)
41. Kirjavainen, M, Sipila, K, Alakangas, E, Savola, T, Salomon, M: Small-scale biomass CHP technologies situation in Finland, Denmark and Sweden: OPET report 12. VTT Processes and Finnish District Heating Association, Espoo (2004)
42. ASME: An evaluation of the cost of incinerating wastes containing PVC. http://files.asme.org/Committees/K&C/TCOB/BRTD/EEW/24116.pdf (2012). Accessed 10 June 2012
43. Peters, MS, Timmerhaus, KD: Plant Design and Economics for Chemical Engineers. McGraw-Hill Book Company, New York (1980)
44. EPA: Full cost accounting in action: case studies of six solid waste management agencies. EPA 530-R-95-041 report. EPA, Washington, DC (2007)
45. Salami, L, et al.: Int Jour Appl Sci Tech. **1**(3), 47–52 (2011)
46. Center for People and Environment: Landfill recovery and use in Nigeria. Report prepared for the Methane-to-Markets program of the US Environmental Protection Agency. https://www.globalmethane.org/Data/347_Landfill.Recovery.and.Use.in.Nigeria(LFGE)Final.Report.pdf (2011). Accessed 3 Sept 2013
47. IEA: IEA Energy Technology Essentials. IEA, Paris (2007)
48. World Bank: Municipal solid waste incineration: World Bank technical guidance report. World Bank, Washington, DC (1999)
49. Kerr, T: Combined Heat and Power. International Energy Agency, Paris (2007)
50. Suberu, MY, Mokhtar, AS, Bashir, N: Potential capability of corn cob residue for small power generation in rural Nigeria. ARPN J Eng Appl Sci. **7**, 1037–1046 (2012)
51. Mohammed, YS, Mokhtar, AS, Bashir, N: Renewable power generation opportunity from municipal solid waste in Lagos metropolis (Nigeria). J Energy Tech Policies. **2**, 1–14 (2012)
52. Mohammed, YS, Bashir, N, Mustafa, MW: Biogenic waste methane emissions and methane optimization for bioelectricity in Nigeria. Renew Sustain Energy Rev. **25**, 643–654 (2013)
53. Mohammed, YS, Mustafa, MW, Bashir, N, Mokhtar, AS: Renewable energy resources for distributed power generation in Nigeria: a review of the potential. Renew Sustain Energy Rev. **22**, 257–268 (2013)

CFD based analysis of flow distribution in a coaxial vacuum tube solar collector with laminar flow conditions

Abdul Waheed Badar*, Reiner Buchholz, Yongsheng Lou and Felix Ziegler

Abstract

A computational fluid dynamics (CFD) analysis has been conducted to find the pressure losses for dividing and combining fluid flow through a tee junction of a solar collector manifold. Simulations are performed for a range of flow ratios and Reynolds numbers, and equations are developed for pressure loss coefficients at junctions. A theoretical model based on successive approximations then is employed to estimate the isothermal and non-isothermal flow distribution in laminar range through a collector consisting of 60 vacuum tubes connected in parallel in a reverse (U-configuration) and parallel (Z-configuration) flow arrangement. The results are in reasonable agreement with the available experimental results for U-configuration. The proposed CFD based strategy can be used as a substitute to setting up and performing costly experiments for estimating junction losses.

Keywords: computational fluid dynamics, junction pressure loss coefficient, dividing manifold, combining manifold, flow distribution

Background

In a typical vacuum tube solar collector design, the fluid flow usually is divided in a manifold to risers connected in parallel. After extracting heat from the attached absorber, the fluid passages are combined in a manifold and the hot fluid is led to the user. Heat gain of a solar collector is affected by the flow rate through the individual risers, and uniform distribution is desirable to achieve the same thermal output from all the tubes for best performance, which is not the case in reality [1]. Both the collector's overall flow rate and flow nonuniformity affect the collector efficiency [2-4].

The risers are generally connected at right angles to the manifold, thus forming a tee joint at each connection which leads to local disturbances. A varying local pressure drop for a flow through such a manifold results in an uneven flow distribution through the whole collector. Wang and Yu [5] developed a numerical model to calculate the isothermal flow distribution using mass and momentum equations at each branch of the header system. Wang and Wu [6] proposed a discrete numerical model and demonstrated the flow and temperature distribution in a flat-plate solar collector for U- and Z-configurations by taking into account the effects of buoyancy and longitudinal heat conduction. Jones and Lior [7] developed a discrete hydrodynamic model, and the resulting nonlinear algebraic equations were solved numerically. For negligible buoyancy effects, they investigated the effects of geometrical parameters of the collector such as the ratio of riser to manifold diameter, number of risers, and length of risers on the flow nonuniformity. Fan et al. [4] carried out CFD calculations to estimate the flow and temperature distribution in a flat-plate solar collector consisting of 16 risers connected in U-configuration, having a tilt of 40°. The influences of flow rate, fluid properties, collector tilt, and collector inlet fluid temperature were investigated. Later, for the same collector, Fan and Furbo [8] studied the buoyancy effects on fluid and temperature distribution by means of CFD calculations. They found that at a certain low flow rate, buoyancy effects are of the same order as the pressure drop in the risers or even larger, which results in a more nonuniform flow distribution. The pattern of flow distribution reverses as compared to the case of high flow rate (25 l/min) with negligible buoyancy effects. Glembin

* Correspondence: awbdr@yahoo.com
Institut für Energietechnik, KT 2, FG Maschinen- und Energieanlagentechnik, Technische Universität Berlin, Marchstraße 18, Berlin 10587, Germany

et al. [2] experimentally investigated the flow distribution in a horizontally placed coaxial vacuum tube solar collector consisting of 60 parallel vacuum tubes connected in U-configuration. They adopted a procedure where flow distribution was calculated from the measured tube outlet temperatures for a range of inlet temperatures and mass flow rates.

An alternative way for the determination of flow distribution is to solve it as a piping network where flow is divided amongst the parallel piping, and each flow path leads to the same pressure loss. The riser components connected to the manifolds interrupt the smooth flow of the fluid in the manifolds and cause additional losses. One of the key parameters in evaluating the flow distribution through such a piping network is the pressure loss at the junctions of dividing and combining manifold. At the junctions, the pressure loss coefficient, in general, depends on the geometry of the component and the Reynolds number, just like the friction factor. However, when the Reynolds numbers are very large, it is usually assumed to be independent of the Reynolds number. Based on the experimental data for turbulent flow, these junction loss coefficients can be found in the literature in the form of analytical or empirical expressions characterized by branch angle, area ratio, and flow ratio [9,10], which are not appropriate to use for laminar flow conditions. On the other hand, Weitbrecht et al. [11] performed experiments on a collector with 10 risers connected in Z-configuration and used the experimentally obtained junction loss coefficients to determine the flow distribution analytically with laminar flow conditions, which form the basis for this paper.

The aim of this work is to demonstrate the alternate use of CFD simulations instead of experiments to estimate the pressure loss coefficients at the junctions for a range of locally changing flow ratios and Reynolds numbers, and using those to calculate the flow distribution theoretically. A theoretical approach as adopted by Weitbrecht et al. [11] is used to estimate the isothermal flow distribution in the horizontally placed collector for both U and Z-configurations. Temperature effects on fluid properties are also taken into account for non-isothermal flow distribution, but the buoyancy effects are neglected for a horizontally placed collector. Modeling results then are compared and validated with those predicted experimentally by Glembin et al. [2]. Moreover, the influence of various other parameters is also investigated.

Methods

Collector specification

The collector considered in this study consists of two modules having a total number of 60 vacuum tubes connected in parallel in U-configuration and placed horizontally as shown in Figure 1a. Each vacuum tube contains a flat absorber sheet of 0.1 m^2 area coated with

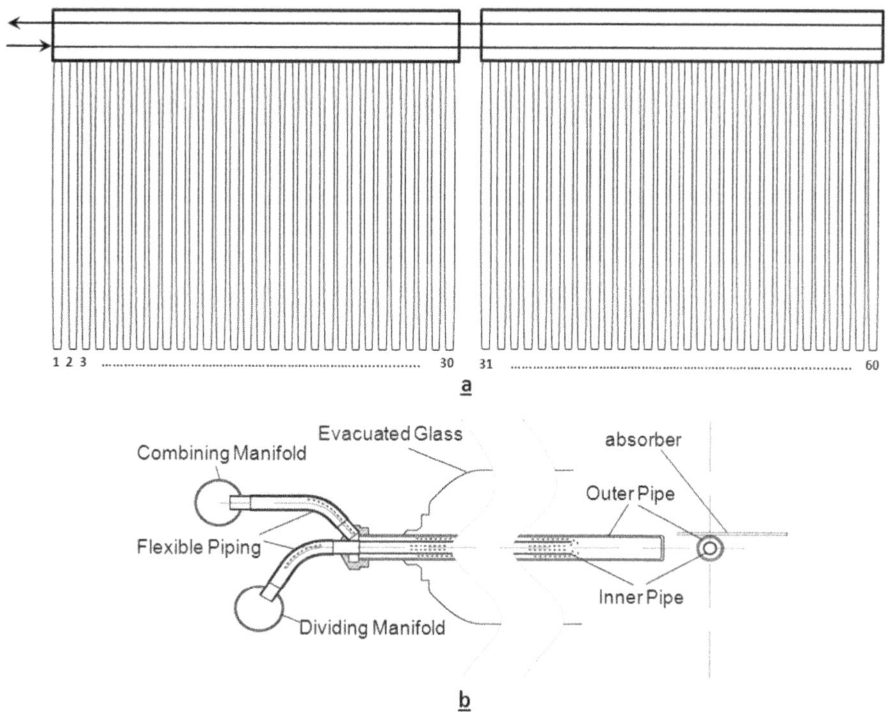

Figure 1 Schematic of the solar collector. (**a**) Collector group of 60 vacuum tubes (**b**) vacuum tube connected to manifold [12].

a selective surface (TiNOX) on a copper substrate and welded to the coaxial copper piping for fluid flow (Figure 1b [12]). A small part of the riser is intruded inside the manifolds, and flexible piping elements are used to connect the coaxial piping to the dividing and combining manifolds as illustrated in Figure 1b [12].

Table 1 summarizes the geometrical parameters of the collector. A mixture of water and propylene glycol [13] is used as a working fluid.

Junction pressure loss coefficients

The main focus of the study is to determine the pressure loss coefficients at the junctions. A part of the manifold attached to the riser forming a tee joint is taken into consideration for the analysis, and the actual flexible corrugated piping elements at inlet and outlet manifolds (Figure 1b) are assumed as straight pipe, as shown in Figure 2. The effect of adjacent branches on each other is not considered.

The loss coefficient k represents the energy dissipation at the junction and is defined as follows:

$$k = \frac{\Delta p}{\rho V_c^2/2}, \tag{1}$$

where Δp is the pressure drop between the upstream and downstream sections of the junction between which the pressure loss under consideration occurs [10]. Therefore, two loss coefficients are defined to characterize a junction loss: one for straight flow (with the index st) and one for side flow (with the index s). In any case, the velocity V_c is the one in the combined flow (see Figure 2).

Based on the idea of Weitbrecht et al. [11], the following relation is used to determine the loss coefficient at the junction:

$$k = \frac{\Delta p_{i-j} - \Delta p_{f,i}}{\rho V_c^2/2}, \tag{2}$$

where Δp_{i-j} represents the pressure loss from CFD simulation and $\Delta p_{f,i}$ is the pressure loss which would occur due to friction only because of the flow from inlet to outlet of the model. It is calculated by standard pressure drop relations. According to Equation 2, equations can be written to represent all the loss coefficients with reference to Figure 2:

$$k_{\mathrm{div,st}} = \frac{\Delta p_{c-\mathrm{st}} - (\Delta p_{f,c} + \Delta p_{f,\mathrm{st}})}{\rho V_c^2/2} \tag{3}$$

$$k_{\mathrm{div},s} = \frac{\Delta p_{c-s} - (\Delta p_{f,c} + \Delta p_{f,s})}{\rho V_c^2/2} \tag{4}$$

$$k_{\mathrm{com,st}} = \frac{\Delta p_{\mathrm{st}-c} - (\Delta p_{f,\mathrm{st}} + \Delta p_{f,c})}{\rho V_c^2/2} \tag{5}$$

$$k_{\mathrm{com},s} = \frac{\Delta p_{s-c} - (\Delta p_{f,s} + \Delta p_{f,c})}{\rho V_c^2/2} \tag{6}$$

while

$$\Delta p_{c-\mathrm{st}} = p_c - p_{\mathrm{st}} \tag{7a}$$

$$\Delta p_{c-s} = p_c - p_s \tag{7b}$$

$$\Delta p_{\mathrm{st}-c} = p_{\mathrm{st}} - p_c \tag{7c}$$

$$\Delta p_{s-c} = p_s - p_c \tag{7d}$$

$$\Delta p_{f,i} = f\frac{L_i}{D_i}\frac{\rho V_i^2}{2} (i = c, s, \mathrm{st}) \tag{8}$$

where $L_c = L_{\mathrm{st}}$, as shown in Figure 2; values of $\Delta p_{c-\mathrm{st}}$, Δp_{c-s}, $\Delta p_{\mathrm{st}-c}$, and $\Delta p_{\mathrm{st}-c}$ are calculated from CFD simulations. The friction factor f can be calculated as follows:

$$f = \frac{64}{Re} \quad For\,\mathrm{Re} < 2,300, \tag{9a}$$

$$\frac{1}{\sqrt{f}} = -1.8 \log\left[\frac{6.9}{\mathrm{Re}} + \left(\frac{\epsilon/D}{3.4}\right)^{1.11}\right]$$
$$\text{For Re} > 2{,}300. \tag{9b}$$

Table 1 Material and geometrical parameters of the solar collector

Description	Specification
Outside diameter of outer riser pipe (mm)	12
Inside diameter of outer riser pipe (mm)	10.4
Length of outer riser pipe (mm)	1,820
Outside diameter of inner riser pipe (mm)	6
Inside diameter of inner riser pipe (mm)	5.3
Length of inner riser pipe (mm)	1,750
Inside diameter of dividing/combining manifold pipe (mm)	22
Spacing between risers (mm)	70
Length of riser intruded inside the manifold pipe (mm)	5

CFD analysis

A 3D meshed model of a manifold connected to a riser is built in GAMBIT (Fluent Inc., Lebanon, NH, USA) as shown in Figure 3b. The mesh influences the accuracy and the simulation time of the solver. Structured meshes have a regular connectivity, which means that each point has the same number of neighbors, while unstructured meshes have irregular connectivity and each point can have a different number of neighbors. The tendency for complex configurations is to use unstructured meshes, which can be automatically generated. However, structured meshes have better accuracy compared with the same amount of unstructured meshes and need less simulation time of solver [14]. Therefore, a structured mesh (hexahedron) is used in this study.

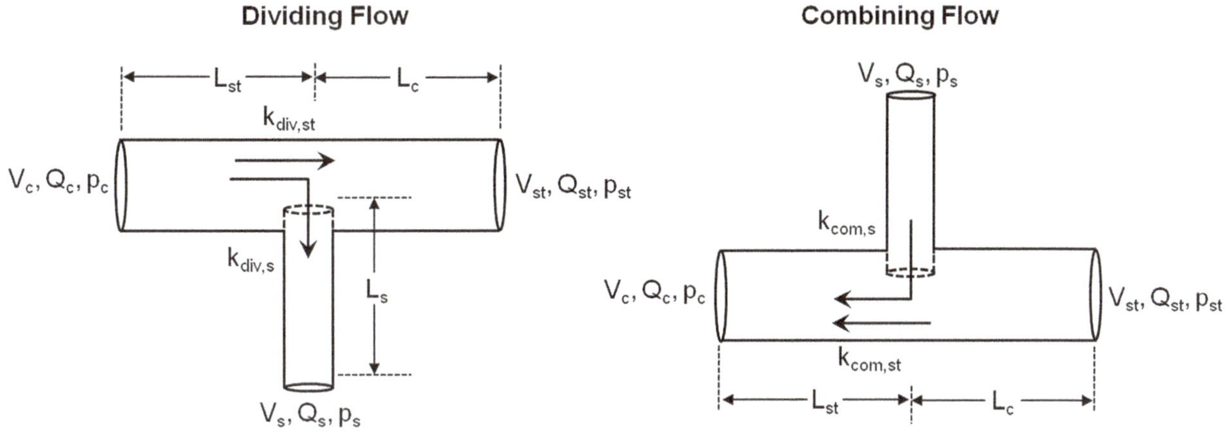

Figure 2 **Loss coefficients at the dividing and combining manifold.**

In order to get the structured mesh, the junction portion of the model is divided into three parts, and each part is meshed separately, as shown in Figure 3a. Initially, part 2 is meshed, then parts 1 and 3 are meshed, and at the end, the remaining straight parts of the manifold are meshed. Figure 3b shows the complete meshed model where a denser mesh is used at the junction than at other parts. The length of the manifold ($L_{st} + L_c$) is 70 mm, and the length of the riser L_s is 30 mm. The effect of mesh density was checked for a mesh density of 7.4E5 and 3.7E6 cells/m^3 which showed a negligible effect on the results. The quality of the mesh was checked based on the criteria of equiangular skew (QEAS) and equisize skew (QEVS). For more than 90% of the volume, the QEAS and QEVS are lower than 0.4, which refers to a good quality mesh.

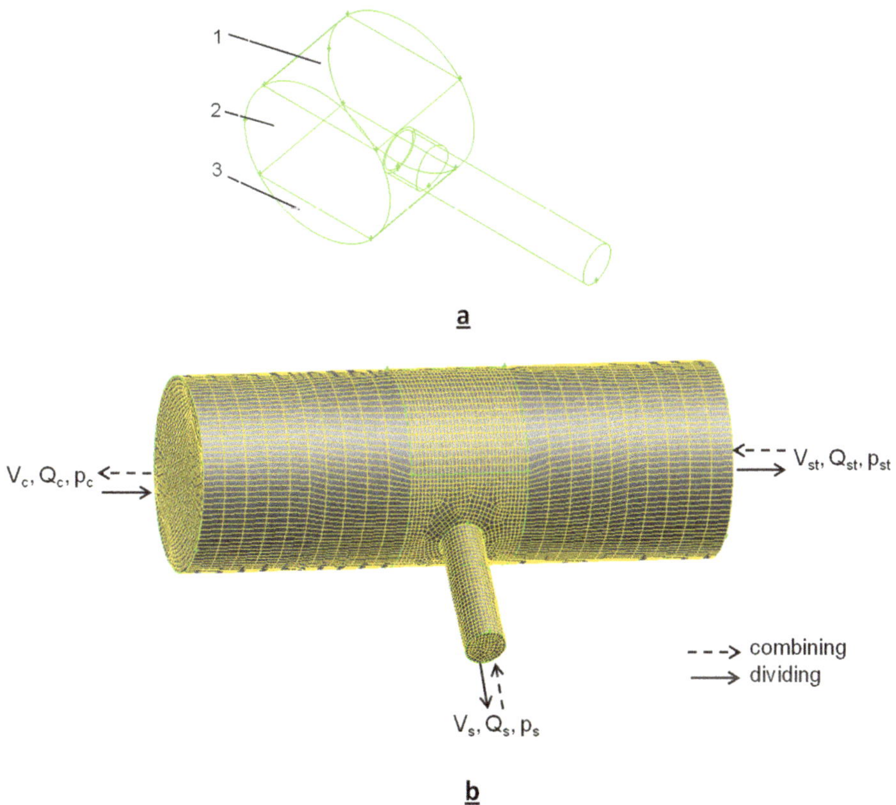

Figure 3 **Schematic meshed model.** (**a**) Portion of junction separated into three parts in order to have a regular mesh. (**b**) Meshed model of the manifold and riser, with higher mesh density at the junction.

The CFD code used was FLUENT 6.3.2 (Fluent Inc., Lebanon, NH, USA). The meshed model in GAMBIT is exported to FLUENT. Computations based on steady flow, pressure based, and implicit formulation are performed using CFD simulations. In this study, three viscous models are selected according to the Reynolds numbers for laminar, transition, and turbulent flows:

1. Re < 2300 laminar model
2. 2,300 ≤ Re ≤ 4,000 *k*-omega standard model (transitional flow)
3. Re > 4,000 *k*-epsilon standard model (enhanced wall treatment).

The following empirical equation is used to define the turbulence intensity:

$$I = 0.16\mathrm{Re}^{-1/8}. \tag{10}$$

Fluid properties (i.e., $\rho = 992$ kg/m^3, $\mu = 0.0010912$ kg/m s) at 80°C are used. For dividing flow with one inlet and two outlets, a uniform velocity inlet boundary condition is used for the inlet, and outflow boundary condition is used for two outlets. Outflow boundary condition is better than the pressure outlet to be used here because the ratio of flow discharge can be fixed in this boundary condition. Hence, it is possible to calculate loss coefficients for a range of riser to main manifold flow ratios.

For combining flow, there are two inlets and one outlet. The velocity inlet boundary condition is used for both inlets corresponding to a certain riser to the main manifold flow ratio. The outlet boundary condition is set to pressure outlet.

For running the program, some more specific information is required. The SIMPLE algorithm is used for pressure-velocity coupling. For faster convergence, the first-order upwind method is opted for discretization of the momentum equation, which is acceptable when the flow is aligned with the grid [15]. The standard discretization scheme is selected for the pressure. Default values for all under-relaxation factors were applied. Convergence criteria selected for the continuity, momentum, and *k*-epsilon equations are less than 5×10^{-5}, 5×10^{-5}, and 1×10^{-6}, respectively.

Hence, for both dividing and combining flow, a number of simulations were performed for a range of Reynolds numbers and riser to manifold flow ratios (Q_s/Q_c) to get the total pressure drop for the straight (manifold) and side (riser) flow. Based on area weighted average, the total pressures at the inlet and outlet faces of the manifolds then are used to calculate the loss coefficients from Equations 3 to 6.

Figure 4 shows the simulation results of the junction loss coefficients for the straight flow in the dividing manifold $k_{\mathrm{div,st}}$. It can be seen that $k_{\mathrm{div,st}}$ shows a variation with Re but negligible dependency on the considered range of flow ratios $Q_{s,i}/Q_{c,i}$. A curve is drawn to represent the trend of the resultant data which are given the form of the following equation:

$$k_{\mathrm{div,st}} = -0.219\ln(\mathrm{Re}) + 2.148 \tag{11}$$

Figure 5 illustrates the variation of the junction loss coefficients for the side flow in the dividing manifold $k_{\mathrm{div},s}$, where it can be seen that $k_{\mathrm{div},s}$ varies both with Re and ratio $Q_{s,i}/Q_{c,i}$. In order to develop an equation for $k_{\mathrm{div},s}$ with respect to Re and $Q_{s,i}/Q_{c,i}$, a procedure is

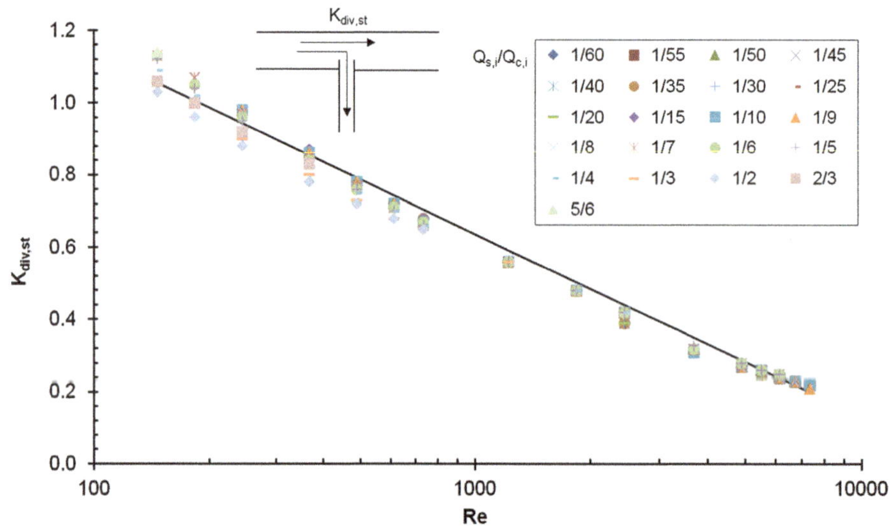

Figure 4 Variation of $k_{\mathrm{div,st}}$ with Re and $Q_{s,i}/Q_{c,i}$ and the estimated trend line.

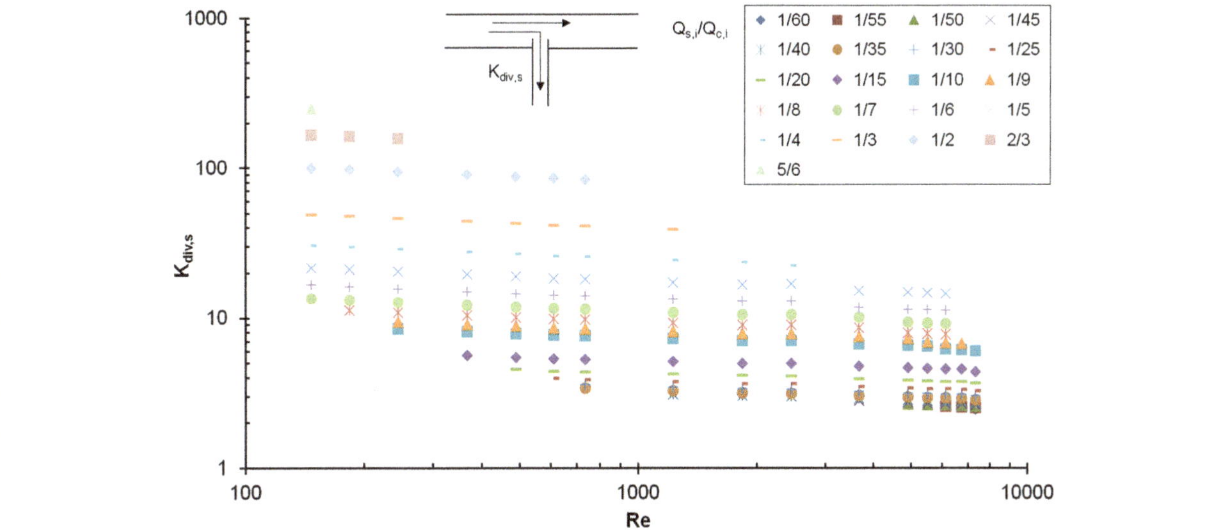

Figure 5 Variation of $k_{div,s}$ with Re and $Q_{s,i}/Q_{c,i}$.

adopted, where in the first step, each data series of $k_{div,s}$ corresponding to $Q_{s,i}/Q_{c,i}$ is fitted in a logarithmic expression in the following form:

$$k_{div,s} = a\ln(\mathrm{Re}) + b, \tag{12}$$

In the second step, coefficients a and b are plotted against the flow ratio $Q_{s,i}/Q_{c,i}$ as shown in Figure 6, which are expressed in the form of following polynomial equations to fit the trend:

$$a = -34.57\left(\frac{Q_{s,i}}{Q_{c,i}}\right) - 1.921\left(\frac{Q_{s,i}}{Q_{c,i}}\right) - 0.12, \tag{13}$$

$$b = 494\left(\frac{Q_{s,i}}{Q_{c,i}}\right)^2 + 40.71\left(\frac{Q_{s,i}}{Q_{c,i}}\right) + 3.08. \tag{14}$$

The combination of Equations 12, 13, and 14 leads to the following equation:

$$k_{div,s} = \left[-34.57\left(\frac{Q_{s,i}}{Q_{c,i}}\right)^2 - 1.921\left(\frac{Q_{s,i}}{Q_{c,i}}\right) - 0.12\right]\ln(\mathrm{Re}) + \left[494\left(\frac{Q_{s,i}}{Q_{c,i}}\right)^2 + 40.71\left(\frac{Q_{s,i}}{Q_{c,i}}\right)^2 + 3.08\right]. \tag{15}$$

Figures 7 and 8 show the junction loss coefficients for the straight and side flow in the combining manifold ($k_{com,st}$ and $k_{com,s}$). Like $k_{div,s}$, the loss coefficients for combining flow ($k_{com,st}$, $k_{com,s}$) also show variation both with Re and $Q_{s,i}/Q_{c,i}$. Notably, the values of $k_{com,s}$ increase significantly with decreasing Reynolds numbers and increasing $Q_{s,i}/Q_{c,i}$. Actually, $k_{com,s}$ becomes negative at high Re and low Q_s/Q_c,

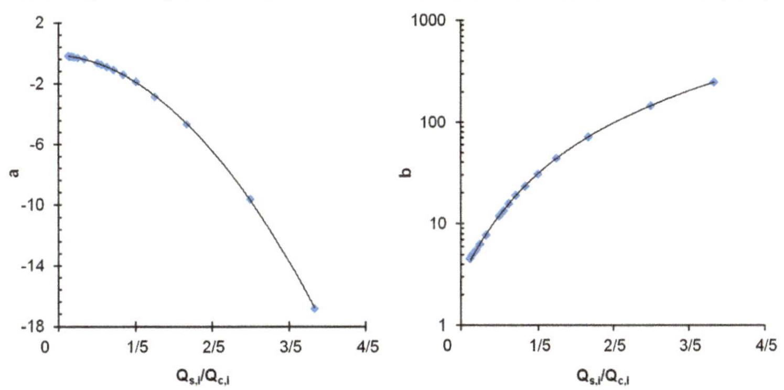

Figure 6 Coefficients a and b versus flow ratio $Q_{s,i}/Q_{c,i}$ and the fitting line.

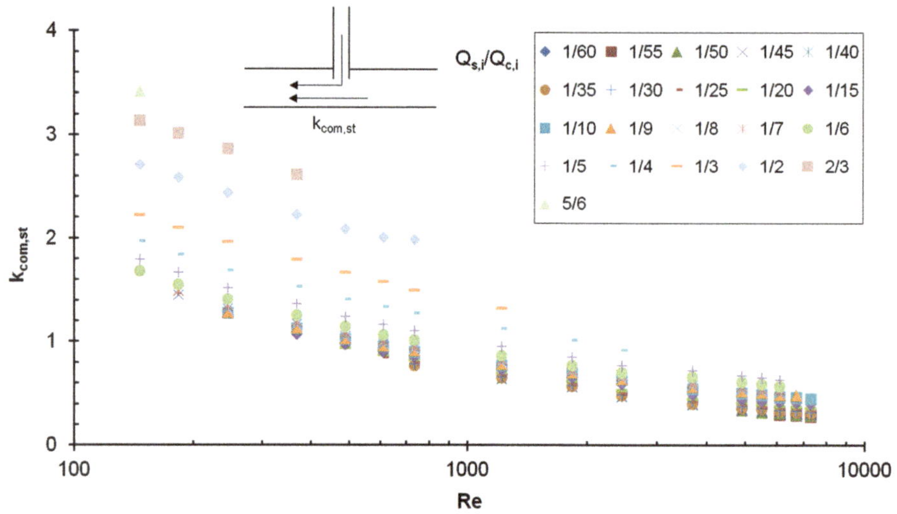

Figure 7 Variation of $k_{com,st}$ with Re and $Q_{s,i}/Q_{c,i}$.

which means there is a pressure increase in the riser at the combining junction. It is because of the fact that the velocity of the fluid in the combining manifold is much higher than the velocity of the fluid coming from the riser, and this may result in negative values of loss coefficients (also stated in the work of Idelchik [9]). By following the same method as before, the following equations are developed for $k_{com,st}$ and $k_{com,sh}$ to fit the data trend:

$$k_{com,st} = 8.919\left(\frac{Q_{s,i}}{Q_{c,i}}\right)^{0.165} \cdot \mathrm{Re}^{\left[0.169\left(\frac{Q_{s,i}}{Q_{c,i}}\right)-0.306\right]}, \quad (16)$$

$$k_{com,s} = \left[-88.64\left(\frac{Q_{s,i}}{Q_{c,i}}\right)^2 + 1.954\left(\frac{Q_{s,i}}{Q_{c,i}}\right) - 0.086\right]\ln(\mathrm{Re}) + \left[908.8\left(\frac{Q_{s,i}}{Q_{c,i}}\right)^2 + 13.381\left(\frac{Q_{s,i}}{Q_{c,i}}\right) - 0.752\right]. \quad (17)$$

The maximum mean deviation of the loss coefficients between the developed equations and the values from simulation is ±6%.

Formulation of the theoretical model

Now after estimating the junction loss coefficients for the prescribed range of turbulent to laminar flow, the flow distribution through the collector consisting of 60 parallel passages can be determined theoretically by solving it as a piping network, connected either in U- or Z-configuration.

U-configuration

Figure 9 illustrates the flow network in the form of junction and frictional resistances through a parallel piping connected in U-configuration. The actual riser consisting of concentric piping is represented by a circular pipe

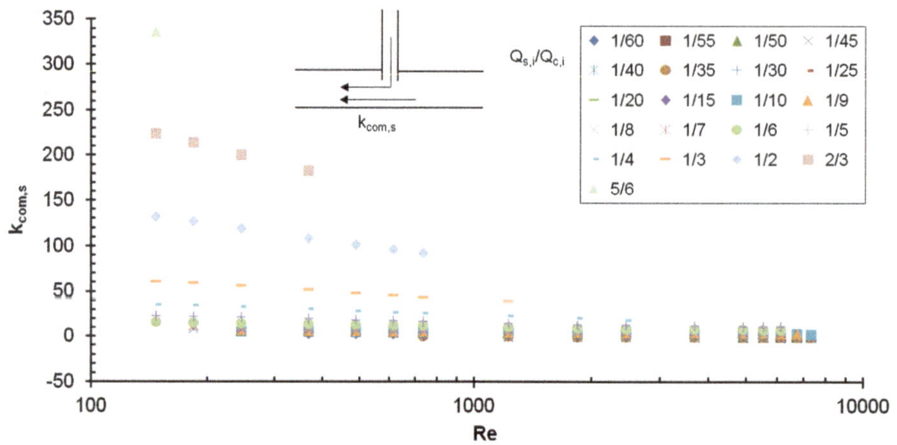

Figure 8 Variation of $k_{com,s}$ with Re and $Q_{s,i}/Q_{c,i}$.

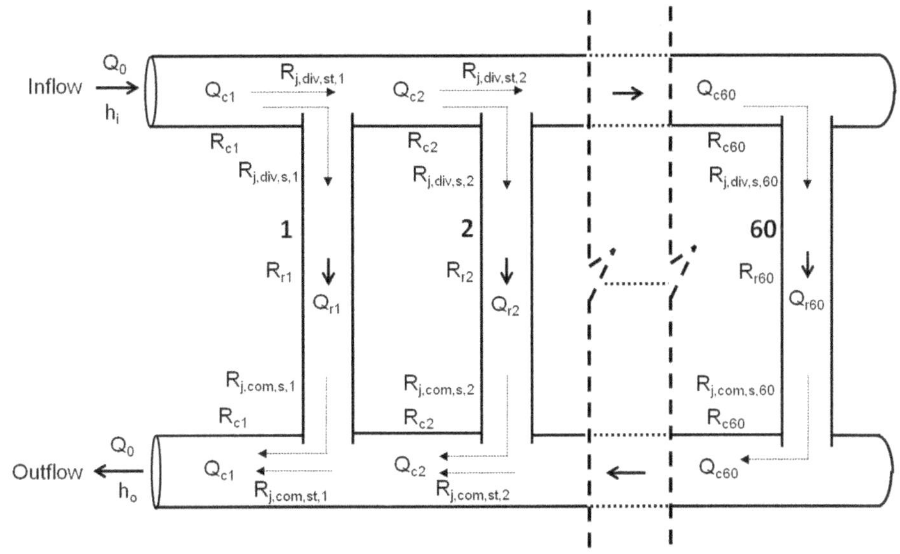

Figure 9 Schematic illustration of the flow scheme and resistances in a U-type configuration.

having equivalent resistance R_r. For each flow route through every riser from inlet to outlet, an energy equation can be written by taking into account the frictional resistance of the straight sections and the junction resistances at the branches.

For instance, it can be seen from Figure 9 that the fluid flow path through riser 1 encounters two frictional resistances through straight sections in the dividing and combining manifolds (R_{c1} and R_{c2}), two junction resistances for the side flow ($R_{j,\text{div},s,1}$ and $R_{j,\text{com},s,1}$), and one frictional resistances through straight section in the riser 1 R_{r1}. Thus, the overall head loss and flow through riser 1 are expressed as follows:

$$h_i - h_o = \left(2R_{c1} + R_{j,\,\text{div},s,1} + R_{j,\text{com},s,1}\right) Q_{c1}{}^2 + R_{r1} Q^{r1r1}{}_{r1}{}^2, \tag{19}$$

$$Q_{r1} = \sqrt{\frac{(h_i - h_o) - \left(2R_{c1} + R_{j,\text{div},s,1} + R_{j,\text{com},s,1}\right)Q_{c1}^2}{R_{r1}}}. \tag{20}$$

Similarly, in general, the flow rate through any other nth riser Q_{rn} (where, $n = 2, 3...60$) can be written as follows:

$$Q_{rn} = \sqrt{\frac{(h_i - h_o) - \sum_{i=1}^{n-1}\left[\left(2R_{ci} + R_{j,\text{div},\text{st},i} + R_{j,\text{com},\text{st},i}\right)Q_{ci}^2\right] - \left(2R_{cn} + R_{j,\text{div},s,n} + R_{j,\text{com},s,n}\right)Q_{cn}^2}{R_{rn}}}, \tag{21}$$

where frictional and junction resistances are as follows:

$$R_{c/r} = \frac{f_{c/r} \cdot l_{c/r}}{2gd_{c/r}A_{c/r}^2}, \tag{22}$$

$$R_{j,\text{div}/\text{com},s/\text{st}} = \frac{k_{\text{div}/\text{com},s/\text{st}}}{2gA_{c/r}^2}. \tag{23}$$

The indices in Equations 22 and 23 such as $R_{c/r}$ indicate the alternative use of the equation for manifold c or riser r. The friction factors f for a circular duct (flow in the first part of the risers) is defined by Equation 9, and for annulus flow in the coaxial piping (flow in the second part - back flow - of the risers), f is given as follows:

$$f = \frac{96}{\text{Re}}. \tag{24}$$

The flow at each non-branch section of the dividing or combining manifold is expressed as follows:

$$Q_{c1} = Q_o, \tag{25a}$$

$$Q_{c(i+1)} = Q_{ci} - Q_{ri}\,(i = 1, 2, \ldots, 59), \tag{25b}$$

$$Q_{c60} = Q_{r60}. \tag{25c}$$

A program is developed to solve Equations 20 to 25 in an iterative manner. For the solution of the model equations, the essential parameters such as the junction loss coefficients in Equation 23 are given by Equations 11, 15, 16, and 17 using the local Reynolds number Re and flow ratio Q_s/Q_c.

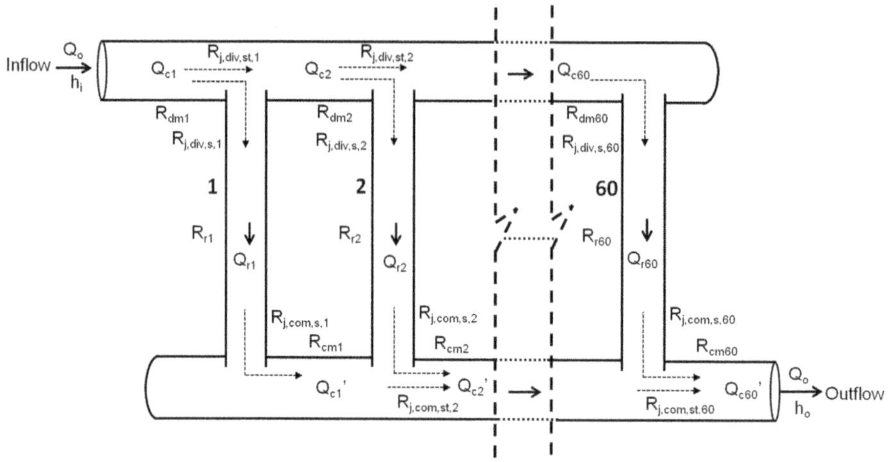

Figure 10 Schematic illustration of the flow scheme and resistances in a Z-type configuration.

Z-configuration

Figure 10 illustrates the flow network in the form of junction and frictional resistances through a parallel piping connected in Z-configuration.

For instance, it can be seen from Figure 10 that the fluid flow path from inlet to outlet through riser 1 encounters frictional resistances in dividing manifold R_{dm1}, riser R_{r1}, and combining manifolds (R_{cm1}, R_{cm2},. . ., R_{cm60}) and resistances due to junctions in dividing and combining manifolds ($R_{j,\mathrm{div},s,1}$, $R_{j,\mathrm{com},s,1}$, $R_{j,\mathrm{com,st},2}$,. . . $R_{j,\mathrm{com,st},60}$). Thus, the overall head loss and flow through riser 1 are expressed as follows:

$$h_i - h_o = \left(R_{\mathrm{dm1}} + R_{j,\mathrm{div},s,1}\right)Q_{c1}^2 + \left(R_{j,\mathrm{com},s,1} + R_{\mathrm{cm1}}\right)Q_{c1}'^2 + R_{r1}Q_{s1}^2 + \sum_{i=2}^{60}\left[\left(R_{\mathrm{cm}i} + R_{j,\mathrm{com,st},i}\right)Q_{c1}'^2\right], \quad (26)$$

The flow at each non-branch section of the dividing or combining manifold is expressed as follows:

$$Q_{c1} = Q_{c60}' = Q_o, \quad (29a)$$

$$Q_{c(i=1)} = Q_{ci} - Q_{ri}\,(i = 1, 2 \ldots, 59), \quad (29b)$$

$$Q_{c(i+1)}' = Q_{ci}' - Q_{ri}\,(i = 1, 2 \ldots, 59). \quad (29c)$$

As for the U-configuration, a program is developed to solve Equations 27 to 29 in an iterative manner for Z-configuration, by employing the same correlations for the junction loss coefficients.

Model validation

The validity of the proposed CFD based model is checked by comparing the model results with the experimentally obtained flow distribution for the same collector by Glembin et al. [2]. Their procedure was based on temperature measurements at the inlet and exit of each collector tube and using the correlation between the mass flow rate and the temperature change of the

$$Q_{r1} = \sqrt{\frac{(h_i - h_o) - \left(R_{dm1} + R_{j,div,s,1}\right)Q_{c1}^2 - \left(R_{j,com,s,1} + R_{cm1}\right)Q_{c1}'^2 - \left(\sum_{i=2}^{60}\left[\left(R_{cmi} + R_{j,st,com,i}\right)Q_{ci}'^2\right]\right)}{R_{s1}}}. \quad (27)$$

In general, the flow rate through any other nth riser Q_{rn} (where $n = 2, 3\ldots, 60$) can be written as follows:

$$Q_{rn} = \sqrt{\frac{(h_i - h_o) - \sum_{i=1}^{n}\left(R_{\mathrm{dm}i}Q_{ci}^2\right) - \sum_{i=1}^{n-1}\left(R_{j,\mathrm{div,st},i}Q_{ci}^2\right) - R_{j,\mathrm{div},s,n}Q_{sn}^2 - \sum_{i=n}^{60}\left(R_{\mathrm{cm},i}Q\prime_{ci}^2\right) - \sum_{i=n}^{60}\left(R_{j,\mathrm{com,st},i}Q\prime_{ci}^2\right) - R_{j,\mathrm{com},s,n}Q\prime_{cn}^2}{R_{rn}}}. \quad (28)$$

fluid while flowing through each collector tube. The experiments were performed for a mass flow range of 200 to 500 kg/h (or $M = 78$ to 31 kg/m^2 h) and an inlet temperature T_{fi} range of 20°C to 80°C. A flow ratio ϵ was calculated, which is used here to compare our modeling results. The flow ratio ϵ is defined as the actual flow Q_{ri} through a riser i to the averaged flow Q_{av} through all risers. It is given as follows:

$$\epsilon = \frac{Q_{ri}}{Q_{av}} = \frac{Q_{ri}}{Q_o/60}. \tag{30}$$

However, these data have not been reported for each experiment but only the average distribution over all experiments.

To account for the temperature effects on fluid properties such as viscosity and density of the fluid, the arithmetic mean of the measured inlet-outlet temperate for each tube (also taken from the work of Glembin et al. [2]) is used to calculate the viscosity and density of the fluid in the respective risers and outlet manifold sections. The estimated junction loss coefficients have shown negligible variation with the temperature.

The details of all the chosen flow rates and inlet temperatures (in the described ranges) of the conducted experiments are not reported in the article of Glembin et al. [2]; therefore, a total of four extreme points of operation corresponding to a flow rate of 78 and 31 kg/m^2 h and inlet temperatures of 20°C and 80°C are selected to compute the flow distribution numerically.

Figure 11 shows the calculated flow distributions for the considered range along with the net averaged flow distributions from the experiment. It can be seen that for a certain inlet temperature ($T_{fi} = 80$°C or 20°C), the overall flow rate has some influence on the flow distribution, i.e., nonuniformity decreases with decreasing flow rate (as shown by Fan et al. [4] and Weitbrecht et al. [11]). However, on the other hand, for a certain mass flow rate ($M = 78$ or 31 kg/m^2 h), the change of inlet temperature from 80 to 20°C shows a larger impact on the calculated flow distribution. With decreasing temperature, viscosity increases and hence the Reynolds number decreases, which results in higher pressure drop for laminar flow through the risers, and this leads to more uniform flow distribution. This variation of Reynolds number in each riser for the above four cases is also shown in Figure 12.

From the simulation, it can be concluded that the use of an average flow distribution as done by Glembin et al. [2] is not really sensible. However, as the averaged experimental curve falls well in between the simulated flow distribution, a reasonable agreement can be stated. Deviations mainly stem from the Reynolds number dependency of the flow distribution. Other reasons for disagreement are the strong sensitivity of the junction loss coefficients at very low Reynolds numbers, geometrical simplification of the actual branch region for the CFD simulation (see Figures 1b and 2), the complexity of the fluid flow in the real conditions, the assumption of fully developed flow in the whole collector, the unknown error margin of the measurements, and the details regarding the tested range of inlet fluid temperatures (between 20°C and 80°C) for the derived experimental flow distribution curve.

Results and discussion

We know that the geometry of the manifold (i.e., simplification of the actual branch region involving flexible

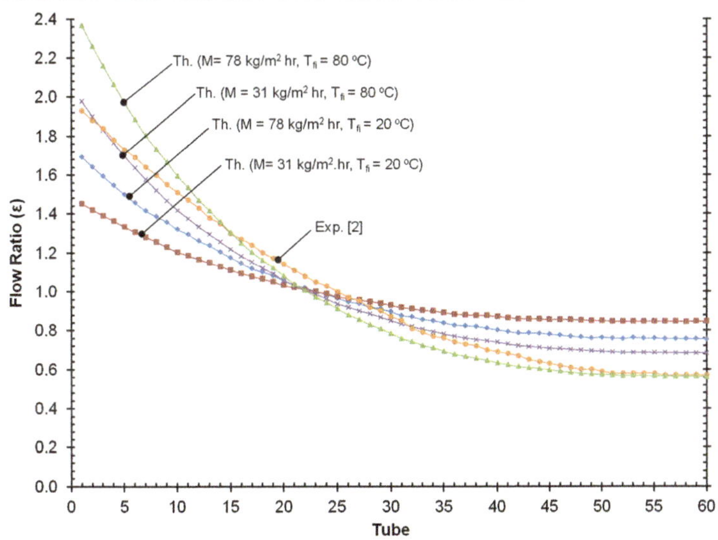

Figure 11 Measured and calculated non-isothermal flow distribution for U-configuration.

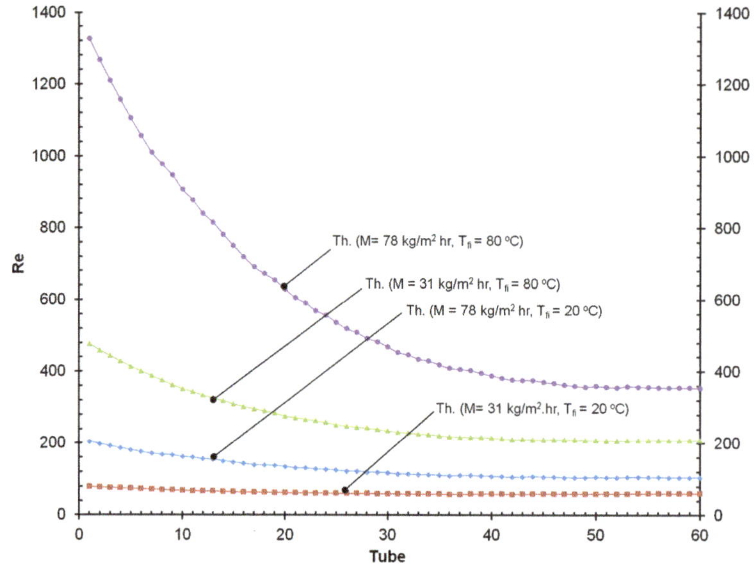

Figure 12 Variation of the Reynolds number in the inner circular pipe of each riser.

corrugated piping, chosen lengths of riser, and manifold as shown in Figures 1, 2, and 3) and boundary conditions which were used for the FLUENT simulation approximates but may not describe the real conditions which prevail inside the manifold at each branch region due to the influence of adjacent branches and the manner in which the flow is changing (decreasing or increasing) in the manifolds. The worse outcome of these assumptions and simplifications could be that estimated values of loss coefficients are too small or too large (if still assuming, the trend of k's as function of Re and flow ratio Q_s/Q_c remains the same as shown in Figures 4, 5, 6, 7, and 8). Therefore, a sensitivity analysis is performed, where the loss coefficients at the junctions are varied over a certain range to see their effect on the estimated flow distribution. For that purpose, isothermal flow distribution for $M = 78$ kg/m^2 h based on junction loss coefficients ($k_{div,st}$, $k_{div,s}$, $k_{com,st}$, and $k_{com,s}$) given by Equations 11, 15, 16, and 17 is used as the reference, and the influence of increasing or decreasing the loss coefficients for straight ($k_{div,st}$ and $k_{com,st}$) and side flow ($k_{div,s}$ and $k_{com,s}$) is examined. It is also useful to determine which loss coefficients have a critical influence on the flow distribution.

Figures 13 and 14 show the resulting flow distribution by varying the junction loss coefficients for the side flow ($k_{div,s}$ and $k_{com,s}$) and straight flow ($k_{div,st}$ and $k_{com,st}$) from the reference case in the range of −40% to +40%, respectively. It can be seen that side flow coefficients

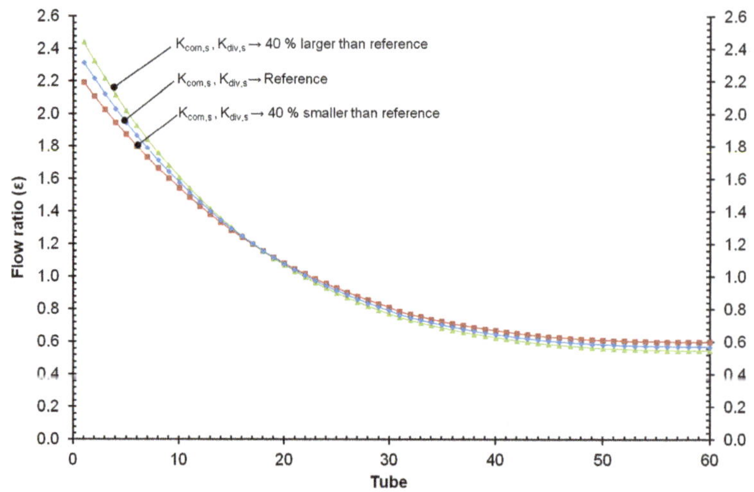

Figure 13 Isothermal flow distribution at $M = 78$ kg/m^2 h for U-configuration by varying $k_{div,s}$ and $k_{com,s}$. The distribution is in the range of −40% to 40% of the reference.

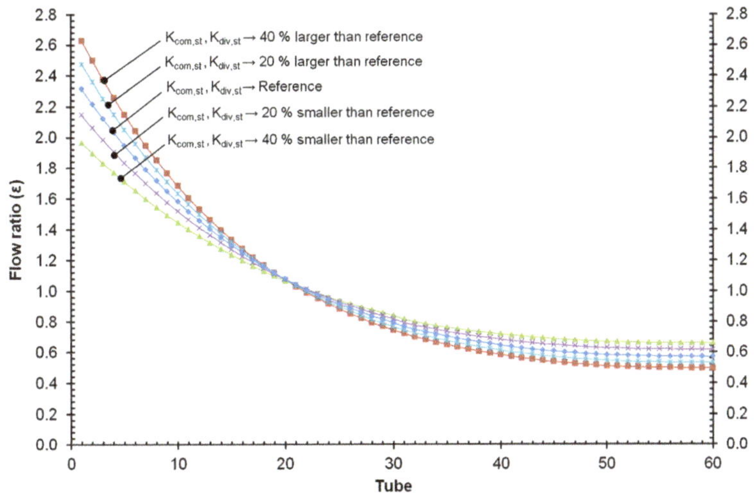

Figure 14 Isothermal flow distribution at $M = 78$ kg/m^2 h for U-configuration by varying $k_{div,st}$ and $k_{com,st}$. The distribution is in the range of −40% to 40% of the reference.

($k_{div,s}$ and $k_{com,s}$) have a marginal effect on the flow distribution by increasing or decreasing it in the range of −40% to +40% (Figure 13), while the straight flow coefficients result in a higher deviation from the reference case in the same range (Figure 14), but still, the deviation is not so drastic from the reference. This can be explained in a way that, for example, a fluid particle which is flowing through the 30th riser has to encounter 29 straight flow junction losses ($k_{div,st}$ and $k_{com,st}$) in the dividing and combining manifolds and only two side flow junction losses ($k_{div,s}$ and $k_{com,s}$) in the 30th riser (see Figure 9). It can be concluded that straight flow junction loss coefficients are more critical in finding the correct flow distribution.

The CFD-based model is now used to investigate various aspects of the flow distribution mechanism. Figure 15 illustrates the comparison of the calculated flow distribution based on the junction loss coefficients from CFD simulations at low Reynolds numbers and the available correlations for turbulent flows from Idelchik [9] and Bassett et al. [10]. It is seen in Figure 15 that the resulting flow distribution curves for Idelchik [9] and Bassett et al. [10] are almost identical but differ appreciably from the present case. The main reason for this deviation is the low values of the junction loss coefficients of the straight flow ($k_{div,st}$ and $k_{com,st}$) from the correlations of Idelchik [9] and/or Bassett et al. [10] depicted in Figure 16 for the encountered range of Reynolds number from 70 to 7,000 (corresponding to total mass flow of 78 kg/m^2 K) in the manifolds.

Figure 17 shows the flow distribution for increasing length of the riser while keeping the geometrical

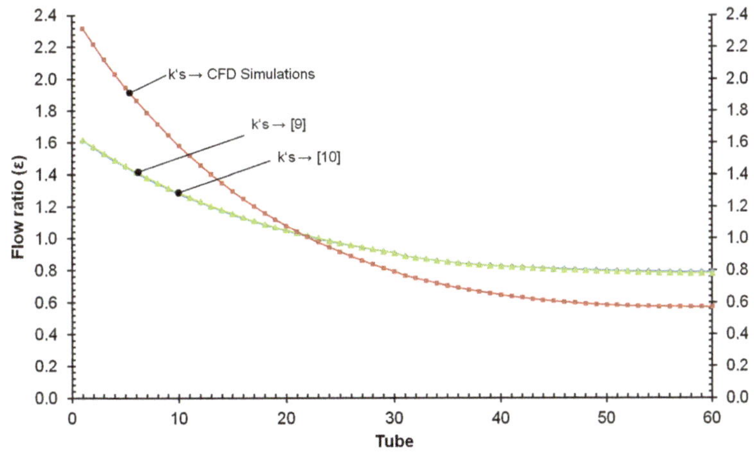

Figure 15 Comparison of isothermal flow distribution (U-configuration) for junction loss coefficients. Based on CFD simulations, Idelchik [9] and Bassett et al. [10] at $M = 78$ kg/m^2 h.

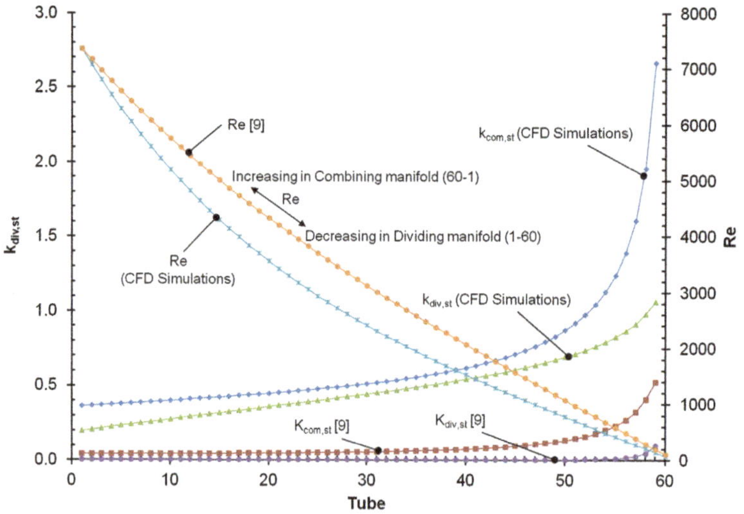

Figure 16 Variation of $k_{div/com,st}$ at each branch of dividing and combining manifold. Based on CFD simulations Idelchik [9] for U-configuration at $M = 78$ kg/m^2 h.

parameters of the manifolds the same (see Table 1). It is obvious that flow uniformity increases with increasing riser length from 1.82 to 8 m, which leads to a well-known principle regarding flow distribution in manifolds, i.e., higher pressure drop in the risers as compared with the manifolds results in more uniform flow distribution.

Finally, the flow distribution is also calculated for the Z-configuration, and comparison is made for the same collector connected in the U- and Z-configurations (Figure 18). For the Z-configuration, the flows are higher at the ends and lower in the middle tubes; they show a similar trend as predicted by McPhedran et al. [1] for 60 tubes and also in good agreement with the experimental results of Glembin et al. [16]. The Z-configuration results in more uniform flow (ϵ is near to unity). The minimum flow ratio ϵ and thus the minimum riser flow rate reached in the case of the Z-configuration (0.8) is 20% higher than for the case of the U-configuration (0.6). However, even in Z-configuration, the flow distribution is not at all uniform.

Conclusions

A CFD analysis is used to estimate the junction losses at the tee junctions of a collector manifold. A simplified model of the junction is built and simulated in FLUENT for a range of Reynolds numbers and riser-to-manifold flow ratios. The resulting junction loss coefficients have

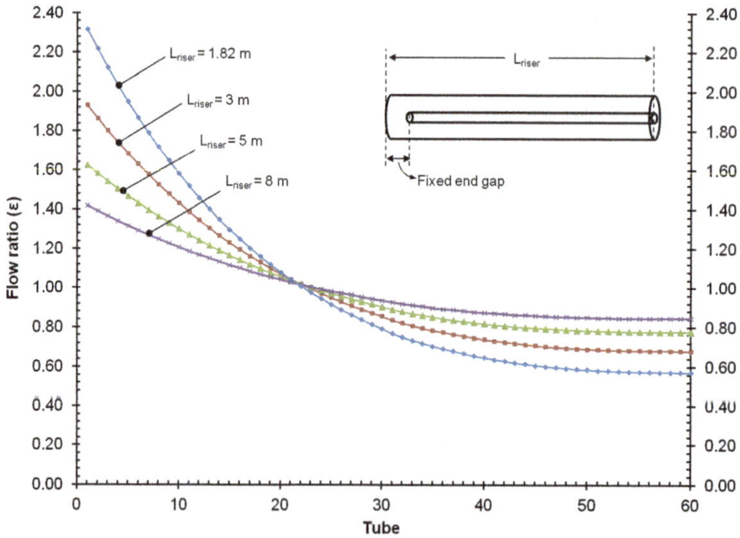

Figure 17 Isothermal flow distribution with increasing riser length for U-configuration at $M = 78$ kg/m^2 h.

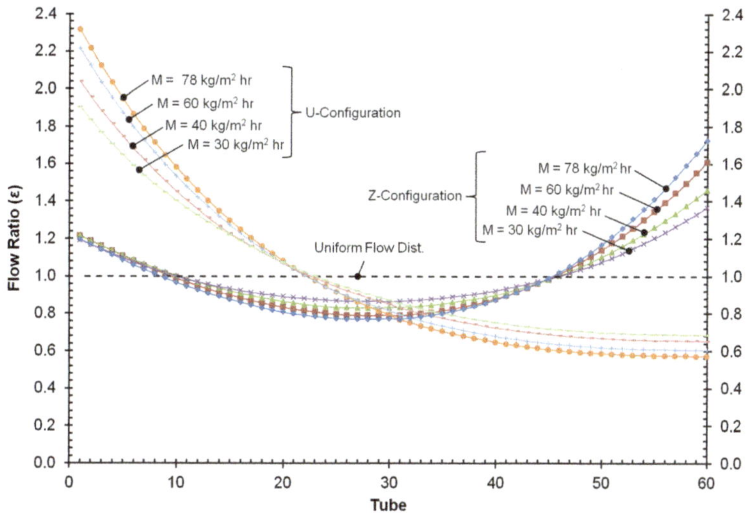

Figure 18 Isothermal flow distribution curves for U- and Z-configuration.

shown a strong dependency on the flow rate at low Reynolds numbers.

The variable loss coefficients are implemented in a theoretical model to predict the flow distribution in a coaxial vacuum tube solar collector arranged in U- and/or Z-configurations. The model is validated with the experimental results for the same collector in U-configuration. The model agrees reasonably well (but not perfectly) to the experiments. The model can be used to predict flow distribution for any number of risers in the prescribed range of Reynolds numbers. Flow uniformity decreases with increasing flow rate and temperature. Parallel flow (Z-configuration) results in more but not perfectly uniform flow than the reverse flow (U-configuration). The proposed CFD-based method can replace the expensive and time-consuming procedure of setting up experiments for estimating junction losses.

Abbreviations

Nomenclature

D: Diameter (m); *F*: moody friction factor (–); $h_i - h_o$: head loss (m); *I*: turbulence intensity; *K*: pressure loss coefficient (–); $k_{com,s}$: local pressure loss coefficient for combining side flow (–); $k_{com,st}$: local pressure loss coefficient for combining straight flow (–); $k_{div,s}$: local pressure loss coefficient for dividing side flow (–); $k_{div,st}$: local pressure loss coefficient for dividing straight flow (–); *L*: length (m); *P*: pressure (Pa); Δp: pressure drop (Pa); Δp_{c-s}: pressure drop between the inlet manifold and the riser in dividing flow (Pa); Δp_{c-st}: pressure drop between the inlet manifold and the straight outlet manifold in dividing flow (Pa); Δp_f: pressure drop due to friction (Pa); $\Delta p_{f,th}$: theoretical calculated pressure drop due to friction (Pa); $\Delta p_{f,c}$: pressure drop in combined manifold due to friction (Pa); $\Delta p_{f,s}$: pressure drop in riser due to friction (Pa); $\Delta p_{f,st}$: pressure drop in straight manifold due to friction (Pa); Δp_{s-c}: pressure drop between the riser and the outlet manifold in combing flow (Pa); Δp_{sim}: pressure drop calculated from simulation (Pa); Δp_{st-c}: pressure drop between the straight inlet manifold and the outlet manifold in combing flow (Pa); *Q*: volume flow (m^3/s); Q_o: total volume flow (m^3/s); *R*: frictional resistance (s^2/m^5); R_c: frictional resistance for the flow through dividing or combining manifold in U-configuration (s^2/m^5); R_{dm}: frictional resistance for the flow through dividing manifold in Z-configuration (s^2/m^5); R_{cm}: frictional resistance for the flow through combining manifold in Z-configuration (s^2/m^5); $R_{j,div,s/st}$: junction resistance for the side or straight flow in dividing manifold (s^2/m^5); $R_{j,com,s/st}$: junction resistance for the side or straight flow in combining manifold (s^2/m^5); Re: Reynolds number (–); T_{fi}: Temperature of fluid at inlet (°C); *V*: Velocity (m/s).

Greek symbols

ϵ: Roughness (m) (Equation 9b); ϵ: flow ratio (–) (Equation 30); μ: dynamic viscosity (kg/m s); ρ: density (kg/m^3).

Subscripts

av: Averaged; *c*: combined; *cr*: combining manifold in Z-configuration; *r*: riser; *s*: side; st: straight.

Competing interests

The authors declare that they have no competing interests.

Authors' contributions

AWB developed and programmed the numerical model for calculating the flow distribution and drafted the manuscript. RB collected the data, conducted the initial study, carried out initial analysis, and participated in the design and coordination of the study. YL participated in the design of the study, constructed the model geometry for simulation, performed simulations, and developed equations to be used in the numerical model. FZ checked the whole work critically and edited and corrected the draft manuscript. All authors read and approved the manuscript.

References

1. McPhedran, R.C., Mackey, D.J.M., McKenzie, D.R., Collins, R.E.: Flow distribution in parallel connected manifolds for evacuated tubular solar collectors. Aust. J. Phys. **36**, 197–219 (1983)
2. Glembin, J., Rockendorf, G., Scheuren, J.: Internal thermal coupling in direct-flow coaxial vacuum tube collectors. Solar Energy **84**, 1137–1146 (2010)
3. Viller, N.M., López, J.M.C., Munoz, F.D.: Numerical 3-D heat flux simulations on flat plate solar collectors. Solar Energy **93**, 1086–1092 (2009)
4. Fan, J., Shah, L.J., Furbo, S.: Flow distribution in a solar collector panel with horizontally inclined absorber strips. Solar Energy **81**, 1501–1511 (2007)
5. Wang, X.A., Yu, P.: Isothermal flow distribution in header systems. Solar Energy **7**, 159–169 (1989)
6. Wang, X.A., Wu, L.G.: Analysis and performance of flat-plate solar collector arrays. Solar Energy **45**, 71–78 (1990)
7. Jones, G.F., Lior, N.: Flow distribution in manifold solar collectors with negligible buoyancy effects. Solar Energy **52**, 289–300 (1994)
8. Fan, J., Furbo, S.: Buoyancy effects on thermal behaviour of a flat-plate solar collector. J Solar Energy Eng **130**(1–12), 021010 (2008)

9. Idelchik, I.E.: Handbook of Hydraulic Resistance, 2nd edn. Springer, Washington (1986)
10. Bassett, M.D., Winterbone, D.E., Pearson, R.J.: Calculation of steady flow pressure loss coefficients for pipe junctions. Proceedings of the institution of mechanical engineers, part C. J. Mech. Eng. Sci. **215**, 861–881 (2001)
11. Weitbrecht, V., Lehmann, D., Richter, A.: Flow distribution in solar collectors with laminar flow conditions. Solar Energy **73**, 433–441 (2002)
12. Buchholz, R.: Hydraulische Untersuchung einer Vakuumröhren Kollektoranlage in Hinsicht auf einen möglichen partiellen Stilstand und damit verbundenen leistungseinbußen. Institut für Energietechnik, Technische Universität, Berlin (2005)
13. Tyfocor, L.S.: Technische Information. P. TYFOROP CHEMIE GmbH, Hamburg (1999). http://www.tyfo.de
14. Filipiak, M.: Mesh Generation. Edinburgh parallel computing centre, the University of Edinburgh, Edinburgh (1996)
15. Fluent Inc: Fluent 6.3 UDF Manual. Fluent Inc, Lebanon (2006)
16. Glembin, J., Eggert, D., Rockendorf, G., Scheuren, J.: Evaporation in solar thermal collectors during operation - reasons and effects of partial stagnation. J Solar Energy Eng **133**(1–9), 041003 (2011)

Methanolysis optimization of sesame (*Sesamum indicum*) oil to biodiesel and fuel quality characterization

Eriola Betiku[1*] and Tunde Folorunsho Adepoju[2]

Abstract

Statistical approach was employed to optimize biodiesel production from sesame oil in this work. Precisely, response surface methodology was applied, and the effects of four variables, *viz.* reaction temperature, catalyst amount, reaction time, and methanol/oil molar ratio, and their reciprocal interactions were determined. Central composite rotatable design was used to generate a total of 30 individual experiments, which were designed to study the effects of these variables during alkali-catalyzed methanolysis of sesame oil. A statistical model predicted the highest conversion yield of sesame biodiesel to be 99.71% at the following optimized variable conditions: reaction temperature of 63°C, catalyst amount of 1.04 wt.%, and methanol/oil molar ratio of 6.24, with a reaction time of 51.09 min. Using these variables under experimental condition in four independent replicates, an actual biodiesel yield of 98.36% was accomplished. The fuel properties of biodiesel produced were found to be within the ASTM D6751 and DIN EN 14214 biodiesel specifications.

Keywords: Biodiesel, Sesame oil, Methanolysis, Optimization, Response surface methodology

Background

The need for alternative fuel, which can solve the problems of finite nature of fossil fuels and their environmental concerns, is imperative. About 98% of carbon emissions result from fossil fuel combustion [1]. The continued and increasing use of diesel derived from fossil oil intensifies air pollution and magnifies the global warming problem caused by CO_2 and other particulate matters [2]. Biodiesel, which is renewable in nature, non-toxic, and environmentally friendly with higher octane number than diesel, is the most promising candidate. It can be defined as the alkyl esters of long-chain fatty acids derived from renewable feedstock like vegetable oils or animal fats, obtained by transesterifying oil or fat with an alcohol [3].

Despite the food vs. fuel challenge facing the world, both edible and non-edible vegetable oil have been successfully used to produce biodiesel with superior qualities to diesel. These feedstocks include soybean oil [4], moringa oil [5], sunflower oil [6], rapeseed oil [7], palm oil [8], jatropha oil [9], camelina oil [10], neem oil [11,12], karanja oil [13], and sesame oil [14]. The ideal vegetable oil for biodiesel must be readily available, its plant should be easy to cultivate, and its composition must include a high percentage of monounsaturated fatty acids (C16:1, C18:1), a low proportion of polyunsaturated acids (C18:2, C18:3) and a controlled amount of saturated fatty acids (C16:0, C18:0) [15].

Transesterification of oil or fat to produce a high yield of biodiesel is typically investigated by optimizing the following reaction variables: alcohol/oil molar ratio, catalyst concentration, reaction temperature, and reaction time involved in the process. Both the effects of the variables and their reciprocal interactions have been evaluated using response surface methodology (RSM). RSM is a useful statistical tool, which has been applied in research for optimizing various processes including pretreatment of the *Jatropha curcas* fruit hull for enzymatic hydrolysis [16], bioremediation of contaminated soil [17], and transesterification reaction of vegetable oils. Transesterification of *Moringa oleifera* oil to biodiesel was optimized via RSM by Rashid et al. [5], whereas Tiwari et al. [9] optimized biodiesel production from

* Correspondence: ebetiku@oauife.edu.ng
[1]Biochemical Engineering Laboratory, Department of Chemical Engineering, Obafemi Awolowo University, Ile-Ife, Osun State 220005, Nigeria
Full list of author information is available at the end of the article

Table 1 Factors and their levels for composite central design

Variable	Symbol	Coded factor levels				
		−2	−1	0	1	2
Reaction temperature (°C)	X_1	50	55	60	65	70
Catalyst amount (wt.%)	X_2	0.7	0.8	0.9	1.0	1.1
Reaction time (min)	X_3	40	45	50	55	60
Methanol/oil molar ratio	X_4	4	5	6	7	8

jatropha oil with high free fatty acids using RSM. Jeong et al. [18] optimized the transesterification process of animal fat ester, while Yuan et al. [7] used RSM for optimization of waste rapeseed oil with high FFA to biodiesel. Also, Zhang et al. [19] applied RSM for the optimization of *Zanthoxylum bungeanum* seed oil transesterification to biodiesel using CaO as catalyst. Optimization of transesterification variables for biodiesel production from cottonseed oil using RSM has also been reported [20].

Sesamum indicum, commonly called beniseed in Africa but globally known as sesame, is a fast-growing crop whose oilseed can be utilized for biodiesel production through transesterification process. It is an important oilseed crop cultivated in many parts of the world. Its seed is composed of 50% to 52% oil, 17% to 19% protein, and 16% to 18% carbohydrate [21]. Fatty acid compositions of the oil are mostly oleic (32.7% to 53.9%), linoleic (39.3% to 59%), palmitic (8.3% to 10.9%), and stearic (3.4% to 6.0%) acids [22]. However, the fatty acid composition of this oil varies considerably among the

Table 2 CCRD and experimental, predicted, and residual values for five-level four-factor response surface analysis

Standard order	X_1 (°C)	X_2 (wt.%)	X_3 (min)	X_4	Experimental value % (*w*/*w*)	Predicted value % (*w*/*w*)	Residual values % (*w*/*w*)%
1	−1	−1	−1	−1	56.68	56.80	−0.12
2	1	−1	−1	−1	66.05	65.73	0.32
3	−1	1	−1	−1	79.06	79.08	−0.02
4	1	1	−1	−1	82.77	82.50	0.27
5	−1	−1	1	−1	63.86	64.05	−0.19
6	1	−1	1	−1	77.67	77.56	0.11
7	−1	1	1	−1	80.44	80.67	−0.23
8	1	1	1	−1	88.78	88.66	0.12
9	−1	−1	−1	1	83.55	83.57	−0.02
10	1	−1	−1	1	85.56	85.28	0.28
11	−1	1	−1	1	88.72	88.78	−0.06
12	1	1	−1	1	85.26	84.97	0.29
13	−1	−1	1	1	89.23	89.45	−0.22
14	1	−1	1	1	95.86	95.74	0.12
15	−1	1	1	1	88.78	89.00	−0.22
16	1	1	1	1	89.94	89.77	0.17
17	−2	0	0	0	77.75	77.14	0.61
18	2	0	0	0	86.10	86.86	−0.76
19	0	−2	0	0	65.13	65.19	−0.06
20	0	2	0	0	81.42	81.51	−0.09
21	0	0	−2	0	82.53	82.93	−0.40
22	0	0	2	0	95.22	94.97	0.25
23	0	0	0	−2	67.10	67.16	−0.06
24	0	0	0	2	94.95	95.04	−0.09
25	0	0	0	0	92.87	92.87	0.00
26	0	0	0	0	92.87	92.87	0.00
27	0	0	0	0	92.87	92.87	0.00
28	0	0	0	0	92.87	92.87	0.00
29	0	0	0	0	92.87	92.87	0.00
30	0	0	0	0	92.87	92.87	0.00

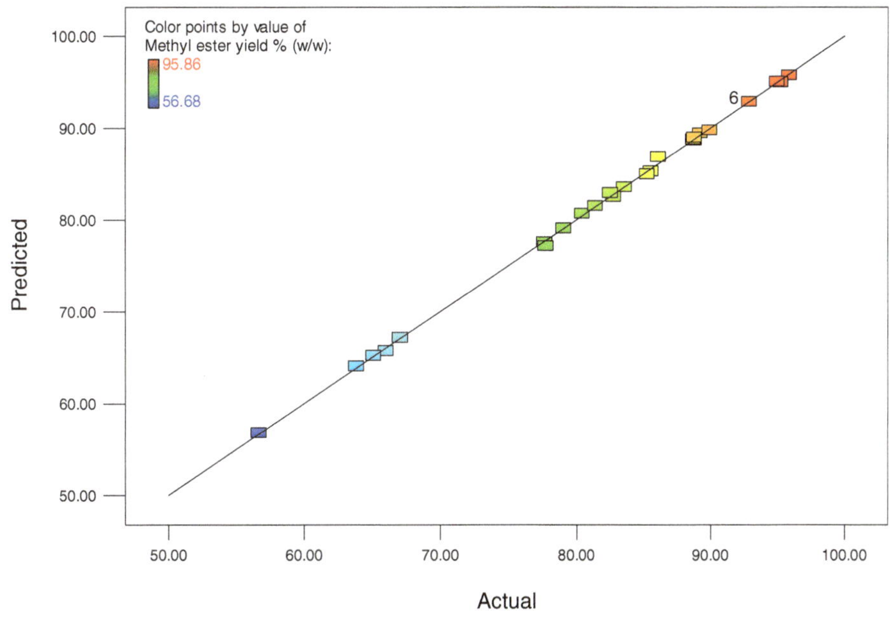

Figure 1 Parity plot showing the distribution of actual vs. predicted values of sesame biodiesel yield.

different varieties worldwide [22,23]. Sesame oil has emerged as a potential feedstock for biodiesel production. Although the transesterification of this seed oil from Turkey and Pakistan cultivars to biodiesel has been reported [14,24,25], the process has not yet been optimized using a statistical approach such as response surface methodology, and the maximum yield reported was 92% [24].

In this present study, an effort was made to optimize the process conditions for transesterification of sesame oil from a Nigeria cultivar to biodiesel. RSM with central composite rotatable design (CCRD) was applied, and the effect of four variables (reaction temperature, catalyst amount, reaction time, and methanol/oil molar ratio) at five levels and their reciprocal interactions were evaluated.

Methods

Materials

S. indicum samples were collected from Lokoja, Nigeria. Chaffs were separated from the oilseeds by winnowing. The cleaned oilseeds were milled to powder by grinding with a plate machine. All chemicals and reagents used for this work were of analytical grades.

Extraction of sesame oil

Soxhlet apparatus and *n*-hexane as solvent were used for the oil extraction. Initially, the apparatus was charged with a known weight of the oilseed powder in a muslin cloth placed in a thimble of the apparatus. A round bottom flask containing a known volume of *n*-hexane was fixed to the end of the apparatus, and a condenser was tightly fixed at the bottom end of the extractor. The whole setup was heated up in a heating mantle at a temperature of 70°C. The excess solvent in the oil was recovered by distillation.

Experimental design of biodiesel production

In this work, the central composite experimental design was employed to optimize the transesterification of sesame oil to biodiesel. A five-level-four-factor design was applied, which generated 30 experimental runs. These included 16 factorial points, 8 axial points, and 6 central points to provide information regarding the interior of the experimental region, making it possible to evaluate

Table 3 Test of significance for every regression coefficient of CCRD

Source	Sum of squares	*df*	Mean square	*F* value	*p* value
X_1	141.17	1	141.17	1,192.78	<0.0001
X_2	399.11	1	399.11	3,364.87	<0.0001
X_3	217.74	1	217.74	1,835.80	<0.0001
X_4	1,166.08	1	1,166.08	9,831.27	<0.0001
X_1X_2	30.44	1	30.44	256.66	<0.0001
X_1X_3	20.95	1	20.95	176.66	<0.0001
X_1X_4	52.16	1	52.16	439.80	<0.0001
X_2X_3	32.06	1	32.06	270.33	<0.0001
X_2X_4	291.47	1	291.47	2,457.40	<0.0001
X_3X_4	1.87	1	1.87	15.77	0.0012
X_1^2	202.54	1	202.54	1,707.61	<0.0001
X_2^2	653.17	1	653.17	5,506.87	<0.0001
X_3^2	26.34	1	26.34	222.05	<0.0001
X_4^2	237.47	1	237.47	2,002.10	<0.0001

the curvature effect. Selected variables for the transesterification process from the sesame oil were reaction temperature (X_1), catalyst amount (X_2), reaction time (X_3), and methanol/oil molar ratio (X_4). The coded levels of the independent variables are given in Table 1. The experiments were randomized to minimize the effects of unexplained variability in the observed response due to extraneous factors.

Sesame biodiesel production procedure

The base-catalyzed reaction was carried out according to the work of Schinas et al. [26]. The transesterification reaction of sesame oil was carried out in a glass reactor (1-L capacity). The reactor was charged with a known weight of the oil sample, which was preheated to the desired temperature. A known weight of NaOH pellet (catalyst) was dissolved in a specific volume of methanol and was transferred into the preheated oil on the hot plate with a magnetic stirrer. The resulting mixture was transferred into a separating funnel for glycerol and methyl ester separation. The glycerol was allowed to separate by gravity for 24 h. Two phases separated clearly, the less dense methyl ester at the top and the denser phase of glycerol at the bottom. The glycerol was tapped off the bottom of the separating funnel, leaving behind methyl ester in the separating funnel. The methyl ester was washed with distilled water to remove residual catalyst, glycerol, methanol, and soap. The methyl ester was further dried over a heated $CaCl_2$ powder. The final product was the sesame biodiesel, and the yield was calculated gravimetrically as described in Equation 1:

$$Beniseed\ biodisel\ yield\ \%\ (w/w) = \frac{Weight\ of\ biodiesel\ produced}{Weight\ of\ Beniseed\ oil\ used} \tag{1}$$

Statistical data analysis

Sesame biodiesel production data were analyzed statistically using response surface methodology, in order to fit the quadratic polynomial equation generated by the Design Expert software version 8.0.3.1 (Stat-Ease Inc., Minneapolis, MN, USA). To correlate the response variable to the independent variables, multiple regressions were used to fit the coefficient of the polynomial model of the response. The quality of the fit of the model was evaluated using a test of significance and analysis of variance (ANOVA). The fitted quadratic response model is described by

$$Y = b_0 + \sum_{i=1}^{k} b_i X_i + \sum_{i=1}^{k} b_{ii} X_i^2 + \sum_{i<j}^{k} b_{ij} X_i X_j + e, \tag{2}$$

where Y is the response variable (sesame biodiesel yield), b_o is the intercept value, b_i (i = 1, 2,. . ., k) is the first-order model coefficient, b_{ij} is the interaction effect, and b_{ii} represents the quadratic coefficients of X_i, and e is the random error.

Table 4 ANOVA of regression equation

Source	Sum of squares	*df*	Mean square	*F* value	*p* value
Model	3,234.82	14	231.06	1,948.06	< 0.0001
Residual	1.78	15	0.12		
Lack of fit	1.78	10	0.18	0.28	0.9589
Pure error	0.000	5	0.000		
Totals corrected for the mean	3,236.60	29			

For the mean square and *F* and *p* values, R^2 = 99.95% and R^2 (adj) = 99.89%.

Oil and fuel properties

Fuel properties, namely, moisture content, density at 15°C, kinematic viscosity at 40°C, iodine value, acid value, higher heating value, flash point, pour point, cloud point, cetane number, diesel index, API, and aniline point of both sesame oil and biodiesel, were determined following standard methods and compared with American and European standards (ASTM D 6751 and DIN EN 14214).

Analysis of fatty acid compositions of sesame biodiesel

Fatty acid profile of the sesame biodiesel was determined using gas chromatography (HP 6890 powered with HP ChemStation Rev. A 09.01 (1206) software, Agilent Technologies Inc., Santa Clara, CA, USA). The sesame biodiesel sample (50 mg) was esterified for 5 min at 95°C

Table 5 ANOVA for response surface quadratic model for intercept

Factors	Coefficient estimate	*df*	Standard error	95% CI low	95% CI high	VIF
Intercept	92.87	1	0.14	92.57	93.16	-
X_1	2.43	1	0.070	2.28	2.58	1.00
X_2	4.08	1	0.070	3.93	4.23	1.00
X_3	3.01	1	0.070	2.86	3.16	1.00
X_4	6.97	1	0.070	6.82	7.12	1.00
X_1X_2	−1.38	1	0.086	−1.56	−1.20	1.00
X_1X_3	1.14	1	0.086	0.96	1.33	1.00
X_1X_4	−1.81	1	0.086	−1.99	−1.62	1.00
X_2X_3	−1.42	1	0.086	−1.60	−1.23	1.00
X_2X_4	−4.27	1	0.086	−4.45	−4.08	1.00
X_3X_4	−0.34	1	0.086	−0.53	−0.16	1.00
X_1^2	−2.72	1	0.066	−2.86	−2.58	1.05
X_2^2	−4.88	1	0.066	−5.02	−4.74	1.05
X_3^2	−0.98	1	0.066	−1.22	−0.84	1.05
X_4^2	−2.94	1	0.066	−3.08	−2.80	1.05

with 3.4 mL of the 0.5 M KOH in dry methanol. The mixture was neutralized using 0.7 M HCl, and 3 mL of 14% boron triflouride in methanol was added. The mixture was heated for 5 min at a temperature of 90°C to achieve complete methylation process. The fatty acids were thrice extracted from the mixture with redistilled *n*-hexane. The content was concentrated to 1 μL for gas chromatography analysis, and 1 μL was injected into the injection port of gas chromatography (GC).

Results and discussion

Properties of the sesame seed oil used as feedstock

The analysis of the seed oil showed that it has a moisture content of 0.09%, specific gravity of 0.88, and

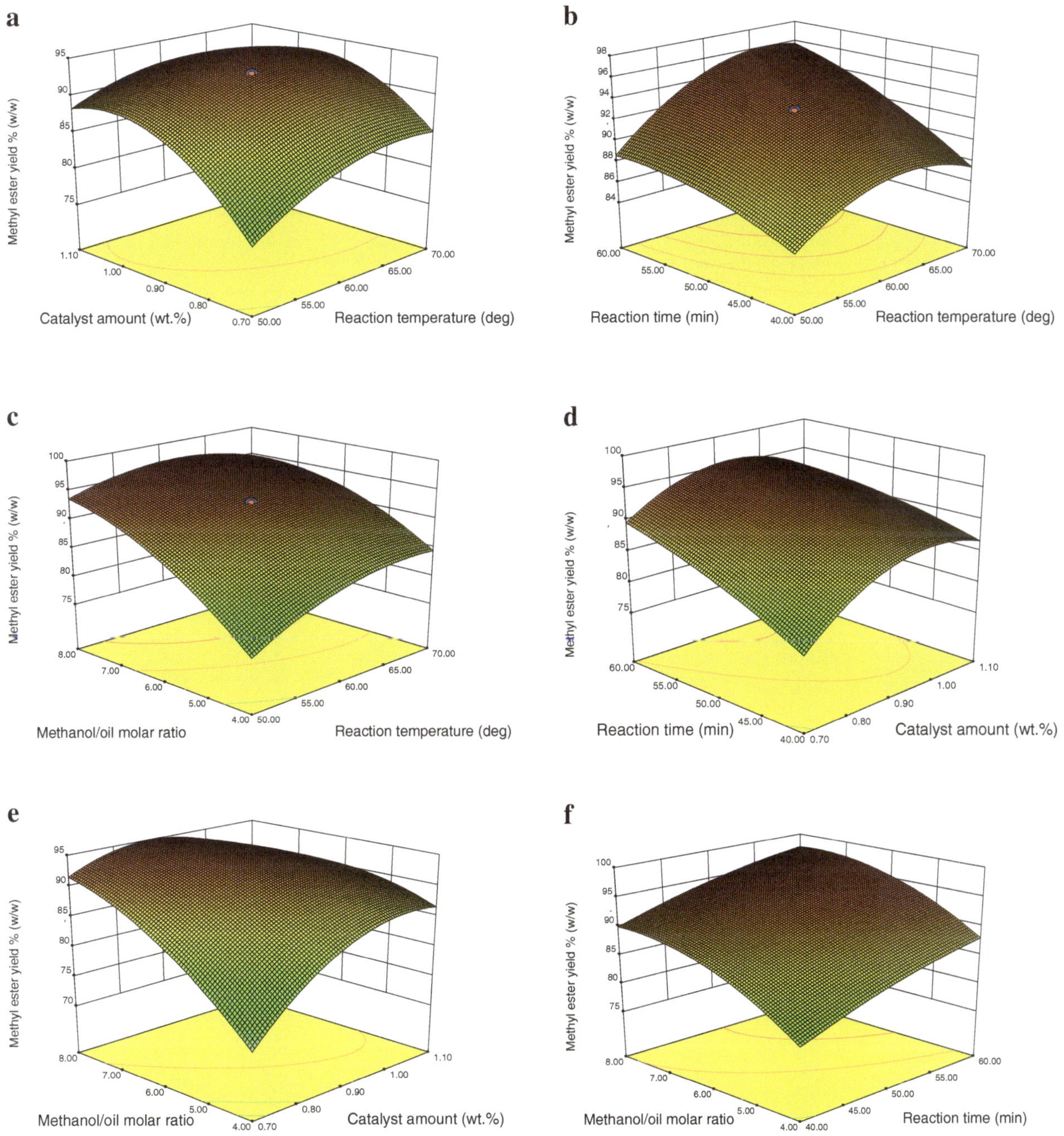

Figure 2 Surface plots for conversion of sesame seed oil to biodiesel. (**a**) catalyst amount and reaction temperature, (**b**) reaction time and reaction temperature, (**c**) methanol/oil molar ratio and reaction temperature, (**d**) reaction time with catalyst amount, (**e**) methanol/oil molar ratio and catalyst amount, and (**f**) methanol/oil molar ratio and reaction time.

kinematic viscosity of 31.39 mm^2/s. The acid value of the oil was 0.50 mg KOH/g oil, while the iodine value was 108 g I_2/100 g oil. Whereas the saponification value and the peroxide value of the oil were 190 mg KOH/g oil and 7.80 meq O_2/kg oil, respectively, its higher heating value and cetane number were 40.02 MJ/kg and 50.73, respectively. These results are within earlier reported values in the literature [14,27].

Optimization of sesame biodiesel production

The coded factors considered in this work with experimental results, predicted values, as well as the residual values obtained, are shown in Table 2. The parity plot showed a satisfactory correlation between the actual and predicted values of sesame biodiesel yield (Figure 1). Table 3 showed the results of the test of significance for every regression coefficient. The results revealed that the p value of the model terms was significant, i.e., $p < 0.05$. In this particular case, the four linear terms (X_1, X_2, X_3, X_4), six cross products (X_1X_2, X_1X_3, X_1X_4, X_2X_3, X_2X_4, X_3X_4), and the four quadratic terms (X_1^2, X_2^2, X_3^2, and X_4^2) were all remarkably significant model terms at 95% confidence level. Based on the large F values and low corresponding p values, all the linear terms had very strong effects on the biodiesel yield. However, methanol/oil molar ratio X_4 (F value = 9831.27) was the most significant variable followed by the catalyst amount X_2 (F value = 3364.87), followed by reaction time X_3 (F value = 1835.80), and lastly, reaction temperature X_1 (F value = 1192.78). In order to minimize error, all the coefficients were considered in the design. The model coefficients and probability values, i.e., coded values, are shown in Table 4. The model also proved suitable for adequate representation of the real relationship among the selected independent variables (X_1, X_2, X_3, X_4). Results of the ANOVA of the regression equation are presented in Table 5. The model F value of 3,234.83 with a low probability value ($p < 0.0001$) implied a high significance for the regression model [7]. The goodness of fit of the model was checked by the coefficient of determination (R^2), which should be at least 0.80 for the good fit of a model [28]. The R^2 value of 0.9995 indicated that the sample variation of 99.95% for biodiesel was attributed to the independent variables, and only 0.05% of the total variations are not explained by the model. The value of the adjusted determination coefficient (adjusted R^2 of 0.9989) was also very high, supporting a high significance of the model [29,30], and all p values were less than 0.0001 except for X_3X_4 with a p value of 0.0012, which indicated the model proved suitable for the adequate representation of the actual relationship among the selected variables. The lack-of-fit term of 0.9589 was not significant relative to the pure error. The final equation in terms of coded factors for the central composite response surface quadratic model is expressed in Equation 3.

$$\begin{aligned} Y(w/w\ \%) = {} & 92.87 + 2.43X_1 + 4.08X_2 \\ & + 3.01X_3 + 6.97X_4 - 1.38X_1X_2 \\ & + 1.14X_1X_3 - 1.81X_1X_4 \\ & - 1.42X_2X_3 - 4.27X_2X_4 \\ & - 0.34X_3X_4 - 2.72X_1^2 - 4.88X_2^2 \\ & - 0.98X_3^2 - 2.94X_4^2 \end{aligned} \quad (3)$$

All the linear terms and cross-product term X_1X_3 had a positive effect on the sesame biodiesel yield, while the rest of the surface quadratic model terms had negative effects on the yield. The low values of standard error observed in the intercept and all the model terms demonstrated that the regression model fits the data well, and the prediction was good (Table 5). The variance inflation factor (VIF) obtained in this study showed that the center points are orthogonal to all other factors in the model.

Generally, the 3D response surface plots are graphical representations of the regression equation for the optimization of the reaction variables, and they are presented in Figure 2. The curvatures' nature of the three-dimensional surfaces in Figure 2a,c,f suggested mutual interaction of catalyst amount with reaction temperature, methanol/oil molar ratio with reaction temperature, and methanol/oil molar ratio with reaction time, respectively. On the other hand, the curvatures' nature of the three-dimensional surfaces in Figure 2b,d,e indicated

Table 6 Properties of sesame biodiesel in comparison to biodiesel standards

Parameters	Sesame oil	Sesame biodiesel	ASTM D6751	EN 14214
Moisture content %	0.09	0.014	<0.03	0.02
Kinematic viscosity (mm^2/s) at 40°C	31.39	4.03	1.9 to 6.0	3.5 to 5.0
Iodine value (g I_2/100 g)	108	86.72	-	120 max
Acid value (mg KOH/g)	0.50	0.25	0.50 max	0.5 max
Density (kg/m^3) at 25°C	833	873	-	860 to 900
Higher heating value (MJ/kg)	40.20	41.35	-	-
Flash point (°C)	-	180	>130	>120
Pour point (°C)	-	6	-	-
Cloud point (°C)	-	18	-	-
Diesel index	55.80	69.17	50.40	-
API	29.30	30.03	36.95	-
Cetane number	50.73	59.80	47 min	51 min
Aniline point (°F)	190.42	230.33	331.00	-

moderate interactions of reaction time with reaction temperature, reaction time with catalyst amount, and methanol/oil molar ratio with catalyst amount, respectively.

The optimal values of the independent variables selected for the transesterification of sesame biodiesel were obtained by solving the regression equation (Equation 2) using the Design Expert software. The optimal conditions for this process were statistically predicted as $X_1 = 63°C$, $X_2 = 1.04$ wt.%, $X_3 = 51.09$ min, and $X_4 = 6.24$. The predicted sesame biodiesel yield under the above set of conditions was $Y = 99.71\%$. In order to verify the prediction of the model, the optimal conditions were applied to three independent replicates, and the average biodiesel yield obtained was 98.63%, which is well within the predicted value for the model equation.

Sesame biodiesel quality characterization and fuel properties

The results of GC analysis revealed that sesame biodiesel is mainly composed of methyl esters of linoleic acid (45.17%) and oleic acid (39.02%), palmitic acid (8.47%) followed by 5.15% stearic acid, 0.93% behenic acid, 0.77% arachidic acid, 0.15% linolenic acid, 0.09% myristic acid, 0.04% palmitoleic acid, 0.03% arachidonic acid, and 0.01% erucic acid. These observations further showed that the sesame biodiesel produced under the optimal condition was within ASTM D 6751 specifications for free and total glycerol contents. In addition, Table 6 shows the results of the evaluation of sesame seed oil and fuel properties of the sesame biodiesel. Most of the fuel properties of sesame biodiesel from this study compared well with ASTM D6751 and EN 14214 biodiesel specifications.

Conclusions

The results obtained in this study using response surface methodology to determine the effects of four reaction variables, namely, oil/methanol molar ratio, reaction temperature, catalyst concentration, and reaction time on sesame biodiesel yield via methanolysis, indicate that the biodiesel produced had fuel properties that satisfied both the ASTM D 6751 and EN 1424 standards. The statistical model predicted the optimal conditions for the selected transesterification variables as reaction temperature of 63°C, catalyst amount of 1.04 wt.%, reaction time of 51.09 min, and methanol/oil molar ratio of 6.24 with an actual sesame biodiesel yield of 98.63%. Hence, the work established the usefulness of RSM for optimum conversion of sesame (*S. indicum*) seed oil to biodiesel and also suggests that sesame oil could be effectively used as feedstock for biodiesel production.

Competing interests

Authors declare that they have no competing interests.

Authors' contributions

EB conceived and designed the research work. He is the major contributor in the preparation of the manuscript. TFA was involved in the laboratory work, and he drafted the manuscript. Both authors read and approved the final manuscript.

Authors' information

EB is a senior lecturer in the Department of Chemical Engineering, Obafemi Awolowo University, Ile-Ife, Nigeria. He holds B.Sc. and M.Sc. degrees in Chemical Engineering from Obafemi Awolowo University. He also received a Ph.D. degree in Biotechnology from Technical University Carolo-Wilhelmina, Braunschweig, Germany in 2005. He was a DAAD scholar and a guest scientist at the National Research Centre for Biotechnology, Germany between April 2001 and December 2004, respectively. His research focus is in the area of biofuel development and bioconversion of agricultural crops and wastes to useful products. TFA is a lecturer in the Department of Chemical Engineering, Landmark University, Nigeria. He holds a M.Sc. degree in Chemical Engineering from Obafemi Awolowo University, Ile-Ife, Nigeria.

Acknowledgments

E. Betiku gratefully acknowledged the equipment donation by the World University Service, Germany and provision of relevant literature by the DAAD. The authors also thank A.K. Omole and O.E. Aluko for their technical support.

Author details

[1]Biochemical Engineering Laboratory, Department of Chemical Engineering, Obafemi Awolowo University, Ile-Ife, Osun State 220005, Nigeria.
[2]Department of Chemical Engineering, Landmark University, P.M.B 1001, Omu-Aran, Kwara State, Nigeria.

References

1. Balat, M: Potential alternatives to edible oils for biodiesel production - a review of current work. Energ. Convers. Manage. **52**, 1479–1492 (2010)
2. Shay, EG: Diesel fuel from vegetable oils: status and opportunities. Biomass Bioenerg. **4**(4), 227–242 (1993)
3. Knothe, G: Biodiesel and renewable diesel: a comparison. Prog. Energ. Combust. **36**, 364–373 (2010)
4. Aransiola, EF, Betiku, E, Layokun, SK, Solomon, BO: Production of biodiesel by transesterification of refined soybean oil. Int. J. Biol. Chem. Sci. **4**, 391–399 (2010)
5. Rashid, U, Anwar, F, Ashraf, M, Saleem, M, Yusup, S: Application of response surface methodology for optimizing transesterification of *Moringa oleifera* oil: biodiesel production. Energ. Convers. Manage. **52**, 3034–3042 (2011)
6. Berrios, M, Siles, J, Martin, M, Martin, A: A kinetic study of the esterification of free fatty acids (FFA) in sunflower oil. Fuel **86**, 2383–2388 (2007)
7. Yuan, X, Liu, J, Zeng, G, Shi, J, Tong, J, Huang, G: Optimization of conversion of waste rapeseed oil with high FFA to biodiesel using response surface methodology. Renew. Energ. **33**, 1678–1684 (2008)
8. Margaretha, YY, Prastyo, HS, Ayucitra, A, Ismadji, S: Calcium oxide from Pomacea sp. shell as a catalyst for biodiesel production. Int. J. Energy Env. Eng. **3**, 33 (2012)
9. Tiwari, AK, Kumar, A, Raheman, H: Biodiesel production from jatropha oil (*Jatropha curcas*) with high free fatty acids: an optimized process. Biomass Bioenerg. **31**, 569–575 (2007)
10. Wu, X, Leung, DYC: Optimization of biodiesel production from camelina oil using orthogonal experiment. Appl. Energy **88**(11), 3615–3624 (2011)
11. Muthu, H, SathyaSelvabala, V, Varathachary, T, Kirupha Selvaraj, D, Nandagopal, J, Subramanian, S: Synthesis of biodiesel from neem oil using sulfated zirconia via transesterification. Braz J Chem Eng **27**(4), 601–608 (2010)
12. Aransiola, EF, Betiku, E, Ikhuomoregbe, DIO, Ojumu, TV: Production of biodiesel from crude neem oil feedstock and its emissions from internal combustion engines. Afr. J. Biotech. **11**(22), 6178–6186 (2012)
13. Meher, LC, Dharmagadda, VSS, Naik, SN: Optimization of alkali-catalyzed transesterification of *Pongamia pinnata* oil for production of biodiesel. Bioresour Technol **97**(12), 1392–1397 (2006)
14. Saydut, A, Duz, MZ, Kaya, C, Kafadar, AB, Hamamci, C: Transesterified sesame (*Sesamum indicum* L.) seed oil as a biodiesel fuel. Bioresour. Technol **99**(14), 6656–6660 (2008)

15. Wang, R, Hanna, MA, Zhou, W-W, Bhadury, PS, Chen, Q, Song, B-A, Yang, S: Production and selected fuel properties of biodiesel from promising non-edible oils: *Euphorbia lathyris* L., *Sapium sebiferum* L. and *Jatropha curcas* L. Bioresour. Technol. **102**(2), 1194–1199 (2011)
16. Marasabessy, A, Kootstra, AMJ, Sanders, JPM, Weusthuis, RA: Dilute H_2SO_4-catalyzed hydrothermal pretreatment to enhance enzymatic digestibility of *Jatropha curcas* fruit hull for ethanol fermentation. Int. J. Energy Env. Eng. **3**, 15 (2012)
17. Agarry, SE, Ogunleye, OO: Box-Behnken design application to study enhanced bioremediation of soil artificially contaminated with spent engine oil using biostimulation strategy. Int. J. Energy. Env. Eng. **3**, 31 (2012)
18. Jeong, GT, Yang, H-S, Park, DH: Optimization of transesterification of animal fat ester using response surface methodology. Bioresour Technol **100**, 25–30 (2009)
19. Zhang, J, Chen, S, Yang, R, Yan, Y: Biodiesel production from vegetable oil using heterogenous acid and alkali catalyst. Fuel **89**(10), 2939–2944 (2010)
20. Fan, X, Wang, X, Chen, F: Biodiesel production from crude cottonseed oil: an optimization process using response surface methodology. The Open Fuels Energy Sci. J. **4**, 1–8 (2011)
21. Ustimenko-Bakumovsky, GV: Plant Growing in the Tropics and Subtropics. Mir Publishers, Moscow (1983)
22. Yeramanos, DM, Hemstreet, S, Saleeb, W, Huszar, CK: Oil content and composition of the seed in the world collection of sesame introductions. J. Am. Oil Chem. Soc. **49**, 20–23 (1972)
23. Brar, GS: Variations and correlations in oil content and fatty acid composition of sesame. Ind. J. Agric. Sci. **52**, 434–439 (1982)
24. Ahmad, M, Khan, MA, Zafar, M, Sultana, S: Environment-friendly renewable energy from sesame biodiesel. Energy Sources, Part A **32**(2), 189–196 (2010)
25. Ahmad, M, Ullah, K, Khan, MA, Ali, S, Zafar, M, Sultana, S: Quantitative and qualitative analysis of sesame oil biodiesel, Energy Sources. Part A. **33**(13), 1239–1249 (2011)
26. Schinas, P, Karavalakis, G, Davaris, C, Anastopoulos, G, Karonis, D, Zannikos, F, Stournas, S, Lois, E: Pumpkin (*Cucurbita pepo* L.) seed oil as an alternative feedstock for the production of biodiesel in Greece. Biomass Bioenerg **33**, 44–49 (2009)
27. Elleuch, M, Besbes, S, Roiseux, O, Blecker, C, Attia, H: Quality characteristics of sesame seeds and by-products. Food Chem **103**(2), 641–650 (2007)
28. Guan, X, Yao, H: Optimization of viscozyme L-assisted extraction of oat bran protein using response surface methodology. Food Chem **106**, 345–351 (2008)
29. Akhnazarova, S, Kefarov, V: Experiment Optimization in Chemistry and Chemical Engineering. Mir Publishers, Moscow (1982)
30. Khuri, AI, Cornell, JA: Response Surfaces: Design and Analysis. Marcel Dekker, New York (1987)

15

Wind tunnel testing and numerical simulation on aerodynamic performance of a three-bladed Savonius wind turbine

Khandakar Niaz Morshed[1], Mosfequr Rahman[2*], Gustavo Molina[2] and Mahbub Ahmed[3]

Abstract

The purpose of this research work is to investigate experimentally and computationally the feasibility of improving the performance of the vertical-axis Savonius wind turbine. The authors first performed a series of wind tunnel investigations on semi-cylindrical three-bladed Savonius rotor scale models with different overlap ratios and without overlap. These experiments were conducted in front of a low-speed subsonic wind tunnel at different Reynolds numbers. Pressures around the concave and convex surfaces of each blade, as well as the static torque for the rotor models, were measured. Using these experimental data, the authors calculated aerodynamic characteristics such as drag coefficients, static torque coefficients, and power coefficients. The authors then performed computational fluid dynamics (CFD) simulations using the commercial CFD software FLUENT and GAMBIT to analyze the static rotor aerodynamics of those models. The experimental and computational results were then compared for verification. Three different models with different overlap ratios were designed and fabricated for the current study to find the effect of overlap ratios. The results from the experimental part of the research show a significant effect of overlap ratio and Reynolds number on the improvement of aerodynamic performance of the Savonius wind turbine. At higher Reynolds number, the turbine model without overlap ratio gives better aerodynamic coefficients, and at lower Reynolds number, the model with moderate overlap ratio gives better results.

Keywords: Savonius wind turbine, Aerodynamic performance, Torque coefficient, Power coefficient, Angle of rotation, Overlap ratio

Background

Background of Savonius wind turbine

Vertical-axis wind turbines (VAWTs) include both a drag-type configuration, such as the Savonius rotor, and a lift-type configuration, such as the Darrieus rotor. The simplest type of vertical-axis wind turbine is the Savonius rotor, the operation of which depends on the difference in drag force when the wind strikes either the convex or concave part of its semi-cylindrical blades. Savonius rotors are good at self-starting and work independently of wind direction. However, its efficiency is relatively lower than that of the lift-type VAWTs. Due to its simple design and low construction cost, Savonius rotors are primarily used to pump water and generate wind power on a small scale, and its large starting torque makes it suitable for starting other types of wind turbines that have inferior starting characteristics, such as the Darrieus rotor and Gyro mill [1]. Recently, some generators with high torque at low rotational speed, suitable for small-scale wind turbines, have been developed, suggesting that Savonius rotors may yet be used to generate electric power [1].

Wind turbine aerodynamics must be designed for optimal output to exploit the wind energy in a specific location. Diaz et al. [2] analyzed the drag and lift coefficients of a Savonius wind turbine to quantify the aerodynamic performance of the rotor. They found that maximum efficiency, in terms of power coefficient, occurs at a tip speed ratio of $\lambda = 1$, and the drag coefficient decreases sharply when the tip speed ratio increases or decreases

* Correspondence: mrahman@georgiasouthern.edu
[2]Mechanical Engineering Department, Georgia Southern University, Statesboro, GA 30460, USA
Full list of author information is available at the end of the article

from this value. They also found that the most important region of Savonius rotor operation occurs at a tip speed ratio around $\lambda = 1$, where the lift coefficient remains as a constant 0.5. Sawada et al. [3] studied the mechanism of rotation of a Savonius rotor with two semi-cylindrical blades and found that a rotor with a gap ratio of 0.21 produces positive static torque at all angles. They also found that lift force contributes significantly to dynamic torque, while the rotor angle is between $\alpha = 240°$ and $\alpha = 330°$. Aldoss and Obeidat [4] used the discrete vortex method to analyze the performance of two Savonius rotors running side-by-side at different separations. They compared their computational results on torque and power coefficients with their experimental results for verification. Fujisawa and Gotoh [5] studied the aerodynamic performance of a Savonius rotor by measuring pressure distribution on the blade surfaces at various rotor angles and tip speed ratios. Torque and power performance, evaluated by integrating the pressure, were in close agreement with direct torque measurements.

Rahman et al. [6-8] experimentally studied aerodynamic characteristics, such as the torque and drag coefficients, of a three-bladed Savonius rotor model by measuring the pressure difference between the convex and concave surfaces of each semi-cylindrical blade of the stationary rotor at different rotor angles and the variation of the separation point with the increase of rotor angle. They used the static coefficients for dynamic prediction and compared the findings in terms of power coefficients for different tip speed ratios with experimental results for the two-bladed Savonius rotor. Rahman et al. [9] conducted both experimental investigations and computational fluid dynamic (CFD) simulations to establish the feasibility of improving the performance of a simple, three-bladed Savonius VAWT. The normal drag coefficient, tangential drag coefficient, and torque coefficient were calculated both experimentally and numerically, and the results were compared. In each case, the calculations matched well. The numerical results were more accurate and gave positive values for combined drag coefficients and the total static torque coefficient.

Gupta et al. [10] compared a three-bucket Savonius wind turbine with a three-bucket Savonius-Darrieus wind turbine. They found that the power coefficient of the combined turbine decreases as the overlap ratio increases. The maximum power coefficient of 51% was found where there was no overlap. They claimed that the combined rotor without overlap, which showed 51% efficiency, was the highest efficiency of a Savonius wind turbine at any overlap condition under these test conditions. Altan et al. [11] did some experimental studies to improve the performance of the Savonius wind turbine using a curtain. They placed the curtain arrangement in front of the rotor in a configuration capable of preventing the negative torque that affects the convex blade surface of the Savonius wind turbine.

Sargolzaei and Kianifar [12] simulated a Savonius wind turbine using artificial neural networks (ANNs) to estimate power ratio and torque. They experimentally investigated seven prototype Savonius wind turbines and compared the experimental results with their predicted ANN results. Their predicted results were in good agreement with their experimental results. They found that increased wind speed causes torque increase. For all their models, they found that maximum torque was at 60° and minimum torque was at 120°. Altan and Atilgan [13] numerically simulated their experimental work using FLUENT 6.0 and GAMBIT 2.0. Their model was two-dimensional, and they used a standard k-ε turbulence model. To calculate pressure and velocity distribution, they used a semi-implicit method for pressure-linked equation (SIMPLE) analysis algorithm. By comparing the numerical and experimental results, they concluded that the curtain improved the performance of Savonius wind turbines.

Saha et al. [14] fabricated a two-stage Savonius wind turbine by inserting valves on the concave side of the blades. They compared its performance to a conventional Savonius wind turbine and found that with valves on a three-bladed turbine, the power coefficient was higher compared to a two-bladed turbine for both semi-circular and twisted blades. Without valves, air strikes the blades and rotates them in a negative direction. Saha et al. also varied the number of stages in a Savonius wind turbine and found that while the power coefficient increased from one to two stages, it decreased from two to three stages due to increased inertia. They tested the twisted blades of one, two, and three stages and found that the three stages had a better power coefficient, and the twisted blades showed better performance.

To decrease the variation in static torque in conventional Savonius rotors at a 0° to 360° rotor angle, Kamoji and Kedare [15] tested a helical rotor with a twist of 90°. They conducted experiments in an open-jet wind tunnel at gap ratios of 0.0, 0.05, and 0.08 to study the effect of gap ratio and Reynolds number on its performance and evaluated static torque, dynamic torque, and power coefficients. They compared its performance with and without a shaft between the end plates at different gap ratios. A helical rotor without a shaft was also compared with the performance of the conventional Savonius rotor. They found that all helical rotors have a positive coefficient of static torque at all rotor angles, but the rotors with a shaft had a lower power coefficient than those without. The coefficient of power of the rotor without a shaft with a 0.0 gap ratio was marginally less than the conventional Savonius rotor.

Gupta et al. [16] investigated the performance of two-bladed Savonius turbine with five overlaps of 16.2%, 20%, 25%, 30%, and 35%. Among them, 16.2% overlap condition showed maximum power extraction. The pressure drop across the rotor from upstream to downstream as well as the maximum pressure difference across the returning bucket is displayed in the same condition. Qasim et al. [17] worked with impeller scoop-frame type with movable vanes wind turbine (VAWT). The objective was to maximize the drag factor by closing the vanes on convex shape and opening when air hits the concave part. Due to the movement of vanes for and against the wind, a higher drag factor is worked on the impeller scoop-frame type with movable vanes and has higher efficiency than flat vanes.

Ghatage and Jyeshtharaj [18] have done an experiment by changing the shape of the blade as well as the blade number. They have studied with both regular curved blade and twisted curved blade. The experiment concluded that the two blades with twist enhance the efficiency of turbine. In their experiment, the 30°-twisted two-bladed turbine gave the better power coefficient. It can be concluded that the twisted blade attributes relatively higher drag over the turbine surface.

Kumbernuss et al. [19] studied two-staged Savonius-type turbines with different number of blades, the shape of the blades, the overlap ratio, and the phase shift angle. The wind turbine was tested under four different wind speeds of 4, 6, 8, and 10 m/s. There were three turbines with the overlap ratios of 0, 0.16, and 0.32. The overlap ratio of 0.16 produced the better performance among the three, followed by the 0.32 overlap ratio. At lower and higher air velocities, the larger and smaller phase shift angles, respectively, will produce better performance of the turbines.

Carrigan et al. [20] had the objective to introduce and demonstrate a fully automated process for optimizing the airfoil cross section of a vertical-axis wind turbine. The objective was to maximize the torque while enforcing typical wind turbine design constraints such as tip-speed ratio, solidity, and blade profile. This work successfully demonstrated a fully automated process for optimizing the airfoil cross section of a VAWT.

Researchers from different parts of the world have been investigating the aerodynamic characteristics of Savonius wind turbines and trying to identify the optimum design in order to achieve better performance compared to horizontal-axis wind turbines. Although much research has been going on experimentally and numerically on Savonius wind turbine performance improvement, there are few to no comprehensive studies using both experimental and numerical methods for various gap ratios at different Reynolds numbers. The primary goal of the present study is to investigate the aerodynamic characteristics of three-bladed Savonius wind turbines in order to contribute to the performance improvement of vertical-axis Savonius wind turbines. To achieve this goal, the authors designed and fabricated Savonius wind turbine scale models with no overlap ratio and two different overlap ratios, measured the pressure distribution around the Savonius turbine rotor models, and calculated the drag coefficients. Static torque was measured using the subsonic wind turbine for all models at varying angles of rotation, the mesh was generated numerically around all turbine models using GAMBIT, and fluid flow fields around the models were solved using k-ε turbulence model of FLUENT. Pressure contours, velocity contours, and torque were determined at various Reynolds numbers. A detail of the experimental and computational procedure of this research work can be found in the thesis work done by one of the authors [21].

Methods

Experimental measurement

Subsonic wind tunnel

A subsonic wind tunnel was designed and built to conduct the experimental measurement of this research as shown in Figure 1. The wind tunnel is 12-ft (3.66 m) long and consists of a converging mouth entry, honeycomb Section Background, test section, fan section, rectangle section, honeycomb Section Methods, converging diverging section, and rectangular exit section.

Mathematical expressions

The following equations were used to calculate various experimental aerodynamic coefficients corresponding to different Reynolds numbers (Re) and tip speed ratios (λ). Savonius rotor model dimensions and wind tunnel measured data were used as inputs for these equations.

Figure 1 Subsonic wind tunnel.

Rotor area:

$$A = DH. \tag{1}$$

Overlap ratio:

$$\mathrm{OR} = \frac{a}{D}. \tag{2}$$

Aspect ratio:

$$\mathrm{AR} = \frac{H}{d}. \tag{3}$$

Angular velocity:

$$\omega = \frac{2\pi N}{60}. \tag{4}$$

Reynolds number:

$$Re = \frac{VD}{\nu}. \tag{5}$$

Tip speed ratio:

$$\lambda = \frac{\omega D}{2V} \tag{6}$$

Torque coefficient:

$$C_\mathrm{q} = \frac{T}{\frac{1}{4}\rho ADV^2}. \tag{7}$$

Power coefficient:

$$C_\mathrm{p} = \frac{P}{\frac{1}{2}\rho AV^3} = \frac{T\omega}{\frac{1}{2}\rho AV^3} = C_\mathrm{q} \times \lambda. \tag{8}$$

Normal drag force:

$$F_\mathrm{n} = \int_0^\pi \Delta p \frac{d}{2} \cos\varnothing d\varnothing = \sum_{i=1}^{17} \Delta p_i \frac{d}{2} \cos\varnothing_i \Delta\varnothing_i. \tag{9}$$

Tangential drag force:

$$F_\mathrm{t} = \int_0^\pi \Delta p \frac{d}{2} sin\varnothing d\varnothing = \sum_{i=1}^{17} \Delta p_i \frac{d}{2} \sin\varnothing_i \Delta\varnothing_i. \tag{10}$$

Normal drag coefficient:

$$C_\mathrm{n} = \frac{F_\mathrm{n}}{\frac{1}{2}\rho V^2 A}. \tag{11}$$

Tangential drag coefficient:

$$C_\mathrm{t} = \frac{F_\mathrm{t}}{\frac{1}{2}\rho V^2 A}. \tag{12}$$

Drag force measurement

The pressure distribution around the concave and convex surfaces of each blade was measured experimentally using a semi-cylindrical three-bladed Savonius VAWT model with overlap distance, $a = 25$ mm, between the adjacent blades as shown in Figure 2. The rotor model was made of stainless steel with each blade diameter, $d = 125$ mm, height, $H = 300$ mm, and rotor diameter, $D = 225$ mm. The overlap ratio (OR) was 0.11, and no shaft was used through the rotor model. The whole rotor was mounted on an iron frame using two separate shafts and bearings at the two ends. The convex and concave surface pressures of each blade were measured at 17 tapping points using 1.5-mm outer diameter and 10-mm-long copper tubes which were press fitted into 17 tapping holes. Those tapping points were located at the mid-plane of each blade to measure the pressure at every 10° interval on the blade surface. The copper tubes were connected to 17 pressure transducers (PX277, Omega Engineering Inc., Stamford, CT, USA) through the 2-mm PVC tubes. Pressures were measured statically at every 30° interval of

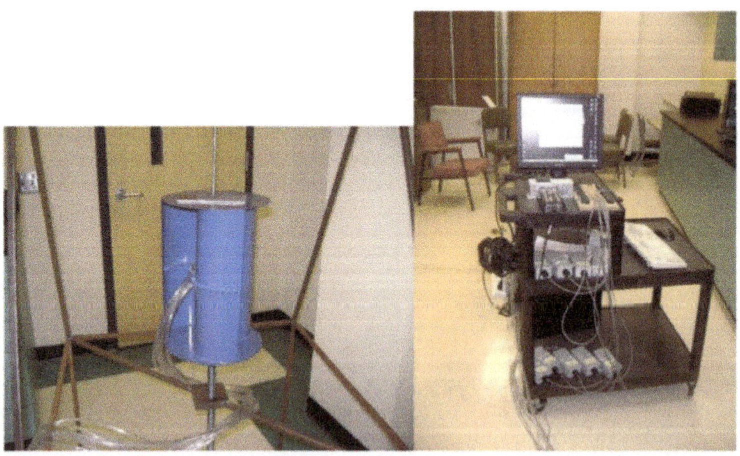

Figure 2 Setup of Savonius rotor model with pressure transducer data acquisition system for drag force measurement.

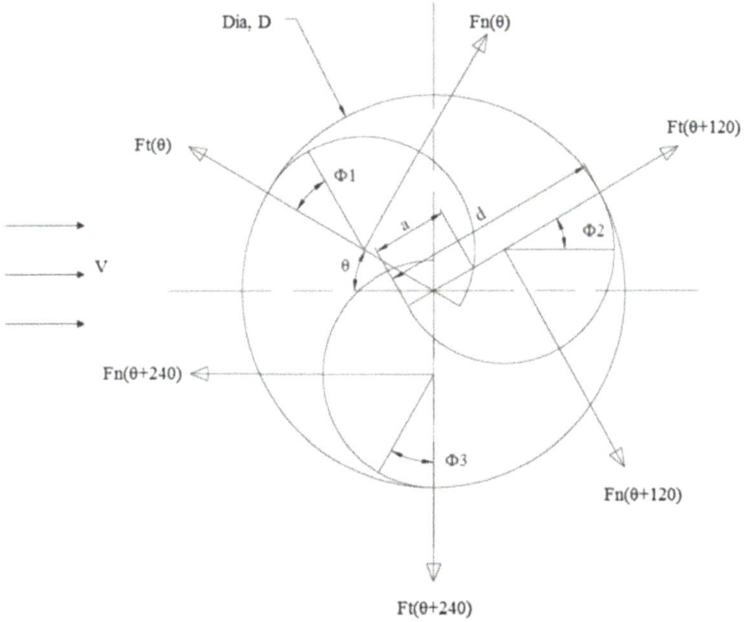

Figure 3 Schematic of the rotor model cross section. Showing the normal and tangential drag forces on each blade.

rotor angle. A personal computer equipped with a data acquisition system was used to record and edit the pressure data. Average wind speed during this experiment was 9.61 m/s. The Reynolds number based on rotor diameter was 1.47×10^5. The normal and tangential drag forces on each blade of the Savonius rotor model were calculated using the measured pressure difference between the concave and convex surfaces of the blades using Equations 9 and 10. Figure 3 shows the cross section of the rotor with the normal and tangential drag force directions. Tangential and normal drag coefficients were then calculated using Equations 11 and 12.

Three-bladed Savonius rotor models

To observe the effect of the OR (the ratio between the distance of the two adjacent blades and the rotor diameter) and Reynolds number on the aerodynamic characteristics of the Savonius rotor, three different rotor models with and without overlap ratios were designed and physically fabricated. Figure 4 shows the three Savonius rotor models with three different ORs. Model 1 was designed without any overlap between the adjacent blades and fabricated with three semi-cylindrical blades of diameter, $d = 127$ mm, and height, $H = 300$ mm. The blades of model 1 were made of acrylic and set 120° apart. The overall diameter of the rotor model 1 was $D = 248$ mm without any central shaft. Model 2 was designed with an overlap distance between the adjacent blades, $a = 25$ mm but having the same blade diameter and height as model 1, and was fabricated with the same number of blades 120° apart. The overall rotor diameter of model 2 was $D = 216$ mm with OR = 0.12. Model 3 was designed and fabricated with an overlap distance between adjacent blades, $a = 50$ mm,

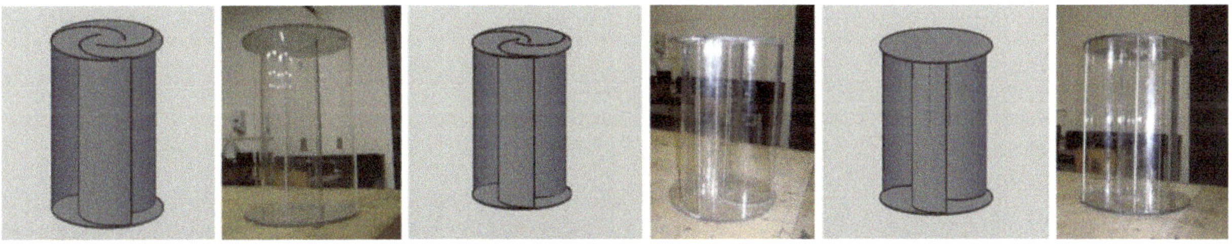

Model 1 Model 2 Model 3

Figure 4 3-D and fabricated views of three Savonius rotor models.

Figure 5 Setup of wind tunnel and Savonius rotor model for static torque measurement.

with the same blade diameter, height, and number of blades set 120° apart as model 1 and model 2. The overall rotor diameter of model 3 was $D = 192$ mm and OR = 0.26. These three models were tested in front of the subsonic wind tunnel for various Reynolds number flow conditions.

Static torque measurement

The experiment was carried out at three different wind speeds: $V = 9.66$, 8.23, and 7.33 m/s. The Reynolds numbers based on the rotor diameter varied from 9.94×10^4 to 1.6×10^5. Experiments were carried out, and data were recorded at room temperature. Static torque (T) for the three different models of the Savonius wind turbine was measured using a static torque meter (TQ-8800 model, Lutron Electronic Enterprise Co., Ltd., Taipei, Taiwan) at three different wind speeds. Torque meter output was in pound-inch which was then converted into Newton-meter. Rotational speed (N) was measured using a non-contact photo tachometer. Equation 4 was used to calculate the angular velocity from the rotational speed.

Savonius wind turbine is a drag-type VAWT where the lift forces are considered to be negligible. Figure 5 shows the experimental setup for torque measurement of a Savonius wind turbine model. When the wind strikes the blade surfaces of the model, two components of drag force are generated on each blade surface. Normal drag force (F_n) acts perpendicularly on the blade surface, and the tangential drag force (F_t) acts tangentially on each blade, as shown in Figure 3. The pressure difference between the concave and convex surfaces on each blade produces these tangential and normal drag forces. These components of drag force are responsible for torque generation within the turbine shaft and can be measured using a torque meter. Equation 7 is used to calculate the torque coefficient from the measured torque value. The power coefficient can be calculated from the measured torque and angular velocity of the rotor using Equation 8.

Numerical investigation

Numerical model selection

For the selection of the numerical model from the CFD code FLUENT 2D, a NACA 4412 airfoil was numerically

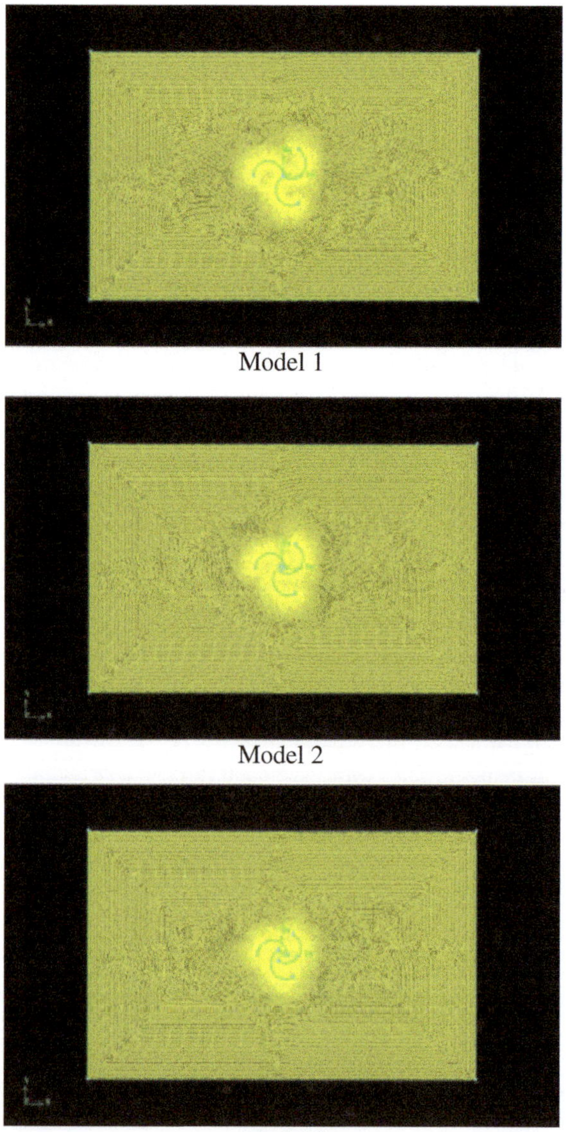

Figure 6 Generated mesh using GAMBIT.

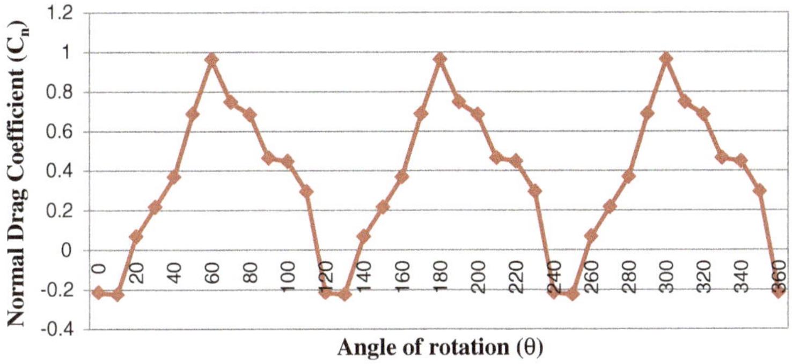

Figure 7 C_n **versus angle of rotation (**θ**) for a three-blade combined effect.**

examined at a different angle of attack and compared with established research results. Using the flow simulation results from both the inviscid model and the *k-ε* turbulence model, lift coefficient was calculated and then compared with published [22] NACA 4412 airfoil results. Comparing the inviscid model and the *k-ε* turbulence model results with established published results, it was found that the *k-ε* turbulence model gave more accurate results than the inviscid model. Therefore, the *k-ε* turbulence model was selected for numerical modeling of the Savonius wind turbine.

Numerical procedure

The *k-ε* turbulence model was used for the computational flow simulation around the Savonius rotor models with different overlap ratios. Commercially available software FLUENT was used to solve the turbulent flow field, and GAMBIT was used for mesh generation around the rotor models. Numerical simulation provides the pressure and velocity values at all nodal points of flow domain around the rotating blades. Figure 6 shows the 2-D mesh generated using GAMBIT within a computational domain around three-bladed Savonius wind turbine models in which the position of the three blades were 0°, 120°, and 240°. The size of the computational domain was 1.6 m × 1.4 m, and the total number of nodes was around 39,992. These computer-generated meshes were then exported into FLUENT for post-processing. The flow of air within the domain around the rotor model was assumed to be turbulent, and the effects of molecular viscosity were considered negligible. Also, the end effects of the turbine have been considered negligible. The simplest 'complete models' of turbulence are two-equation models in which the solution of two separate transport equations allows the turbulent velocity and length scales to be independently determined.

The standard *k-ε* turbulence model in FLUENT was used for the analysis of turbulent flow around rotor models. The pressure–velocity coupling is achieved using the well-known SIMPLE method by Patankar [23]. Turbulence kinetic energy (k) and turbulence dissipation rate (ε) first-order upwind scheme was chosen for the momentum equation solution. The standard *k-ε* turbulence model [24] is a semi-empirical model based on model transport equations for k and its ε. The model transport equation for k was derived from the exact

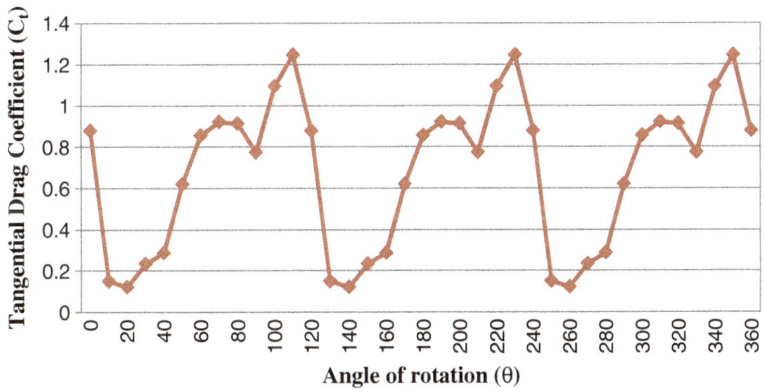

Figure 8 C_t **versus angle of rotation (**θ**) for the three-bladed rotor.**

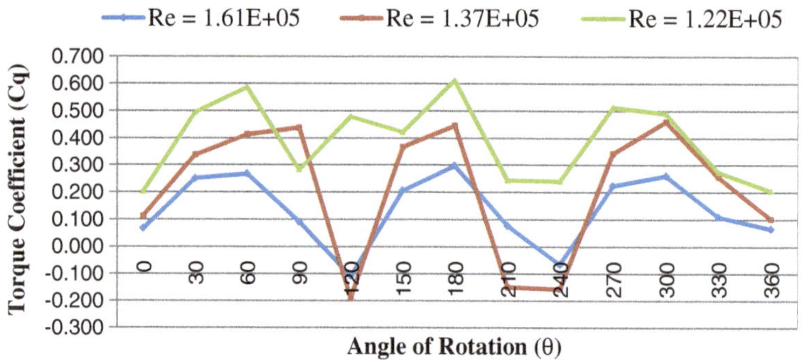

Figure 9 C_q **versus angle of rotation (*θ*) for model 1.**

equation, while the model transport equation for ε was obtained using physical reasoning and bears little resemblance to its mathematically exact counterpart.

The turbulence kinetic energy, k, and its rate of dissipation, ε, were obtained from the following transport equations:

$$\frac{\partial}{\partial t}(\rho k) + \frac{\partial}{\partial x_i}(\rho k u_i) = \frac{\partial}{\partial x_j}\left[\left(\mu + \frac{\mu_t}{\sigma_k}\right)\frac{\partial k}{\partial x_j}\right] + G_k + G_h - \rho\varepsilon - Y_M + S_k, \tag{13}$$

$$\frac{\partial}{\partial t}(\rho \varepsilon) + \frac{\partial}{\partial x_i}(\rho \varepsilon u_i) = \frac{\partial}{\partial x_j}\left[\left(\mu + \frac{\mu_t}{\sigma_\varepsilon}\right)\frac{\partial \varepsilon}{\partial x_j}\right] + c_{1\varepsilon}\frac{\varepsilon}{k}(G_k + C_{3\varepsilon}G_b) - C_{2\varepsilon}\rho\frac{\varepsilon^2}{k} + S_\varepsilon. \tag{14}$$

In these equations, G_k represents the generation of turbulence kinetic energy due to the mean velocity gradients; G_b, the generation of turbulence kinetic energy due to buoyancy; Y_M, the contribution of the fluctuating dilatation in compressible turbulence to the overall dissipation rate; and $C_{1\varepsilon}$, $C_{2\varepsilon}$, and $C_{3\varepsilon}$, constants. σ_k and σ_ε are the turbulent Prandtl numbers for k and ε, respectively. S_k and S_ε are user-defined source terms. The turbulent (or eddy) viscosity, μ_t, is computed by combining k and ε as follows:

$$\mu_t = \rho C_\mu \frac{k^2}{\varepsilon}, \tag{15}$$

where C_μ is a constant.

Boundary conditions were assigned with symmetric top and bottom, the left side was open with inlet free-stream velocity, and the right side was open with an atmospheric pressure outlet. Inlet air velocity was considered the same as the experimental values, i.e., 9.66, 8.23, and 7.33 m/s, and air density was considered at 1.2 kg/m^3. The blades were considered as moving walls and their rotational velocity was provided from the rpm measured during the experiment. The convergence of the sequential iterative solution is achieved when the sum of the absolute differences of the solution variables between two successive iterations falls below a pre-specified small number, which was chosen as 1×10^{-5} in

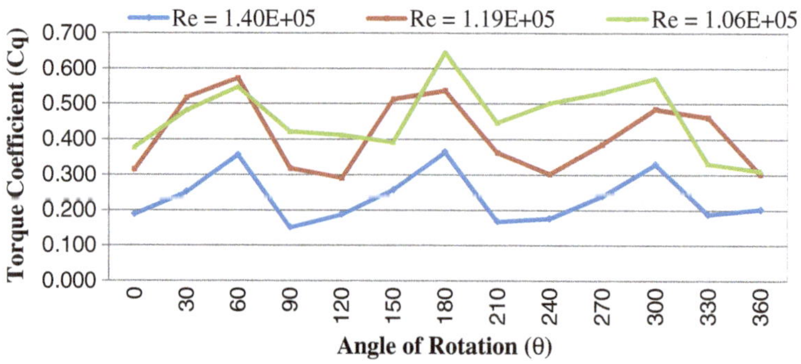

Figure 10 C_q **versus angle of rotation (*θ*) for model 2.**

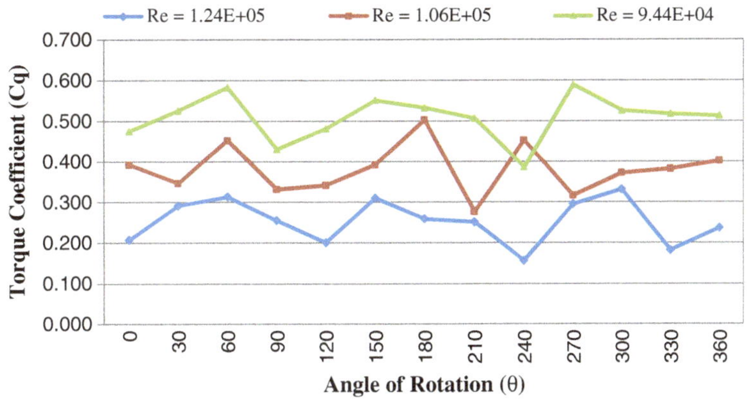

Figure 11 C_q **versus angle of rotation (*θ*) for model 3.**

this study. For all models using *k*-*ε* turbulence model convergence criteria, 1×10^{-5} was set and tested for continuity, *x* velocity, *y* velocity, *k*, and *ε*.

Results and discussion

Experimental results

Normal and tangential drag coefficients

Normal drag coefficient (C_n) variation with the change in angle of rotation (*θ*) for the three-bladed Savonius wind turbine model is shown in Figure 7. A three-blade combined effect at every 10° interval from 0° to 360° is shown in this plot. Normal drag coefficient increases with the increase of rotor angle from 0° to 60° and then decreases with the increase of rotor angle up to 100°. Normal drag coefficient is responsible for torque generation in the rotor model. The same pattern of normal drag coefficient repeats from 120° to 230° and from 240° to 350°.

Tangential drag coefficient (C_t) with change in rotor angle (*θ*) for every 10° interval from 0° to 360° is shown in Figure 8. The figure shows that a sharp drop occurs from 0° to 10° and then sharp increase occurs from 10° to 40°. Again, a sharp drop occurs in the tangential drag coefficient from 40° to 90°, and a sharp increase occurs from 90° to 120°. However, for every angle of rotation, the tangential drag coefficient remains positive which is a very important factor for producing thrust in the rotor model. The same pattern of tangential drag coefficient repeats from 120° to 230° and from 240° to 350° of angle of rotation.

Torque coefficient variation for three individual Savonius VAWT models

Torque coefficient of the Savonius wind turbine model 1 was calculated for three different Reynolds number. Figure 9 shows torque coefficient (C_q) variation with the increase of angle of rotation (*θ*). Torque coefficient was calculated for combined blade effect at every 30° interval from 0° to 360°. Three Reynolds numbers for model 1 were 1.61×10^5 (for wind speed 9.66 m/s), 1.37×10^5 (for wind speed 8.23 m/s), and 1.22×10^5 (for wind speed 7.33 m/s). For every Reynolds number, the values of

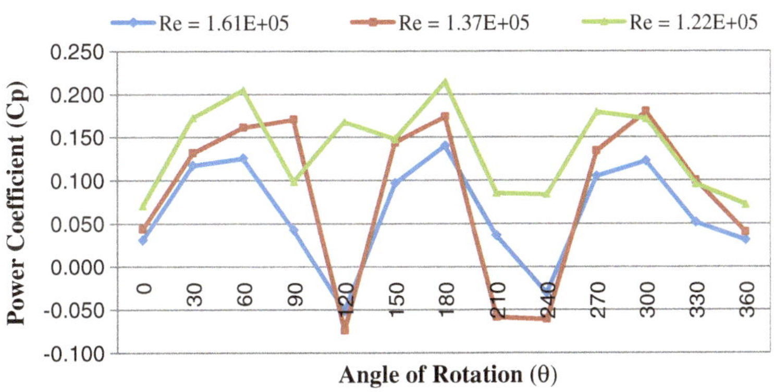

Figure 12 C_p **versus angle of rotation (*θ*) for model 1.**

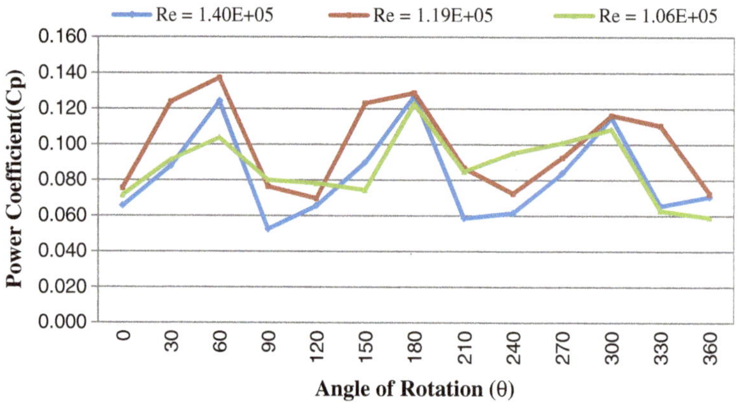

Figure 13 C_p **versus angle of rotation (*θ*) for model 2.**

torque coefficient increase from 0° to 60° and then start to decrease from 60° to 120°. The same pattern repeats for the rotor angle from 120° to 210° and from 240° to 330°. For Reynolds number, 1.00×10^5, the value of torque coefficient becomes negative at 120°, 210°, and 240°. It is desired to remove the negative torque for all rotor positions, as this negative torque causes reverse rotation which can reduce power output. Figure 10 shows C_q variation with the angle of rotation (θ) for rotor model 2. Similarly, the combined blade effect on torque coefficient with 30° interval from 0° to 360° was calculated. Reynolds numbers for model 2 were 1.40×10^5 (for wind speed 9.66 m/s), 1.19×10^5 (for wind speed 8.23 m/s), and 1.06×10^5 (for wind speed 7.33 m/s). From the figure, it can be seen that the torque coefficient increases from 0° to 60° and decreases at 90°, and again increases at 120° (except at $Re = 1.06 \times 10^5$). There was no negative torque coefficient for this model. The same pattern repeats for the rotor angle from 120° to 210° and from 240° to 330°. Figure 11 shows C_q variation with the increase of angle of rotation (θ) for model 3. Likewise other two models, combined blade effect on torque coefficient at every 30° interval from 0° to 360° was calculated. Reynolds numbers for model 3 were 1.24×10^5 (for wind speed 9.66 m/s), 1.06×10^5 (for wind speed 8.23 m/s), and 9.44×10^5 (for wind speed 7.33 m/s). For $Re = 1.24 \times 10^5$, 1.06×10^5, and 9.44×10^5, the pattern of the graph looks similar. The torque coefficient increases from 0° to 60° then decreases at 90°, and again increases at 120° (except for $Re = 1.24 \times 10^5$). The same pattern repeats for the rotor angle from 120° to 210° and from 240° to 330°.

Power coefficient variation for three individual Savonius VAWT models

Power coefficient (C_p) was calculated using the relationship between C_p and C_q which is $C_p = C_q \times \lambda$ at three different *Re* for all three models. Figure 12 shows C_p variation with angle of rotation (θ) from 0° to 360° for model 1. Trends of the plots are similar for $Re = 1.61 \times$

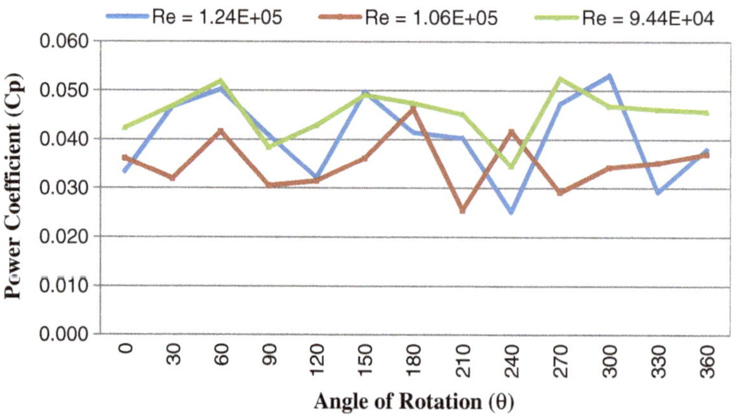

Figure 14 C_p **versus angle of rotation (*θ*) for model 3.**

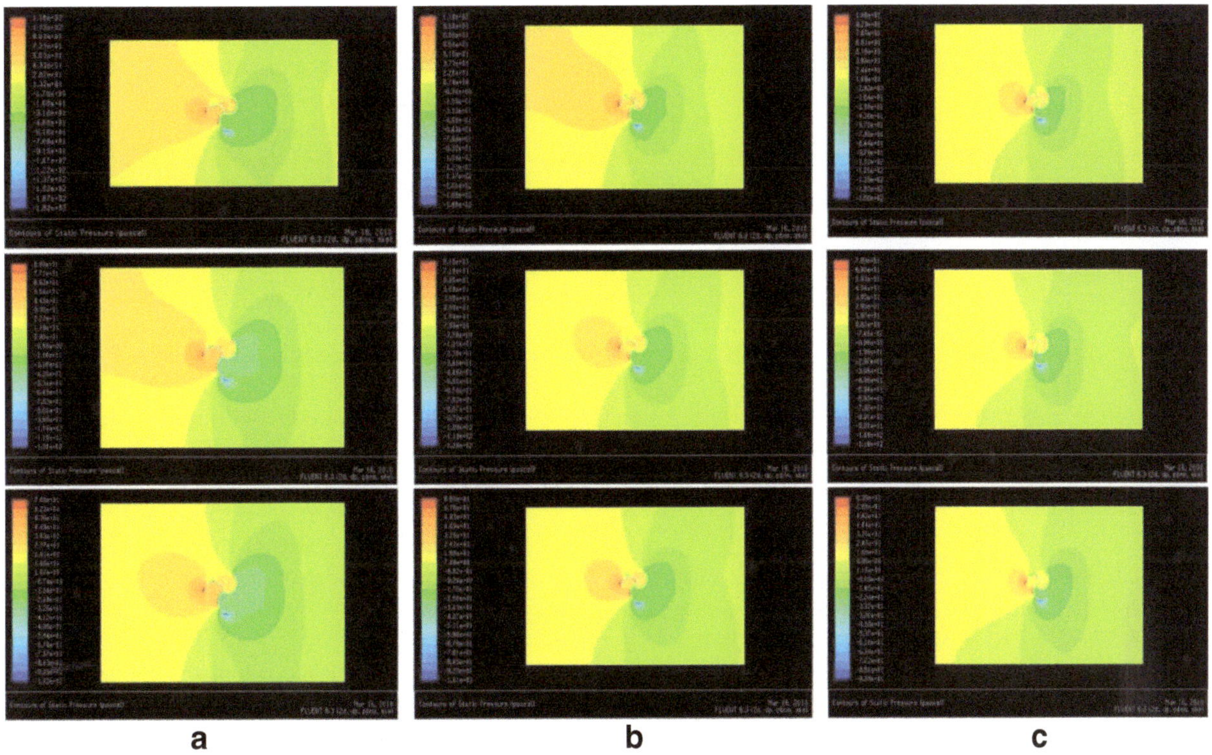

Figure 15 Pressure contours around Savonius rotor (a) model 1, (b) model 2, (c) model 3 for all three Reynolds numbers. (Reynolds numbers in descending order from top to bottom).

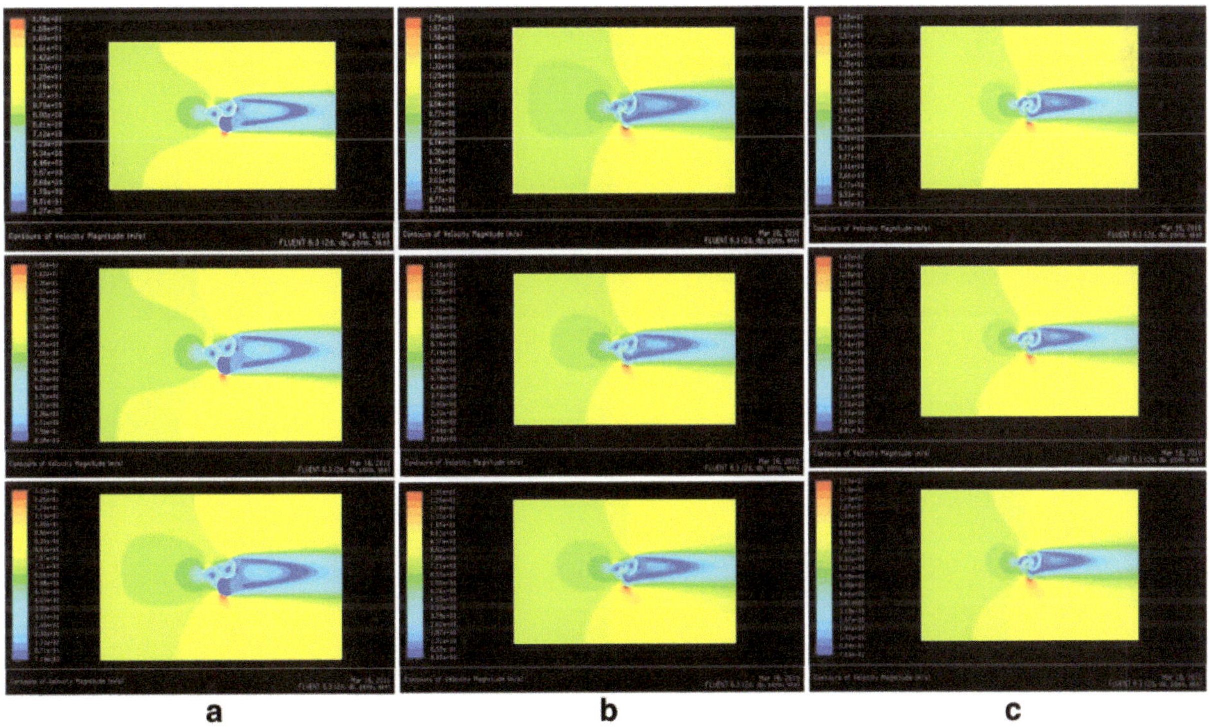

Figure 16 Velocity contours around Savonius rotor (a) model 1, (b) model 2, (c) model 3 for all three Reynolds numbers. (Reynolds numbers are in descending order from top to bottom).

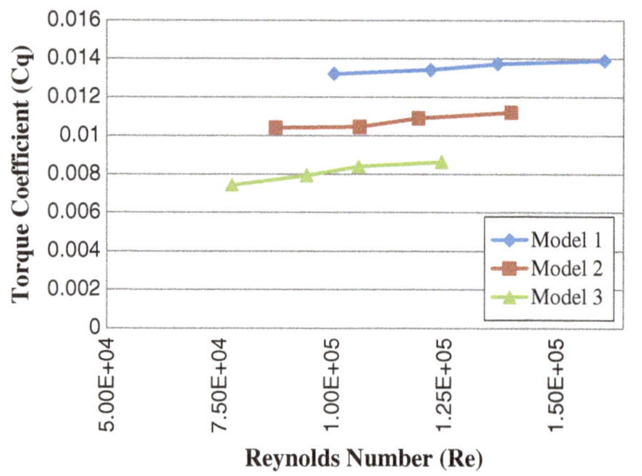

Figure 17 Numerical C_q versus *Re* for three models.

10^5, 1.37×10^5, and 1.22×10^5. Power coefficient was negative at 120°, 210°, and 240° for $Re = 1.37 \times 10^5$ and at 120° and 240° for $Re = 1.61 \times 10^5$. For this model, better power coefficient variation occurred at $Re = 1.22 \times 10^5$. Figure 13 shows C_p variation with angle of rotation (θ) at three different *Re* for model 2. There is no negative power coefficient for this model at any Reynolds number. Figure 14 shows C_p variation with the change of angle of rotation (θ) from 0° to 360° for model 3. For this model, power coefficient variation follows the similar trend for all Reynolds number, increasing from 0° to 60° and then decreasing up to 120° then repeats from 120° to 230° and from 240° to 330°.

Numerical results

Pressure contours for three models at three different Reynolds numbers

Pressure contours generated from numerical simulation of model 1, model 2, and model 3 for three different Reynolds numbers are shown in Figure 15a,b,c, respectively. For all these cases, higher pressure values were found at the convex side of the first blade Savonius rotor model. Negative pressure region was developed from the convex side of blade 2 to some portion of the convex side of blade 3. This negative pressure is creating pressure difference between the concave and convex surfaces that eventually rotates the turbine blades.

Velocity contours for three models at three different Reynolds numbers

Contours of velocity magnitude for Savonius rotor model 1, model 2, and model 3 at three different Reynolds numbers are shown in Figure 16a,b,c, respectively. Patterns of the contours are almost the same for different Reynolds numbers; the only exception is a slight variation in velocity magnitude. Once the wind strikes the turbine blades, the velocity starts to decrease at the trailing edge of the Savonius wind turbine model, but after some distance travel, the turbine blades start to regain their velocity. Higher velocity region was created at the top and bottom sides of the wind turbine model.

Numerical torque coefficient

Figure 17 shows the numerically calculated C_q variation with different *Re* for three different models. With the increase of Reynolds number, the torque coefficient slightly increases for all three models. Model 1 gives better torque coefficient compared to the other two models.

Comparison of numerical and experimental power coefficients

Numerical C_p was calculated by multiplying the numerical C_q and λ. Figures 18, 19 and 20 show the comparison of numerically and experimentally calculated C_p of the three Savonius rotor models with the increase of λ. Converged solutions of the power coefficient values were considered at all tip speed ratios for numerical results, whereas the power coefficient at four rotor positions 0°, 30°, 60°, and 90° were considered for experimental values. Combined blade effect was considered for both experimental and numerical calculations. Figure 18

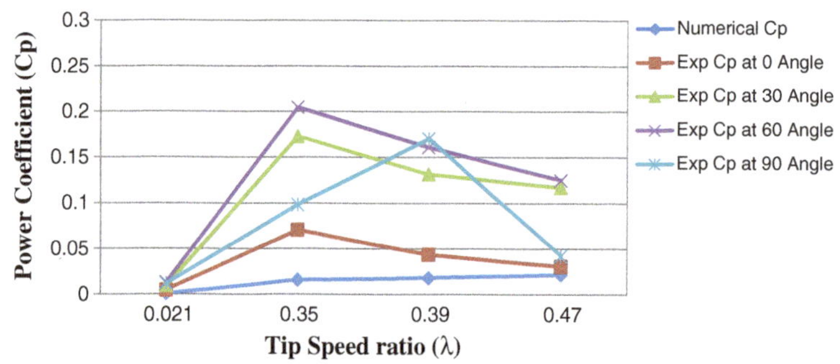

Figure 18 C_p versus λ for model 1.

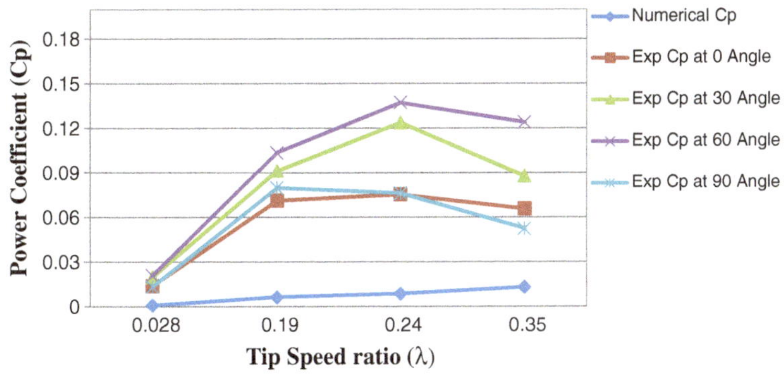

Figure 19 C_p **versus** λ **for model 2.**

shows that for model 1 experimental power coefficient at rotor position 0° is very close to the numerical results. However, the deviation is huge for rotor position 60°. Disturbance from the surrounding environment causes a sudden power coefficient increase for the experimental value at different rotor positions, whereas for numerical calculation of power coefficient, the boundary effect has to consider for simulation purpose which causes a reduction in converged numerical power coefficient value. Similar pattern of power coefficient variation is observed for model 2 and model 3 as shown in Figures 19 and 20. However, the magnitude of the C_p decreases for model 2 and model 3 for both numerical and experimental cases.

Error analysis

Normal drag coefficient, tangential drag coefficient, torque coefficient, and power coefficient were calculated both experimentally and numerically and compared. Experimental power coefficient matches well with the numerical results. The experimental results are slightly deviated from the numerical results. In both computational and experimental cases, there can be some possible sources of errors, but computational results are still more towards the ideal case except some assumptions that were made for the model formulation. In the experimental case, more possibilities of errors can be found because of equipment and human imperfection. This could be the reason for lower values of coefficients in the experimental case.

Conclusions

Three different three-bladed Savonius wind turbine scale models with different overlap ratios (model 1, no overlap; model 2, overlap ratio 0.12; and model 3, overlap ratio 0.26) were designed and fabricated for the current study. Aerodynamic characteristics of these models were experimentally investigated using the subsonic wind tunnel. Experimental investigation was performed at different Reynolds numbers. Numerical investigation was also performed to determine torque and power coefficients using GAMBIT and FLUENT. The current study shows that lower Reynolds number gave better C_q variation with the increase of the angle of rotation for each model. Model 2 demonstrates better experimental C_q for all three different wind speeds (9.66, 8.23, and 7.33 m/s). For model 1 with $Re = 1.22 \times 10^5$, model 2 with $Re = 1.19 \times 10^5$, and model 3 with $Re = 9.94 \times 10^4$, the experimental C_p shows higher and positive values compared to other Reynolds numbers. Model 2 shows the better experimental C_p at wind speeds of 9.66 and 8.23 m/s. However, for wind speed 7.33 m/s, model 1 shows the better C_p. Power coefficient calculated from the numerical method shows that it is always increasing with the

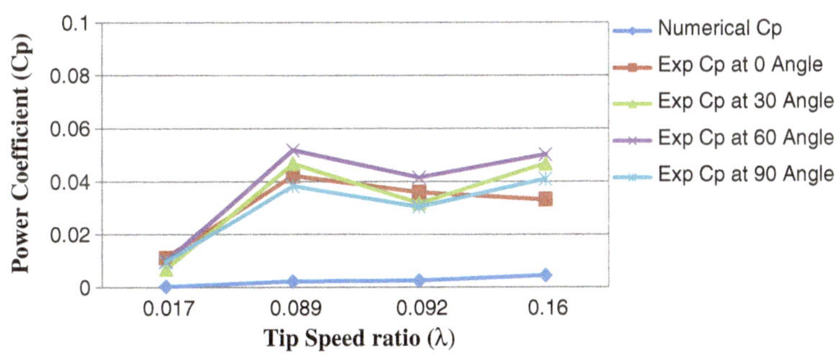

Figure 20 C_p **versus** λ **for model 3.**

increase of tip speed ratio. For model 1, numerical power coefficient matches well with the corresponding experimental values at 0° rotor position.

Abbreviations

a: overlap distance between two adjacent blades (mm); *A*: rotor area (m^2); AR: aspect ratio; C_n: normal drag coefficient; C_p: power coefficient; C_q: torque coefficient; C_t: tangential drag coefficient; *d*: blade diameter (mm); *D*: overall rotor diameter (mm); F_n: normal drag force (N); F_t: tangential drag force (N); *H*: rotor height (mm); *k*: turbulence kinetic energy; *N*: rotational speed (rpm); OR: overlap ratio: ratio of overlap distance between two adjacent blades and rotor diameter (OR = *a* / D); *P*: power (W); *Re*: Reynolds number; *T*: torque (N.m); *V*: wind velocity (m/s); *v*: kinematic viscosity (m^2/s); ρ: air density (kg/m^3); θ: angle of rotation (°); Φ: angular position of the pressure tapping points on three blades (°); ω: angular velocity (rad/s); λ: tip speed ratio; Δp: pressure difference (Pa); ε: turbulence dissipation rate.

Competing interests

The authors declare that they have no competing interests.

Authors' contributions

KNM carried out both experimental and computational investigations on the performance improvement of the three-bladed Savonius wind turbine with varying overlap ratios under the supervision of MR. GM and MA were also in the research group who gave their inputs time to time. All authors have their contributions in different percentages in developing the manuscript of this paper. All authors read and approved the final manuscript.

Acknowledgments

The authors acknowledge the funding and facility support from the Mechanical Engineering Department of Georgia Southern University. They would also like to thank Mr. Andrew Michaud for helping in fabricating different Savonius wind turbine models.

Author details

[1]Mechanical Engineering Department, Colorado State University, Fort Collins, CO 80523, USA. [2]Mechanical Engineering Department, Georgia Southern University, Statesboro, GA 30460, USA. [3]Biochemistry, Chemistry & Physics, Southern Arkansas University, Magnolia, AR 71753, USA.

References

1. Hayashi, T, Li, Y, Hara, Y, Suzuki, K: Wind tunnel test on a three stage out phase Savonius rotor. JSME International Journal **48**(1), 9–16 (2005)
2. Diaz, F, Gavalda, J, Massons, J: Drag and lift coefficients of the Savonius wind machine. J. Wind. Eng. **1991**(15), 240–246 (1991)
3. Sawada, T, Nahamura, M, Kamada, S: Blade force measurement and flow visualization of Savonius rotors. Bulletin of JSME **29**, 2095–2100 (1986)
4. Aldoss, TK, Obeidat, KM: Performance analysis of two Savonius rotors running side by side using the discrete vortex method. Wind. Eng. **11**, 79–88 (1987)
5. Fujisawa, N, Gotoh, F: Experimental study on the aerodynamic performance of a Savonius rotor. ASME J Sol Energ Eng **116**, 148–152 (1994)
6. Rahman, M, Islam, MQ, Islam, AKMS: Prediction of dynamic characteristics of a three bladed Savonius rotor. In: Proceeding of the third international conference on fluid mechanics and heat transfer. Dhaka, Bangladesh. 15–16 December 1999
7. Rahman, M: Torque and drag characteristics of a three bladed Savonius rotor. Mechanical Engineering Department, Bangladesh University of Engineering and Technology, Bangladesh (1999). M.Sc. thesis
8. Rahman, M, Islam, MQ, Islam, AKMS: Aerodynamic characteristics of a three bladed Savonius rotor. In: Proceeding of the 2nd international seminar on renewable energy for poverty alleviation. IEB, Bangladesh. 26–27 November 1999
9. Rahman, M, Morshed, KN, Lewis, J, Fullar, M: Experimental and numerical investigations on drag and torque characteristics of three-bladed Savonius wind turbine. In: Proceedings of 2009 ASME International Mechanical Engineering Congress and Exposition (IMECE 2009). Lake Buena Vista, Orlando. 13–19 November 2009
10. Gupta, R, Biswas, A, Sharma, KK: Comparative study of a three bucket Savonius rotor with a combined three-bucket Savonius–three-bladed Darrieus rotor. Renew Energy **33**, 1974–1981 (2008)
11. Altan, BD, Atilgan, M, Ozdamar, A: An experimental study on improvement of a Savonius rotor performance with curtaining. Exp. Therm. Fluid. Sci. **32**, 1673–1678 (2008)
12. Sargolzaei, J, Kianifar, A: Modeling and simulation of wind turbine Savonius rotor using artificial neural networks for estimation of the power ratio and torque. Simulat. Model Pract. Theor. **17**, 1290–1298 (2009)
13. Altan, BD, Atilgan, M: An experimental and numerical study on the improvement of the performance of Savonius wind rotor. Energ. Convers. Manag. **49**, 3425–3432 (2008)
14. Saha, UK, Thotla, S, Maity, D: Optimum design configuration of Savonius rotor through wind tunnel experiment. J. Wind. Eng. Ind. Aerod. **96**, 1359–1375 (2008)
15. Kamoji, MA, Kedare, SB: Wind tunnel tests on a single stage helical Savonius rotor. In: 5th AIAA International Energy Conversion Engineering Conference. St. Louis, Missouri. 25–28 June 2007
16. Gupta, R, Das, R, Gautam, R, Deka, SS: CFD analysis of a two bucket Savonius rotor for various overlap conditions. ISESCO J. Sci. Tech. **8**(13), 67–74 (2012)
17. Qasim, AY, Usubamatov, R, Zain, ZM: Investagation and design impeller type vertical axis wind turbine. Australian J Basic Appl Sci **5**(12), 121–126 (2011)
18. Ghatage, SV, Jyeshtharaj, BJ: Optimisation of vertical axis wind turbine: CFD simulations and experimental measurements. Can. J. Chem. Eng. **90**, 1186–1201 (2012)
19. Kumbernuss, J, Chen, J, Yang, HX, Lu, L: Investigation into the relationship of the overlap ratio and shift angle of double stage three bladed vertical axis wind turbine (VAWT). J. Wind Eng. Ind. Aerod. **107**, 57–75 (2012)
20. Carrigan, TJ, Brian, HD, Zhen, XH, Bo, PW: Aerodynamic shape optimization of a vertical-axis wind turbine using differential evolution. International Scholarly Research Network ISRN Renewable Energy **2012**, 1–16 (2012)
21. Morshed, KN: Experimental and numerical investigations on aerodynamic characteristics of Savonius wind turbine with various overlap ratios. Georgia Southern University (2010). M.S. thesis
22. Pinkerton, RB: Calculated and measured pressure distribution over the midspan section of the NACA 4412 airfoil. NACA report no. **563**, 365–380 (1936)
23. Patankar, SV: Numerical Heat Transfer and Fluid Flow. Hemisphere Publishing Corporation, Washington, D.C. (1980)
24. Launder, BE, Spalding, DB: Lectures in Mathematical Models of Turbulence. Academic Press, London (1972)

Simulation on the effect of the blade number on the rotational characteristic on a horizontal axis river current turbine

Aditya Rachman*, Ridway Balaka, Jenny Delly and Yuspian Gunawan

Abstract

The power system based on the horizontal axis turbines has been extensively employed to harness the kinetic energy of the moving fluid to generate the clean renewable energy in several domains, such as wind, river, and sea. The characteristic of the rotational speed is one of the prominent aspects in designing the power based on the horizontal axis turbine. This is the characteristic that can determine the quantity of the electricity production in the generator on the power technology system. The number of blades is another important aspect in designing the technology. It is believed that the number of blades not only determines the aerodynamic performance, the construction, and the production cost but also the turbine rotational speed characteristics. This study investigates the effect of the blade number on the rotational speed characteristic of a horizontal axis river current turbine. The methodology of generating the investigation is a parametric study using the blade element momentum theory. The result shows that the turbines with low blade number have high rotational speed characteristics while the turbines with high blade number have low rotational speed characteristics. The very low attack angle which yields into a low lift coefficient is one of the factors responsible for the inability of the turbines with high blade number to operate at high rotation. A further discussion on the results on this investigation relating to the hydrodynamic behavior is explored in this study. It is recommended that when a turbine is decided to have a high blade number, the transmission ratio should be high to obtain a high rotation in the generator.

Keywords: River turbine; Design; Blade number; Rotational operation

Background

The role of energy in the economic development in all nations is so indispensible [1,2]. The industry, transportation, trade, and communication will be posing massive challenges without the support of energy. Unfortunately, many regions, especially in developing nations, still seem to pose the problem of energy access. The topography and geography characteristic and the condition of a dispersed population on a region are some main obstacles for an effort on the distribution of energy [3,4].

The power generation based on the river current turbine is one of the clean decentralized renewable technology potential to supply the energy to the regions where the conditions of the topography, geography, and population become the obstacles for the distribution of energy. The working principle of this technology is based on the conversion of the kinetic energy of the river flow into electricity. This technology requires no level difference (head) and less civil works and causes less detrimental environmental impacts than the conventional hydroelectric power [5].

There are two main classes of the turbine configurations: horizontal axis and vertical axis turbines. Some papers [6-10] show the success of several projects in generating the electricity from water flow, such as rivers, tides, and ocean currents using the horizontal and vertical axis turbines. The horizontal turbines have several beneficial features than the vertical turbines. Some of these features are higher efficiency, easier start-up, less torque fluctuations, and higher rotation speeds [11]. A study [11] reports a survey on research, development, and demonstration on the water current turbine technology which is conducted presently or had been dismissed from the stage of the proof concept to the

* Correspondence: aditya_rchmn@yahoo.com
Department of Mechanical Engineering of Haluoleo University, Anduonohu Kendari 93232 Indonesia

commercialization. The results on this survey show that the horizontal axis turbines are more frequently employed than the vertical turbines.

In the mechanism of energy conversion in a modern horizontal turbine, the fluid kinetic energy is intercepted by the rotor blades to produce lift, as shown in Figure 1. This force is utilized to generate a torque, which is used to drive the generator shaft to produce electricity [12,13].

The mechanism of the energy production in the generator is based on the production of the electromotive force (EMF). Faraday's law states that the EMF will increase when a permanent magnet is moved relatively to a conductor, or vice versa, to give the effect of the flux change. The amount of EMF is proportional to the rate of the flux change.

The rotational speed characteristic of the generator shaft plays an essential role in determining the rate of the flux change. This is because the shaft drives one of the two components inside the generator, the conductor or the magnet, to generate the relative movement in order to produce the flux change. The role of the rotational speed characteristic of the turbine is also salient in determining the rate of the flux change. This is because generally the shaft of the generator is connected to the shaft of the turbine via a transmission system. When the turbine rotational speed is high, the rotational speed on the generator shaft is possibly high, making the rate on the flux change to be high and vice versa. Thus, in the design stage of the power technology based on the horizontal axis river current, the turbine rotational speed characteristics should conform to that of the generator rotation speed; otherwise, the rate on the flux change will fail to meet its requirement, and thus, the energy production in the generator is highly possibly unsatisfactory, resulting in an inefficient power technology.

Instead of the aerodynamic performance, the manufacturing cost, and the construction, the design parameter of the blade number seems to have a connection to the rotational speed characteristics of the turbine. Studies [14-18] indicate the role of the rotational speed on the effect of the blade number on the performance of horizontal axis wind turbines. These studies show that the turbines with high blade number perform better at low rotation speeds, while those with lower number obtain high performance at high rotation speeds.

In the water current domain, several papers [6-10,19-21] show some projects on the power technology based on horizontal axis turbines, in some fields such as tidal streams, ocean currents, and river flows, conducted by several companies and universities. Most of these papers provide the detailed information of the number of blades utilized in the projects. However, it seems that the explanation on the connection between the blade number and the turbine rotational speed cannot be found in these papers. Some of the papers seem to focus on the effect of the blade number on the hydrodynamic performance.

A study [8] shows a series of pilot projects on horizontal axis river current turbines by the Northern Territory University, Australia. The projects show that the change in the blade number is always accompanied by the change on the transmission ratio. Turbines with higher blade number require a higher transmission ratio than that of the turbines with lower blade number. This may indicate that the change on the transmission ratio is conducted to maintain the generator rotational speed when the number of blades is varied. A reference [13] explains the relation between the tip speed ratio (TSR), a ratio of the turbine rotational speed and the ambient flow velocity, and the solidity of rotor, a ratio of the area of the blades and the rotor swept area. A rotor with a

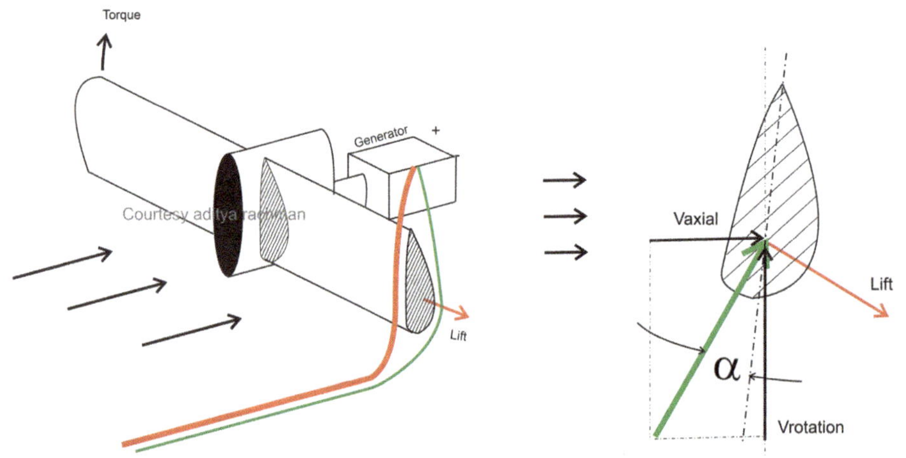

Figure 1 The mechanism of the energy conversion in a modern horizontal turbine. The lift force and the velocity in a blade. Illustrated by Rachman adopted in [12,13].

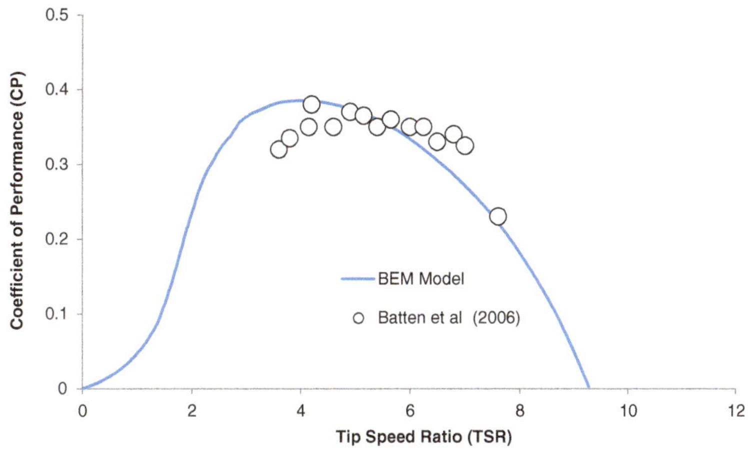

Figure 2 The results in the validation of the BEM model to the experimental results in [21].

high speed will have less blade area than the rotor of a slower machine. Another reference [12] explains that a low solidity rotor is intended to obtain a high rotational speed in order to reduce the gearbox ratio. These explanations can assert the indication in the previous references regarding the connection of the rotational speed and the blade number. As the lower bladed turbine will obviously have a less blade area than that of the turbine with high blade number, the rotational speed of the lower bladed turbine will be higher than that of the turbine with high bladed number. Although some of these references can indicate more relationship between the turbine rotational speed and the number of blades, a comprehensive discussion of the phenomenon seems to be less explored.

The knowledge on the behavior on the connection between the blade number and the turbine rotational speed can assist the designer in obtaining an appropriate decision in the design stage, especially in selecting the specification of the generator and the transmission ratio when the blade number is decided, in order to contribute in obtaining a technology with a desirable performance outcome. This study aims to investigate the effect of the blade number on the turbine rotational speed characteristic of the horizontal axis river current turbine. This study also attempts to explore more comprehensively the reasons behind the results of the investigation.

Methods

This investigation is conducted by a parametric study using the mathematical model of the Blade Element Momentum (BEM).

Blade element momentum model

In wind domain, there are many approaches, theories, and models to evaluate the performance of the horizontal axis turbine. The model of BEM is one of the oldest methods and remains to be the most widely used

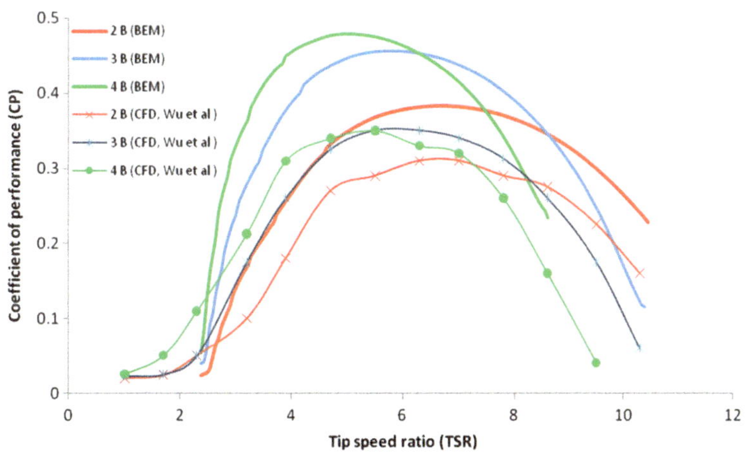

Figure 3 Coefficient of performance versus tip speed ratio in the variation of blade number.

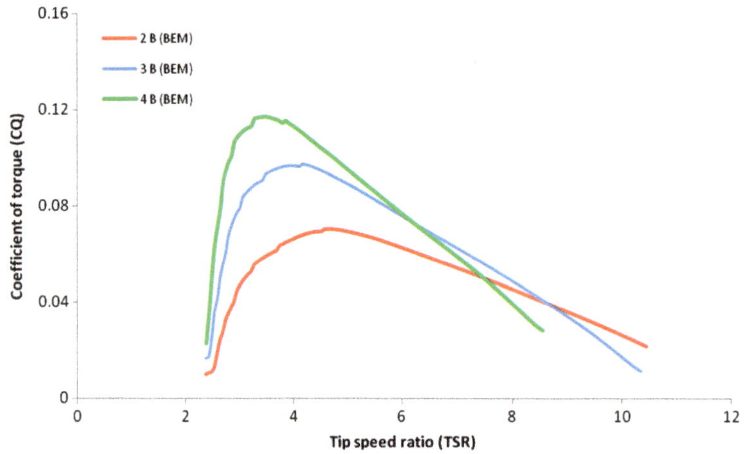

Figure 4 **Coefficient of torque versus tip speed ratio in the variation of blade number.**

method [22]. This model allows the user to evaluate the force (lift) acting on the blades, the torque in the shaft, and the power at the blade with different wind speeds [23].

It is possible to apply this model into the water domain. One of the reasons is the similarity in the working principle of the horizontal wind turbine and that of the water current turbine [24]. The mechanism of the water current turbine is to extract the kinetic energy of the flow of water, similar to the mechanism in the wind turbine. According to [25], the mathematical equation of the water current turbines is analogous to the wind turbines.

The development of BEM model is based on the combination of the momentum theory and the blade element theory [12,13]. The momentum theory is developed form the Bernoulli equation and the momentum balance equation. Both equations can be applied into not only air but also water. The blade element theory is based on the equation of lift and drag forces on the cross section of a blade. In water domain, the lift and drag force mechanism has been applied in several applications such as the boat propeller and the wing on the boat hull.

Validation of BEM model

To validate the BEM model, before being utilized in the investigation, a result from a parametric study using the BEM model will be compared to the experimental result obtained from a study in [21]. In this parametric study, the blade number is set to be three with the diameter of 0.8 m. The blade angle is 25° at the hub and 5° at the tip (twisted). The hydrofoil type is NACA 638xx. The water velocity is 1.54 m/s. These data are similar to those in [21].

It is shown that the data obtained from the BEM model and those of the experimental result in [21] show a good agreement (see Figure 2). It can be seen that the curve resulted from the BEM model is close to some of the data in the experimental result.

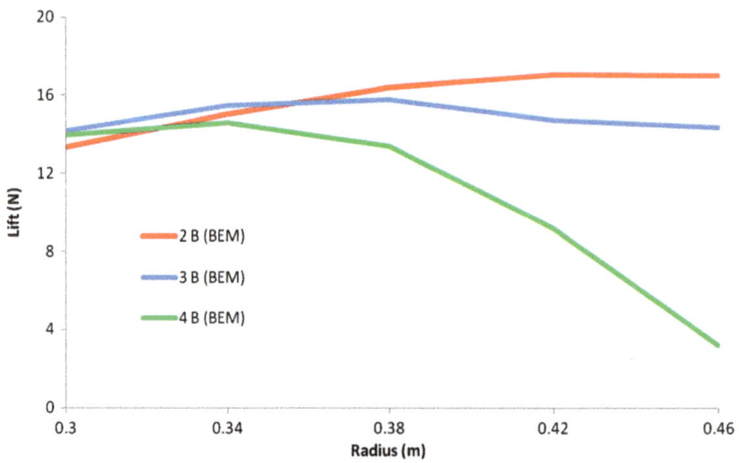

Figure 5 **Lift forces versus radius (> 0.3 m) in different blade numbers in TSR 11.**

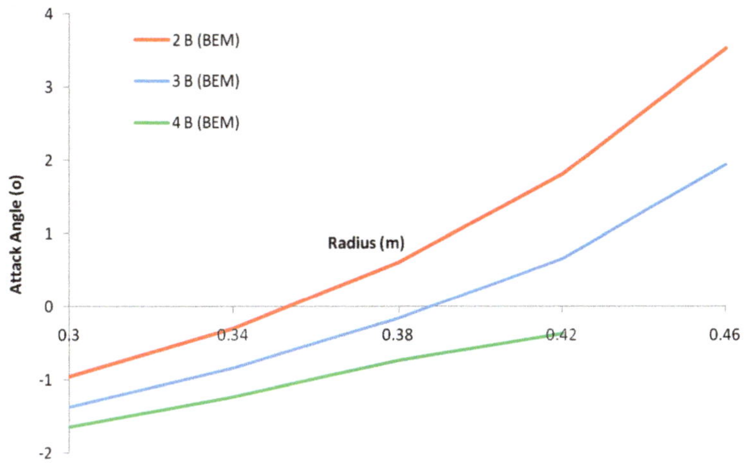

Figure 6 Attack angles versus radius (> 0.3 m) in different blade numbers in TSR 11.

Data of investigation

In the investigation on the effect of the blade number to the rotational speed characteristic of the river turbine employing BEM model, the input parameter of the turbine radius is 0.3 m and the hub radius is 0.07 m. The hydrofoil type is NACA 63415. The data of the drag and lift coefficients of this hydrofoil are obtained from [26]. The blade width is 0.14 m at the hub and 0.02 m at the tip (tapered). The blade angle is 23° at the hub and 0° at the tip (twisted). The water velocity is 2 m/s. The rotational speed of the turbine is varied from 0 to 400 RPM. The number of blades is varied by two, three, and four.

The result of this parametric study is presented in a graph illustrating the variation of the rotational speed of the turbine blade, represented by TSR and the blade number on the coefficient of performance generated. To validate the result of this study, it compares to the computational results obtained in [20]. To obtain additional information that can be used to explain the reasons of the investigation results, this study presents the data of the lift force, the axial rotor flow velocity, and the attack angle. These additional data are obtained during the calculation of the coefficient of performance using the BEM model.

MATLAB program is employed to calculate the coefficient of performance and the additional data as required by the iteration process in some of stages of the BEM calculation.

Results and discussion

The results on the investigation using BEM model show a good agreement to the computational results in [20]. Even the amount of the coefficient of performance is different; both numerical results show a similar tendency of the higher rotational speed range of the turbine with lower blade number than that of the turbine with higher blade number. This is because the power is still possible to be generated at the high rotational speed for the low-bladed turbine.

The turbine with two blades possibly operates at TSR of more than 11, while the turbine with four blades

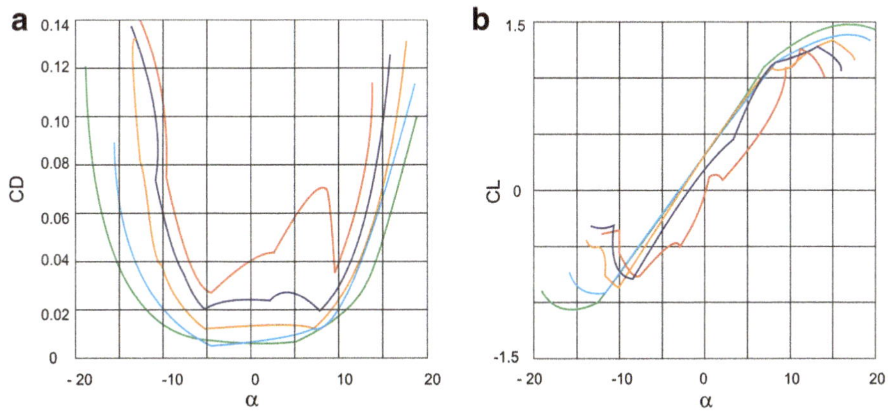

Figure 7 The coefficients of (a) drag (b) lift of NACA 63415.

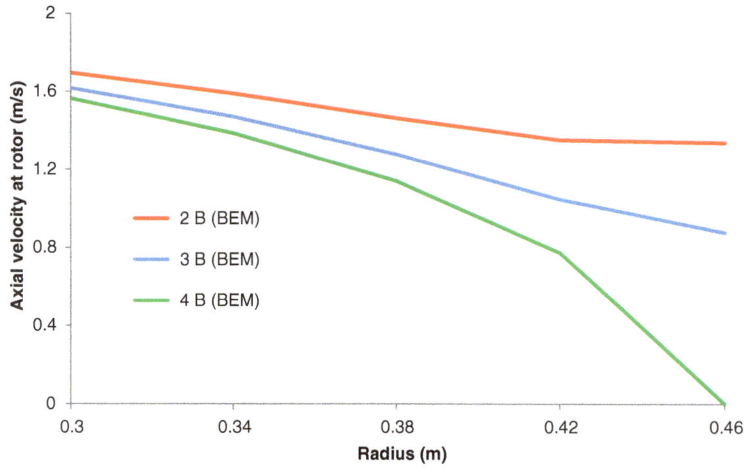

Figure 8 **Axial velocities versus radius (> 0.3 m) in different blade numbers in TSR 11.**

seems impossible to operate at TSR of more than 11 (Figure 3). The turbine with two blades still possibly produces the power at TSR of more than 11 as the torque can still be generated at a corresponding rotation condition (Figure 4). In contrary, the torque generated for the turbine with four blades at TSR of more than 11 seems to be highly impossible; thus, the turbine seems to be impossible to generate satisfactorily the power beyond TSR of 11.

At a high rotational speed (TSR 11), the lift forces in high radius for the turbine with four blades tend to be very low (see Figure 5). This results in the low torque generated of the turbine at the high rotational speed. For the lower bladed turbines, the lift forces in high radius are still relatively high at the high rotational speed, making the turbines maintain the torque at corresponding rotational speed.

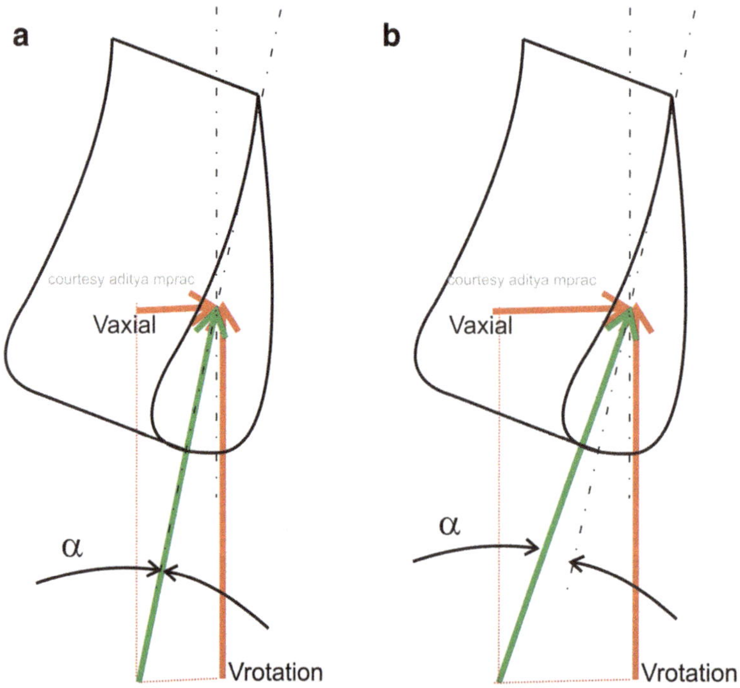

Figure 9 **The condition of attack angle and axial velocity at a high rotational speed.** **(a)** High-bladed turbine. **(b)** Low-bladed turbine.

At corresponding rotational speed condition, the attack angles of the turbine with four blades in high radius are very low (negative) (see Figure 6). This results in the low lift coefficient. This is due to the value of the lift coefficient for the NACA 63415 that would be low, even negative, at the very low attack angle (see Figure 7b). As the lift coefficient is low or even negative, the lift generated will be low or possibly be negative. For the turbines with lower blade number, the attack angles for the high radius are relatively not too low. This possibly makes the lift still possible to exist, thus making the turbines to still generate the torque at the high rotational speed.

At TSR of 11, the axial velocities at rotor in the high radius for the turbine with four blades are much lower than those of the turbines with less blade number (see Figure 8). This low axial velocity may cause the low attack angle. The following paragraph explains the relation of the low axial velocity at rotor and the low attack angle for the high-bladed turbine at the high rotational speed.

In the BEM model, the amount and the direction of the wind acting on the turbine blade depend on the sum vector of the axial velocity and the blade rotational speed. As both velocities (the axial velocity and the rotational speed) are perpendicular with each other, in the condition of the very low axial velocity and the high rotational speed, the position of the velocity acting on the turbine blade tends to be closely parallel to the rotational speed. This causes a low attack angle (see Figure 9).

The condition of the low axial velocity at the high rotational speed for the turbine with high blade number could be the blockade effect created by the number of blade area interacted with fluid. For the turbines with lower blade number, the blockade effect is relatively less. This makes the lower bladed turbines possible to obtain a relatively higher attack angle at the high rotational speed than that of the high-bladed turbine.

Conclusion

This study has investigated the effect of the blade number on the rotational speed characteristic of a horizontal axis river current turbine. Following are some conclusions that can be drawn.

The small number of blades allows the turbine to operate at the high rotational speed; in contrast, it is impossible for the turbine with high blade number to operate at the high rotation. One of the factors responsible is the inability of the high-bladed turbine to generate the power at the high rotational speed. This is because the attack angle at corresponding rotational speed is very low, resulting in low lift coefficient. In contrary, for the turbine with less blade number, it is possible to obtain a relatively more moderate attack angle at the high rotational speed, to allow the turbine obtaining the high lift coefficient at the high rotation speed.

The low rotational speed characteristic for the high-bladed turbine requires a high gearing ratio for the mechanical transmission system. As a result, it is possible to obtain high looses as more friction surfaces can exist caused by the requirement of the multi-gearing system. This may, thus, reduce the overall river power performance. In addition, the high ratio may add the dimension of the transmission on the power based on river current; thus, it may increase the weight which in turn influences the construction consideration.

Competing interests

The authors declare that they have no competing interests.

Authors' contributions

AR developed a mathematical code for BEM Calculation in MATLAB. AR and RB conducted the parametric study, collected the references and the informations, and drafted the manuscript. JD and YG corrected the draft manuscript. All authors read and approved the final manuscript.

Acknowledgements

AR developed a mathematical code for BEM Calculation in MATLAB. AR and RB conducted the parametric study, collected the references and the informations, and drafted the manuscript. JD and YG corrected the draft manuscript. All authors read and approved the final manuscript. In addition: Thank to the head of computer laboratory of Mechanical Engineering Department of Haluoleo University due to MATLAB program facility.

References

1. Oyedepo, S.O.: Efficient energy utilization as a tool for sustainable development in Nigeria. Int J Energy Environ Eng **3**, 11 (2012)
2. Kaundinya, D.P., Balachandra, P., Ravindranath, N.H.: Grid-connected versus stand-alone energy systems for decentralized power. Renew Sustain Energy Rev power rev literature **13**(8), 2041–2050 (2009)
3. Lal, S., Raturi, A.: Techno-economic analysis of a hybrid mini-grid system for Fiji islands. Int J Energy Environ Eng **3**, 10 (2010)
4. Dasuki, A.S., Djamin, M., Lubis, A.Y.: The strategy of photovoltaic technology development in Indonesia. Renew Energy **22**(1–3), 321–326 (2001)
5. Khan, M.J., Iqbal, M.T., Quaicoe, J.E.: River current energy conversion systems: Progress, prospects and challenges. Renew Sustain Energy Rev **12**, 2177–2193 (2007)
6. Guney, M.S., Kaygusuz, K.: Hydrokinetic energy conversion systems: a technology status review. Renew Sustain Energy Rev **14**, 2996–3004 (2010)
7. Sornes, K.: Small-scale water current turbines for river applications. (2010). http://www.zero.no/publikasjoner/small-scale-water-current-turbines-for-river-applications.pdf. Accessed 1 January 2012
8. Anyi, M., Kirke, B.: Evaluation of small axial flow hydrokinetic turbines for remote communities. Energy Sustain Dev **14**, 110–116 (2010)
9. Li, D., Wang, S., Yuan, P.: An overview of development of tidal current in China: energy resource, conversion technology and opportunities. Renew Sustain Energy Rev **14**, 2896–2905 (2010)
10. Rourke, F.O., Boyle, F., Reynolds, A.: Marine current energy devices: current status and possible future applications in Ireland. Renew Sustain Energy Rev **14**, 1026–1036 (2010)
11. Khan, M.J., Bhuyan, G., Iqbal, M.T., Quaicoe, J.E.: Hydrokinetic energy conversion systems and assessment of horizontal and vertical axis turbines for river and tidal applications: a technology status review. Appl Energy **86**, 1823–1835 (2009)
12. Burton, T., Sharpe, D., Jenkins, N., Bossanyi, E.: Wind Energy Handbook. Wiley, Chichester (2001)
13. Manwell, J.F., McGowan, J.G., Rogers, A.L.: Wind Energy Explained: Theory, Design and Application, University of Massachusetts, Amherst, John Wiley & Son Ltd. (2002)
14. Nishizawa, Y.: An experimental study of the shapes of rotor for horizontal-axis small wind turbines. In: Ibrahim, A.-B. (ed.) Wind Turbines, vol. 978-953-307-221-0.

InTech (2011). http://www.intechopen.com/books/wind-turbines/an-experimental-study-of-the-shapes-of-rotor-for-horizontal-axis-small-wind-turbines. Accessed 1 February 2012
15. Rachman, A.: Parametric study on a horizontal axis ducted wind turbine. University of Wollongong, Dissertation (2010)
16. Atmadi, S., Fitroh, A.J.: Analisis and Optimasi Sudu SKEA 5 KW untuk Pemompaan. J Teknologi Dirgantara **8**, 108–115 (2010)
17. Wang, S.H., Cheng, S.H.: Blade number effect for a ducted wind turbine. J Mechanic Sci Technol **221**, 1984–1992 (2008)
18. Swanson, J., Visser, K.: Improving the efficiency of small wind turbines by optimizing blade number and solidity. (2002). http://www.clarkson.edu/reu/archives/2002/abstracts/Swanson.pdf
19. Rachman, A., Balaka, R., Delly, J.: A mathematical model for predicting the performance of a horizontal axis river current turbine with consideration of the blade parameters. Proceeding International Conference and Exhibition Sustainable Energy and Advanced Materials (ICESEAM 2011), Solo, 3–4 October 2011. UNS Indonesia 2011, Solo Indonesia (2011). ISBN 2088–8864:66
20. Wu, H., Chen, L., Yu, M., Li, W., Chen, B.: On design and performance prediction of the horizontal-axis water turbine. Ocean Eng **50**(15), 23–30 (2012)
21. Batten, W.M.J., Bahaj, A.S., Molland, A.F., Chaplin, J.R.: Experimentally validated numerical, method for the hydrodynamic design of horizontal axis tidal turbines. Ocean Eng **34**, 1013–1020 (2006)
22. Liu, S., Janajreh, I.: Development and application of an improved blade element momentum method model on horizontal axis wind turbines. Int J Energy Environ Eng **3**, 30 (2012)
23. Lanzafame, R., Messina, M.: Horizontal axis wind turbine working at maximum power coefficient continuously. J Renew Energy **35**, 301–306 (2010)
24. Clark, R.H.: Elements of Tidal-Electric Engineering. Wiley, Pistcataway (2007)
25. Li, Y., Sander, S.M.: Numerical analysis of the characteristics of vertical axis tidal current turbines. J Renew Energy **35**, 435–442 (2010)
26. Aerofoil Tool: NACA 63-415 AIRFOIL (n63415-il). (2013). http://airfoiltools.com/airfoil/details?airfoil=n63415-il

Analysis and optimization of current collecting systems in PEM fuel cells

Peiwen Li[1*], Jeong-Pill Ki[1,2] and Hong Liu[1]

Abstract

This paper presents analytical and experimental studies on optimization of the gas delivery and current collection system in a proton exchange membrane (PEM) fuel cell for the objective of reducing ohmic loss, thereby achieving higher power density. Specifically, the dimensions of current collection ribs as well as the rib distribution were optimized to get a maximized power density in a fuel cell. In the modeling process, the power output from a fixed area of membrane is calculated through analysis of an electrical circuit simulating the current from electrochemical reaction flowing to the current collectors. Current collectors of two-dimensional ribs and three-dimensional pillars were considered. Analyses found that three-dimensional pillars allow higher power density in a PEM fuel cell. Considering the mass transfer enhancement effect, three-dimensional pillars as current collectors in gas flow field may be a good choice if the fuel cell operates at low current density and there is no liquid water blocking the flow channels. The analyses did not consider the existence of liquid water, meaning the current density is not very high. The study concluded that decreasing the size of both the current collector and its control area yields a significant benefit to a higher power density. A preliminary experimental test in a PEM fuel cell has verified the conclusion of the analytical work.

Keywords: Proton-exchange-membrane fuel cell, Optimization, Current collectors, Analysis

Background

As environmentally friendly power sources, proton exchange membrane (PEM) fuel cells have received increased attention in recent years. Emphasis of research has either been placed on producing high power density with adequate energy conversion efficiency for mobile power source applications or on high energy efficiency with adequate power density for stationary power source applications. One of the big hurdles to the widespread commercialization of fuel cells is due to the high cost and insufficiently high power densities.

One of the important approaches to having high power density in a fuel cell is to reduce the internal losses. The internal losses can always be attributed to the three types of polarizations [1]: activation polarization, concentration polarization, and ohmic polarization. As these polarizations increase, the overpotential can become too large for the fuel cell to produce a high power density with reasonably high energy efficiency [2-4]. The minimization of polarization will largely depend on the mechanism that causes each specific type of polarization. The activation polarization is caused by the slowness of the electrochemical reactions taking place on the surface of the electrodes. Consequently, a proportion of the available ideal voltage is sacrificed in the electrochemical reaction. The concentration polarization is due to mass transfer resistance from bulk flow to the surface of the electrodes when fuel and oxygen are consumed and water is generated. The ohmic polarization is the straightforward resistance to the flow of electrons through the material of the electrodes and the various interconnections as well as the resistance to the flow of protons through the electrolyte.

The reduction of activation polarization is largely related to membrane and catalyst properties. The electrical potential loss due to ohmic and concentration polarization is strongly related to the configurations of the current conduction and collection elements [5,6]. These elements also significantly affect the mass transfer process since the current collectors are ribs or walls that guide the flow of reactant and product species in a fuel

* Correspondence: peiwen@email.arizona.edu
[1]Department of Aerospace and Mechanical Engineering, University of Arizona, Tucson, AZ, 85721, USA
Full list of author information is available at the end of the article

cell [7,8]. The total of ohmic polarization and concentration polarization may take a large proportion in the overall losses of fuel cell if the flow channels are not well designed or if the contact resistances between cell components are large [9,10].

A qualitative description of the control and optimization of ohmic loss is worth reviewing here. On an electrode/electrolyte layer with a fixed area, using small current collectors can expose a larger membrane area to reactant gases, therefore extracting greater current from the electrochemical reaction. However, a smaller current collector has a small contact area to electrodes and also makes the pathway for current conducting along the in-plane direction of electrode layers longer. This increases the ohmic loss. To strike the best balance between the two competing effects, an optimal arrangement of the current collectors on a bipolar plate is needed. This raises issues about optimization of the size of a current collector and the area from which the current due to electrochemical reaction is collected [11,12]. Obviously, the objective of this optimization is to obtain a maximum power density on an electrode/electrolyte membrane in a fixed area.

The minimization of the fuel cell voltage losses due to mass transfer and ohmic resistances will largely depend on the design of gas flow fields and current collecting system on the bipolar plates. The bipolar plates perform functions including the electrical connection between cells, supply of reactant gases, water and heat management, and support for membrane electrode assembly (MEA). Improvement in the design of current collecting and flow delivery system is therefore important to the high efficiency and maximized power density in fuel cells.

Various bipolar plate designs have been tested and evaluated by researchers worldwide in the field of PEM fuel cells [13,14]. The considered factors include the following: properties of materials [15], cost benefit analysis [16], the optimization of flow field [17], etc. The essential objective of these analyses is to achieve a maximum power output at low cost of materials of bipolar plates.

In this study, the analyses for the optimization of a current collection system will consider two different structures of current collectors. The first type is a two-dimensional (2D) rib design which is very common in proton exchange membrane fuel cell (PEMFC) products so far; the second type is a three-dimensional (3D) pillar design. The analysis will find optimal sizes of ribs and pillars in a given area of membrane to meet the objective of maximum power density. The same optimization analysis will be carried further for different sizes of the given area of the membrane. Using these results, optimal sizes and ratios of the current collectors and their control areas will be recommended for fuel cell design.

There are several published papers that present optimization of the flow channels and current collecting ribs using computational fluid dynamics (CFD) methods [6,10,13]. The models and methods used in these papers are CFD-based and inconvenient for a quick design optimization as might be incidentally required. To avoid the heavy-load numerical computations, the current paper presents a model that can be easily implemented. It considered typical flow channel and current collector structures as they possess the common features of current collection processes in PEM fuel cells. The model and results presented are of significance to the practical design of fuel cells when heavy CFD analysis is unavailable.

Methods

Structure of current collectors

The long gas channels and two-dimensional ribs shown in Figure 1 are typical characteristics of a bipolar plate in a fuel cell. From the mass transfer point of view, this flow field design has disadvantages in promoting the mixing of flows. To mitigate this shortcoming, 3D pillars, shown in Figure 2, for current collection and mass transfer enhancement have been proposed [18] for solid oxide fuel cells by one of the current authors. In order to give a comprehensive study of the current collectors, the present physical model will analyze both 2D and 3D current collecting elements in the gas delivery field of PEM fuel cells.

Analytical model

To analyze the current collecting system, it is convenient to consider just one single collector which collects the current of the electrochemical reaction in a controlled area. An electrical circuit will establish the relationship

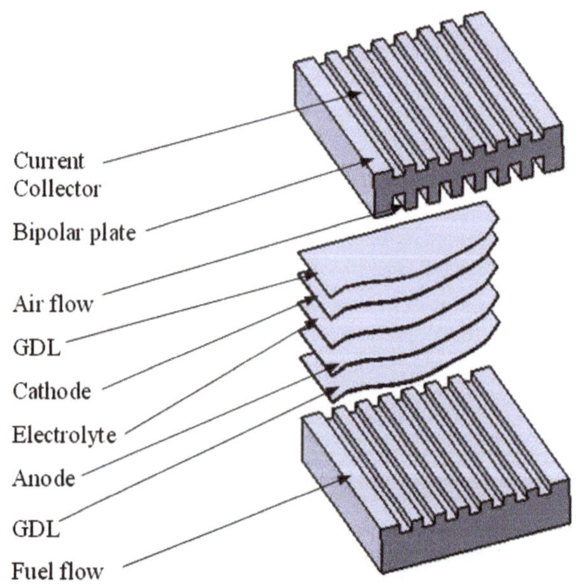

Figure 1 Two-dimensional current collectors on a bipolar plate.

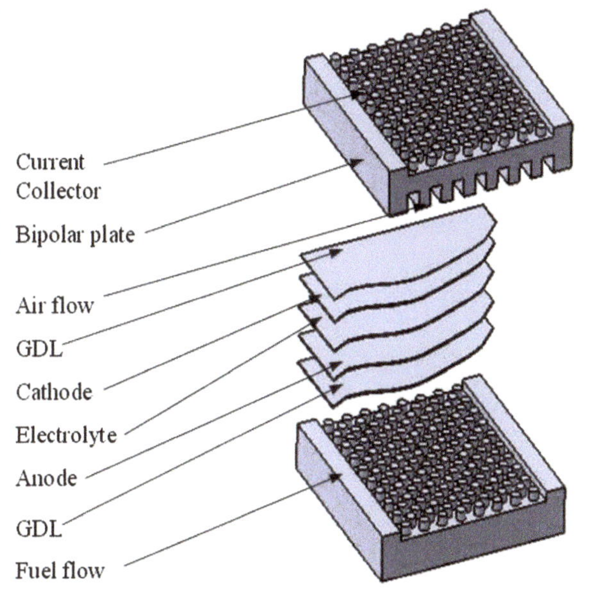

Figure 2 Three-dimensional current collectors on a bipolar plate.

of the relevant parameters to the power density. These parameters include the dimensions of current collectors and control area, the fuel cell operating temperature, and properties of the membrane and electrodes. The effect of current collection area and current collector size on the power density will be discussed in order to find optimal dimensions.

Figures 3 and 4, respectively, show the physical models of a single current collector and the electrical circuits for 2D-type and 3D-type collectors in a PEMFC. Resistances in the anode, cathode, electrolyte, and gas diffusion layers (GDL) are considered. Contact resistances of the gas diffusion layer to electrodes and gas diffusion layer to current collectors are also considered. Tables 1 and 2 give the properties of all components such as thicknesses, resistivities (at 25°C), and contact resistances. In the gas diffusion layer and electrodes, only resistances in the in-plane direction are considered to be significant. For proton conduction in the electrolyte, resistance is in the through-plane direction.

At a given current density for an area (with a width of d_o for 2D case and a diameter of d_o in 3D case) of the MEA layer, the voltage difference between the cathode-side and the anode-side current collectors can be found through analysis of the circuit. Given conditions include the Nernst potentials and the resistances from contact and conduction. The potentials on every node in the circuit can be related using equations from Kirchhoff's laws.

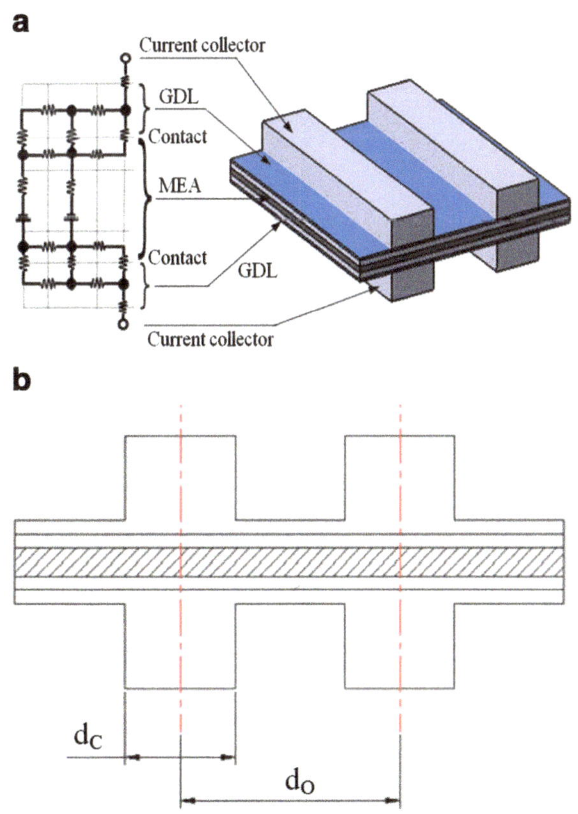

Figure 3 2D current collectors and current collection circuit. (**a**) Circuit for the current collection; (**b**) front view of the current collectors.

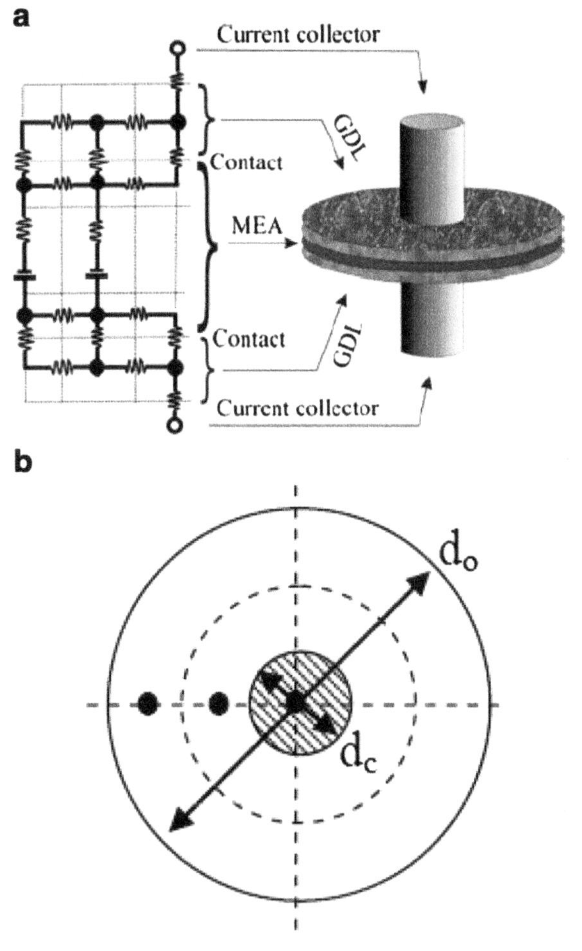

Figure 4 3D current collectors and current collection circuit. (**a**) Circuit for the current collection; (**b**) top view of grid allocation.

Consequently, potentials are available when the equations from all nodes are jointly solved. With potentials known, the power output from the electrode/electrolyte layer of a fixed area is obtained.

The analysis of the current collection may be based on given voltages (at anode-side $V = 0$, and at cathode-side $V = V_{cell}$). The total current leaving the cathode is calculated once the potentials at all grids in the circuit are obtained [19].

Table 1 Parameters of simulated PEMFC components showing contact resistances

Studied contact resistances(Ω cm^2)		
Anode-GDL	Cathode-GDL	Current collector-GDL
0.001	0.001	0.001
0.005	0.005	0.005
0.010	0.010	0.010
0.015	0.015	0.015

Table 2 Parameters of simulated PEMFC components showing thicknesses and resistivities

	Anode	Cathode	Electrolyte	GDL
Resistivity (Ω cm)	1.8868	1.8868	9.091	0.0127
Thickness (μm)	10	10	100	100

The electromotive forces (emf) in the circuit shown in Figures 3a and 4a are

$$E = E^o - \eta_{conc} - \eta_{act}, \tag{1}$$

where the standard state emf, E^o, is based on the low heating value for the electrochemical reaction in a PEM fuel cell; η_{conc} is the concentration polarization. The activation polarization η_{act} is based on equations taken from literature [20]:

$$\eta_{act} = \frac{RT}{0.5F} \ln\left[\frac{ip^c}{i_{O_2} p^c_{O_2}}\right] + \frac{RT}{F} \ln\left[\frac{ip^a}{i_{H_2} p^a_{H_2}}\right], \tag{2}$$

where i is the current density; i_{H_2} (1,000 A/m^2 [10]) and i_{O_2} (100 A/m^2 [20]) are the exchange current densities of the anode and cathode sides, respectively; p^a is the pressure on the anode side, and p^c is that on the cathode side; $p^a_{H_2}$ and $p^c_{O_2}$ are the partial pressures of hydrogen on the anode side and oxygen on the cathode side, respectively; T is the temperature of the electrochemical reaction. Table 3 lists several cases identified by the control area of a current collector.

The optimization and fuel cell performance are studied at temperatures of 25°C, 50°C, 60°C, and 75°C. The temperature is used to calculate the ideal electromotive force in Equation 1. Also, the protonic conductivity in the membrane varies depending on the temperature,

$$\sigma(T) = (0.005139\lambda - 0.00326) \exp\left[1,268\left(\frac{1}{303} - \frac{1}{273 + T}\right)\right], \tag{3}$$

where $\lambda > 1$. This parameter is specifically defined as the number of moles of water per number of sulfonic sites in the membrane:

Table 3 The size of control area (d_o)

Number	3D type (mm)	2D type (mm)
a	1.5	1.5
b	1.8	1.8
c	2.0	2.0
d	3.0	3.0
e	4.0	4.0
f	6.0	6.0

Table 4 The values of protonic conductivity and resistivity depending on the operating temperature

Temperature (°C)	Protonic conductivity (1/Ωcm)	Resistivity (Ωcm)
25	0.1100	9.0910
50	0.1529	6.5401
60	0.1720	5.8128
75	0.2027	6.5401

$$\lambda = \frac{\text{number of molecules}}{\text{number of sulfonic sites}}. \tag{4}$$

In this simulation, the water content was considered to be approximately 14 [20], assuming that the membrane is well hydrated. Protonic conductivity was calculated by Equation 3 to consider the effect of temperature. Some values of the protonic conductivity at different temperatures are listed in Table 4.

Results and discussions

The voltage and current density relationships obtained for a fixed area centered by a current collector are presented in the following sections. The concentration polarization is not considered in this optimization analysis. Because of this, the results of the maximum power density obtained in the study will thus be considered as the asymptotic power under the selected operating conditions.

Optimization of the size of 3D current collector/pillar in the center of a fixed control area

This optimization selected an operating temperature of 25°C for analysis. The contact resistance of $R_{cont} = 0.001$ Ω cm^2 between current collectors, gas delivery layers, and electrodes is assumed. Cylindrical current collectors, as shown in Figure 4, are investigated first. The cases of control areas of diameters 2.0, 4.0, and 6.0 mm were considered.

The cell voltages and power densities on a control area centered by a current collector are shown against the current densities in Figure 5. It is clear that current collectors of different sizes will result in different maximum power densities. An optimal current collector diameter is found to be 0.72 mm for the case of a control area with diameter of 2 mm, which can result in a maximum power density of 1.1233 W/cm^2. Results from the cases of control areas of 4.0 mm and 6.0 mm show similar scenarios, although the optimal sizes of current collectors are different in each case.

It is also found that a higher optimized power density is available from a smaller control area. The maximum power density at the optimal current collector size of $d_c = 0.72$ mm in the case of $d_o = 2$ mm is higher than that

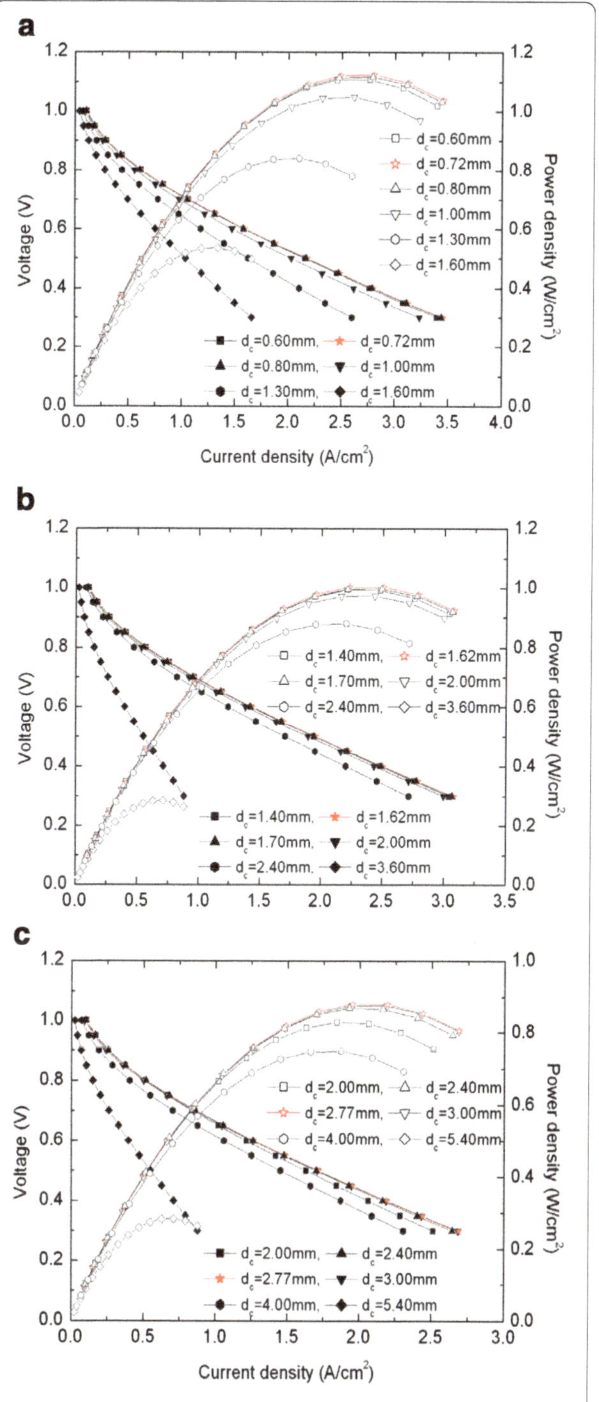

Figure 5 Optimum current collector size in a fixed control area. (**a**) $d_o = 2.0$ mm, (**b**) $d_o = 4.0$ mm, and (**c**) $d_o = 6.0$ mm.

obtained in the case of $d_c = 2.77$ mm when $d_o = 6$ mm. This feature is observed even when different contact resistances, ranging from $R_{cont} = 0.001$ to $R_{cont} = 0.015$, are applied. These results strongly suggest that densely distributed small-sized 3D-type current collectors in a gas delivery field are preferable. To give a clear

Table 5 The optimum current collector sizes and maximum power density

Contact resistance (Ω cm^2)	Size of control area d_o(mm)	Optimum size of collector d_c(mm)	Maximum power density (W/cm^2)
0.001	1.5	0.53	1.1457
	1.8	0.64	1.1329
	2.0	0.72	1.1233
	3.0	1.14	1.0673
	4.0	1.62	1.0038
	6.0	2.77	0.8797
0.005	1.5	0.70	0.8487
	1.8	0.84	0.8441
	2.0	0.91	0.8396
	3.0	1.44	0.8193
	4.0	1.96	0.7924
	6.0	3.12	0.7300
0.010	1.5	0.78	0.6836
	1.8	0.94	0.6810
	2.0	1.05	0.6791
	3.0	1.59	0.6674
	4.0	2.14	0.6524
	6.0	3.33	0.6153
0.015	1.5	0.83	0.5836
	1.8	0.99	0.5817
	2.0	1.11	0.5803
	3.0	1.67	0.5724
	4.0	2.24	0.5624
	6.0	3.45	0.5370

comparison of the obtained results, Table 5 includes the optimal current collector size and the maximum available power density in different control areas. Figure 6 shows the results given in Table 5.

Firstly, the contact resistance has a significant effect on the maximum available power density. With the decrease of the contact resistance, the maximum output power density increases. The contact resistance is related to many parameters such as clamping force, elastic modulus, thickness of MEA, the pressure of fuel and oxidant, temperature, and humidity [21]. Of these various parameters, the clamping force is relatively important. If the clamping force is unreasonably small, the interfacial contact resistance between the gas diffusion layer and current collector will increase and the maximum power output will decrease [22,23]. The curves of d_o versus d_c in Figure 6 show that ideal current collector sizes exist for the specific contact resistance. This result reminds fuel cell manufacturers that the assembly of PEM fuel cell components should be carefully considered in order to attain the optimal compaction pressure for the appropriate contact resistance. If the contact resistance varies significantly, the effect of optimization of the current collectors may not be maintained.

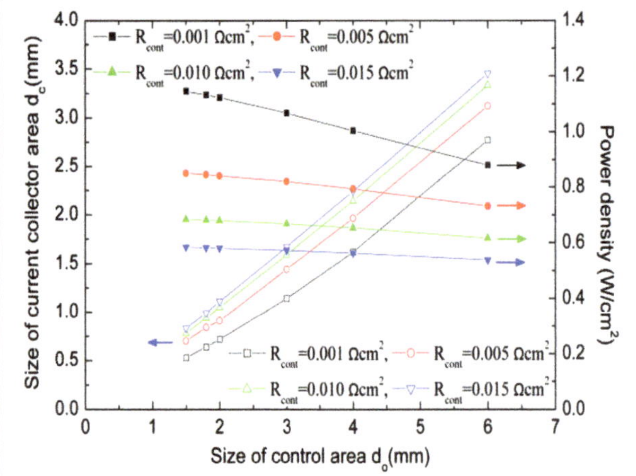

Figure 6 **Optimum values of power density and d_o versus d_c at cases with different contact resistances.**

Optimization of the size of 2D rib current collectors

The two-dimensional current collection ribs, shown in Figure 1, are commonly adopted in commercially available PEM fuel cells. At a contact resistance of $R_{cont} = 0.001\ \Omega\ cm^2$ and electrochemical reaction temperature of 25°C, the optimal current collectors and maximum power densities are obtained for control areas (denoted by d_O) of 1.5, 1.8, 2.0, 3.0, 4.0, and 6.0 mm. As shown in Figure 7, results similar to those from the 3D-type current collectors have been obtained; small control

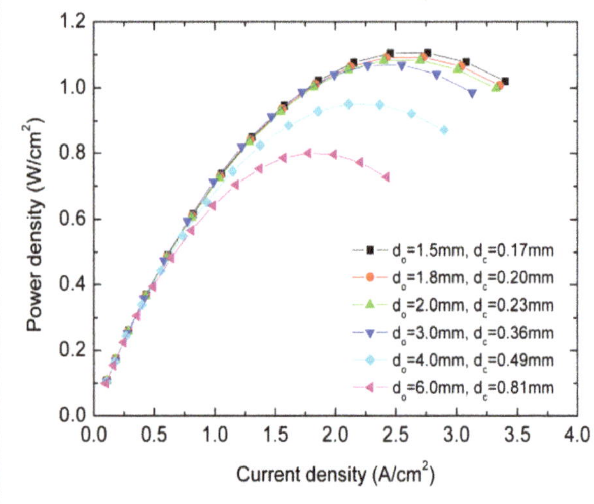

Figure 7 **Power density versus current density from optimized current collector at different control areas.**

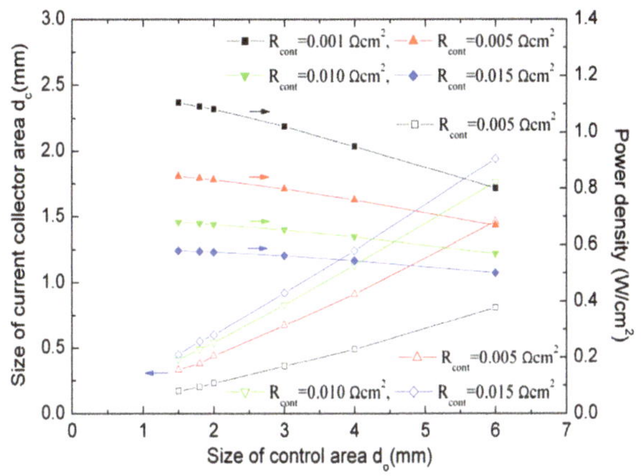

Figure 8 Optimum sizes of 2D current collection ribs and power density for different control areas.

areas and optimized current collectors are advantageous due to the corresponding higher power densities.

A similar investigation for 2D current collection ribs at different contact resistances is also made. Again, for all the cases, the power densities increase when the control area decreases and the current collector width is optimized. Contact resistances significantly affect the maximum output power.

The analytical results of 2D current collecting ribs from Figure 8 are also given in Table 6 for the convenience of comparison between different cases.

Table 6 Optimum current collector sizes and maximum power density at different contact resistances for 2D ribs

Contact resistance (Ω cm^2)	Control area size (mm)	Optimum size (mm)	Maximum power density (W/cm^2)
0.001	1.5	0.17	1.1056
	1.8	0.20	1.0917
	2.0	0.23	1.0817
	3.0	0.36	1.0209
	4.0	0.49	0.9495
	6.0	0.81	0.8003
0.005	1.5	0.33	0.8429
	1.8	0.38	0.8357
	2.0	0.44	0.8307
	3.0	0.67	0.7983
	4.0	0.91	0.7583
	6.0	1.46	0.6681
0.010	1.5	0.41	0.6792
	1.8	0.49	0.6750
	2.0	0.54	0.6718
	3.0	0.83	0.6522
	4.0	1.13	0.6273
	6.0	1.76	0.5685
0.015	1.5	0.45	0.5800
	1.8	0.55	0.5769
	2.0	0.60	0.5747
	3.0	0.92	0.5610
	4.0	1.24	0.5434
	6.0	1.94	0.5007

Comparison of the optimization result of 3D current collectors and 2D current collectors

It is interesting and of significance to compare the maximum available power densities based on geometrically different types of current collectors. Table 7 shows the maximum power densities obtained from the analysis for 3D cylindrical and 2D rib-type current collectors. The

Table 7 Comparison of maximum power density obtained from using 3D-type and 2D-type current collectors

Contact resistance (Ω cm^2)	Size of control area d_o (mm)	Maximum power density for 3D collectors (W/cm^2)	Maximum power density for 2D collectors (W/cm^2)
0.001	1.5	1.1457	1.1056
	1.8	1.1329	1.0917
	2.0	1.1233	1.0817
	3.0	1.0673	1.0209
	4.0	1.0038	0.9495
	6.0	0.8797	0.8003
0.005	1.5	0.8487	0.8429
	1.8	0.8441	0.8357
	2.0	0.8396	0.8307
	3.0	0.8193	0.7983
	4.0	0.7924	0.7583
	6.0	0.7300	0.6681
0.010	1.5	0.6836	0.6792
	1.8	0.6810	0.6750
	2.0	0.6791	0.6718
	3.0	0.6674	0.6522
	4.0	0.6524	0.6273
	6.0	0.6153	0.5685
0.015	1.5	0.5836	0.5800
	1.8	0.5817	0.5769
	2.0	0.5803	0.5747
	3.0	0.5724	0.5610
	4.0	0.5624	0.5434
	6.0	0.5370	0.5007

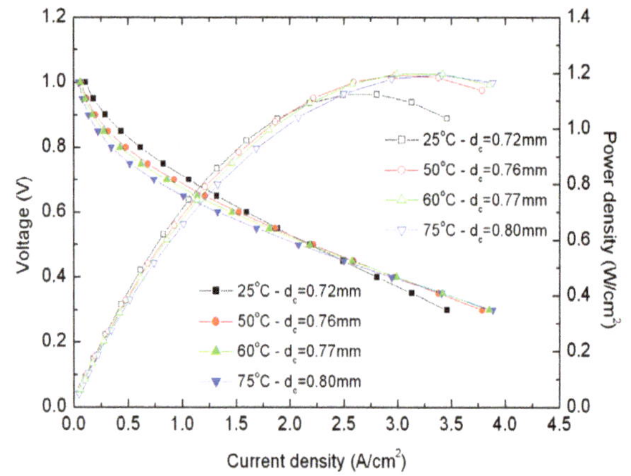

Figure 9 Optimum power density at different temperatures for 3D pillar current collectors (R_{cont} = 0.001 Ω cm^2, d_o = 2.0 mm).

definition for the dimensions of d_o is given in Figures 3 and 4 for 3D and 2D current collectors, respectively. A general trend is identified from this comparison: the maximum power density from using the 3D cylindrical current collector is higher than that from using 2D ribs. This conclusion is drawn without considering the liquid water effect in PEM fuel cells. About 4 % to 7 % of improvement of the maximum power density may be available if 3D cylindrical current collectors are used instead of 2D ribs. Note that for Table 7, the optimum sizes of the current collectors for every case can be found in Tables 5 and 6.

Applicability of optimization effect at different operating temperatures

The fuel cell operating temperatures affect the performance in a complicated manner. Temperature affects the mass transfer polarization since the gas diffusivities are functions of temperature. Temperatures also affect the ideal electromotive force, protonic conductivity of the membrane, and activation polarization. Since the concentration polarization is not considered in the optimization analysis in this work, results regarding the effect of temperature on the maximum available power density are only preliminary studies.

Figure 9 shows the optimized power densities and voltages against current densities for cases of operating temperatures from 25°C to 75°C. The contact resistance is 0.001 Ω cm^2, and the control area size is d_0 = 2.0 mm for a cylindrical current collector. Similarly, Figure 10 indicates the same type of results for 2D rib current collectors. The values for d_c given in Figures 9 and 10 are optimized sizes of current collectors under the different temperatures. Since the increase of temperature will result in an increase of protonic conductivity, higher optimum power density may be available when the operating temperature is high.

Experimental results

In the experimental work by Hental et al. [24], a similar phenomenon as revealed in the present optimization analysis was observed in the development of new materials of current collectors for PEMFCs. They reported that using smaller current collecting ribs and gas channels elicited higher power density in PEM fuel cells. This agrees with the conclusion obtained in the present analysis.

In this study's accompanying experimental work, a test for PEM fuel cells was conducted at room temperatures (25°C) to verify the conclusions from the analytical work. The tested fuel cells have the same membrane, but the current collector size and control area in hydrogen and oxygen flow fields on the bipolar plates are different. One has large-sized 3D current collectors, and the other has

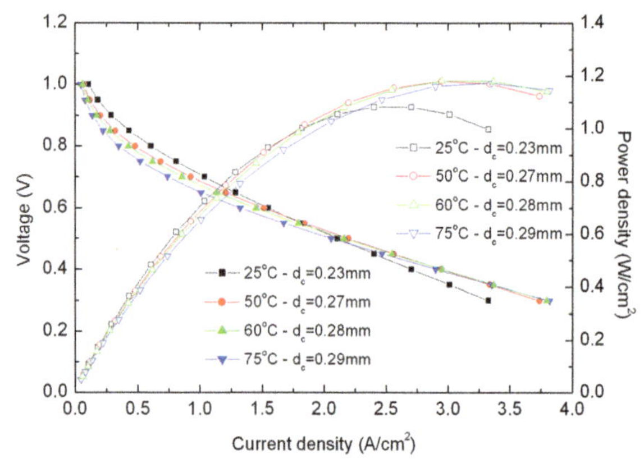

Figure 10 Optimum power density at different temperatures for 2D rib current collectors (R_{cont} = 0.001 Ω cm2, d_o = 2.0 mm).

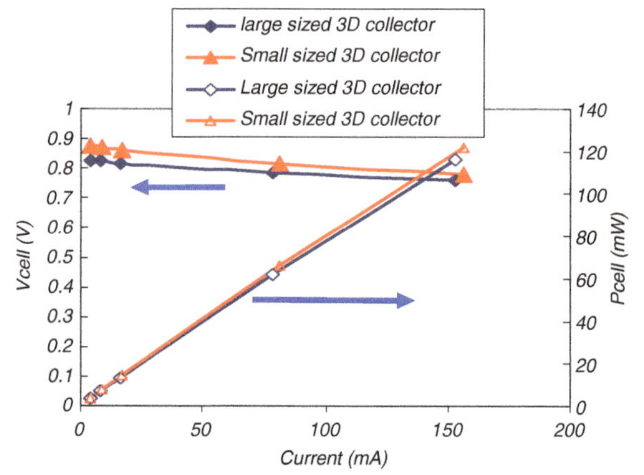

Figure 11 Experimental test for fuel cells at room temperature (25°C). Oxygen was supplied from free convection airflow.

small-sized 3D current collectors. Oxygen with no humidification is provided through natural convective airflow. As shown in Figure 11, a higher voltage output is obtained for small 3D current collectors. A comprehensive experimental test will be carried out in the near future.

Conclusions

Analysis and experimental work was conducted to optimize the flow field and current collectors in PEM fuel cells. From the analysis, the dimensions of current collecting ribs and the rib distribution can be optimized in order to have maximum power density in a fuel cell. The analytical models consider current collectors of two-dimentional ribs and three-dimensional pillars. Upon optimization, using three-dimensional pillars in a flow field can obtain higher power density in a PEM fuel cell. This conclusion is drawn without considering the liquid water effect in PEM fuel cells.

It is also found that higher power densities can be obtained using smaller current collectors with optimized distribution in a flow field. Considering the mass transfer enhancement effect, three-dimensional pillars, as current collectors in gas flow field, are recommended. However, when liquid water exists in a PEMFC, two-dimensional ribs are usually better as the liquid can be excluded with more ease than in the case of three-dimensional pillars. A preliminary, yet not comprehensive, experimental test at low current density without liquid water in PEM fuel cells has found better performance when small sized 3D current collecting pillars and optimal distribution are adopted in a flow field.

Competing interests

The authors declare that they have no competing interests.

Acknowledgment

The support from the Office of Naval Research of the USA and the University of Tennessee SimCenter under the contract #8500011366 is gratefully acknowledged.

Author details

[1]Department of Aerospace and Mechanical Engineering, University of Arizona, Tucson, AZ, 85721, USA. [2]Current address: Department of Mechanical and Aerospace Engineering, University of Texas at Arlington, Arlington, TX, 76019, USA.

Authors' contributions

PL conceived the concept and procedures of the optimization analysis and drafted the manuscript. JK carried out computations and analysis. HL checked equations and analysis, and updated and revised the manuscripts. All authors read and approved the final manuscript.

Authors' information

PL earned his Ph.D. (1995) degree in energy and power engineering area from Xi'an Jiaotong University, China. He is currently an assistant professor in the Department of Aerospace and Mechanical Engineering at the University of Arizona, USA. He has been interested in heat transfer enhancement in industrial processes, turbulence drag reduction, gas turbine cooling technologies, the multi-physics transport phenomena in fuel cells and electrolysis cells, and concentrated solar thermal power systems. Professor PL is an active member of the American Society of Mechanical Engineers. He is also a reviewer of papers for more than 10 technical journals.

JK received his M.S. degree (2007) from the Department of Aerospace and Mechanical Engineering at the University of Arizona in 2007. He is currently a Ph.D. student in the Department of Mechanical and Aerospace Engineering at the University of Texas at Arlington, USA. His research focuses on thermal/fluid sciences in energy system, modeling and system dynamics of solid oxide fuel cell and gas turbine (GT) hybrid systems. His Ph.D. dissertation work is the integrated modeling approach of solid oxide fuel cell-based combined heat and power (CHP) system.

HL received her bachelor degree (2005) in energy and power engineering area from Xi'an Jiaotong University, China. She is currently a Ph.D. student in the Department of Aerospace and Mechanical Engineering at the University of Arizona, USA. Her research focuses on numerical analysis of fluid flow and heat transfer in industrial processes and energy systems. Her Ph.D. dissertation work is experimental and numerical study of the multi-physics transport phenomena in internal reforming solid oxide fuel cells.

References

1. Rowe, A, Li, X: Mathematical modeling of proton exchange membrane fuel cells. J. Power Sourc. **102**, 82–96 (2001)
2. Srinivasan, S, Mosdale, R, Stevens, P, Yang, C: Fuel cells: reaching the era of clear and efficient power generation in the twenty-first century. Annu. Rev. Energ. Environ. **24**, 281–328 (1999)
3. Gardner, FJ: Thermodynamic processes in solid oxide and other fuel cells. Proc. Inst. Mech. Engrs. **211**, 367–380 (1997)
4. Suzuki, K, Iwai, H, Kim, JH, Li, PW, Teshima, K: Solid oxide fuel cell and micro gas turbine hybrid cycle and related fluid flow and heat transfer. In: The 12th International Heat Transfer Conference, Grenoble France 18–23 August 2002
5. Roshandel, R, Arbabi, F, Karimi Moghaddam, G: Simulation of an innovative flow-field design based on a bio inspired pattern for PEM fuel cells. Renew. Energ. **41**, 86–95 (2012)
6. Ferng, YM, Su, A: A three-cell CFD model used to investigate the effects of different flow channel designs on PEMFC performance. Int. J. Hydrogen Energ. **32**, 4466–4476 (2007)
7. Yan, W-M, Yang, C-H, Soong, C-Y, Chen, F, Mei, S-C: Experimental studies on optimal operating conditions for different flow field designs of PEM fuel cells. J. Power. Sourc. **160**, 284–292 (2006)
8. Wang, X-D, Duan, Y-Y, Yan, W-M, Peng, X-F: Effects of flow channel geometry on cell performance for PEM fuel cells with parallel and interdigitated flow fields. Electrochim. Acta **53**, 5334–5343 (2008)
9. Manso, AP, Marzo, FF, Garmendia Mujika, M., Barranco, J., Lorenzo, A.: Numerical analysis of the influence of the channel cross-section aspect ratio on the performance of a PEM fuel cell with serpentine flow field design. Int. J. Hydrogen Energ. **36**, 6795–6808 (2011)
10. Lobato, J, Canizares, P, Rodrigo, MA, Pinar, JF, Mena, E, Ubeda, D: Three-dimensional model of a 50 cm^2 high temperature PEM fuel cell. Study of the flow channel geometry influence. Int. J. Hydrogen Energ. **35**, 5510–5520 (2010)
11. Li, P-W, Ki, JP: Analysis and optimization of current collecting systems in PEM fuel cells, Paper No. FUELCELL2007-25060. In: Proceedings of FUELCELL2007, Fifth International Conference of Fuel Cell Science, Engineering and Technology, New York, 18–20 June 2007
12. Goebel, SG: Impact of land width and channel span on fuel cell performance. J. Power Sourc. **196**, 7550–7554 (2011)
13. Hashemi, F, Rowshanzamir, S, Rezakazemi, M: CFD simulation of PEM fuel cell performance: effect of straight and serpentine flow fields. Math. Comput. Model. **55**, 1540–1557 (2012)
14. Peng, L, Lai, X, Liu, D, Hu, P, Ni, J: Flow channel shape optimum design for hydroformed metal bipolar plate in PEM fuel cell. J. Power Sourc. **178**, 223–230 (2008)
15. Yen, C-Y, Liao, S-H, Lin, Y-F, Hung, C-H, Lin, Y-Y, Ma, C-C: Preparation and properties of high performance nanocomposite bipolar plate for fuel cell. J. Power Sourc. **162**, 309–315 (2006)
16. Jayakumar, K, Pandiyan, S, Rajalakshmi, N, Dhathathreyan, KS: Cost-benefit analysis of commercial bipolar plates for PEMFC's. J. Power Sourc. **161**, 454–459 (2006)
17. Hontanon, E, Escudero, MJ, Bautista, C, Garcia-Ybarra, PL, Daza, L: Optimisation of flow-field in polymer electrolyte membrane fuel cells using

computational fluid dynamics techniques. J. Power Sourc. **86**, 363–368 (2000)

18. Li, PW, Chen, SP, Chyu, MK: Novel gas distributors and optimization for high power density in fuel cells. J. Power Sourc. **140**, 311–318 (2004)
19. Li, PW, Schaefer, L, Chyu, MK: Multiple processes in solid oxide fuel cells. In: Sunden B, Faghri M (eds.) Transport Phenomena in Fuel Cells, pp. 1–42. WIT Press, Boston (2005) 2005:1–42.
20. Shimpalee, S, Dutta, S: Numerical prediction of temperature distribution in PEM fuel cells. Numer. Heat Tran. **38**, 111–128 (2000)
21. Scherer, GG: Interfacial aspects in the development of polymer electrolyte fuel cells. Solid State Ionics **94**, 249–257 (1997)
22. Mishra, V, Yang, F, Pitchumani, R: Measurement and prediction of electrical contact resistance between gas diffusion layers and bipolar plate for applications to PEM fuel cells. ASME J. Fuel Cell Sci. Technol. **1**, 2–9 (2004)
23. Lee, WK, Ho, CH, Zee, JWV, Murthy, M: The effect of compression and gas diffusion layers on the performance of a PEM fuel cell. J. Power Sourc. **84**, 45–51 (1999)
24. Hental, PL, Lakeman, BJ, Mepsted, GO, Adcock, PL, Moore, JM: New material for polymer electrolyte membrane fuel cell current collectors. J. Power. Sourc. **80**, 235–241 (1999)

18

Evaluation of food waste disposal options in terms of global warming and energy recovery: Korea

Mi Hyung Kim[1*], Han Byul Song[2], Yuleum Song[3], In Tae Jeong[4] and Jung Wk Kim[1]

Abstract

This study is aimed to evaluate and compare environmental impacts of various food waste management systems: anaerobic digestion, co-digestion with sewage sludge, and volume reduction using a garbage dryer followed by incineration from generation to final disposal. An environmental credit using life cycle assessment was employed to compare by-products. The entire disposal process, namely discharge, collection, transportation, treatment, and final disposal, was included in the system boundary. The functional unit was 1 tonne of food waste from households for each treatment option. Global warming potential of the category indicator was selected to assess the environmental impact of food waste disposal options. The net global warming potential (environmental credit) of the options (wet based) was 33 kg of carbon dioxide equivalent (CO_2-eq) for anaerobic digestion and −315 kg of CO_2-eq for incineration by renewable energy production as electricity, thermal energy, and primary materials avoided. We found that dryer-incineration option was an available alternative for food waste recycling in a metropolitan area in Korea.

Keywords: Food waste, Life cycle assessment, Global warming potential, Renewable energy, Environmental impact

Background

Energy recovery from wastes has a great potential for reducing CO_2 emissions and thereby resulting in greenhouse gas (GHG) emissions. Food wastes can be utilized as a substitution for traditional nature resource and energy under the situation of energy exhaustion and sustainable crisis nowadays. Due to the large amount and proportion in the municipal solid waste (MSW), food waste is expected to play an important role in the low-carbon economy and recycling-based society [1,2].

Anaerobic digestion for food wastes has been commonly used to produce energy from organic wastes [3-5]. It was reported that 1 m^3 of biogas had 21 MJ of energy by anaerobic digestion, and it could generate 2.04 kW h of electricity considering the 35% of generation efficiency [6]. Co-digestion of food wastes has been useful for energy recovery by increasing methane production from the sludge digestion in a sewage treatment plant (STP) [7]. It was known that the anaerobic co-digestion of sewage sludge and highly rich organics such as food wastes or animal manure could increase biogas production [8,9]. Another research evaluated five food waste disposal options by life cycle assessment (LCA) and found that global warming potential (GWP) of the options were 273 kg of carbon dioxide equivalent (CO_2-eq) for anaerobic home composting, 82 kg of CO_2-eq for co-disposal with MSW, 52 kg of CO_2-eq for centralized composting, 13 kg of CO_2-eq for food waste processor system, and 0.3 kg of CO_2-eq for aerobic home composting [10]. GWP of each scenario was produced from 182 kg of food waste (wet based).

LCA is an effective environmental tool dealing with the complex interaction between the environment and a product or activity. Though LCA aims to be science-based, it involves a number of technical assumptions and value choices. In spite of its limitations, the environmental impacts obtained from LCA analysis are described as potential impacts because they are not specified in time and space. LCA has been conducted as a useful tool for a full cycle of MSW management, including collection, transport, waste treatment, infrastructure for waste treatment facilities, and production of energy and ancillary materials consumed [11].

* Correspondence: mhkim9@snu.ac.kr
[1]Department of Environmental Planning, Graduate School of Environmental Studies, Seoul National University, San 56-1, Sillim-dong, Gwanak-gu, Seoul 151-742, Republic of Korea
Full list of author information is available at the end of the article

The geographical boundary of the study was Jungnang-gu, which is formed by the three mountains and located in Northeastern Seoul Metropolitan City; it is a separated area with large populations that are divided into five districts. The area of 18.53 km^2 occupies 3% of Seoul. Data were collected from the government of the district and plants in the study area.

The objectives of this study were to evaluate and compare the environmental impacts of food waste management systems including anaerobic digestion, co-digestion with sewage sludge, and incineration of dried food waste from the time of generation to final disposal. LCA methodology was used to analyze environmental impacts of the systems from the perspective of global warming. Tax issues in carbon emission regulations associated with climate change are to be an important matter of international concern. All countries of the world make an effort to mitigate climate change as activities that reduce GHG emissions. One of the options is to produce renewable energy from wastes that would result in GHG reduction [12]. Therefore, GWP of the result of characterization was finally selected as an environmental indicator to evaluate the impacts of global warming. Each GHG value is quantified as equalization of carbon dioxide by Intergovernmental Panel on Climate Change (IPCC) indicators [13]. The GHGs contain carbon dioxide (CO_2), methane (CH_4), nitrous oxide (N_2O), chlorofluorocarbon (CFCs), hydrofluorocarbon (HFCs), perfluorocarbon (PFCs), and sulfur hexafluoride (SF_6).

Methods

Study areas

In Korea, the direct landfilling of raw food wastes was banned in 2005 in order to mitigate the shortage of landfill site, to protect groundwater and soil contamination, and to promote recycling of food wastes as an available resource [14,15]. These efforts resulted in a highly effective recycling rate in food wastes (e.g., over 94% of food waste recycled in the nation in 2006). Approximately 13,372 tonnes of food waste was generated per day as of 2006 in Korea, 261 tonnes of which (2% of daily generation) was landfilled; 509 tonnes (4%), incinerated; 12,603 tonnes (94%), recycled (Table 1). The study area, Jungnang District, is located in Northeastern Seoul. It consists of a residential area (57.8%), green area (40.4%), commercial area (1.7%), and semi-industrial area (0.1%). The district, which had 174,475 households (421,422 residents) in 2012, generated 81 tonnes of food waste from single houses (9.8%), town houses (34.1%), apartments (35.3%), and restaurants (13.6%) per day as of 2008, 65.5 tonnes of which (81% of daily generation) were recycled for animal feed and 15 tonnes (19%) for compost.

Statistical data showed that the treatment methods of food wastes in Korea have been concentrated on feed manufacturing for animals and composting. These facilities for treating food wastes by these methods have been usually located in a rural area such as the Gyeonggi province; therefore, the food waste generated in the metropolitan area such as in Seoul has to be generally passed to other districts and has been causing problems for transportation, odor, hygiene concerns, and adverse environmental impacts. It is recommended that food waste treatment and disposals should be covered within their own district. It also requires the use of the most effective treatment system that minimizes environmental risks.

Modeling approach

In this study, the function was defined as food waste disposal, and the functional unit was defined as 1 tonne of food waste from households for each treatment option. For data calculation, the reference flow was defined as 1 tonne of food waste.

In the present study, the system boundary for LCA describes the time from initial food waste generation at the household to its final disposal. To calculate GWP of food waste treatment options, five stages were defined and investigated; they are separate discharge, separate collection, transportation, treatment, and resource/energy production of disposal from food waste generation to final disposal. Food wastes were collected and transported to a

Table 1 Food waste treatment types and generation in Korea (tonnes/day (%))

	Year					
	2001	**2002**	**2003**	**2004**	**2005**	**2006**
Total amount	11,237	11,397	11,398	11,464	12,976	13,372
Landfilling	3,856 (34.3)	3,345 (29.3)	2,836 (24.9)	1,607 (14.0)	356 (2.7)	261 (2.0)
Incineration	1,003 (8.9)	922 (8.1)	844 (7.4)	541 (4.7)	516 (4.0)	509 (3.8)
Recycling sum	6,378 (56.8)	7,130 (62.6)	7,718 (67.7)	9,316 (81.3)	12,104 (93.3)	12,603 (94.2)
Animal feed	3,524 (55.3)	3,526 (49.5)	3,832 (49.6)	4,434 (47.6)	5,110 (42.2)	5,703 (45.2)
Compost	2,598 (40.7)	3,259 (45.7)	3,391 (43.9)	3,955 (42.4)	5,759 (47.6)	5,660 (44.9)
Others[a]	256 (4.0)	345 (4.8)	495 (6.4)	927 (10.0)	1,235 (10.2)	1,240 (9.8)

[a]Others include anaerobic digestion and co-digestion with sewage sludge. Adapted from the Ministry of Environment (South Korea) [14].

treatment facility. An avoided impact analysis was applied to energy and resource recovery. Electricity and steam can be produced by converting waste to captured biogas; the steam was used for warming the digester. The produced steam was excluded because energy usage such as electricity or fuel for processes was already not counted as the same amount as the steam from biogas. The decision rules for the inclusion of input and output materials into the system boundary were cumulative weight, energy, and relevance to the environmental aspects, all suggested by ISO/TR 14049 [16]. The decision rule applied was to include up to 99% of the total product weight in the system boundary. Energy and hazardous substances were all included into the system boundary regardless of their quantities.

The environmental impacts of each stage were analyzed using LCA methodology and an indicator model. The life cycle inventory tool, Total 3.0, developed by Korea Eco-Products Institute (Seoul, South Korea), was used.

Scenario construction and process description

Three different scenarios of food waste disposal were demonstrated. Each system was to treat food wastes to stabilize and sterilize the final goal of waste reduction and disposal by environment-friendly methods. Therefore, in the study, all scenarios had a consideration on final disposal. The scenarios were defined as follows:

- *Scenario 1 anaerobic digestion.* Biogas was produced by hydrolysis, acidification, and methane fermentation.
- *Scenario 2 co-digestion with sewage sludge.* Pretreated food waste was anaerobically digested with primary sludge in the STP.
- *Scenario 3 dryer-incineration.* Food wastes were dried by a machine after discharging at a household kitchen and moved to a facility for energy recovery.

Scenarios 1 and 2 were presented as potential waste management methods that are not commonly used in Korea but could meet current demands for renewable energy recovery from biomass and GHG reduction. Scenario 3 may be a possible option. It has been reported that various garbage dryers have been used in households in Korea as well as in many countries.

Anaerobic digestion

Food wastes were injected to the input hopper for pretreatment after weighing. Vinyl, animal born, and debris were sorted for the shredding process. The next step was that the pretreated food wastes were supplied to an anaerobic digester. Captured biogas was converted to steam. This steam was used for warming the digester. About 75%, 5%, 10%, and 10% of the input mass output were wastewater, screenings, final sludge, and biogas, respectively. This final sludge was used as an ingredient in manufacturing a fertilizer [17]. Produced biogas was burnt in engines that turn generators to make electricity. About 35% of the input energy outputs as electric energy, and the rest of the heat from the engine cooling water was used to warm the digesters. The produced steam was recycled within the system, and electricity was sold at the national power exchange (Figure 1).

Co-digestion with sludge

The process was classified with two subunit processes, pretreatment and co-digestion. Food wastes should be pretreated by sorting vinyl and animal born before injection to an anaerobic digester in the STP to promote organic decomposition and stabilization. Co-digestion with food wastes took advantage of the increasing biogas due to high concentrated organic input in the STP. Food wastes could be supported as a carbon source when advanced treatment technologies are applied to a plant. Food wastes were injected to a hopper and moved to a centrifuge through shredding and sorting processes. The cakes generated from the centrifuge were moved to a landfill site for final disposal, and other screenings from sorting processes were moved to a resource recovery facility for incineration. Pretreated food wastes were highly rich organic liquids that looked like pulp because screenings and centrifuged cakes were removed. The liquids were delivered to an anaerobic digester in the STP; it was mixed with primary sewer sludge and was stabilized. Biogas was produced through hydrolysis, acidification, and methane fermentation processes (Figure 2).

Dryer-incineration

After the moisture contained in food wastes was removed using a dryer, the dried food wastes were discharged with MSW. MSW including dried food wastes were collected by a garbage truck and stored at a transfer station. The wastes were moved to a resource recovery facility for incineration and burnt at 850°C to approximately 1,100°C. The high-temperature gas produced steam which operated a steam turbine and then produced electricity. The steam was passed to a heat exchanger, and it supported the residential area's hot water for room heating. Air emissions and wastewater were discharged during the process. Wastewater was moved to the STP after pretreatment. Fly ash and bottom ash were moved to a landfill for final disposal.

Allocation and avoided impact approach

ISO 14041 mention that allocation should be applied following these stepwise procedure [18,19]:

Step 1 Whenever possible, allocation should be avoided by dividing the unit process to be allocated into two or more subprocesses and collecting the input and output

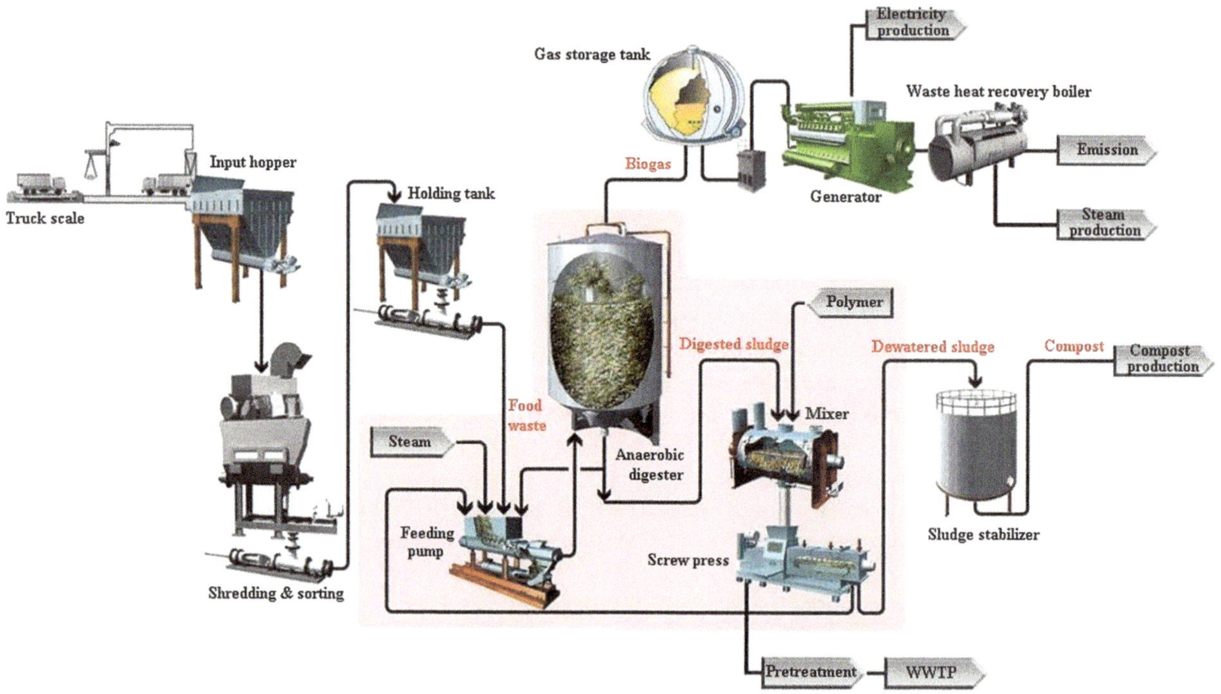

Figure 1 Schematic diagram of an anaerobic digestion process (Courtesy of Seohee Construction, Seoul, Korea; http://www.seohee.co.kr).

data related to these subprocesses or by expanding the product system to include the additional functions related to the coproducts.

Step 2 Where allocation cannot be avoided, the inputs and outputs of the system should be partitioned between its different products or functions in a way which reflects the underlying physical relationships between them.

Step 3 Where physical relationship alone cannot be established or used as the basis for allocation, the

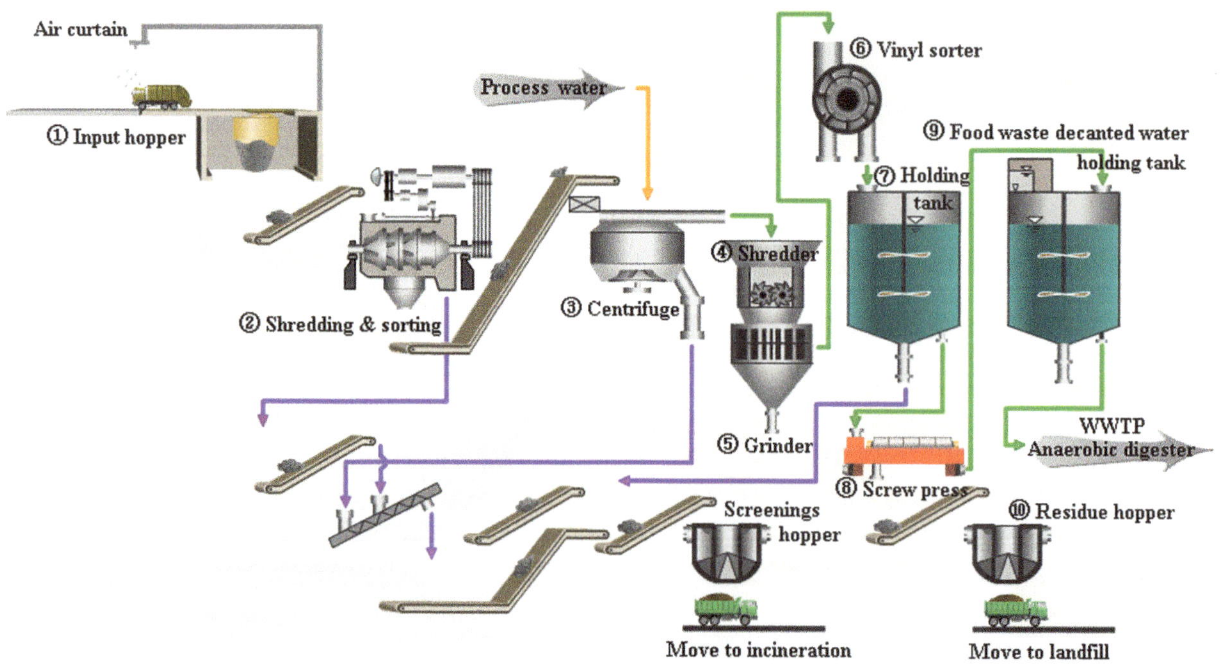

Figure 2 Schematic diagram of a co-digestion process. (Courtesy of Daegu Metropolitan City Environmental Installations Corporation, Daegu, South Korea; http://www.dgeic.or.kr).

inputs should be allocated between the products and functions in a way which reflects other relationships between them (i.e., economic value of the products).

An expanded system boundary was applied for the comparison of different by-products of each scenario. The most appropriate method for the quantification of environmental credit (or benefit) that resulted from the recycling of by-products is the system expansion method as shown in Figure 3. By-products generated from the product manufacturing process can be used as raw materials for other product systems, resulting in reduction of the use of virgin raw materials in the other product systems. It has been generally accepted that the use of recycled materials is environmentally preferable to that of virgin raw materials. This is because the environmental loads associated with the processing of recycled materials are less than those associated with the extraction and processing of virgin raw materials [20]. In the anaerobic digestion scenario, produced biogas primarily used to warm a digester then to a surplus biogas converts to electric energy. In the co-digestion scenario, produced biogas is completely used to warm a digester. Therefore, the system had no surplus energy to use outside of the system boundary. In the dryer-incineration scenario, produced heat energy converted to steam is completely used for residential room heating. An environmental credit which resulted from recycling of food wastes could be analyzed and quantified by avoided impact analysis. Therefore, these food waste recycling systems could reduce negative impacts to the environment as compared with new commercial product manufacturing.

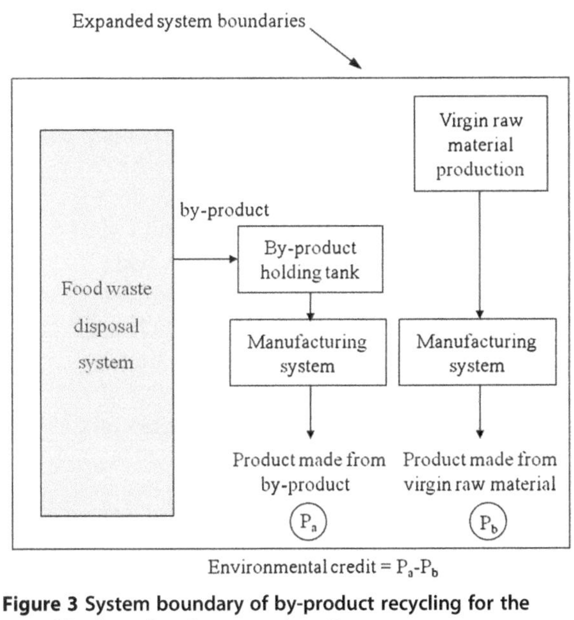

Figure 3 System boundary of by-product recycling for the quantification of environmental credit.

Assumptions

There were some assumptions:

Most of the discharged sludge from the STP have been dumped to the ocean. However, it should be restricted to protect the ocean's environment. An environmental impact was simulated under the assumption that sludge was managed by incineration.

Biosolids from anaerobic digestion was to be a raw material for the manufacturing of a fertilizer.

Screenings were delivered to a landfill for the final disposal.

Final wastes were burnt at the Nowon Energy Recovery Facility in case of incineration or buried at the Sudokwon landfill site in case of landfilling.

When additives such as coagulants and chemicals for pH adjusting were added, the distance for transportation was 8 km, which is based on a study that over 10 km of distance was not cost-effective [21].

Results and discussion

Data collection

Data quality should be characterized by quantitative and qualitative aspects. Data collection was based on site inspections. In the present study, time-related coverage was from 2005 to 2009; however, plant data were collected with the age of data within the latest year, specifically 2008. Literature data were collected with the time scope within the last 5 years. Data were collected by plant survey at a facility in Sainggok, Busan City for anaerobic digestion, in Shincheon, Daegu City for co-digestion, and in Nowon, Seoul metropolitan for dryer-incineration to satisfy the expanded system boundaries of each scenario as shown in Figure 4. Up- and downstream databases, which were connected with technosphere, were taken by the Ministry of the Environment and the Ministry of the Knowledge Economy, Korea. In the case of non-existing data, data inventories were directly constructed with site inspection data, or overseas inventory supported by SimaPro 7.1 was used. The analysis was conducted in accordance with the methods stipulated in ISO 14040 [22,23].

Life cycle inventory analysis

Discharge stage

Dryer-incineration. Food wastes were dried using a garbage dryer at the discharge stage by a discharger. Two types of the popular garbage dryers which dominated at the market share were installed in ten of the sample houses and were tested by residents. Under the data from the user surveys, 75.9% weight reduction in efficiency resulted from one operation on 1 day and in the next, which is 3.35 min of continuous activity. Electricity consumption was averaged at 648.884 kW h/tonne of

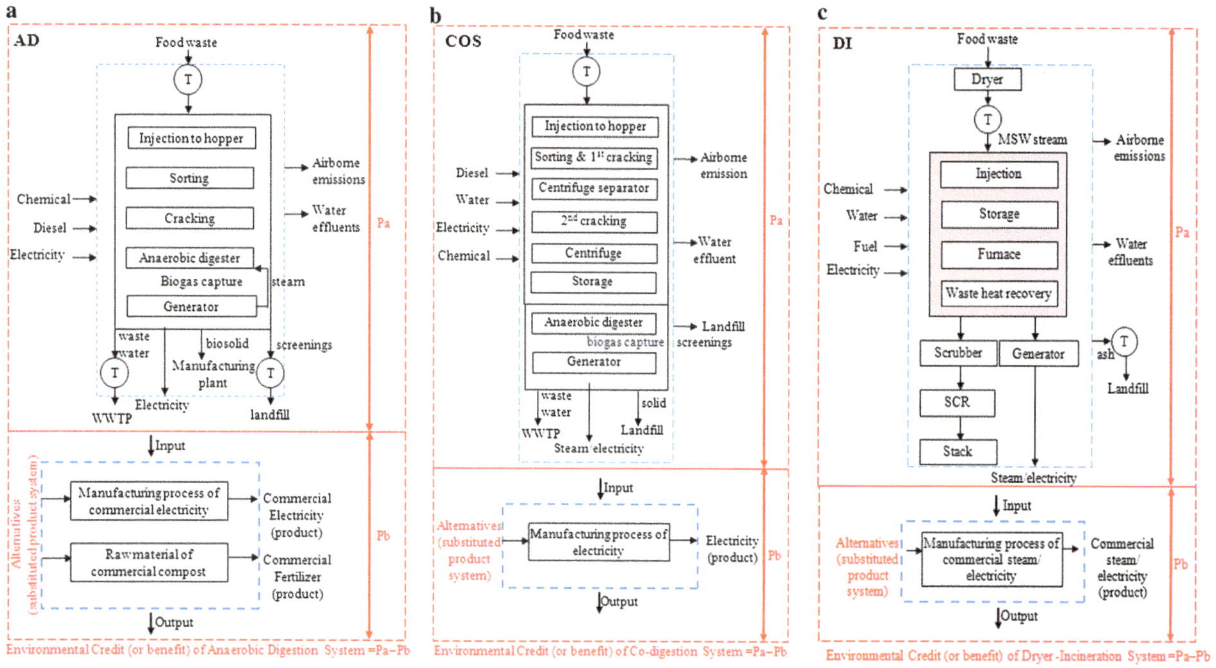

Figure 4 Expanded system boundary of scenarios (T, transportation). (**a**) Anaerobic digestion, (**b**) co-digestion, and (**c**) dryer-incineration.

food waste using both stirring warm air drying and warm air circulation drying.

Separate collection stage

Food wastes were kerbside collected by a 5-tonne-capacity garbage truck and transported to a transfer station. The distance for collection depended on the collection systems. However, the round-trip distance was accepted, and the average distance of 94.4 km was directly measured in Jungnang District, the study region. Fuel efficiency was 4.8 km/L, and the load was 5 tonnes/truck. In the dryer-incineration scenario, about 0.24 tonne of dried food waste was collected.

Transportation stage

The study area, Jungnang, has had its own transfer station. Food wastes were transported from a transfer station to each facility by an 11-tonne truck. The distances from the transfer station to each facility were as follows: around 5.21 km to the anaerobic digestion facility for the anaerobic digestion scenario and around 3.97 km to the STP for the co-digestion scenario. In the case of dryer-incineration scenario, around 8.42 km was the calculated distance to the Nowon energy recovery facility; however, the weight of the dried food wastes was around 0.24 tonne.

Treatment stage

- *Anaerobic digestion.* The target facility treated 200 tonnes of food waste by anaerobic digestion per day. Approximately 14.29 tonnes of biogas and 22 tonnes of biosolids were produced for a day. About 40 tonnes of processed water was recycled. Around 200 tonnes of wastewater and 21.2 tonnes of screenings were produced. The quality of wastewater was 5,440 mg/L of BOD, 4,520 mg/L of SS, 2,400 mg/L of T-N, and 19 mg/L of T-P. The wastewater was pretreated before moving to the STP. For this process, 13,645 kW h of electricity usage was required. Produced biogas was converted to 48,735 kW h of electricity and 40.79 tonnes of steam per day. Steam was used to warm the digester. Electricity typically covered 13,645 kW h of daily usage for the process operation, and a surplus of 35,089 kW h was sold at Korea Power Exchange. Air emissions from the combustion of generated biogas were calculated by IPCC emission factors. Produced CO_2 was not counted on GHG because it had a biogenic origin. It was investigated that 242.64 kW h of electricity, 4,280,000 kcal/h of steam, and 0.11 tonne of biosolids were produced from 1 tonne of food waste. In this study, surplus electricity and the biosolids were analyzed by an avoided impact approach. The gate-to-gate data of the scenario is shown in Table 2.
- *Co-digestion.* The selected facility treated about 200 tonnes of food waste per day with mixing sewage sludge in the STP. The co-digestion process was divided into two major sections, pretreatment and co-digestion. The pretreatment covered sorting,

Table 2 Gate-to-gate data for the scenarios at treatment stage

		Materials	Anaerobic digestion	Co-digestion	Dryer-incineration	Unit
Input	Materials	Food waste	1.00E + 03	1.00E + 03	2.41E + 02	kg
		Chemicals	2.70E + 02	3.11E + 01	1.53E + 00	kg
		Process water	-	-	1.08E + 01	kg
	Energy	Diesel	-	-	1.12E − 01	kg
		Liquefied natural gas	-	-	6.22E + 01	MJ
		Electricity	6.82E + 01	3.68E + 01	7.91E + 00	kW h
Output	By-products	Electricity	2.43E + 02	-	-	kW h
		Biosolids	1.10E + 02	-	-	kg
		Waste heat recovery	-	-	2.16E + 00	Gcal
		Wastewater/leachate	1.00E + 03	-	3.25E + 01	kg
		Sludge	-	1.41E + 02	-	kg
		Screenings	1.10E + 02	3.59E + 02	-	kg
		Fly ash	-	-	3.81E − 01	kg
		Bottom ash	-	-	2.49E + 00	kg
	Air emission	CO_2	-	-	7.66E + 02	kg
		CH_4	5.48E − 02	6.37E − 03	4.70E − 02	kg
		N_2O	7.32E − 03	8.50E − 04	-	kg
		NO_x	1.83E − 01	2.12E − 02	6.25E − 01	kg
		CO	1.83E + 00	2.12E − 01	5.69E − 01	kg
		MVOC	9.12E − 02	1.06E − 02	-	kg
		Particles	-	-	6.35E − 02	kg
		Dioxin	-	-	2.63E − 06	kg

shredding, and dewatering. Pretreated food wastes, which were in pulp status, were injected to an anaerobic digester with primary sewage sludge. On the basis of plant data from July 2007 to June,2008, an average of 193 tonnes of food waste per day was pretreated, and then 293 m^3 of pulp-phase food waste and 1,206 m^3 of sewage sludge per day were co-digested by anaerobic digestion. Finally, 11,150 m^3 of biogas per day was produced. An average of 3,700 kW h of electricity per day was required. Around 213 tonnes of processed water per day was needed to move the pulp-phase pretreated food waste easily to the anaerobic digester. Approximately 77 tonnes (around 40% of the daily treated food waste) of screenings and dewatered cakes per day were sorted and finally landfilled. In this system, pretreated food wastes (flow rate at 250 m^3/day, volatile solid (VS) 39 kg/L) and the first sludge from a primary sedimentation tank (flow rate 1,300 m^3/day, VS 12.9 g/L) were mixed in the anaerobic digester. Sewage sludge and pretreated food wastes were injected to the digester in a ratio of 5:1, and the ratio of volatile solid concentration of sludge and food wastes was 1:3. Biogas production in 2006 was 7,236 m^3 from sewage sludge alone; this increased to 9,745 and 9,559 m^3 of biogas in 2007 and 2008, respectively, when co-digestion with food wastes was introduced. It was represented that co-digestion with the food waste system produced around 34% of biogas more than sludge digestion alone.

o *Allocation for biogas quantification.* Biogas production was calculated by allocation. In 2008, a daily increase, around 2,323 m^3, was produced from 196 tonnes of food waste; therefore, it was estimated that 11.85 m^3 of biogas was produced from 1 tonne of food waste. It was 13.32 m^3 in 2007 (Table 3). It was converted to 8.92 kg of weight based from the

Table 3 Biogas production from the co-digestion

Year	Total biogas production (m^3/day)	Food waste input (kg/day)	Biogas production from food waste (m^3/kg)	Note
2006	7.24E + 03	0.00E + 00	0.00E + 00	Sewage sludge only
2007	9.75E + 03	1.88E + 05	1.33E − 02	Sewage sludge plus food wastes
2008	9.56E + 03	1.96E + 05	1.19E − 02	Sewage sludge plus food wastes

average value of 12.59 m^3. CO_2-eq of 186 kg was generated as GHG from co-digestion treatment process. GHG production from the combustion of biogas, which was generated for this process, was calculated using IPCC emission factors. If generated CO_2 was a biogenic origin, then this CO_2 was excluded on the counted GHG.

o *Allocation for electricity quantification.* Electricity usage was calculated by allocation. According to the visiting surveys at selected STPs, the station for pumping to a grit chamber, facility operation (including settler and clarifier), and aerator operation needed 27.75%, 8.86%, and 37.46% of total electricity usage, respectively. In the case of the sludge treatment process, dewatering, sludge thickening, and deodorization needed 19.89%, 4.54%, and 1.50% of electricity, respectively. Therefore, 74% and 26% of electricity were used for sewage treatment process and sludge treatment process, respectively. Another selected STP showed that 50% of electricity was consumed for aeration, and 32% and 9% of electricity was consumed by the facility operation and pump station, respectively. Only 8% of electricity was used for anaerobic digestion. Therefore, it was assumed that an average of 17,511 kW h (17%) was used for sludge treatment by anaerobic digestion. Allocation was followed by flux rate, therefore, it was estimated that 3,502 kW h, which was one fifth of electricity, was consumed for anaerobic digestion per day. The facility treated around 196 tonnes of food waste a day; therefore, it was assumed that daily electricity for anaerobic digestion per 1 tonne of food waste was 17.87 kW h. In conclusion, 18.88 kW h, which was already calculated for pretreatment of 1 tonne of food waste, and 17.87 kW h, which was calculated by allocation here for anaerobic digestion, were counted for this scenario. The sum of two electric values, 36.75 kW h, was estimated as electricity usage for 1 tonne of food waste treatment by co-digestion.

We calculated that 36.75 kW h of electricity was used and 8.92 kg of biogas was produced for the treatment of 1 tonne of food waste by allocation. Steam, which was converted from the captured biogas, was used for digester warming. The environmental impact of the recycled processed water and steam was not considered because it was already counted on the system. It was estimated that 141 kg of sludge was discharged from 1 tonne of food waste. The gate-to-gate data is shown in Table 2.

- *Dryer-incineration.* Reduction efficiency was 75.9% of food waste; therefore, the water containing food wastes after drying with the use of a garbage dryer reduced in weight from 1 to 0.24 tonne (Table 2). Volume was also reduced; however, ultimate constituents were not changed. Produced fly ash and bottom ash were transported to the Sudolwon landfill site for final disposal. As the ash from complete incineration contained no organic carbon when it arrived at the landfill, the burial of ash would generate no landfill gas (LFG). Carbon was not completely removed by incineration, but it can be assumed that the landfill of incinerator ash resulted in no LFG emissions [24]. Therefore, we estimated that GHG from the ash at the landfill was not produced; however, GHG by transportation was added. Around 364 kg of CO_2, 7.99 kg of CH_4, and 2.75 kg of N_2O were produced. The produced CO_2 was non-biogenic and was produced from energy consumption. However, CO_2 by incineration of food waste was not counted as GHG because it was biogenic in origin. The caloric value (wet based) of 1 tonne of dried food waste was 2.92 Gcal, and energy recovery would be 9,047 MJ if the average recovery rate of 74% was applied (Table 4). Therefore, it was estimated that around 657 kg of GHG was reduced.

Disposal stage

The distances of transportation from facilities to final disposal sites were calculated. It was assumed that the screenings were landfilled and the sludge was incinerated. The distances from each facility to the landfill site and resource recycling (incineration) facility were calculated respectively by GIS. One-way distances were 46 and 11 km for anaerobic digestion, 47 and 12 km for co-digestion from the facility to the landfill and resource recycling facility, respectively, and the distance was approximately 50 km between the landfill and incineration plant.

Life cycle impact assessment analysis

Characterized impact on the global warming impact category was assessed using GWP for 100 years [25].

Anaerobic digestion

The indicator, GWP, represented that the treatment and disposal stages mainly contributed to the environment.

Table 4 Energy recovery rate of the dryer-incineration

Factor	Value	Recovery rate
Low-heating value of dried food waste (wb[a])	2.92 Gcal/tonne	74%
Energy recovery	9,047 MJ[b]	
GHG reduction	657 kg CO_2/tonne	

[a]Wet based; [b]1 Gcal = 4186.8 MJ.

Directly discharged CO_2 for digestion was biogenic in origin; therefore, it was not counted as GHG. In addition, produced methane was captured. However, the treatment stage still contributed to GWP the most because chemicals and electricity usage were sources of GHG. Screenings were finally landfilled, and wastewater was treated chemically or biologically. These final disposals also contributed to global warming. This anaerobic digestion system treated 1 tonne of food waste and produced 0.07 tonne of biogas. The biogas was converted to 244 kW h of electricity and 204 kg of steam. About 68.23 kW h of electricity was primarily used for the facility operation. Surplus electricity of 175 kW h was sold at Korea Power Exchange. Produced steam was used for warming a digester. About 110 kg of biosolids was also produced. It was sold as a raw material of compost. For this scenario, 211 kg of CO_2-eq of GHG were discharged from 1 tonne of food waste. Around 8.8, 0.7, 131, and 70.1 kg of CO_2-eq of GHG were produced in the collection, transportation, treatment, and disposal stages, respectively (Table 5).

Co-digestion

The indicator, GWP, represented that disposal stages mainly contributed to the environment. Screenings, which were produced for the pretreatment process, were disposed by landfilling. LFG was the major reason of global warming in the co-digestion scenario. For this scenario, 259 kg of CO_2-eq of GHG were discharge from 1 tonne of food waste. Around 8.8, 0.5, 37.7, and 212 kg of CO_2-eq of GHG were produced in the collection, transportation, treatment, and disposal stages, respectively (Table 5). Landfilling of the screenings was contributed to GWP the most in this scenario.

Dryer-incineration

Electricity was consumed for the garbage dryer operation. Produced CO_2 in the incineration process was biogenic in origin; therefore, it was excluded from the GHG counting. In addition, combusted waste heat was recovered and converted to steam. This system treated 1 tonne of food waste and produced 2.16 Gcal of heat energy. It was sold and used in the residential area for room heating with hot water heating. For this scenario, 342 kg of CO_2-eq of GHG were discharged from 1 tonne of food waste. Around 321, 8.8, 1.1, 10.3, and 0.7 kg of CO_2-eq of GHG were produced in the discharge, collection, transportation, treatment, and disposal stages, respectively (Table 5).

Avoided impact analysis

Avoided impact was analyzed in order to calculate efficiencies of GHG reduction relevant to alternatives. The commercial production of steam and electricity was analyzed by LCA. The result showed that 0.04 and 0.50 kg of CO_2-eq of GHG were produced when 1 MJ of steam and 1 kW h of electricity were manufactured, respectively. About 178 kg of CO_2-eq by anaerobic

Table 5 GHG emissions (kg CO_2-eq/tonne) at each stage

		Anaerobic digestion	Co-digestion	Dryer-incineration
Discharge (household)		0.00E + 00	0.00E + 00	3.21E + 02
Electricity		-	-	3.21 E + 02
Collection (diesel)		8.81E + 00	8.81E + 00	8.81E + 00
Transportation To Facility (diesel)		6.71E − 01	5.10E − 01	1.08E + 00
Treatment process		1.32E + 02	3.77E + 01	1.03E + 01
Direct GHG emission		3.42E + 00	3.97E − 01	4.90E + 00
Electricity		3.38E + 01	1.82E + 01	3.92E + 00
LNG		-	-	5.99E − 01
Water		-	-	1.23E − 03
Diesel		-	-	4.30E − 02
Additive	Transportation	1.87E − 02	1.87E − 02	1.15E − 03
	Chemical	9.37E + 01	1.91E + 01	8.38E − 01
Final disposal		7.01E + 01	2.12E + 02	6.83E − 01
Transportation to disposal site		9.49E − 01	3.14E + 00	3.30E − 01
Electricity		1.34E + 00	-	-
Solid		6.40E + 01	2.09E + 02	-
Wastewater		3.83E + 00	-	3.53E − 01
Total		2.11E + 02	2.59E + 02	3.42 E + 02

Table 6 Alternative effects of scenarios

Scenario	1-tonne food waste treatment options			Avoided impact		
	By-product		GWP (Pa)	CO_2 reduction (Pb)	Environmental credit (Pa-Pb)	Unit
Anaerobic digestion	Electricity	1.74E + 02(kWh)	2.11E + 02	8.46E + 01	3.32E + 01	kg
	Biosolids	1.10E − 01(tonne)		9.32E + 01		
Co-digestion	NA	NA	2.59E + 02	NA	NA	-
Dryer-incineration	Waste heat	9.05E + 03(MJ)	3.42E + 02	6.57E + 02	−3.15E + 02	kg

digestion and 657 kg of CO_2-eq by dryer-incineration from 1 tonne of food waste, respectively, were calculated by avoided impact analysis. In the case of dryer-incineration, GHG was reduced by waste heat recovery; however, GWP was still high through the full stages. It was because a garbage dryer operation required electricity usage.

This comparison of three different food waste disposal scenarios found that the GWP of each scenario before the offset of generated electricity is around 211 kg of CO_2-eq for anaerobic digestion, 259 kg of CO_2-eq for co-digestion, and 342 kg of CO_2-eq for dryer-incineration from 1 tonne of food waste, respectively, all based on the wet weight. GWP of dryer-incineration in the treatment stage was low because CO_2 production from incineration was biogenic in origin; however, GWP in the discharge stage was the highest because electricity usage for a waste dryer operation in the discharge stage contributed to GWP. Anaerobic digestion, co-digestion, and dryer-incineration scenarios were higher in GWP in that order.

Around 110 kg of biosolids, 244 kW h of electricity, and 0.2 tonne of steam were produced by the anaerobic digestion scenario. Biosolids were used as a raw material in the composting industry. Steam was not included in analyzing the by-product because it was recycled in the system boundary. Around 70 kW h of produced electricity was used for facility operation; therefore, 174 kW h of electricity could be sold in a market. Incineration scenario produced 2.16 Gcal of waste heat. It was converted to steam and supplied to residential areas. Steam of the co-digestion scenario was recycled within the system boundary. Therefore, steam production was excluded in the list of by-products. Environmental credit was 33 kg of CO_2-eq for anaerobic digestion and −315 kg of CO_2-eq for dryer-incineration from 1 tonne of food waste, respectively (Table 6).

In the present study, the indicator, GWP, was used for environmental impact analysis. We analyzed six impact categories, ADP, AP, EP, GWP, ODP, and POCP for the scenarios as shown in Table 7. The results show that GWPs of all scenarios affect the environment the most although POCP of dryer-incineration is high.

We analyzed the B/C ratio of the scenarios and found that it was 0.11 for anaerobic digestion and 0.26 for dryer-incineration in the previous research [26]. We

Table 7 Environmental impact for each impact category of scenarios

	Characterization (a)				Normalization reference (b)		Normalized result (a/b)		
	AD	COS	DI	Unit	Factor	Unit	AD	COS	DI
ADP	1.05	2.80	4.02	1/year	24.9	kg/person-years^2	4.22	1.12	1.61
	E + 00	E − 01	E − 01				E − 02	E − 02	E − 02
AP	5.56	1.84	3.32	kg of SO_2-eq/kg	39.8	kg of SO_2-eq/person-years	1.40	4.61	8.33
	E − 01	E − 01	E − 01				E − 02	E − 03	E − 03
EP	1.79	2.92	1.06	kg of PO_4^{3-}-eq/kg	13.1	kg of PO_4^{3-}-eq/person-years	1.37	2.23	8.08
	E + 00	E − 02	E − 01				E − 01	E − 03	E − 03
GWP	2.12	2.60	3.42	kg of CO_2-eq/kg	5530	kg of CO_2-eq/person-years	3.83	4.70	6.18
	E + 03	E + 03	E + 03				E − 01	E − 01	E − 01
ODP	1.28	4.72	1.04	kg of CFC11-eq/kg	0.0407	kg of CFC-eq/person-years	3.10	1.20	2.60
	E − 05	E − 06	E − 05				E − 04	E − 04	E − 04
POCP	2.23	1.01	1.01	kg of C_2H_4-eq/kg	10.3	kg of C_2H_4-eq/person-years	2.16	9.81	9.76
	E − 01	E − 01	E + 03				E − 02	E − 03	E + 01

AD, anaerobic digestion; COS, co-digestion with sludge; DI, dryer-incineration; ADP, abiotic resource depletion potentials; AP, acidification potentials; EP, eutrophication potentials; GWP, global warming potentials; ODP, ozone depletion potentials; POCP, photochemical oxidant creation potentials.

could not find it for co-digestion because the by-product of co-digestion was recycled within the system boundary.

Conclusions

This study evaluated three food waste disposal systems: anaerobic digestion, co-digestion, and incineration of dried food waste, using LCA from the perspective of global warming and energy/resource recovery. Since landfilling of food wastes was banned in 2005, animal feeding and composting have been the major practice for treating them, and anaerobic digestion and co-digestion were the minor ones. Incineration is an available method for energy recovery from municipal solid wastes. Incineration of raw food waste has been practiced for increasing the heating value in a few local governments in Korea; however, it provokes a lot of controversy because of much moisture and salinity constituent. The three scenarios were evaluated though they are not major disposal methods in Korea because we concentrated in energy recovery from food wastes to suggest alternatives.

GWP, which is an indicator to evaluate environmental impact, was acquired in five stages: discharge, collection, transportation, treatment, and disposal stages. We found that the GWPs were estimated to be 211 kg of CO_2-eq for anaerobic digestion, 259 kg of CO_2-eq for co-digestion, and 342 kg of CO_2-eq for dryer-incineration from 1 tonne of food waste.

From analyzing the by-products from each scenario, 174 kW h of electricity and 110 kg of raw materials for compost from anaerobic digestion and 9,050 MJ of heat from dryer-incineration were produced. An avoided impact showed that by-products from 1 tonne of food waste could reduce GHG; for example, 178 kg of CO_2-eq by anaerobic digestion and 657 kg of CO_2-eq by incineration might offset the global warming potential of each system under the assumption that the by-products were completely utilized. Therefore, the environmental credit of the three options based on the wet weight of food waste was found to be 33 kg of CO_2-eq for anaerobic digestion and −315 kg of CO_2-eq for dryer-incineration from 1 tonne of food waste by electricity, thermal energy generated, and primary materials avoided. We found that dryer-incineration is the best available alternative for producing renewable energy in Korea. In the case of the co-digestion scenario, future studies need to increase biogas production. In the dryer-incineration scenario, it is available to acquire heat energy although energy consumption is high at the discharge stage. It could be one of the practical treatment methods under reducing energy requirements.

The limitations and future studies suggested in this study are as follows: Jungnang District in Seoul City, one of the facilities currently operated for each scenario in the study area, was selected and evaluated; however, the data from the selected facilities may not represent national or global trends.

Competing interests

The authors declare that they have no competing interests.

Authors' contributions

MHK carried out the study plan, analyzed environmental impacts, and drafted the manuscript. HBS collected data and drafted the manuscript, tables, and figures. YS participated in the literature review, reviewed the manuscript, and corrected an error. ITJ developed the methodology and participated in the analysis of the results. JWK participated in the analysis of the results. All authors read and approved the final manuscript.

Authors' information

MHK is an engineer. This research is a part of her PhD dissertation. HBS received her BS in Chemical Engineering from Soongsil University. YS is currently a doctorate degree student in the School of Medicine, University of Missouri. ITJ got a PhD from Ajou University and is an associate researcher at Korea Environmental Industry and Technology Institute. JWK was the PhD adviser for MHK at Seoul National University.

Acknowledgments

The authors would like to give their special thanks to Prof. Dong-Soo Lee, Prof. Je-Yong Yoon, Dr. Mann-Young Kim, and Dr. Kee-Young Yoo for their helpful advice.

Author details

[1]Department of Environmental Planning, Graduate School of Environmental Studies, Seoul National University, San 56-1, Sillim-dong, Gwanak-gu, Seoul 151-742, Republic of Korea. [2]Department of Chemical Engineering, Soongsil University, Sangdo-ro 369, Dongjak-gu, Seoul 156-743, Republic of Korea. [3]Department of Molecular Microbiology and Immunology, School of Medicine, University of Missouri, Columbia, MO 65201, USA. [4]Carbon Management Office, Korea Environmental Industry and Technology Institute, 215 Jinheung-ro, Bulgwang-dong, Eunpyeong-gu, Seoul 122-706, Republic of Korea.

References

1. Pandyaswargo, AH, Onoda, H, Nagata, K: Energy recovery potential and life cycle impact assessment of municipal solid waste management technologies in Asian countries using ELP model. Int J Energy Environ Eng **3**, 28 (2012)
2. Anglani, N, Muliere, G: The impact of renewable energy and energy efficient technologies, what to choose in case of limited supportive actions: a case study. Int J Energy Environ Eng **2**(4), 83–94 (2008). 2008
3. Zhang, R, El-Mashad, HM, Hartman, K, Wang, FG, Liu, G, Choate, C, Gamble, P: Characterization of food waste as feedstock for anaerobic digestion. Bioresource Technol **98**(4), 929–935 (2007)
4. Lai, CM, Ke, GR, Chung, MY: Potentials of food wastes for power generation and energy conversion in Taiwan. Renew Energ. **34**(8), 1913–1915 (2009)
5. Khoo, HH, Lim, TZ, Tan, RBH: Food waste conversion options in Singapore: environmental impacts based on an LCA perspective. Sci Total Environ **408**(6), 1367–1373 (2010)
6. Murphy, JD, McKeogh, E, Kiely, G: Technical/economic/environmental analysis of biogas utilization. Appl Eng. **77**(4), 407–427 (2004)
7. Nam, JJ, Yoon, YM, Lee, YH, So, KH, Kim, CH: Life cycle assessment of greenhouse gas emissions from livestock and food wastes co-digestive biogas production system. Kor J Environ Agric. **27**(4), 406–412 (2008)
8. Gómez, X, Cuetos, MJ, Cara, J, Morán, A, Garcia, AI: Anaerobic co-digestion of primary sludgy and the fruit and vegetable fraction of the municipal solid wastes: Conditions for mixing and evaluation of the organic loading rate. Renew Energ. **31**(12), 2017–2024 (2006)
9. Kabouris, JC, Tezelm, U, Pavlostathism, SG, Engelmannm, M, Dulaneym, J, Gillette, RA, Todd, AC: Methane recovery from the anaerobic codigestion of municipal sludge and FOG. Bioresource Technol. **100**(15), 3701–3705 (2009)

10. Lundie, S, Peters, GM: Life cycle assessment of food waste management options. J Cleaner Prod. **13**(3), 275–286 (2005)
11. Zhao, W, van der Voet, E, Zhang, Y, Huppes, G: Life cycle assessment of municipal solid waste management with regard to greenhouse gas emissions: case study of Tianjin, China. Sci Total Environ **407**(5), 1517–1526 (2009)
12. Eddine, BT, Salah, MM: Solid waste as renewable source of energy: current and future possibility in Algeria. Int J Energy Environ Eng **3**, 17 (2012)
13. Eggleston, S, Buendia, L, Miwa, K, Ngara, T, Tanabe, K (eds.): 2006 Guidelines for National Greenhouse Gas Inventories. Intergovernmental Panel on Climate Change, Institute for Global Environmental Strategies, Hayama (2006)
14. Ministry of Environment (South Korea): Master Plan of Food Waste Reduction and Recycling. Ministry of Environment, Gwacheon (1998)
15. Ministry of Environment (South Korea): Quality Standard Establishment of Energy Products from Organic Waste. Ministry of Environment, Gwacheon (2008)
16. International Organization for Standardization/TR: Environmental Management - Life Cycle Assessment - Examples of Application of ISO 14041 to Goal and Scope Definition and Inventory Analysis. ISO/TR 14049:2000(E), Geneva (2000)
17. Korres, NE, Singh, A, Nizami, AS, Murphy, JD: Is grass biomethane a sustainable transport biofuel? Biofuel Bioprod Bior. **4**(3), 310–325 (2010)
18. International Organization for Standardization: Environmental Management - Life Cycle Assessment - Goal and Scope Definition and Inventory Analysis. ISO 14041:1998(E), Geneva (1998)
19. Ekvall, T, Finnveden, G: Allocation in ISO 14041 - a critical review. J Cleaner Prod. **9**(3), 197–208 (2001)
20. Lee, KM, Park, PJ: Estimation of the environmental credit for the recycling of granulated blast furnace slag based on LCA. Resour Conserv Recy. **44**(2), 139–151 (2005)
21. Kim, MH, Kim, JW: Comparison through a LCA evaluation analysis of food waste disposal options from the perspective of global warming and resource recovery. Sci Total Environ **408**(19), 3998–4006 (2010)
22. International Organization for Standardization: Environmental Management - Life Cycle Assessment - Principles and Framework. ISO14040:2006(E)., Geneva (2006)
23. International Organization for Standardization: Environmental Management - Life Cycle Assessment - Requirements and Guidelines ISO 14044:2006(E)., Geneva (2006)
24. McDougall, F, White, P, Franke, M, Hindle, P: Integrated Solid Waste Management: A Life Cycle Inventory, 2nd edn. Blackwell Science Inc, Oxford (2001)
25. Guinée, JB: Handbook on Life Cycle Assessment: Operational Guide to the ISO Standards. Kluwer Academic Publishers, Alphen aan den Rijn (2002)
26. Kim, MH, Song, YE, Song, HB, Kim, JW, Hwang, SJ: Evaluation of food waste disposal options by LCC analysis from the perspective of global warming: Jungnang case, South Korea. Waste Manag **31**, 2112–2120 (2011)

Effect of storage parameters on stability of Jatropha-derived biodiesel

Purabi Mazumdar[1], Venu Babu Borugadda[2], Vaibhav V Goud[1,2*] and Lingaraj Sahoo[1,3*]

Abstract

Jatropha curcas L. seeds as a raw material for biodiesel production is a rapidly growing interest over the world because of its high oil content, ecological adaptability, and excellent fuel properties. Though there is an increase in productivity of biodiesel, showing solution for future energy insecurity, there still remains some concern for commercialization due to its susceptibility to degradation during long storage. The aim of this paper is to investigate the effect of temperature and ambient condition on Jatropha biodiesel storage. An experiment was conducted for a period of 12 months, where Jatropha biodiesel stored in three groups at different temperatures (4°C, 25°C, and 35°C) and environmental conditions (exposed in dark, light, and air). At regular intervals, the samples were taken out to analyze acid value, density, kinematic viscosity, and thermogravimetric profile to monitor the quality of biodiesel. Analysis showed that acid value, density, kinematic viscosity, and the onset temperature of volatilization and distillation increases with the increase in storage time of biodiesel samples. However, Jatropha biodiesel stored at 35°C, in contact with ambient air and light showed highest degradation compared to those which were stored at 25°C and 4°C. Among all the parameters studied, high temperature and air exposure are the two most potent parameters which accelerate the degradation process. Along with that, light exposure had mild but significant effect on Jatropha biodiesel degradation over a long storage period.

Keywords: Biodiesel, *Jatropha curcas*, Transesterification, Free fatty acids, TGA

Background

Biodiesel is a monoalkyl ester of long-chain fatty acid derived from vegetable oil or animal fat [1]. It constitutes the most reliable renewable source for substituting diesel. The finite nature of fossil fuel reserves and growing green house gas emission have promoted research on alternative fuels [2,3]. Biodiesel has advantages over conventional diesel fuels because of its renewable, environmentally benign and biodegradable nature. The main disadvantage is that biodiesels are severely prone to degradation than conventional diesel fuels during long storage [4]. Storage stability is a critical issue regarding the successful commercialization of biodiesel in fuel market. Storage stability is the ability of a fuel to resist changes in its physicochemical characteristics brought by interaction with its environment. In the presence of air or oxygen, biodiesel will be hydrolyzed into alcohol and acid. The presence of alcohol reduces flash point, and the presence of acid increases the total acid value. All these make biodiesel unstable during storage. But in order to introduce biodiesel in the transport industry, it should meet accepted fuel standards and quality assurance [5].

There are various reports in the literature on the storage, oxidation stability, and the effect of antioxidant concentration on biodiesel. The effect of different synthetic and natural antioxidants on oxidation stability of biodiesel produced from rapeseed oil, sunflower oil, used frying oil, beef tallow, and soya bean oil has been reported [6,7]. Long time storage stability are also investigated on biodiesels synthesized from rapeseed oil, used frying oil, high oleic sunflower oil, high and low erucic *Brassica carinata* oil [5,8,9]. Dunn [10] examined the oxidative stability of soy-bean oil methyl esters by analyzing oil stability index Polavka et al. [11] studied the oxidation stability of rapeseed oil and waste frying oil methyl esters using differential thermal analysis and

* Correspondence: vvgoud@iitg.ernet.in; ls@iitg.ernet.in
[1]Center for Energy, Indian Institute of Technology Guwahati, Guwahati, Assam 781039, India
[3]Department of Biotechnology, Indian Institute of Technology Guwahati, Guwahati, Assam 781039, India
Full list of author information is available at the end of the article

Rancimat (Metrohm AG, Herisau, Switzerland). Most of the studies on the storage and oxidation stability are carried out on biodiesel derived from edible oil. However, the studies on biodiesel derived from non-edible oil seeds are scanty [12-14]. Das et al. [12] analyzed the effect of light, air exposure, and antioxidants namely propyl gallate, butylatedhydrox-yanisole, and butylatedhydroxytoluene on long-term storage stability of karanja oil methyl ester (KOME). Their study showed that oxidative stability of KOME decreases with the increase in storage time and increases with higher concentration of antioxidant levels.

Among non-edible oil crop, recently, Jatropha is in the top priority in national biodiesel program because of its high oil content and promising fuel properties. Several researchers have performed experiments for efficient Jatropha biodiesel production, physicochemical characterization [15,16], and engine performance [17]. Sarin et al. [18] monitored the effect of metal contaminants and antioxidant concentrations on the storage stability of Jatropha biodiesel. However, no proper study has been conducted on the effect of different storage parameters like temperature and exposure to light and air on Jatropha methyl ester (JME) storage. The aim of this work is to investigate the influence of different parameters like temperature and exposure to light and air on Jatropha biodiesel degradation. Acid value, density, kinematic viscosity (KV), and thermogravimetric analysis (TGA) results for methyl esters were analyzed for a period of 12 months and were compared with the initial values to monitor the changes in the quality of methyl esters. The study will provide knowledge on the factor influencing degradation of Jatropha biodiesel, which will be immensely helpful in maintaining the quality of Jatropha biodiesel over long-term storage.

Methods

Materials

Jatropha seeds were collected from Assam, Northeast India. The seeds were separated from the fruit mechanically and cleaned manually to remove all foreign materials. Then, the seeds were dried under similar temperature (60°C) and humidity conditions to reach constant weight. The oil was extracted from grinded kernels in Soxhlet apparatus using hexane as per the standard American Oil Chemical Society procedure for 8 h. The extract was concentrated in Rotavapor (Sigma-Aldrich, MO, USA). All the other chemicals and reagents like methanol, ethanol, n-hexane, potassium hydroxide, sodium hydroxide, and phenolphthalein indicator were analytical reagent grade and were purchased from M/s Merck (Merck & Co., Inc., NJ, USA).

Preparation of biodiesel from Jatropha oil

Due to low-acid value of the native oil, single-step direct transesterification procedure was followed. JME was synthesized by refluxing the oil at 60°C, employing 1:6 molar ratio of oil to methanol for 1.5 h. KOH (1 wt%) was used as the catalyst. The mixture was stirred using magnetic beads (400 rpm). After completion of the reaction, the mixture was cooled to room temperature and was transferred to a separating funnel, leading to the separation into two phases. The bottom glycerol layer was discarded. The top ester layer was washed several times with warmed Milliporewater (EMD Millipore Corporation, Billerica, MA, USA) to remove all the trace of catalyst, glycerol, and soap. The washed methyl ester was further purified under vacuum on a rotary evaporator.

The storage condition and analysis

The acid value, density, KV, and TGA analysis were performed and recorded as 0 month values after the preparation of Jatropha methyl ester. Then, the methyl ester samples were stored for 12 months in three individual groups at different storage conditions. The groups were made on the basis of different storage temperatures: 4°C, 25°C, and 35°C. Each group consists of three subgroups: in the first subgroup, the samples were kept in sealed and in dark environment. In the second subgroup, the samples were kept in a sealed environment but exposed to light, while in the third one, the samples were kept but exposed to both light and ambient air. The samples from each group (in three replicates) were analyzed periodically at the intervals of 6 months for different fuel quality parameters such as acid value, density, KV, and TGA.

Analysis

Acid value

Acid value denotes free fatty acid content of oil. It is defined as the amount of potassium hydroxide (in mg) required for neutralizing 1 g of oil sample. It was calculated according to AOCS Official method (Te TA-64, 1997), by titrating solution of JME in neutral ethanol with 0.5 N KOH. Phenolphthalein solution was used as an indicator for titration. Acid value was calculated using the following expression:

$$\text{Acid value (mg KOH/g)} = \text{Volume of the titrant (in ml)} \times N \times 56.10/\text{Mass of the sample(in g)},$$

where N is the normality of accurately standardized sodium hydroxide solution.

Density

The density of a material is defined as mass per unit volume. The densities of the stored JME samples were determined at 25°C, using specific gravity bottle, according to ASTM D4052 method. Density was calculated using the following expression:

$$\text{Density}\left(\text{kg/m}^3\right) = \text{Mass (kg)/Volume}\left(\text{m}^3\right).$$

Kinematic viscosity

KV is the measure of resistance of a fluid flow against gravity and is determined as the ratio of dynamic viscosity to the density. Dynamic viscosity was measured for stored JME samples with Rheometer (RheoStress RS1, Thermo Electron, Germany) at 40°C. At three different shear rates (50, 100, and 150 s^{-1}), the dynamic viscosities were measured, and the average values were considered. KV was calculated using the following expression:

$$\text{KV}\left(\text{m}^2/\text{s}\right) = \text{Dynamic viscosity(pa.s or kg/m.s)/Density}\left(\text{kg/m}^3\right).$$

Thermogravimetric analysis

The thermogravimetric profile of the JME samples was obtained using Thermogravimetric Analyzer (TGA 851e/LF/1100, Mettler, Switzerland). Conventional TGA was performed on 6 ± 0.5 mg samples under high purity nitrogen atmosphere at a flow rate of 40 ml/min. The samples were heated from ambient temperature to 900°C and at constant heating rate of 10°C/min.

Results and discussion

The degradation behavior of JME samples on long storage under different conditions was assessed by monitoring the changes in their physicochemical properties such as acid value, kinematic viscosity, onset temperature of volatilization, and distillation temperature of thermal decomposition with respect to their initial value at 0 month.

Effect of temperature on biodiesel storage

Temperature plays an important role in the degradation of methyl ester. High temperatures accelerate the process of oxidation which results in the increase in density and viscosity of the fuel [19]. The acid value is one of the promising methods to monitor the degradation of methyl ester. Figure 1a,b,c represents the change in the acid value of the JME samples stored under different conditions. The methyl ester degradation shows direct correlation with acid value. Therefore, increase in acid value with storage time indicates gradual degradation of the methyl ester sample. According to ASTM D6751, the standard maximum allowable limit of the acid value for biodiesel is around 0.5 mg KOH/g. At the beginning

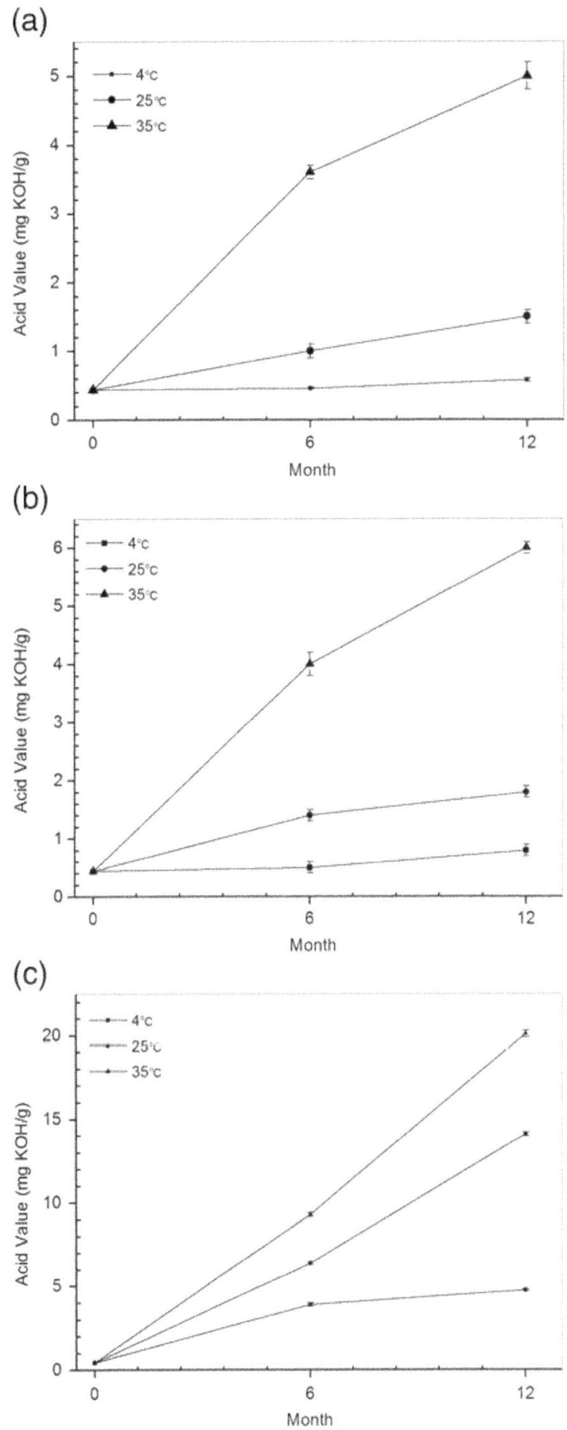

Figure 1 Change in the acid value. The Jatropha biodiesel was kept under different temperatures (4°C, 25°C, and 35°C) during 12 months of storage. (**a**) Sealed (devoid of light and air), (**b**) sealed and exposed to light, and (**c**) with air and exposed to light.

(i.e., at 0 months), the acid value of the sample was measured to be 0.44 mg KOH/g which is well below the maximum allowable limit. After 6 months of storage, the acid value of the methyl ester sample stored at 35°C was found to be the highest (9.3 mg KOH/g), whereas the sample stored at 4°C showed the lowest value (3.9 mg KOH/g). Further extending the storage period, JME stored at 4°C showed increase in acid value from 0.44 to 4.8 mg KOH/g. For the samples stored at 25°C, increase in the acid value was found to increase from 0.44 to 14.1 mg KOH/g, but the acid value was found to be increased up to 20.1 mg KOH/g for samples stored at 35°C. This clearly indicates that prolonged exposures of methyl ester to high temperature can deteriorate the biodiesel quality, which was also observed in previous studies with canola, palm, soybean, and sunflower methyl ester [4].

Comparison of density of the stored sample with initial value (i.e., 0 month value) over time also gives an idea about the changes in quality of the biodiesel upon extended storage. Figure 2a,b,c shows that with increase in storage time, the density of JME also increases proportionally. The increase in the density of methyl ester is mainly due to the interaction of peroxide molecules which formed due to the degradation of esters [20]. The density values found to be increased for all JME samples kept under extended storage. The initial density values of the samples for 0 months were measured to be around 0.85 kg/m^3. After 6 months of storage period, maximum increase in density of the biodiesel stored at 35°C was measured to be 1.9 kg/m^3.On the other hand, biodiesel samples stored at 4°C and 25°C, maximum increase in density was observed around 1.15 kg/m^3 and 1.6 kg/m^3, respectively (Figure 2c). Further extending the storage period, the maximum increase in density of the samples stored at 35°C over 12-months period was measured as 2.3 kg/m^3. Whereas, in the case of the other samples stored at 4°C and 25°C,the density increase was from 0.85 to 1.3 kg/m^3 and 0.85 to 1.8 kg/m^3, respectively. The data clearly revealed the influence of higher temperature on methyl ester degradation. The increase in density of biodiesel is the first indication of methyl ester degradation.

Kinematic viscosity is another suitable method for monitoring the oxidative degradation of methyl ester. Kinematic viscosity tends to increase upon further oxidation due to the formation of secondary oxidative product. Figure 3a,b,c represents the change in kinematic viscosity of Jatropha biodiesel over 12 months of storage. At the beginning, the kinematic viscosity of JME was found to be around 6.10 × 10^{-6} m^2/s (0-month value) which was within the specified range of ASTM D6751 standards. During the storage over a period of 12 months, the kinematic viscosity values of all the methyl ester samples were found to be increased. After 6 months of

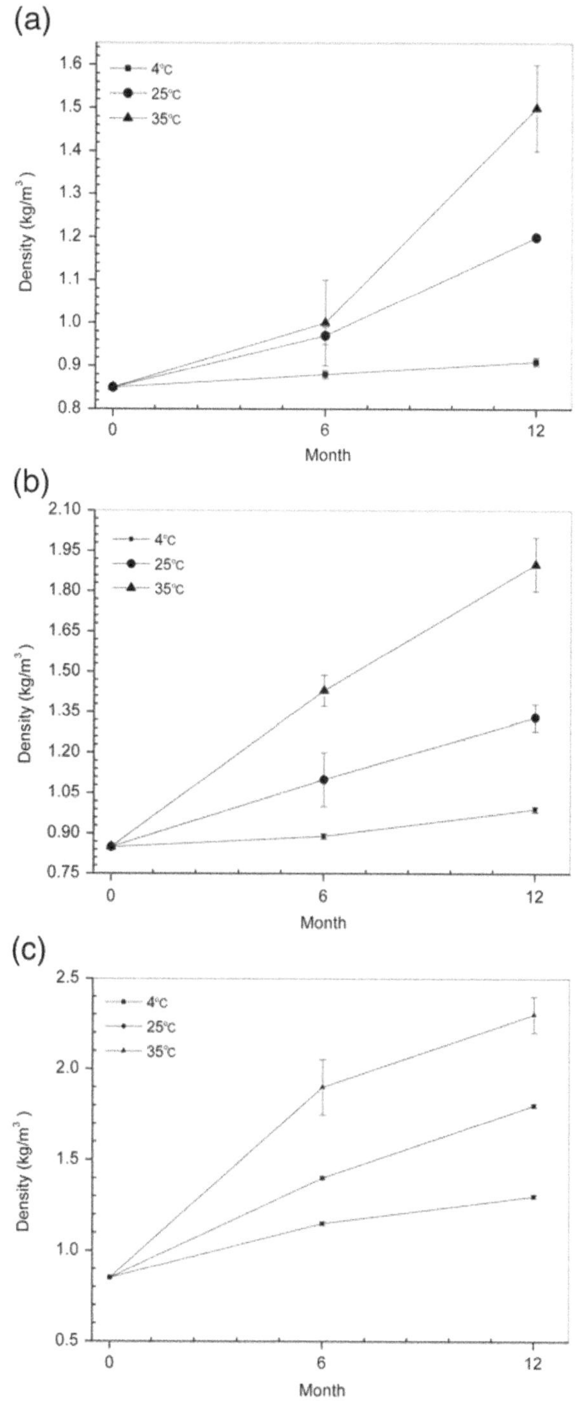

Figure 2 Change in density. The Jatropha biodiesel kept under different temperature (4°C, 25°C and 35°C) during 12 month of storage. **(a)** Sealed (devoid of light and air), **(b)** sealed and exposed to light, and **(c)** with air and exposed to light.

storage, maximum increase in kinematic viscosity of the JME stored at 35°C was measured to be 9.2 × 10^{-6} m^2/s. On the other hand, biodiesel samples which were stored at 4°C and 25°C, maximum increase in density was found to be around 8 × 10^{-6} and 8.5 × 10^{-6} m^2/s, respectively

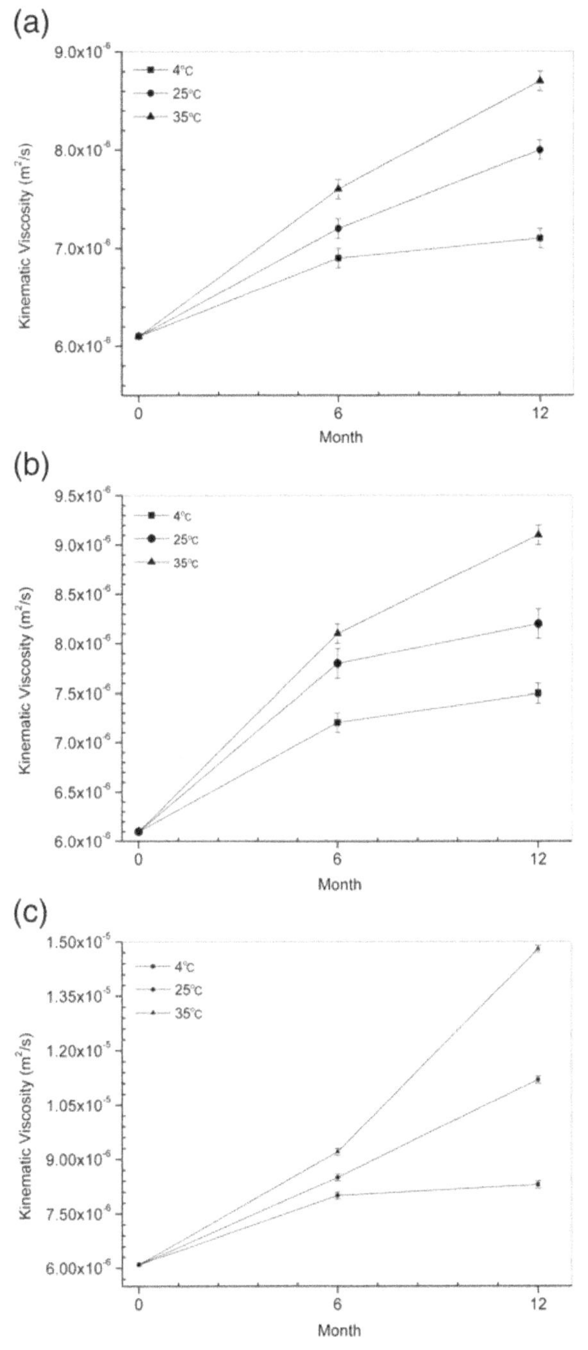

Figure 3 Change in kinematic viscosity of Jatropha biodiesel kept under different temperatures (4°C, 25°C, and 35°C) during 12 months of storage. (**a**) Sealed (devoid of light and air), (**b**) sealed and exposed to light, (**c**) exposed to air and light.

(Figure 3c). Further extending the storage period over 12 months, the JME stored at 35°C showed highest viscosity value of 1.48 × 10^{-5} m^2/s compared to 4°C (8.30 × 10^{-6} m^2/s) and 25°C (1.12 × 10^{-5} m^2/s). This increase in viscosity of the samples might be due to the formation of acids and polymeric compound as a result of oxidation, which further can lead to the formation of gums and sediments [5]. The gum and sediment formation is detrimental to engine operation. The increase in kinematic viscosity is a clear indication of methyl ester degradation which was also observed in stability studies in rapeseed and palm oil biodiesel over longer period of storage [21,22].

Table 1 presents the TGA data of JME stored under different conditions for a period of 12 months. The data shows that with storage time, onset temperature of volatilization and distillation temperature of thermal decomposition also increased. At the beginning (i.e., 0 month), the onset temperature of volatilization of Jatropha biodiesel was around 120°C. In the JME kept at 4°C and 25°C after 6 months of storage, the onset temperature of volatilization has increased to a maximum of up to 139°C and 165°C, respectively. On the other hand, for the JME kept at 35°C after 6 months of storage, the onset temperature of volatilization has increased a maximum of up to 220°C. Further extending the storage period over 12 months, the JME kept at 4°C and 25°C, the onset temperatures of volatilization were found to be increased to maximum of up to 148°C and 173°C, respectively. Whereas, the JME kept at 35°C after 12 months storage, the onset temperature of volatilization has increased a maximum of up to 310°C. The distillation temperature of thermal degradation of JME stored at 35°C is much higher compared to those of the samples which are kept at 25°C and 4°C. The data presented in the Table 1 show that higher temperature speeds up the formation of secondary oxidative product. Exposure to high temperature will help in isomerisation of methylene interrupted poly-unsaturated olefin structure to form more stable high molecular weight polymer [19], which increases thermal stability of biodiesel. Due to the formation of those thermally stable compounds, degraded methyl ester was found to alter the thermal profile compared to initial thermal profile.

Effect of light on biodiesel storage

To study the influence of light on degradation, JME samples were stored in light and as well as in dark conditions. Figure 1a shows the change in acid value of the biodiesel samples stored under dark and sealed conditions, whereas Figure 1b represents acid value change for samples exposed to light and sealed condition. After 6 months of storage, the samples stored at 4°C under dark and sealed conditions were analyzed for acid value. The acid value of the sample kept in dark condition was found to be increased from 0.44 to 0.46 mg KOH/g. Further extending the storage period over 12 months, the increase in acid value of JME was up to 0.58 mg KOH/g. On the other hand, for the sample stored over a period of 6 months at 4°C in light and sealed condition,

Table 1 Influence of storage parameter on volatilization and distillation temperature of Jatropha biodiesel

Samples		Onset temperature for volatilization (°C)			Distillation temperature (°C)					
Temperature (°C)	Storage condition	0 months	6 months	12 months	0 months		6 months		12 months	
	Sealed and dark		125	130	50%	90%	50%	90%	50%	90%
4	Sealed and in light		130	138			228	264	239	275
	In light and air exposure		139	148			241	275	250	298
	Sealed and dark	120	148	155	225	260	255	285	269	319
25	Sealed and in light		153	162			267	301	271	328
	In light and air exposure		165	173			290	340	310	349
	Sealed and dark		179	240			310	350	320	358
35	Sealed and in light		186	280			325	320	336	380
	In light and air exposure		220	310			345	420	390	480

increase in the acid value was measured as 0.44 to 0.50 mg KOH/g. After, 12 months of storage, the acid value further increased up to 0.80 mg KOH/g. In the samples stored at 25°C over a period of 12 months in sealed and dark condition, significant increase in the acid value (i.e., 0.44 to 1.5 mg KOH/g) was observed. Whereas, in the same sample stored in sealed condition but exposed to light, the acid value after 12 months of extended storage increases from of 0.44 to 1.8 mg KOH/g. The acid value of methyl ester after 12 months of storage which was kept at 35°C under dark and sealed condition increases from 0.44 to 5.0 mg KOH/g; whereas, in light and sealed condition, the acid value increased from 0.44 to 6.0 mg KOH/g. From this, it is clear that prolonged exposure of light on JME samples fastens the degradation rate compared to the dark condition. This finding was found to be in agreement with the previously published report on degradation of rapeseed biodiesel, where the sample kept in exposure to light showed higher degradation than in dark condition [21].

Increase in the formation of degraded product increases the density and kinematic viscosity of methyl ester. Figure 2a represents the changes in density of stored sample in sealed and dark condition, and Figure 2b is for the sample in sealed but exposed to light. The density of the biodiesel samples kept in dark condition was increased to a maximum of up to 1.0 and 1.5 kg/m^3 in 6 and 12 months of storage, respectively. On the other hand, the density of the biodiesel samples in sealed and light condition increased to a maximum of up to 1.4 and 1.9 kg/m^3 in 6 and 12 months of storage, respectively. Kinematic viscosity also followed the same pattern. The kinematic viscosity of the biodiesel samples stored in dark and sealed condition has increased to a maximum of up to 7.6 × 10^{-6} and 8.7 × 10^{-6} m^2/s for 6 and 12 months of storage, respectively; whereas, the kinematic viscosity of the biodiesel sample stored in light increased to a maximum of up to 8.1 × 10^{-6} and 9.1 × 10^{-6} m^2/s in 6 and 12 months, respectively. The above result gives clear evidence in support of evolution of higher density and kinematic viscosity of stored methyl ester in light in long periods than in dark condition.

Like the other parameters, the effect of temperature on thermal decomposition has also been studied in detail (Table 1). To understand the sample behavior, thermogravimetric analysis was performed for the samples stored under different conditions. The onset temperature of volatilization of 0 month sample was around 120°C. Whereas, for the sample stored over 12 months in light and sealed conditions, the onset temperature of volatilization increased to a maximum of up to 162°C. But, it remains at 138°C for the sample kept in dark and sealed conditions. The distillation temperature required for thermal decomposition of biodiesel kept in light was higher than the sample kept in dark (Table 1). Therefore, from the TGA results, it can be concluded that although the sample stored in light and dark conditions showed closed pattern of degradation, the degradation rate was significantly higher for the samples stored in light than in the dark condition.

Effect of air exposure on biodiesel storage

Air exposure enhances the degradation of methyl esters. During the oxidation process of methyl esters, it usually forms a radical next to double bond which quickly binds with oxygen in air [23]. This forms peroxide, which immediately forms new radical, and thus, the process of radical auto-oxidation cycle starts which exponentially increases the degradation of methyl esters. Figure 1c shows the change in acid value of the JME sample stored in exposure to air. At the beginning (0 month), the acid value of JME was around 0.44 mg KOH/g. After 6 months of storage, the acid value increased from 0.44 to 3.9 mg KOH/g for the samples which are kept at 4°C and exposed to air; whereas, for the sample stored at the same temperature but under sealed condition, increase was not significant, i.e., 0.5 mg KOH/g (Figure 1b,c). Further extending the storage period over 12 months, the increase in acid value was up to 4.8 mg KOH/g for the sample which was kept in exposure to air; whereas, maximum increase in the acid value was found up to 0.8 mg KOH/g for the sealed sample (Figure 1b,c). The samples which were kept at 25°C in exposure to air, the acid value after 6 months of storage was also found to increase from 0.44 to 6.37 mg KOH/g (Figure 1c). On the other hand, maximum increase in the acid value was up to 1.4 mg KOH/g for air-tight sample (Figure 1b). Over 12 months of storage, the increase in acid value was up to 14.1 mg KOH/g for air-exposed JME sample and 1.8 mg KOH/g for air-tight sample stored at 25°C (Figure 1b,c). On the other hand, the samples that are stored at 35°C and exposed to air, the acid value after 6 months of storage was found to increase to 9.3 mg KOH/g; whereas, in sealed condition, the maximum increase was up to 4 mg KOH/g (Figure 1b,c). In this case, also after 12 months of storage, the increase in acid value was up to 20.1 mg KOH/g for air-exposed sample. For the sealed sample, the maximum increase in acid value was up to 6 mg KOH/g (Figure 1b,c). The results obtained showed that air exposure makes biodiesel more susceptible to degradation. When the stored JME samples were compared, the rate of degradation was found to be significantly less for the air-tight samples than that for the air-exposed samples.

Figure 2c represents the change in density of stored JME but exposed to air for a period of 12 months. Density values were also found to follow the same pattern like the acid value of the stored JME sample. For all the samples, the density value was found to be increased with increased in storage time. After 6 months of storage at 4°C, the density of the air-exposed sample was found to be increased from 0.85 (0 months) to 1.15 kg/m^3 (Figure 2c). On the other hand, no significant change in the density value was observed (i.e., 0.85 (0 months) to 0.89 kg/m^3) for the samples stored in air-tight condition (Figure 2b). After, 12 months of storage, the density increased up to 1.3 kg/m^3 for samples that were kept but exposed to air and reached a maximum of up to 0.99 kg/m^3 for sealed samples (Figure 2b,c). The density values were found to be increased for the samples that were kept in exposure to ambient air than those for the samples in sealed condition for all storage temperature considered in this study (4°C, 25°C, or 35°C). Result obtained showed that air exposure increases the methyl ester degradation over long storage period.

Figure 3c represents the change in kinematic viscosity of Jatropha biodiesel over 12 months of storage period. Kinematic viscosities of all samples were found to be increased with increase in storage period. When the kinematic viscosity of JME samples stored in exposure to air and air-tight condition compared, significant difference was observed. After 6 months of storage, the JME samples that were kept at 4°C in exposure to air, the kinematic viscosity was to increase from 6.1×10^{-6} (initial value) to 8×10^{-6} m^2/s (Figure 3c); whereas, in sealed condition, the value increased to a maximum of up to 7.2×10^{-6} m^2/s (Figure 3b). After 12 months of storage, the increase in kinematic viscosity was up to 8.3×10^{-6} m^2/s for air-exposed sample, and for air-tight sample, a maximum increase was up to 7.5×10^{-6} m^2/s (Figure 3b,c). The samples that were kept at 25°C and 35°C in exposure to air, the kinematic viscosities after 6 months of storage were also found to be increased up to $8.5 \times 10^{-6}$and 9.2×10^{-6} m^2/s, respectively (Figure 3c); whereas, in sealed condition, a maximum increase in kinematic viscosity was up to $7.8 \times 10^{-6}$and 8.1×10^{-6} m^2/s for the samples kept at 25°C and 35°C, respectively (Figure 3b). Similarly, after 12 months of storage, increase in the kinematic viscosity was found to be $1.12 \times 10^{-5}$and 1.48×10^{-5} m^2/s for samples kept in exposure to air at 25°C and 35°C, respectively (Figure 3c). On the other hand, for the same sample kept in sealed condition, the increase in kinematic viscosity was found up to 8.2×10^{-6} and 9.1×10^{-6} m^2/s at 25°C and 35°C, respectively (Figure 3b). The result clearly revealed the detrimental effect of air exposure on the quality of methyl esters over long storage period, which was also observed in long-storage stability of vegetable oil and used frying oil-based biodiesel [5].

The onset temperature of volatilization and distillation temperature of thermal decomposition for all JME samples were also increased with increase in storage time. However, temperature required for thermal decomposition of JME samples exposed to ambient air was higher than that of the sample stored in sealed condition over a period of 12 months (Table 1). The initial onset temperature of volatilization for JME was around 120°C. The onset temperature of volatilization increased up to 148°C for JME samples that were exposed to air (stored at 4°C). It only raised maximum up to 138°C for air-tight sample stored at the same temperature (i.e., 4°C). JME samples stored at 25°C and 35°C also showed similar pattern (Table 1). For the storage period of 12 months, the onset temperature of volatilization increased up to 173°C for air-exposed samples, and a maximum increase was up to 162°C for air-tight samples stored at 25°C. On the other hand, for the sample stored at 35°C over 12 months, the onset temperature of volatilization increased up to 310°C for air-exposed JME samples. For air-tight samples stored at 35°C, maximum increase in onset temperature of volatilization was noted as 280°C. The distillation temperature required for thermal decomposition of biodiesel was also found to be higher for the sample exposed to air than that for the sample kept in sealed condition (Table 1). This increase in thermal stability of degraded biodiesel was mainly due to the decomposition of hydroperoxide which forms high molecular weight secondary oxidative product like aldehydes. Air exposure catalyzes the formation of this thermo stable secondary oxidative product. From the above discussion, it is clear that air exposure was one of the driving factors for worsening the quality of Jatropha biodiesel over long-term storage.

Conclusions

For successful commercialization of Jatropha biodiesel, maintaining the quality of biodiesel during storage period is of considerable significance. The study investigated experimentally the effect of storage parameter on Jatropha biodiesel stability. Results showed that JME quality deteriorates over long storage period. Increase in the value of qualitative parameters such as acidity, density, kinematic viscosity, onset temperature of volatilization, and distillation proves that the process of degradation increases with the increase in storage period. JME deterioration was more pronounced at 35°C compared to 25°C and 4°C. The degradation levels of Jatropha biodiesel stored at 25°C and 4°C were of similar pattern and less than the biodiesel stored at 35°C. Increase in the acid value of biodiesel stored at 4°C (from 0.44 to 4.8 mg KOH/g) and 25°C (from 0.44 to 14.1 mg KOH/g) was found to be moderate, while increase in the acid value was abrupt for the sample stored at 35°C (from 0.44 to 20.1 mg KOH/g). This abrupt increase in the acid value was observed for the sample stored in condition exposed to ambient air. In the analysis of the other two parameters, the density and kinematic viscosity also support this result. The maximum increase in density value was found to be from 0.85 to 1.3 kg/m^3 in case of 4°C storage temperature, from 0.85 to 1.8 kg/m^3 for 25°C, and for 35°C, it increased up to 2.3 kg/m^3 for the sample kept but exposed to air. The kinematic viscosity of biodiesel kept but exposed to air increased from 6.1×10^{-6} m^2/s to maximum of 1.48×10^{-5} m^2/s for sample stored at 35°C, and 1.12×10^{-5} m^2/s for sample stored at 25°C; whereas, in case of the sample stored at 4°C, it only increases slightly up to 8.3×10^{-6} m^2/s. Although light exposure did not show profound effect on biodiesel degradation, slight significant difference was observed in degradation level between the Jatropha biodiesel kept in light and in dark. The TGA data also showed higher onset temperature of volatilization and distillation temperatures for sample kept at 35°C compared to sample kept at 25°C and 4°C;the degradation was significantly higher in air-exposed condition than air-tight condition. Comparing all the data obtained from the analysis, it can be concluded that among all storage parameters, high temperature and air exposure were the two strong factors in fastening the Jatropha biodiesel degradation. Proper care should be taken for these parameters during storage; otherwise, biodiesel may become unusable very soon.

Competing interest

The authors declare that they have no competing interests.

Authors' contributions

PM and VBB carried out the study plan and were involved in the laboratory work. VVG and LS initiated, guided, and supervised the research work. VVG reviewed the revised manuscript and helped in correcting the manuscript. All the authors read and approved the final manuscript.

Authors' information

PM is a Ph.D. scholar at the Center for Energy, Indian Institute of Technology Guwahati. Her focus on research lies in biofuel production and genetic improvement studies in biofuel plants. VBB holds masters degree in Chemical Engineering. He is currently pursuing his Ph.D. at the Department of Chemical Engineering, Indian Institute of Technology Guwahati in the area of biofuels and biolubricants processing. LS is working as professor at IIT Guwahati. His research interests include genetic manipulation of crops for stress tolerance. VVG is working as an assistant professor at IIT Guwahati. His research interests are in the area of biofuel development and bioconversion of agricultural crops and wastes to useful products.

Author details

[1]Center for Energy, Indian Institute of Technology Guwahati, Guwahati, Assam 781039, India. [2]Department of Chemical Engineering, Indian Institute of Technology Guwahati, Guwahati, Assam 781039, India. [3]Department of Biotechnology, Indian Institute of Technology Guwahati, Guwahati, Assam 781039, India.

References

1. Raj, MT, Kandasamy, MKK: Tamanu oil - an alternative fuel for variable compression ratio engine. Int. J. Energy. Env. Eng. **3**, 8 (2012)
2. Betiku, E, Adepoju, TF: Methanolysis optimization of sesame (*Sesamumindicum*) oil to biodiesel and fuel quality characterization. Int. J. Energy Env. Eng. **4**, 9 (2013)
3. Eddine, BT, Salah, MM: Solid waste as renewable source of energy: current and future possibility in Algeria. Int. J. Energy Env. Eng. **3**, 17 (2012)
4. Moser, BR: Influence of extended storage on fuel properties of methyl esters prepared from Canola, palm, soybean and sunflower oils. Renew. Energy **36**, 1221–1226 (2011)
5. Bouaid, A, Martinez, M, Aracil, J: Long storage stability of biodiesel from vegetable and used frying oils. Fuel **86**, 2596–2602 (2007)
6. Mittelbach, M, Schober, S: The influence of antioxidants on the oxidation stability of biodiesel. J. Am. Oil Chem. Soc. **80**(8), 817–823 (2003)
7. Dunn, RO: Effect of antioxidants on oxidative stability of methyl soyate. Fuel Process. Technol. **86**(10), 1071–1085 (2005)
8. Mittelbach, M, Gangl, S: Long storage stability of biodiesel made from rapeseed and used frying oil. J. Am. Oil. Chem. Soc. **78**(6), 573–577 (2001)
9. Bondioli, P, Gasparoli, A, Bella, LD: Biodiesel stability under commercial storage conditions over one year. Eur. J. Lipid Sci. Technol **105**(12), 735–741 (2003)
10. Dunn, RO: Oxidative stability of soybean oil fatty acid methyl esters by oil stability index (OSI). J. Am. Oil Chem. Soc. **82**(5), 381–387 (2005)
11. Polavka, J, Paligova, J, Cvengros, J, Simon, P: Oxidation stability of methyl esters studied by differential thermal analysis and Rancimat. J. Am. Oil Chem. Soc. **82**(7), 519–524 (2005)
12. Das, LM, Bora, DK, Pradhan, S, Naik, MK, Naik, SN: Long-term storage stability of biodiesel produced from Karanja oil. Fuel **88**, 2315–2318 (2009)
13. Diwani, GE, Rafie, SE, Hawash, S: Protection of biodiesel and oil from degradation by natural antioxidants of Egyptian Jatropha. Int. J. Environ. Sci. Tech **6**(3), 369–378 (2009)
14. Jain, S, Sharma, MP: Long term storage stability of *Jatropha curcas* biodiesel. Energy **36**, 5409–5415 (2011)
15. Lu, H, Liu, Y, Zhou, HZ, Yang, Y, Chen, M, Liang, B: Production of biodiesel from *Jatropha curcas* L. oil. Comp. Chem. Eng. **33**, 1091–1096 (2009)

16. Berchmans, HJ, Hirata, S: Biodiesel production from crude Jatropha curcas L. seed oil with a high content of free fatty acids. Bioresour. Technol. **99**, 1716–1721 (2008)
17. Agarwal, D, Agarwal, AK: Performance and emissions characteristics of Jatropha oil (preheated and blends) in a direct injection compression ignition engine. Appl. Therm. Eng. **27**, 2314–2323 (2007)
18. Sarin, A, Arora, R, Singh, NP, Sharma, M, Malhotra, RK: Influence of metal contaminants on oxidation stability of Jatropha biodiesel. Energy **34**, 1271–1275 (2009)
19. Jain, S, Sharma, MP: Stability of biodiesel and its blends: a review. Renew. Sustain. Energ. Rev. **14**, 667–678 (2010)
20. Pattamaprom, C, Pakdee, W, Ngamjaroen, S: Storage degradation of palm-derived biodiesels: its effects on chemical properties and engine performance. Renew. Energ. **37**, 412–418 (2012)
21. Leung, DYC, Koo, BCP, Guo, Y: Degradation of biodiesel under different storage condition. Bioresour. Technol. **97**, 250–256 (2006)
22. Lin, C-Y, Chiu, C-C: Effects of oxidation during long term storage on the fuel properties of palm-oil based biodiesel. Energy Fuel **23**, 3285–3289 (2009)
23. Bora, DK, Das, LM, Babu, MKG: Storage stability of mahua oil methyl ester. J. Sci. Indus. Res. **68**, 149–158 (2009)

Wind profile characteristics and turbine performance analysis in Kano, north-western Nigeria

Oluseyi O Ajayi[1*], Richard O Fagbenle[2], James Katende[3], Samson A Aasa[1] and Joshua O Okeniyi[1]

Abstract

This study analyzed the electricity generation potential from wind at Kano, Nigeria (12.05°N; 08.2°E; altitude 472.5 m; air density 1.1705 kg/m^3). Twenty one years (1987 to 2007) monthly mean wind speed data at a height of 10 m were assessed from the Nigeria Meteorological Department, Oshodi. The data were subjected to different statistical tests and also compared with the two-parameter Weibull probability density function. The outcome shows that the average monthly wind speed ranged from 6.6 to 9.5 m/s. Seasonally, average wind speeds ranged between 6.6 to 8.5 m/s and 7.4 to 9.5 m/s for dry (October to March) and wet (April to September) seasons, respectively. Also, estimated monthly wind power ranged between 3.6 and 12.5 MWh/m^2. The most probable and maximum energy carrying wind speeds were also determined and the two parameters of the Weibull statistics were found to lie between 2.1 ≤ k ≤ 4.9 and 7.3 ≤ c ≤ 10.7, respectively. These results indicate that wind speeds at Kano may be economically viable for wind-to-electricity at and above the height of 10 m. In addition, five practical turbine models were assessed for the site's wind profile, with results suggesting strong economic viability.

Keywords: Wind power; Renewable energy; Weibull; Kano; Nigeria

Background

Globally different sources of energy production which include fossil fuel burning, small and large scale hydro, nuclear power, biomass burning, etc., have made major headlines in national and regional discussions of energy development. Some of these sources indeed have produced adequate amount of energy for community and national electricity production. However, the need to secure the environment while also developing the national economy has driven the trend toward diversification of energy production modules across the globe [1,2]. More so, the finite nature of conventional energy sources renders them unsustainable for long-term energy planning. Based on this, the focus of energy development and planning has shifted to include the non-conventional yet environmentally friendly sources of wind, solar, biomass, geothermal, and other renewable energy sources.

Moreover, before reasonable financing for renewable energy is provided by local or international investors, adequate resource and viability assessments must be made, among many other policy initiatives and derivatives. For Nigeria, the national energy policy and the renewable energy master plan [3,4] stipulate the country's desire to increase electricity production through the development of its alternative energy sources, of which wind, solar and biogas are major. Thus, efforts at measuring and assessing these sources for electricity generation are vital to its energy development. Since wind resources are site specific, wind energy measurements of as many sites as possible are required to have a proper wind classification of a nation. Based on this, several studies have attempted to classify the wind profile characteristics of Nigeria either, by site, region, or nation [5-26]. Ajayi [5,6] and Ohunakin et al. [9] profiled the results of some wind energy assessment studies in Nigeria to include those of pre-2000 and few post-2000 wind databases. On regional basis, different studies exist that characterized the wind profiles of sites in the North West [8,9], North East [7,8,11], North Central [8,10], South West [14] and South East [26] regions, respectively. Some

* Correspondence: oluseyi.ajayi@covenantuniversity.edu.ng
[1]Mechanical Engineering Department, Covenant University, P. M. B. 1023, Ota, Nigeria
Full list of author information is available at the end of the article

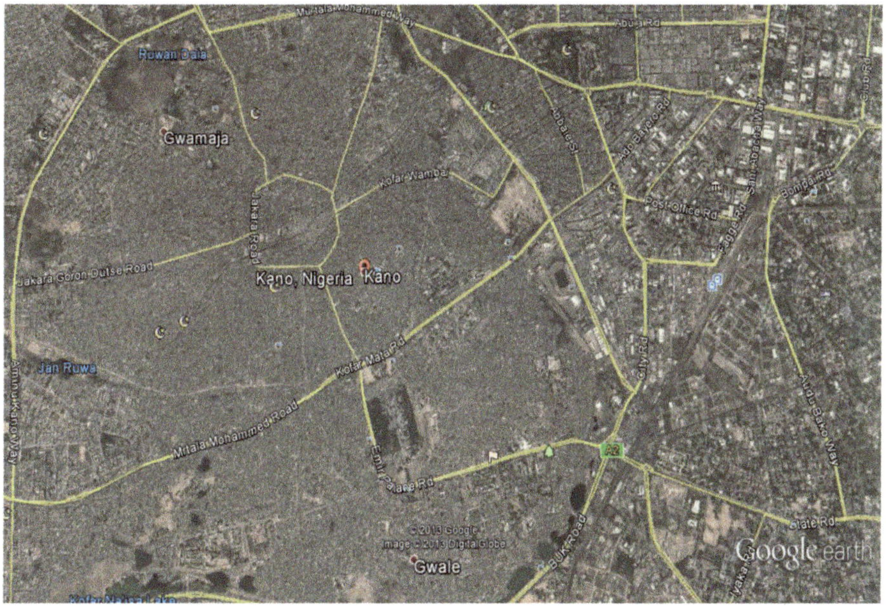

Figure 1 A view of the studied site at Kano, Kano State.

recent studies also exist that profiled the wind characteristics of single sites in Nigeria [12,13,15,17,21,23].

Each of the articles cited above demonstrated that Nigeria can employ wind resources for power generation across studied sites and geopolitical zones. Of interest to this study, however, are articles that focus on sites located in and around Kano [8,9,20,27]. ECN-UNDP [20] reported wind speed results based on the outcome of work carried out by Lahmeyer (International) [22]. The work of Lahmeyer (International), according to Ajayi et al. [12], used 12 months wind speed data to determine the average wind speed for ten selected sites across the country. One of the sites is Funtua (11.52°N; 7.31°E; Altitude 616 m), a place in Kano State. It reported average wind speed of 4.07 at 10 m height. Wind power generation at the site using three different turbines was evaluated to be 116.3, 281.2, and 963.7 MWh/yr at heights 34.5, 42.0, and 44.0 m, respectively. The models are FL 100, and 250, and V52 made by Fuhrlander and Vestas, respectively [12]. The application of this result is however limited by the number of data points (12 months). Also, because of the limited data spread, the monthly and seasonal variations associated with wind speed profiles were not captured. It is worthy of note that wind speed variation are location specific and associated with high variability in time and space; therefore, good results are based on historical data of many years.

Based on the aforementioned, studies by Ohunakin et al. [9] and Ohunakin [8] employed several years' wind speed data to characterize the wind profile of a meteorological site in Kano. The site is of the Nigeria meteorological agency. Moreover, while the studies suggested that the wind profile at the site is suitable for generating electricity, they failed to properly characterize the site wind for turbine sitting and applications. For the sitting and/or application of wind turbines at a site, adequate and robust

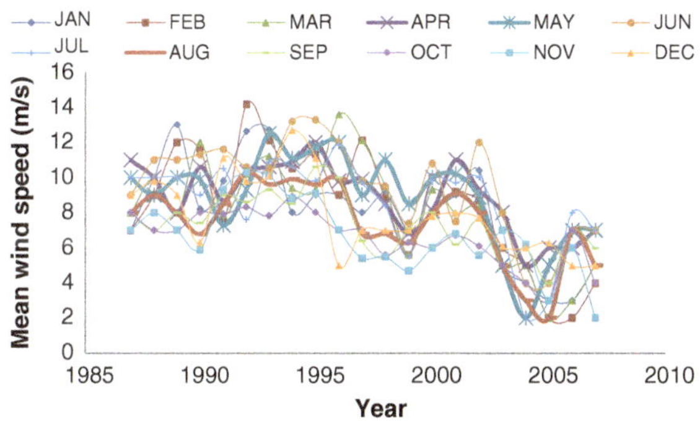

Figure 2 Plot of annual monthly mean wind speeds for the entire period.

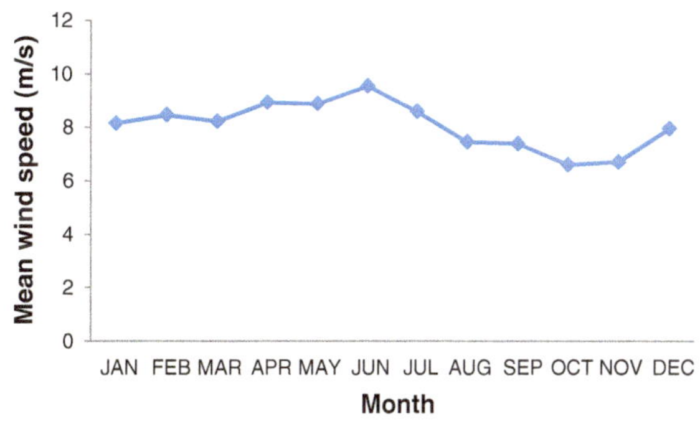

Figure 3 Monthly mean wind speeds for the entire period.

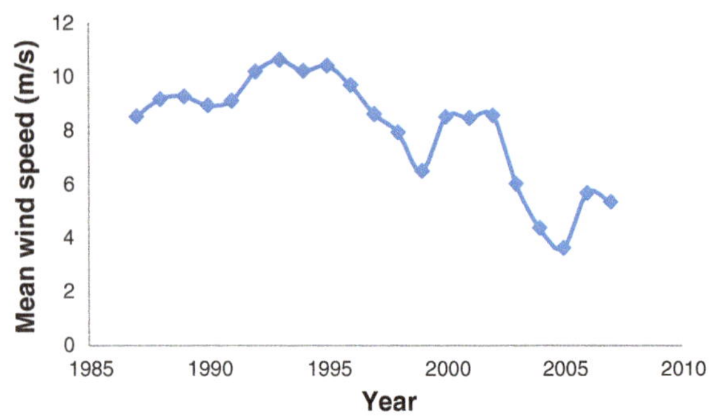

Figure 4 Yearly mean wind speed for the entire period.

site characteristics are required. Worthy of note is the fact that sitting of a wind turbine at a potential site does not just involve collection and analysis of wind data. It should also, of necessity, involve other factors and activities. These include the knowledge and analysis of the prevailing wind directions, surface roughness, topography, upwind obstacles such as trees or buildings, wind shear profile, and/or the influence of terrain contours.

In addition to the aforementioned, Ohunakin et al. [27] worked on the economic analysis of wind energy conversion using levelized cost of energy and present value cost for some sites which include the meteorological site in Kano. However, due to the fact that the work compared potential generation from different sites, it assumed constant generation from the sites across the period spanning 37 years. This places a limitation on the econometric result. The expected value of mean, taken over a period of years, is statistically significant over the actual data. Thus, employing the mean value for the analysis would have given a better and more applicable result because of the variability associated with wind resources. Further to this, the value of the cost for operation and maintenance employed in the study is minimal and not significant with market reality. The focus of this study was therefore to properly classify the wind profile characteristics of the meteorological site for turbine sitting and application. It also aimed at determining the economic viability of employing modern Mega wind turbines to generate wind electricity at the site. Figure 1 gives the view of the site. The study made use of monthly mean wind resource measurements averaged from daily readings to statistically analyze the wind resource potential of Kano for power generation.

Methods

Twenty one years (1987 to 2007) monthly mean wind data for Kano (12.05°N; 08.2°E; altitude 472.5 m; and air density 1.1705 kg/m^3) were sourced from records archived at the Nigeria Meteorological Department, Oshodi, Lagos State, south-west Nigeria. Three-hour daily readings over the period considered were used and subjected to various statistical operations. The data were recorded continuously using a cup generator anemometer at a height of 10 m. Figure 2 gives the whole data spread across the period considered, while Figure 3 gives the 21 years' monthly average distribution of the mean speeds, and Figure 4 gives the average annual distribution of mean speeds for the period.

Statistical Analysis

The two-parameter Weibull statistical distribution was employed to characterize the wind profile of the site. This is because the Weibull two-parameter probability density function (PDF) has been proved to be the adequate and accurate of several statistical distributions [28-33]. The two-parameter PDF is given as:

$$f(v) = \left(\frac{k}{c}\right)\left(\frac{v}{c}\right)^{k-1} \exp\left[-\left(\frac{v}{c}\right)^{k}\right], \quad (1)$$

where $f(v)$ is the probability of observing wind speed, v. k is the dimensionless Weibull shape parameter and c (m/s), is the Weibull scale parameter.

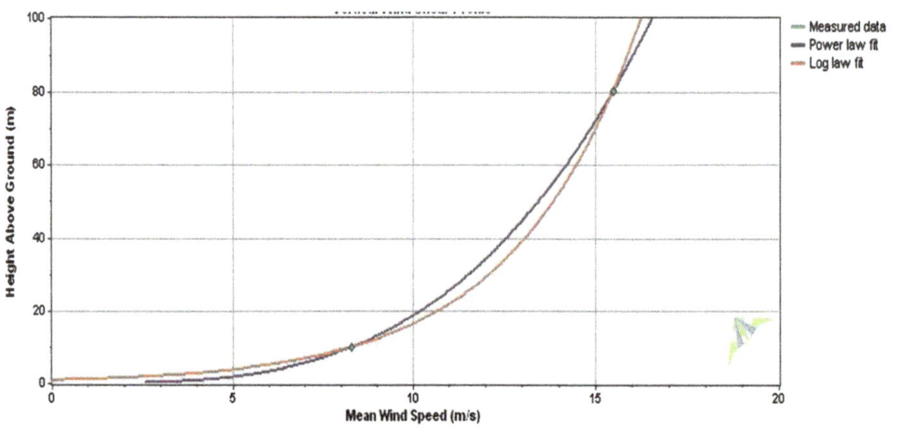

Figure 5 The vertical wind shear profile.

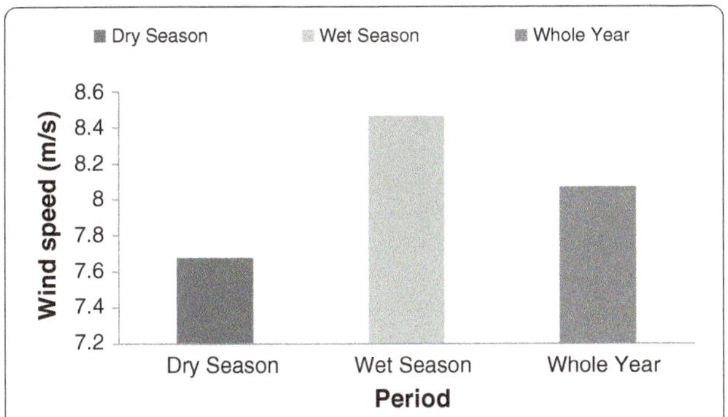

Figure 6 Seasonal and whole-year values of mean wind speed across the period.

The corresponding Weibull cumulative density function (CDF) is given as:

$$F(v) = 1\text{-}\exp\left[-\left(\frac{v}{c}\right)^k\right]. \quad (2)$$

Equation 2 can be translated into a linear form of $y = mx + C$ as:

$$\ln[-\ln[1-F(v)]] = k\ln v - k\ln c,$$

where $y = \ln[-\ln[1 - F(v)]]$, $mx = k \ln v$ and $C = - k \ln c$.

Thus, a plot of $\ln[-\ln[1 - F(v)]]$ on the vertical against $\ln v$ on the abscissa gives the shape (k) and scale (c) parameters of the Weibull distribution by evaluating the slope and intercept on the vertical axis.

The mean value of the wind speed v_m and standard deviation σ for the Weibull distribution are given as [4,5]:

$$v_m = c\Gamma\left(1+\frac{1}{k}\right) \quad (3)$$

and

$$\sigma = \sqrt{c^2\left\{\Gamma\left(1+\frac{2}{k}\right)-\left[\Gamma\left(1+\frac{1}{k}\right)\right]^2\right\}}, \quad (4)$$

where Γ ( ) is the gamma function of ( ).

However, the two wind speeds of utmost interest for wind resource assessment are the wind speed carrying maximum energy (v_{Emax}) and the most probable or modal wind speed (v_{mp}). These were evaluated from:

$$v_{\text{Emax}} = c\left(\frac{k+2}{k}\right)^{\frac{1}{k}} \quad (5)$$

$$v_{\text{mp}} = c\left(\frac{k-1}{k}\right)^{\frac{1}{k}}. \quad (6)$$

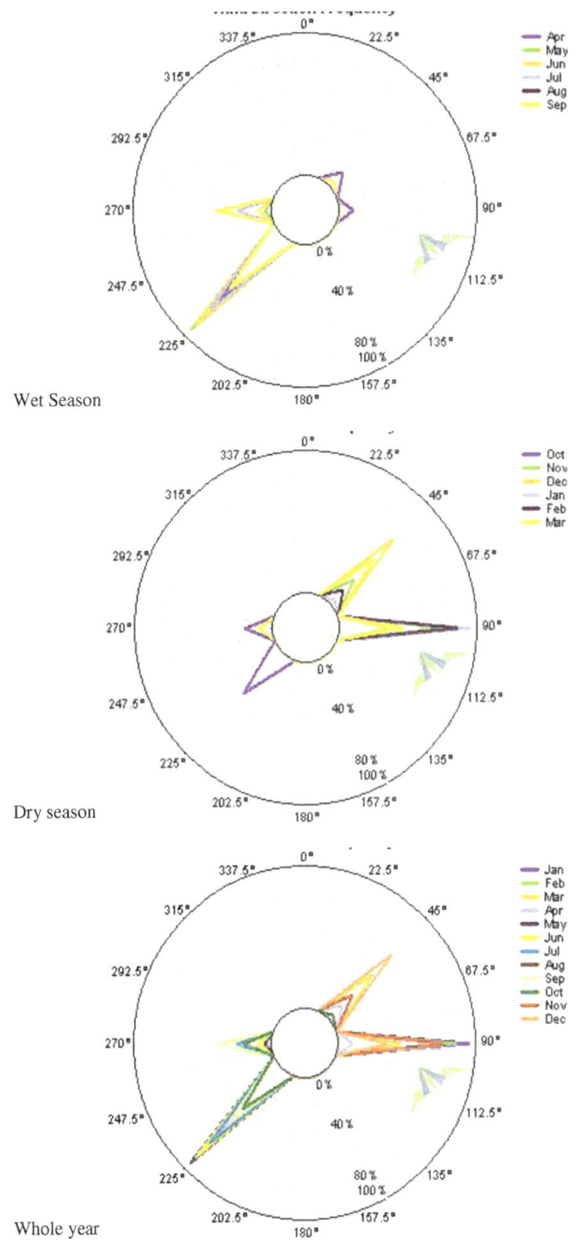

Figure 7 Prevailing wind directions for wet and dry seasons and whole-year analysis. Wet (April to September) and dry (October to March) seasons.

Estimation of wind power density

The available average power in the wind flowing through a rotor blade with swept area, A (m^2) is known to increase as the cube of its mean wind speed, v_m as:

$$P(v) = \frac{1}{2}\rho A {v_m}^3. \quad (7)$$

However, in this study, the wind power per unit area is estimated based on the Weibull PDF as:

Table 1 Some Weibull results and estimation parameters for the whole 21 years

Period	$v_{Weibull}$ (m/s)	k (–)	c (m/s)	$\sigma_{Weibull}$ (m/s)	σ_{actual} (m/s)	R^2	RMSE	COE	K-S P value
Month									
January	8.2	2.7	9.2	3.3	2.9	1.0	0.7	0.9	0.7
February	8.7	2.1	9.8	4.4	3.4	0.9	1.3	0.8	0.6
March	8.4	2.4	9.4	3.7	3.0	0.9	0.9	0.9	0.4
April	8.9	4.9	9.7	2.1	1.9	1.0	0.3	1.0	0.9
May	9.1	2.7	10.2	3.7	2.6	0.9	1.4	0.7	0.5
June	9.6	3.4	10.7	3.2	2.7	0.9	0.8	0.9	0.8
July	8.9	2.4	10.0	3.9	2.5	0.8	1.9	0.4	0.2
August	7.6	2.7	8.5	3.0	2.2	0.9	1.1	0.8	0.4
September	7.4	4.9	8.0	1.7	1.7	1.0	0.2	1.0	1.0
October	6.6	4.1	7.3	1.8	1.6	0.9	0.4	0.9	0.8
November	6.8	2.9	7.6	2.5	2.1	0.9	0.6	0.9	0.6
December	7.9	4.1	8.8	2.2	2.2	0.9	0.5	1.0	0.8
Season									
Dry	7.7	3.0	8.6	2.8	2.7	1.0	0.3	1.0	0.7
Wet	8.5	3.4	9.5	2.8	2.4	1.0	0.6	0.9	0.2
Whole year	8.1	3.2	9.1	2.8	2.6	1.0	0.4	1.0	0.1
1987	8.5	6.9	9.1	1.4	1.3	0.9	0.4	0.9	0.5
1988	9.1	7.5	9.7	1.4	1.3	0.9	0.4	0.9	0.7
1989	9.2	5.1	10.1	2.1	2.0	0.9	0.6	0.9	0.5
1990	8.9	4.5	9.8	2.2	2.1	1.0	0.4	1.0	1.0
1991	9.1	7.1	9.7	1.5	1.4	0.9	0.4	0.9	1.0
1992	10.2	6.2	10.9	1.9	1.8	0.9	0.6	0.9	0.3
1993	10.6	8.1	11.2	1.5	1.4	1.0	0.2	1.0	1.0
1994	10.2	7.0	10.9	1.7	1.6	0.9	0.4	0.9	0.9
1995	10.4	7.5	11.0	1.6	1.5	0.9	0.3	0.9	0.9
1996	9.7	3.9	10.7	2.8	2.5	1.0	0.5	1.0	1.0
1997	8.6	4.3	9.4	2.3	2.2	0.9	0.5	0.9	0.9
1998	7.9	4.5	8.7	2.0	1.8	0.9	0.4	0.9	0.9
1999	6.5	6.9	6.9	1.1	1.0	1.0	0.2	0.9	1.0
2000	8.5	5.7	9.2	1.7	1.5	0.9	0.4	0.9	0.8
2001	8.4	6.2	9.1	1.6	1.5	1.0	0.3	1.0	1.0
2002	8.5	5.2	9.3	1.9	1.8	1.0	0.3	1.0	1.0
2003	6.0	5.0	6.6	1.4	1.2	0.7	0.6	0.8	0.3
2004	4.4	2.8	4.9	1.7	1.5	0.9	0.4	0.9	0.8
2005	3.6	2.6	4.1	1.5	1.5	0.9	0.4	0.9	0.6
2006	5.8	2.4	6.5	2.6	2.0	0.9	1.0	0.7	0.6
2007	5.4	3.0	6.1	2.0	1.6	0.9	0.7	0.8	0.6

$$p(v) = \frac{P(v)}{A} = \frac{1}{2}\rho c^3 \Gamma\left(1 + \frac{3}{k}\right), \tag{8}$$

where, $P(v)$ is the wind power (W), $p(v)$ is wind power density (W/m^2), and ρ is the air density at the site.

Performance of the Weibull distribution model

The accuracy of the Weibull distribution in estimating the site's actual wind speeds was evaluated using the coefficient of determination, R^2, the root mean square error (RMSE) and the Nash-Sutcliffe model coefficient of efficiency (COE) [30,34]. These are given by:

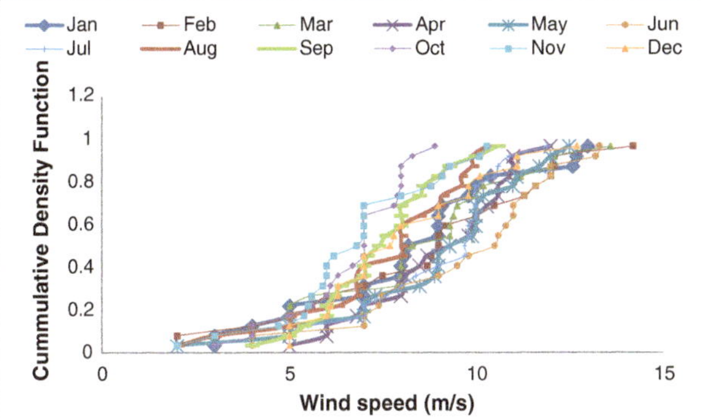

Figure 8 CDF plots of the whole data series.

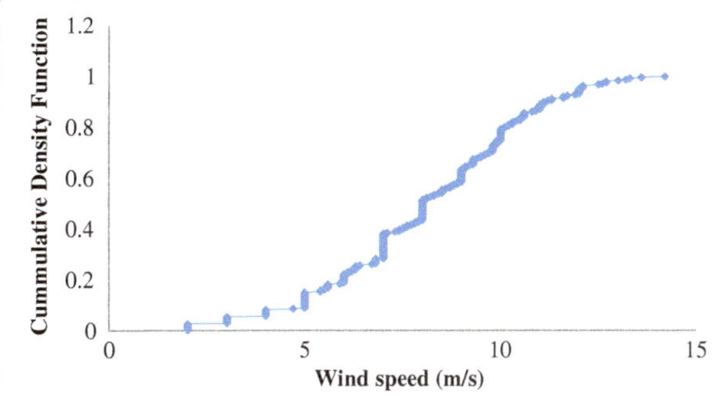

Figure 10 CDF plots of whole-year data series.

$$R^2 = \frac{\sum_{i=1}^{N}(y_i - z_i)^2 - \sum_{i=1}^{N}(x_i - y_i)^2}{\sum_{i=1}^{N}(y_i - z_i)^2}, \tag{9}$$

where

$$R = \sqrt{R^2}; \tag{10}$$

$$RMSE = \left[\frac{1}{N}\sum_{i=1}^{N}(y_i - x_i)^2\right]^{\frac{1}{2}} \tag{11}$$

$$COE = 1 - \frac{\sum_{i=1}^{N}(y_i - x_i)^2}{\sum_{i=1}^{N}(y_i - \bar{x})^2}, \tag{12}$$

where y_i is the ith actual data, x_i is the ith predicted Weibull result, z is the mean of the actual data, and N is the number of observations.

The accuracy of the Weibull results was evaluated by the proximity of the values of R^2 and COE to unity (or 100%) and the proximity to zero of the values of RMSE.

Suitability/goodness of fit test

To ascertain how closely the measured data follow a Weibull distribution, the Kolmogorov-Smirnov (K-S) goodness of fit test [35-39] was employed to measure the absolute difference between the measured distribution function $F^*(x)$ and the Weibull distribution function $F(x)$ [40,41]. The expression for the K-S test is given as:

$$d = d(x_1, \ldots, x_n) = \sqrt{n} \sup_{-\infty < x < \infty} |F * (x) - F(x)|,$$

where n is the sample size. From the value of d, the P value of the K-S goodness of fit could be obtained using the relationship given by:

$$\left.\begin{array}{ll} P = 1 & \text{for } d < 0.22 \\ P = 1 - \frac{\sqrt{2\pi}}{d}\exp\left(\frac{-\pi^2}{8d^2}\right) & \text{for } 0.22 \le d \le 0.80 \\ P = 2\exp(-2d^2) + \exp(-8d^2) - \exp(-18d^2) & \text{for } 0.80 < d \le 3.15 \\ P = 0 & \text{for } d > 3.15 \end{array}\right\}. \tag{13}$$

Therefore, at a significance level $\alpha = 0.05$, the P value can be compared directly with α to test the hypotheses as:

$$H_0 : P \ge \alpha$$
$$H_A : P < \alpha,$$

where H_0 is the null hypothesis, suggesting that the two-parameter Weibull distribution suitably approximates the Kano wind speed profiles. H_A is the alternative hypothesis indicating otherwise.

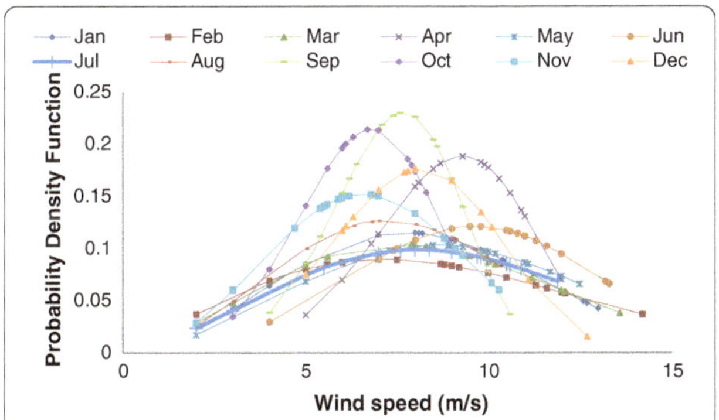

Figure 9 PDF plots of the whole data series.

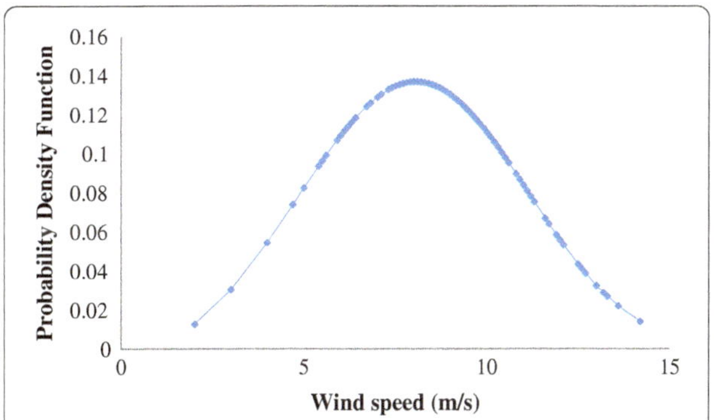

Figure 11 PDF plots of whole-year data series.

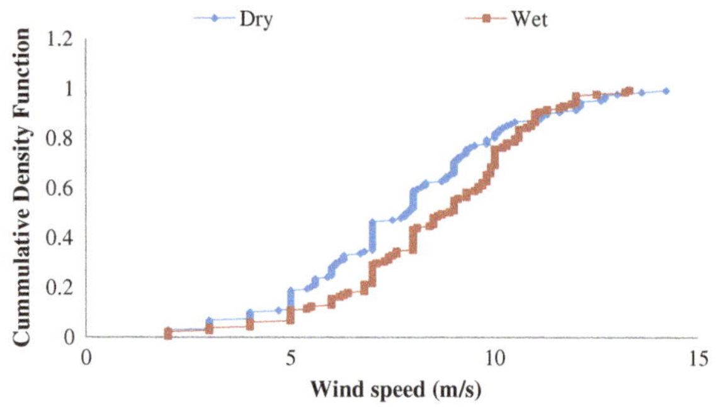

Figure 12 CDF plots of dry and wet seasons.

Simulating the electrical power output from a wind turbine model

The magnitude of wind power that can be produced from a practical wind turbine applied at the site was simulated using Equation 14 [31]:

$$P_e = \begin{cases} 0 & (v < v_c) \\ P_{eR}\dfrac{v^k - v_c^k}{v_R^k - v_c^k} & v_c \leq v \leq v_R \\ P_{eR} & v_R \leq v \leq v_F \\ 0 & v > v_F \end{cases} \tag{14}$$

The average power output ($P_{e,ave}$) from a turbine corresponding to the total energy production and related to the total income/cost analysis was evaluated from:

$$P_{e,ave} = P_{eR}\left\{\frac{e^{-\left(v_c/c\right)^k} - e^{-\left(v_R/c\right)^k}}{\left(v_R/c\right)^k - \left(v_c/c\right)^k} - e^{\left(v_F/c\right)^k}\right\} \tag{15}$$

The capacity factor (CF) of generation is evaluated from:

$$CF = \frac{P_{e,ave}}{P_{eR}}, \tag{16}$$

where P_{eR} = rated electrical power, v_c = cut-in wind speed, v_R = rated wind speed and v_F = cut-off wind speed.

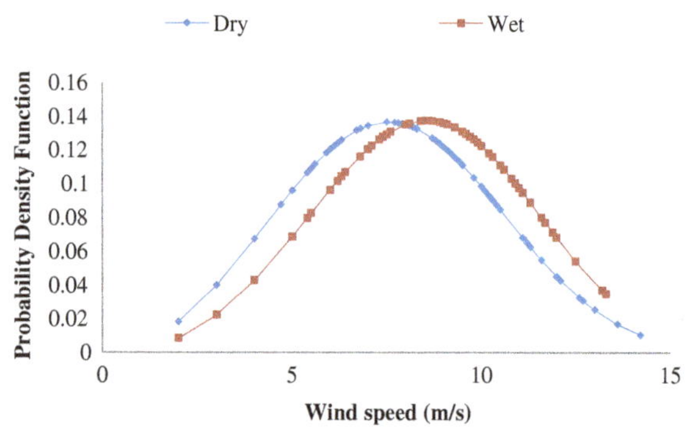

Figure 13 PDF plots of dry and wet seasons.

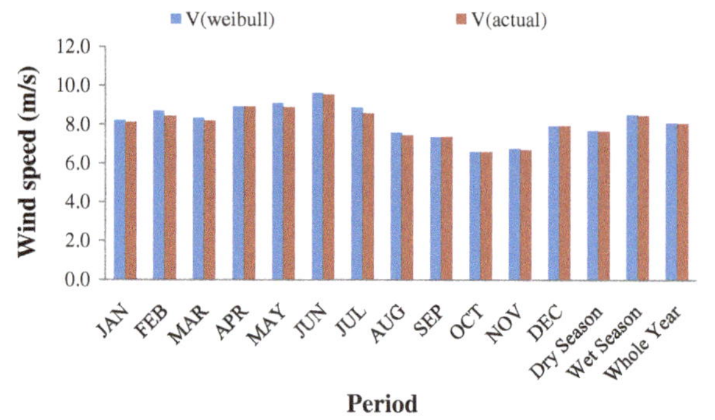

Figure 14 Plots showing the predictive ability of Weibull distribution for monthly and seasonal wind speed distributions.

Results and discussion

Wind profile assessment

Figure 2 indicates that there was a general decline in the values of the wind speed profiles across the years considered while October and November appear to be the months with the least wind supply of the station. The reason behind the drop is not immediately assumable. It is however believed that it may be due to developmental changes around the site. These changes have introduced wind brakes (e.g. tall buildings) that were absent some years back. Furthermore, these changes are notable because of the site's location, which is in the heart of the capital city of Kano. Based on this, the vertical wind shear profile was determined using the Windographer® software [42] and the observable site's characteristics. This gave Figure 5. Thus, even though there was a decline in the observable wind speeds across the periods shown in Figure 2, wind speed profile increased with heights at the site.

The magnitude of the monthly mean measured wind speeds from January to December across the whole 21-year period lay within the range of 2.0 to 14.2 m/s and those of the seasons from dry (October to March) to wet (April to September) every year lay within the range 2.0 to 14.2 and 2.0 to 13.3 m/s, respectively, at the 10-m height. Yearly,

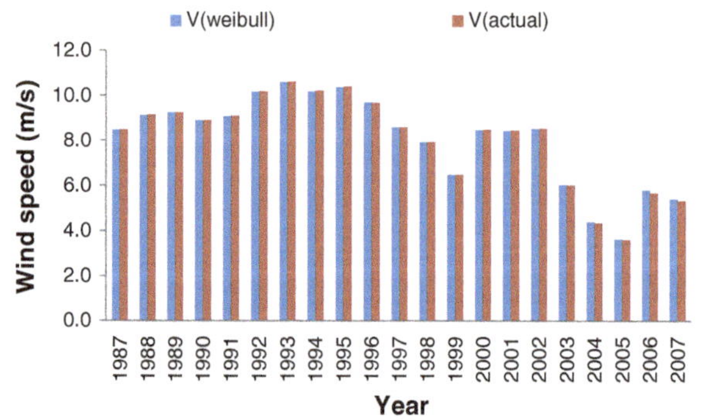

Figure 15 Plots showing the predictive ability of Weibull distribution for yearly mean wind speed distribution.

the mean measured wind speeds from 1987 to 2007 also ranged from 2.0 to 14.2 m/s. However, to determine the fluctuations in the wind speed profiles, the 21-year average of all monthly and yearly mean wind speeds of Figure 2 were estimated and plotted as shown in Figures 3 and 4. Figures 3 and 4 show similar trends with Figure 2 across the months and 2005 experienced the lowest wind speed values, between 2.0 and 6.3 m/s. In addition, the period with the highest energy potential was from January to June, while June appears to have the greatest potential with speed range of 4.0 to 13.3 m/s. The average monthly wind speed range was from 6.6 (in October) to 9.5 m/s (in June). Seasonally, the wet season appears to have more wind energy potential than the dry (Figure 6), while the 21-year average for all of the data was 8.1 m/s. Seasonal mean speed range for dry, was from 6.6 (in October) to 8.5 m/s (in February), while for wet, it ranged from 7.4 (in September) to 9.5 (in June), respectively. These clearly indicate that Kano experiences wind speeds that are viable for economically beneficial wind energy production at 10-m heights and above.

Figure 7 shows the prevailing wind directions across the months and seasons. It shows that the wind directions in the dry period (October to March) vary differently from those of the wet (April to September). The prevailing wind directions tend toward the east during the dry period and south westerly during the wet. Worthy of note however is the fact that as the period tends toward seasonal changes, the wind directions also changes until the prevailing direction is achieved.

The site's characteristics analysis gave wind power class of 7, surface roughness of 0.906 m and a roughness class of 3.83. The wind power class signifies an excellent wind profile [43,44] for the site while the roughness class signifies a site with landscape having many trees and buildings [45].

The result of a Weibull statistical analysis on the wind speed data are displayed in Table 1, and the corresponding CDF and PDF plots for the whole years' data and seasons are shown in Figures 8, 9, 10, 11, 12, and 13. These figures demonstrate that the wind profiles for these periods and whole year data series follow the same cumulative distribution pattern.

Moreover, Figures 8 and 9 reveal that up to 60% of the whole data series were values that ranged from about 7.0 to 10.8 m/s and below, while up to 90% of the data series was values that ranged from about 8.3 to 13.2 m/s.

The dry and wet seasons' Weibull plots of Figure 12 indicate that between 60% and 90% of the dry season's data falls within the range 8.2 to 11.6 m/s and below, while those of the wet season ranged from 9.6 to 11.1 m/s and below. From Table 1, the Kolmogorov-Smirnov P values show that the two-parameter Weibull distribution was adequate for approximating the wind profiles at Kano.

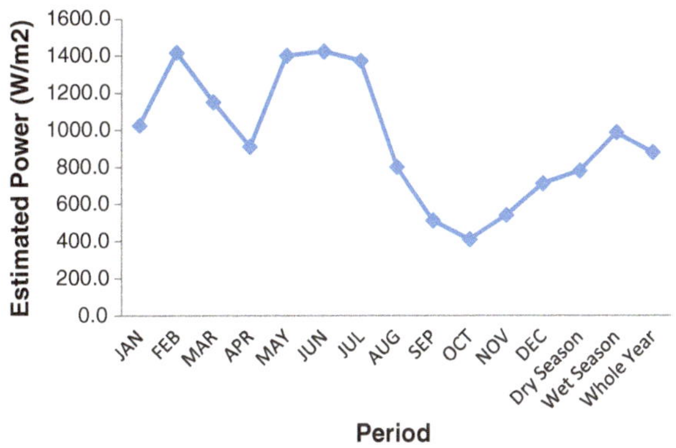

Figure 16 Periodic variation of power density.

The Weibull results (Table 1) were compared with the actual data to assess the degree of convergence. The results are displayed in Figures 14 and 15.

Figures 14 and 15 therefore show that the Weibull statistical distribution predicted the mean wind speed profiles of the station quite adequately. This was corroborated by the associated values of R^2 and COE of Table 1.

However, the RMSE values of Table 1 show some values closer to or greater than 1.0. The reason behind this was the value of the prediction from the Weibull results. This phenomenon is interpreted as when the predicted result is greater than the actual data, the value of RMSE will be closer to or greater than 1.0. Thus, high RMSE values indicate Weibull result greater than actual measured value thus implying over prediction. Overall the, Weibull was adequate at predicting both annual variations and periodic variations of wind profiles going by the R^2, P, and COE values.

On the wind power density across the period and years considered (Figures 16 and 17), June had the highest wind power potential. Also, the period of February, and those between May and July every year can be regarded as the period with very good potential for wind power

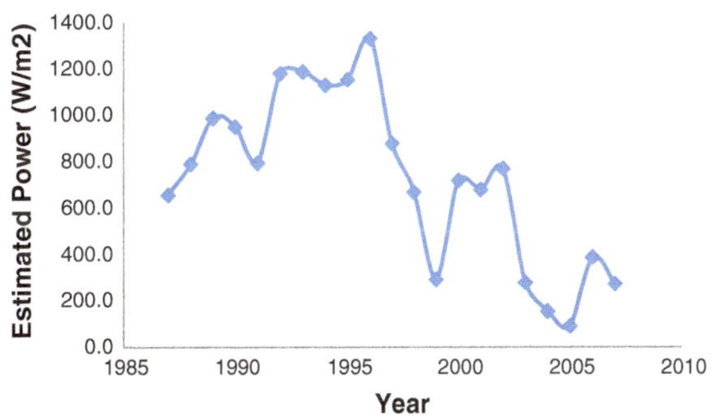

Figure 17 Annual variation of power density.

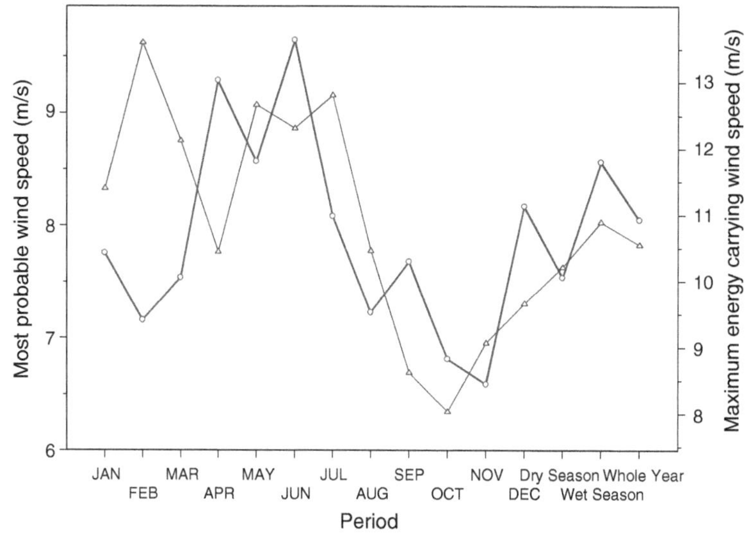

Figure 18 Plot comparing the periodic most probable and maximum energy carrying wind speeds.

harvest. The monthly values ranged between 410.8 (in November) and 1,424.4 W/m^2 (in June), while the dry and wet seasons and whole-year values ranged from 782.9 (dry) to 988.4 W/m^2 (wet). This significant monthly change in power density according to Keyhani et al. [46] underscores the importance of distinguishing different months and periods of the year when a wind power project is assessed or designed to produce maximum power. Furthermore, the values of k and c from the analysis across all period ranged between $2.1 \leq k \leq 4.9$ and $7.3 \leq c \leq 10.7$, respectively. These high values of k and c ($k \geq 2$ and $c \geq 2$) indicate a data spread in a perfectly normal distribution [29], and also show that the data spread exhibits good uniformity with relatively small scatter [47]. The scale parameter, c, further indicates how windy a location under consideration is, whereas the shape parameter, k, indicates how peaked the wind distribution is [46]. Thus, if the wind speeds tends to spike steeply at a certain value, the distribution will have a high k value.

The most probable (v_{mp}) and maximum energy carrying wind speeds (v_{Emax}) analyses for the periods and years are given by Figures 18 and 19. The values of v_{mp} ranged from 6.6 to 9.6 m/s (for January to December), 7.5 m/s (dry season), 8.6 m/s (wet season), and 8.0 m/s (whole years). For the yearly analysis, it ranged from 3.4 to 11.1 m/s. Also, the values of v_{Emax} ranged from 2.1 to 13.6 m/s (for January to December), 10.2 m/s (dry season), 10.9 m/s (wet season), and 10.5 m/s (whole years). For the whole years' it ranged from 5.1 to 11.9 m/s (1987 to 2007), respectively.

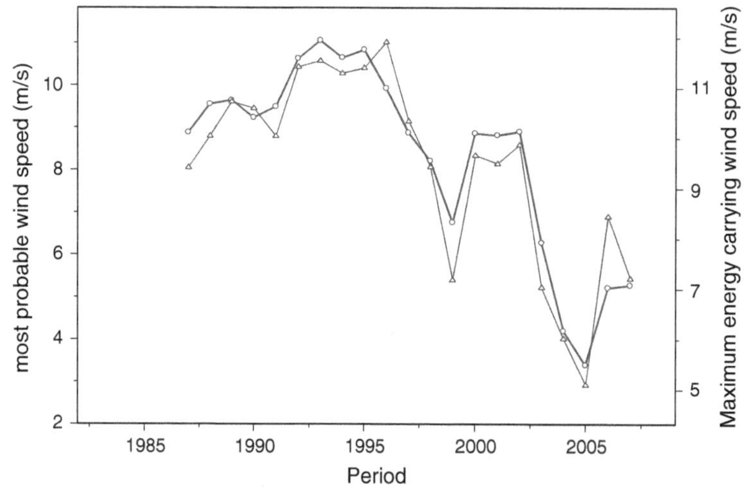

Figure 19 Plot comparing the annual most probable and maximum energy carrying wind speeds.

Table 2 Turbines' technical parameters

Wind machine	v_c (m/s)	v_F (m/s)	v_R (m/s)	P_{eR} (kW)	Hub height (m)	Rotor diameter (m)
GE 1.5sle	3.5	25	14	1,500	65/80	77
GE 1.5xle	3.5	20	11.5	1,500	80	82.5
AV 928	3	25	11.6	2,500	80	93.2
V90	4	25	15	3,000	80	90
SWT-3.6-107	3	25	13	3,600	80	107

Adapting real wind turbines to the Kano site

According to Ajayi et al. [12], installing a wind turbine is capital intensive. Therefore it is worthwhile to first match a turbine parameter to the site's wind profile. Initially, this entails estimating the amount of electrical power that a particular wind turbine will likely generate. The estimated capacity factor (CF) is a pointer to the turbine's generation capacity.

Based on the above, this study employed five different turbines with technical parameters given in Table 2. Equations 14, 15, and 16 were then used to evaluate the power output from each of the turbines. However, because the turbine hub heights are at 80 m, the wind profile characteristics were estimated for this height using Equation 17 [12]. The speeds at this new height were then employed for Weibull re-analyses. It is worthy of note that although it is thought that recent developments may have led to the development of wind brakes around the site's location within the capital city of Kano, the hub heights of practical wind turbines are moreover much more than 10 m. Thus at heights above 20 m, the economic viability of wind energy at the site can be demonstrated.

$$v_{\text{ref}} = v_{10}\left(\frac{h_{80}}{h_{10}}\right)^{\alpha} = v_{10}(8)^{\alpha} = v_{80}, \qquad (17)$$

where $v_{\text{ref}} = v_{80}$ = wind speed at 80 m, v_{10} = wind speed at 10-m height, h_{80} = 80 m height, h_{10} = 10-m height, and α = wind shear coefficient for the sites = 0.143.

The value of α (wind shear coefficient) was used to determine the wind speeds at higher height instead of the roughness length, z_o. This is because, while α is a dynamic value that varies according to a large number of factors which include time of day, season, atmospheric stability, and regional topography, the z_o, obtained from the log law, is only valid under certain assumptions regarding atmospheric stability. Actual values may however deviate from the log law. Thus, it is difficult to argue that the wind profile at the site follow the log law or power law. The power law gives better fit to wider range of speed including high wind speeds. However, the value of α has been reported to be suitable for all sites. Also α is approximately equal to 0.143 for neutral stability and regarded as a reasonable but conservative estimate [48-53]. Thus, the commonest and widely accepted value of α for most sites is 0.143 [12].

The results of the Weibull analyses at 80-m heights for monthly, seasonal, and whole-year analyses are presented in Figures 20 and 21. These reveal that the wind speeds at this height ranged from 9.12 to 12.95 m/s. The values of the scale (c) and shape (k) parameters were $10.22 \leq c \leq 14.42$ and $1.55 \leq k \leq 4.91$, respectively. The range of wind speeds at this height further strengthens the economic viability of wind energy projects and wind farm development at Kano. The capacity factors of deploying the five wind turbines at the site are presented in Figure 22.

Figure 22 shows that the AV 928 turbine produced at the highest CF with values between 51.95 and 78.99%. It is closely followed by the GE 1.5xle and SWT-3.6-107 with the production CF ranged from 48.91 to 74.54%

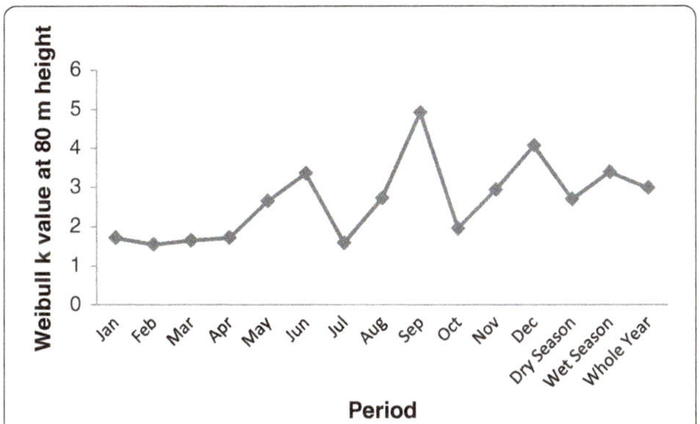

Figure 20 Weibull shape (*k*) parameter for monthly, seasonal, and whole-year analyses at 80-m height.

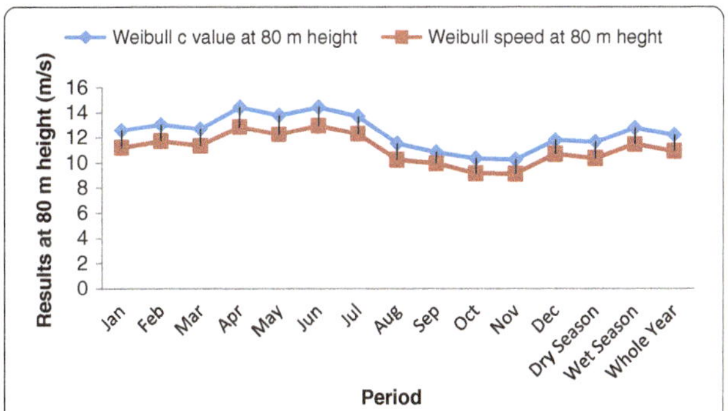

Figure 21 Weibull scale (*c*) parameter and estimated wind speeds for monthly, seasonal, and whole-year analyses. Height at 80 m.

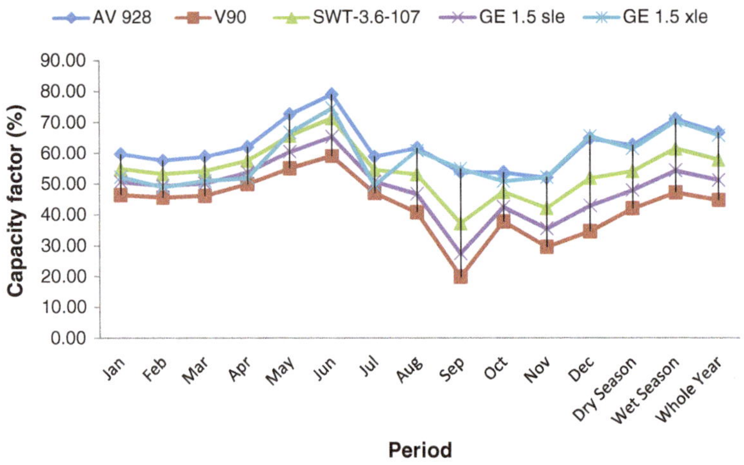

Figure 22 Capacity factors of generation using the turbines at the site.

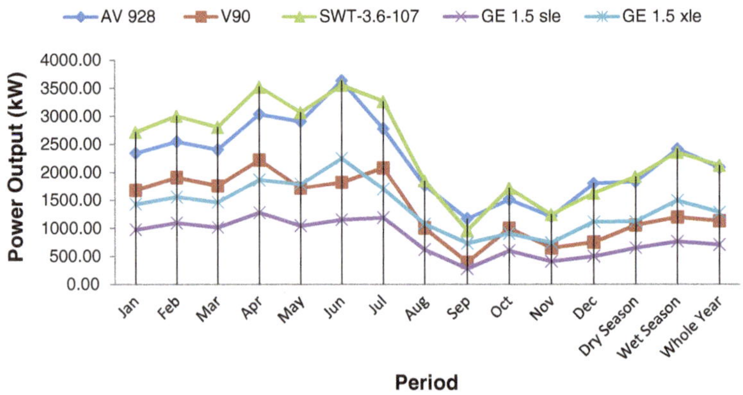

Figure 23 Power output from turbines based on the site's mean wind speed profile at 80-m height.

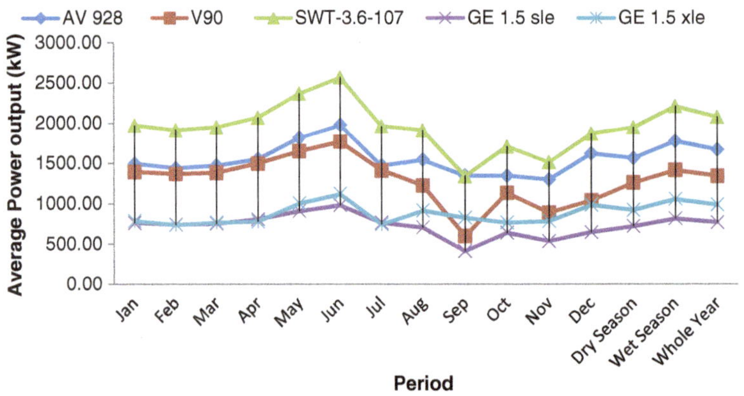

Figure 24 Average power output from turbines based on site's wind speed profile at 80-m height.

Table 3 Range of values from January to December across the period

Item	Wind machine														
	AV 928			V90			SWT-3.6-107			GE 1.5sle			GE 1.5xle		
	P_e (kW)	$P_{e,ave}$ (kW)	*CF* (%)	P_e (kW)	$P_{e,ave}$ (kW)	*CF* (%)	P_e (kW)	$P_{e,ave}$ (kW)	*CF* (%)	P_e (kW)	$P_{e,ave}$ (kW)	*CF* (%)	P_e (kW)	$P_{e,ave}$ (kW)	*CF* (%)
Minimum	1164.27365	1298.52	51.94	391.77	597.08	19.90	957.26	1338.82	37.19	276.35	410.93	27.395465	727.7756	733.7201	48.914673
Maximum	3631.51342	1974.86	78.99	2219.23	1770.06	59.00	3553.76	2569.39	71.37	1273.60	979.64	65.309319	2249.5683	1118.1235	74.541568

P_e, power output, $P_{e,ave}$, average power output, and *CF* capacity factor.

Table 4 Assumptions used for the econometrics analysis [12; 59]

Item	Assumed value
R_C	20.0%
R_{OM}	25.0%
R_I	12.0%
I_R	8.6%
R_{SC}	10.0%
t	20.0 years

Table 5 Assumed turbine model price

Turbine model	Assumed price (€)
GE 1.5sle	1,500,000
GE 1.5xle	1,500,000
AV 928	2,500,000
V90	3,000,000
SWT-3.6-107	3,500,000

and 37.19 to 71.37%, respectively. In term of production capacity, the producible wind powers at the site are presented in Figures 23 and 24. Figures 23 and 24 in comparison with Figure 22 demonstrate that, although the model SWT-3.6-107 was third best in terms of the CF of average production over the rated power, its production capacity is the best. This is closely followed by AV 928. The reason can be adduced to the fact that, for all values greater than the cut-in speed, their power-producing potential is more than the AV 928 and other models. This is due to its power rating (P_{eR}). Consequently, a turbine speed rating (v_R) of between 11.0 and 13.0 m/s, and also cut-in (v_c) and cut-off (v_F) speeds of 3.0 and 25 m/s, respectively, will be valuable for the site. The range of results (minimum and maximum) obtained when the five wind turbines were adapted to the site are presented in Table 3.

Economic benefit of wind power generation at the site

The econometrics analysis was carried out with Equations 18 and 19 [12] using the assumptions stated in Tables 4 and 5. The results are presented in Table 6.

$$C_{pv} = x(1+R_C) + \frac{x}{t}R_{om}\left[\frac{1+I_R}{R_I - I_R}\right] \times \left[1 - \left(\frac{1+I_R}{1+R_I}\right)^t\right] - xR_{sc}(1+R_C)\left(\frac{1+I_R}{1+R_I}\right)^t \quad (18)$$

$$C_{SC/kWh} = \frac{C_{PV}}{\text{Annual}P_{e,ave} \times t}, \quad (19)$$

where C_{PV} = present value cost, x = turbine price, R_C = rate chargeable on turbine price to arrive at the cost for civil/structural works, R_{OM} = rate chargeable on annual turbine price to arrive at the cost for Operation and Maintenance, R_I = prevailing interest rate, I_R = prevailing inflation rate, R_{SC} = rate chargeable on total investment cost, t = turbine life or period of operation of turbine, and $C_{SC/kWh}$ = specific cost per kWh of wind electricity.

Turbine prices have been varying for some years. There is no specific fixed price because of the interplay between demand and supply and other variables which include the cost of materials. Various prices have been quoted around € 900 to 1,300 per kW by suppliers. However, a price of € 1,000/kW was adopted as the assumed rule of thumb [54-56]. An average of 20% of turbine cost price is usually associated with the cost for civil/structural works [57]. The R_{OM} value includes the costs for insurance, regular maintenance, repair, spare parts and administration [58]. The current inflation and interest rates in Nigeria are 8.6% and 12.0% (according to CBN report [59]).

Table 6 shows the model AV 928 provides the lowest production cost per kWh. However, since the model SWT-3.6-107 produces more power per wind speed greater than the cut-in speed, it may seem the preferred choice. Therefore, prior to any conclusion, an economic decision should be made based on whether to compromise the potential for more power output by choosing the lowest cost of power or vice versa [12]. Worthy of note is the fact that the econometrics result gives the specific cost per kWh of generating wind electricity from the site. It is at variance with electricity tariff in Nigeria which is between *N* 4 and *N* 23.71 (€ 0.02 and € 0.15) depending on the location and the cost of living of the area. In order to carry out adequate comparison, the factors such as cost of grid connection,

Table 6 Econometrics analysis of wind electricity generation at the site

Turbine model	Present value cost (€)	Average P_{erave} per annum (from Table 5) × 106 kWh	20 years average P_{erave} ($t \times P_{erave}$) × 10^6 kWh	Specific cost per kWh (€)	Specific cost per kWh (Nigeria naira)
GE 1.5sle	1,978,445	6.7	134	0.015	3.06
GE 1.5xle	1,978,445	8.6	172	0.012	2.38
AV 928	3,297,409	14.6	292	0.011	2.34
V90	3,956,891	11.7	234	0.017	3.50
SWT-3.6-107	4,616,372	18.8	376	0.012	2.54

cost of transmission and distribution, control system cost, and cost incurred from losses in transmission would have to be included in the analysis for Nigeria.

Conclusions

An assessment of the potential of wind power generation at Kano, Nigeria was carried out. Twenty one years of three-hourly mean wind data at 10-m height from the Nigeria meteorological department, Oshodi, Nigeria were assessed and subjected to Weibull two-parameter and other statistical analyses. It was discovered that:

1. The two-parameter Weibull probability distribution adequately predicts the mean wind speed distribution at Kano.
2. From direct data analysis, of 252 recorded monthly mean wind speeds, representing whole 21 years of monthly mean data measurements, the cumulative frequency of mean wind speeds from 4.0 m/s and below was only 23 and that for 5.0 m/s and below was 20, suggesting that monthly mean wind speed values greater than 5.0 m/s were prevalent in Kano. The 21 years' monthly average wind speed variation ranged from 6.6 to 9.5 m/s. Seasonally, the magnitude of mean wind speed ranged from 7.7 (dry) to 8.5 (wet), while the whole year average gave wind speed value of 8.1 m/s, respectively.
3. The cumulative density function of the Weibull statistical distribution revealed that up to 60% of the whole data series were values that ranged from about 7.0 to 10.8 m/s and below, while up to 90% of the data series were values that ranged from about 8.3 to 13.2 m/s. The dry and wet seasons Weibull plots also indicate that between 60% and 90% of the dry season's and wet season's data series were values that ranged from at most 8.2 to 11.6 m/s, and 9.6 to 11.1 m/s respectively.
4. For the wind power density variation based on the Weibull analysis, January to June every year may be considered as the periods with good potential for wind energy harvest. However, the months of February and periods of May and July appeared to have better potential while June had the highest potential for wind energy harvest. The range of monthly wind power density lay between 410.8 and 1424.4 W/m^2.

Competing interests

The authors declare that they have no competing interests.

Authors' contributions

OOA collected the wind data and drafted the manuscript. OOA, SAA, and JOO carried out the comprehensive analysis. ROF and JK supervised the study and approved the original draft. All authors read and approved the final manuscript.

Acknowledgments

Authors appreciate the Proprietor and Management of Covenant University, Ota, Nigeria for the conducive environment made available in the institution for carrying out this research, the CU Centre for Learning Resources (CLR) for useful subscriptions which facilitate access to important reference materials and the CU Centre for System and Information Services (CSIS) for ensuring availability of internet facilities that make accessing the materials possible.

Author details

[1]Mechanical Engineering Department, Covenant University, P. M. B. 1023, Ota, Nigeria. [2]Mechanical Engineering Department, Obafemi Awolowo University, Ile Ife, Nigeria. [3]College of Engineering and Technology, Botswana International University of Science and Technology, Gaborone, Botswana.

References

1. Islam, SM: Increasing wind energy penetration level using pumped hydro storage in island micro-grid system. Int. J. Energy Environ. Eng. **3**(9), 1–12 (2012). doi:10.1186/2251-6832-3-9
2. Boukelia Taqiy Eddine, BT, Salah, MM: Solid waste as renewable source of energy: current and future possibility in Algeria. Int. J. Energy Environ. Eng. **3**(17), 1–12 (2012). doi:10.1186/2251-6832-3-17
3. Federal Ministry of Power and Steel (FMPS): Renewable Electricity Action Programme, International Center for Energy, Environment and Development (ICEED), Abuja. (2006). http://www.iceednigeria.org/workspace/uploads/dec.-2006-2.pdf, Accessed 16 July 2013
4. Federal Ministry of Power and Steel: Renewable Electricity Policy Guidelines, Federal Republic of Nigeria. (2006). http://www.iceednigeria.org/workspace/uploads/dec.-2006.pdf, Accessed 19 July 2013
5. Ajayi, OO: Assessment of utilization of wind energy resources in Nigeria. Ener. Pol. **37**(2), 720–723 (2009)
6. Ajayi, OO: The potential for wind energy in Nigeria. Wind Eng. **34**(3), 303–312 (2010)
7. Ohunakin, OS: Wind resources in North-East geopolitical zone, Nigeria: an assessment of the monthly and seasonal characteristics. Renew. Sustain. Energy Rev. **15**(4), 1977–1987 (2011)
8. Ohunakin, OS: Wind resource evaluation in six selected high altitude locations in Nigeria. Renew. Energy **36**, 3273–3281 (2011)
9. Ohunakin, OS, Adaramola, MS, Oyewola, OM: Wind energy evaluation for electricity generation using WECS in seven selected locations in Nigeria. Appl. Energy **88**, 3197–3206 (2011)
10. Ohunakin, OS: Assessment of wind energy resources for electricity generation using WECS in North-Central region, Nigeria. Renew. Sustain. Energy Rev. **15**, 1968–1976 (2011)
11. Fagbenle, RO, Katende, J, Ajayi, OO, Okeniyi, JO: Assessment of wind energy potential of two sites in North - East, Nigeria. Renew. Energy **36**, 1277–1283 (2011)
12. Ajayi, OO, Fagbenle, RO, Katende, J: Wind profile characteristics and econometric analysis of wind power generation of a site in Sokoto State, Nigeria. Energy Sci. Technol. **1**(2), 54–66 (2011)
13. Ajayi, OO, Fagbenle, RO, Katende, J, Okeniyi Joshua, O, Omotosho, OA: Wind energy potential for power generation of a local site in Gusau, Nigeria. Int. J. Energy Clean Environ. **11**(1–4), 99–116 (2011)
14. Ajayi, OO, Fagbenle, RO, Katende, J: Assessment of wind power potential and wind electricity generation using WECS of two sites in South West, Nigeria. Int. J. Energ. Sci. **1**(2), 78–92 (2011)
15. Ohunakin, OS: Wind characteristics and wind energy potential assessment in Uyo, Nigeria. J. Eng. Appl. Sci. **6**(2), 141–146 (2011)
16. Adejokun, JA: The three-dimensional structure of the inter-tropical discontinuity over Nigeria. Nigerian Meteorol. Services Tech. Note (39) (1966)
17. Adekoya, LO, Adewale, AA: Wind energy potential of Nigeria. Renew. Energy **2**(1), 35–39 (1992)
18. Asiegbu, AD, Iwuoha, GS: Studies of wind resources in Umudike, South East Nigeria – an assessment of economic viability. J. Eng. Appl. Sci. **2**(10), 1539–1541 (2007)
19. Bugaje, IM: Renewable energy for sustainable development in Africa: a review. Renew. Sustain. Energy Rev. **10**(6), 603–612 (2006)
20. ECN-UNDP: Renewable energy master plan: final draft report. (2005). http://www.iceednigeria.org/workspace/uploads/nov.-2005.pdf, Accessed 19 July 2013

21. Fadare, DAA: Statistical analysis of wind energy potential in Ibadan, Nigeria based on Weibull distribution function. Pac. J. Sci. Technol. **9**(1), 110–119 (2008)
22. Lahmeyer (International) Consultants: Report on Nigeria wind power mapping projects. Federal Min. Sci. Technol. , 37–51 (2005)
23. Ngala, GM, Alkali, B, Aji, MA: Viability of wind energy as a power generation source in Maiduguri, Borno State, Nigeria. Renew. Energy **32**(13), 2242–2246 (2007)
24. Ojosu, JO, Salawu, RI: A survey of wind energy potential in Nigeria. Sol. Wind Technol. **7**(2 and 3), 155–167 (1990)
25. Ojosu, JO, Salawu, RI: An evaluation of wind energy potential as a power generation source in Nigeria. Sol. Wind Technol. **7**(6), 663–673 (1990)
26. Oyedepo, SO, Adaramola, MS, Paul, SS: Analysis of wind speed data and wind energy potential in three selected locations in south-east Nigeria. Int. J. Energy Environ. Eng. **3**(7), 1–11 (2012). doi:10.1186/2251-6832-3-7
27. Ohunakin, SO, Oyewola, OM, Adaramola, MS: Economic analysis of wind energy conversion systems using levelized cost of electricity and present value cost methods in Nigeria. Int. J. Energy Environ. Eng. **4**(2), 1–8 (2012). doi:10.1186/2251-6832-4-2
28. Ajayi, OO, Fagbenle, RO, Katende, J, Okeniyi, JO: Availability of wind energy resource potential for power generation of Jos. Nigeria. Frontiers Energ. **5**(4), 376–385 (2011)
29. Carta, JA, Ramı´rez, P, Vela´zquez, S: A review of wind speed probability distributions used in wind energy analysis: Case studies in the Canary Islands: case studies in the Canary Islands. Renew. Sustain. Energy Rev. **13**, 933–955 (2009)
30. Montgomery, DC, Runger, GC: Applied statistics and probability for engineers, 3rd edn. Wiley, New Jersey (2003)
31. Akpinar, EK, Akpinar, S: An assessment on seasonal analysis of wind energy characteristics and wind turbine characteristics. Energ. Convers. Manage. **46**, 1848–167 (2005)
32. Tina, G, Gagliano, S, Raiti, S: Hybrid solar/wind power system probabilistic modelling for long-term performance assessment. Sol. Energ. **80**, 578–88 (2006)
33. Hau, E: Wind turbines - fundamentals, technologies, application, economics, 2nd edn. Springer, New York (2006)
34. Krause, P, Boyle, DP, Bäse, F: Comparison of different efficiency criteria for hydrological model assessment. Adv. Geosci. **5**, 89–97 (2005)
35. Murthy, DNP, Xie, M, Jiang, R: Weibull Models, (Wiley Series in Probability and Statistics). Wiley, Hoboken, New Jersey (2004)
36. Izquierdo, D, Alonso, C, Andrade, C, Castellote, M: Potentiostatic determination of chloride threshold values for rebar depassivation: experimental and statistical study. Electrochim. Acta. **49**, 2731–2739 (2004)
37. Roberge, PR, Roberge, PR: Statistical interpretation of corrosion test results. In: International Corrosion: Fundamentals, Testing, and Protection, ASM Handbook, Vol 13A, pp. 425–429. ASM International, Novelty, Ohio (2003)
38. Roberge, PR: Handbook of Corrosion Engineering. McGraw-Hill Companies, Inc, New York (2000)
39. Omotosho, OA, Okeniyi, JO, Ajayi, OO: Performance evaluation of potassium dichromate and potassium chromate inhibitors on concrete steel rebar corrosion. J. Fail. Anal. Prev. **10**(5), 408–415 (2010)
40. Polyanin, AD, Manzhirov, AV: Handbook of mathematics for engineers and scientists. Chapman & Hall/CRC, Boca Raton (2007)
41. Soong, TT: Fundamentals of probability and statistics for engineers. Wiley, England (2004)
42. Mistaya Engineering, Inc: Windographer. (2013). www.mistaya.ca/windographer/overview.htm, Accessed 22 May 2013
43. WINDEIS: National Renewable Energy Laboratory estimates of wind energy resources on blm-administered lands. http://windeis.anl.gov/documents/fpeis/maintext/Vol2/appendices/appendix_b/Vol2AppB.pdf, Accessed 22 May 2013
44. Danish Wind Industry Association: Wind energy reference manual part 1: wind energy concepts. (2003). http://wiki.windpower.org/index.php/Wind_energy_concepts#roughness, Accessed 22 May 2013
45. Schwartz, M: Wind resource estimation and mapping at the National Renewable Energy Laboratory, pp. 1–6. Paper presented at the ASES Solar '99 Conference, Portland, Maine. (1999). http://www.nrel.gov/docs/fy99osti/26245.pdf, Accessed 22 May 2013
46. Keyhani, A, Ghasemi-Varnamkhasti, M, Khanali, M, Abbaszadeh, R: An assessment of wind energy potential as a power generation source in the capital of Iran, Tehran. Energy **35**, 188–201 (2010)
47. Lipson, C, Sheth, NJ: Statistical design and analysis of engineering experiments. McGraw-Hill, New York (1973)
48. Røkenes, K: Investigation of terrain effects with respect to wind farm siting, p. 199p. Norwegian University of Science and Technology, Doctoral Thesis (2009)
49. Carpman, N: Turbulence Intensity in Complex Environments and its Influence on Small Wind Turbines, p. 55p. Uppsala University, Uppsala, Department of Earth Sciences Geotryckeriet (2011)
50. Kubik, ML, Coker, PJ, Hunt, C: Using meteorological wind data to estimate turbine generation output: a sensitivity analysis, pp. 4074–4081. In World Renewable Energy Congress, Sweden (2011)
51. Gipe, P: Wind power: renewable energy for home, farm, and business. Chelsea Green Publishing, White River Junction, Vermont (2004)
52. Arya, SPS: Introduction to micrometeorology. Academic Press, Inc., Waltham (1988)
53. Freris, LL: Wind energy conversion systems. Prentice Hall, Inc., Upper Saddle River (1990)
54. Ucar, A, Balo, F: Investigation of wind characteristics and assessment of wind-generation potentiality in Uludag-Bursa, Turkey. Appl. Energy **86**, 333–339 (2009)
55. Ahmed, AS: Wind energy as a potential generation source at Ras Benas, Egypt. Renew. Sustain. Energy Rev. **14**, 2167–2173 (2010)
56. Lima, L, Filho, CRB: Investigation of wind characteristics and wind energy assessment in Sao Joao do Cariri (SJC) - Paraiba, Brazil. (2010). http://www.worldenergy.org/documents/congresspapers/46.pdf, Accessed 23 May 2013]
57. EWEA: Wind Energy - the facts: cost and prices, vol. 2. (2003). http://www.ewea.org/fileadmin/eweadocuments/documents/publications/WETF/Facts_Volume_2.pdf, Accessed 23 May 2013
58. Wind Energy: The facts: Operation and maintenance cost of wind generated power. http://www.wind-energy-the-facts.org/en/part-3-economics-of-wind-power/chapter-1-cost-of-on-land-wind-power/operation-and-maintenance-costs-of-wind-generated-power.html, Accessed 23 May 2013
59. Central Bank of Nigeria: Economic indicators. www.cenbank.org, Accessed 27 April 2013

Experimental investigations of butanol-gasoline blends effects on the combustion process in a SI engine

Simona Silvia Merola*, Cinzia Tornatore, Luca Marchitto, Gerardo Valentino and Felice Esposito Corcione

Abstract

Fuel blend of alcohol and conventional hydrocarbon fuels for a spark-ignition engine can increase the fuel octane rating and the power for a given engine displacement and compression ratio. In this work, the influence of butanol addition to gasoline in a port fuel injection, spark-ignition engine was investigated. The experiments were realized in a single-cylinder ported fuel injection spark-ignition (SI) engine with an external boosting device. The optically accessible engine was equipped with the head of a commercial SI turbocharged engine with the same geometrical specifications (bore, stroke and compression ratio) as the research engine. The effect on the spark ignition combustion process of 20% and 40% of n-butanol blended in volume with pure gasoline was investigated through cycle-resolved visualization. The engine worked at low speed, medium boosting and wide-open throttle. Fuel injections both in closed-valve and open-valve conditions were considered. Comparisons between the parameters related to the flame luminosity and the pressure signals were performed. Butanol blends allowed working in more advanced spark timing without knocking occurrence. The duration of injection for butanol blends was increased to obtain a stoichiometric mixture. In open-valve injection condition, the fuel deposits on intake manifold and piston surfaces decreased, allowing a reduction in fuel consumption. BU40 granted the performance levels of gasoline and, in open-valve injection, allowed to minimize the abnormal combustion effects including the emission of ultrafine carbonaceous particles at the exhaust. In-cylinder investigations were correlated to engine out emissions.

Keywords: Optical diagnostics, Cycle-resolved visualization, PFI SI boosted engine, Butanol-gasoline blend, Injection timing

Background

Increasing global concern due to air pollution and to the limited oil reserves has generated much interest in the environmental friendly fuels alternative to petroleum-based fuels, in particular for the transport sector in which the energy consumption depends almost exclusively on fossil fuels. Several countries aim to use sustainable biofuels, which generate a clear and net GHG saving and have no negative impact on biodiversity and land use. In this scenario, butanol has strong potential as a biofuel. Like ethanol, butanol can be produced both by petrochemical and fermentative processes. The production of biobutanol by fermentation for use as a biofuel is generating considerable interest as it offers certain advantages in comparison with bioethanol. These include higher energy content, lower water adsorption and corrosive properties, better blending abilities and the ability to be used in conventional internal combustion engines without the need for modification. Biobutanol can be produced from starch or sugar-based substrates by fermentation (acetone-butanol-ethanol named ABE fermentation process).

However, cost issues, the relatively low yield and sluggish fermentations, as well as problems caused by end product inhibition and phage infections, meant that ABE butanol could not compete on a commercial scale with butanol produced synthetically, and almost all production ceased as the petrochemical industry evolved. However, the increasing interest in the use of biobutanol as a transport fuel induces a number of companies to explore novel alternatives to traditional ABE fermentation, which

* Correspondence: s.merola@im.cnr.it
Istituto Motori-CNR, Via Marconi 8, Naples, 80125, Italy

would enable biobutanol to be produced on an industrial scale.

Regarding the automotive use of biobutanol, the technology to make biobutanol, a nonfood-based biofuel, cost-competitive with fossil fuels isn't here yet, but several companies are working with this target. With respect to gasoline, butanol (or biobutanol) has a number of advantages over other common alcohol fuels such as ethanol and methanol. The energy density of gasoline is about 32 MJ/L, while butanol shows 29.2 MJ/L compared to ethanol, 19.6 MJ/L, and methanol, 16 MJ/L. This makes butanol so close to gasoline that it can allow a straight-across replacement in terms of energy [1]. Butanol is far less hygroscopic than methanol, ethanol and propanol. These lower alcohols are fully miscible with water, whereas butanol has only a modest water solubility. This allows a low-energy intermediate purification step [2]. Butanol is less corrosive than ethanol, can be transported in existing pipelines and is much safer to work with than lower alcohols based on its relatively high boiling point and flashpoint. In comparison with ethanol, the adding of butanol to conventional hydrocarbon fuels for use in a spark-ignition engine can increase fuel octane rating and power for a given engine displacement and compression ratio, thereby reducing fossil fuel consumption and CO_2 emissions [3-5]. Ethanol use has been widely investigated for spark-ignition engines, while few studies have been performed on butanol-gasoline combustion and on butanol-fueled engines [6,7]. Literature is particularly poor with respect to boosted spark-ignition (SI) engine experimental data. Almost all of the studies about butanol-gasoline blends consisted of the evaluation of performance, fuel consumption and exhaust emissions for different engine-operating conditions [6-10]. The in-cylinder process characterizations were principally realized through pressure measurements. These research activities demonstrated that the concentrations of 20% to 40% butanol in gasoline enabled to run the engine at a leaner mixture than gasoline for a fixed performance. These blends offered UHC emissions similar to gasoline, and they increased at higher butanol concentrations. The blends decreased the NO_x emissions to a lower level than with pure gasoline at its leanest mixture. The slight increase in specific fuel consumption (SFC) with the butanol addition was related to the blend's reduced combustion enthalpy. For example, B40 has a 10% lower combustion enthalpy than gasoline, which increases SFC of 10% for stoichiometric and slightly lean mixtures. It was measured that, by adding butanol, the coefficient of variation of indicated mean effective pressure (COV_{IMEP}) was reduced, particularly with lean mixtures, and the fully turbulent combustion phase (10% to 90% MFB) was similar in duration for all blends and pure gasoline. This latter finding showed that butanol has a similar or higher laminar flame speed than gasoline [6,11].

In recent works, the performance of a gasoline engine fuelled with gasoline-butanol blends of different mixing fractions was analyzed. It was demonstrated that butanol is a very promising alternative fuel with great potential for saving energy; a reduction of 14% in brake-specific energy consumption and emissions was observed [12].

Recent experimental investigations conducted using a single-cylinder spark-ignition research engine allowed comparing the performance and emissions of neat n-butanol fuel to that of gasoline and ethanol. It was found that gasoline and butanol are closest in engine performance, with butanol producing slightly less brake torque. Exhaust gas temperature and nitrogen oxide measurements show that butanol combusts at a lower peak temperature. Of particular interest were the emissions of unburned hydrocarbons, which were between two and three times to those of gasoline, suggesting that butanol is not atomizing as effectively as gasoline and ethanol [13].

At the same time, fundamental biobutanol combustion work was carried out; the oxidation of butanol-gasoline surrogate mixtures (85 to 15 vol.%) was studied using a jet-stirred reactor in the work by Dagaut and Togbé [14]. The aim of this paper is better comprehension of in-cylinder phenomena correlated with butanol-gasoline combustion in a SI engine. To this goal, cycle-resolved visualization was performed to follow the flame propagation from the spark ignition to the late combustion phase. The experiments were realized in a single-cylinder ported fuel injection (PFI) SI-boosted engine. The optically accessible engine was equipped with the cylinder head of a commercial SI turbocharged engine with the same geometrical specifications (bore, stroke and compression ratio) of the research engine. Butanol-gasoline blend was tested for several engine operating conditions. Changes in spark timing and fuel injection phasing were considered. Comparison between the parameters related to flame luminosity and to pressure signals were performed, and in-cylinder investigations were correlated to engine out emissions.

Methods

Experimental apparatus

The engine used for the experiments was an optically accessible single-cylinder PFI SI engine. It was equipped with the cylinder head of a commercial SI turbocharged engine with the same geometrical specifications. Further details on the engine are reported in Table 1. The head had four valves and a centrally located spark plug. The injection system was the same as the commercial one with a ten-hole injector. An external air compressor was used to simulate boosted conditions of intake air pressure and temperature

Table 1 Specifications of the single-cylinder ported fuel injection engine

Engine specifications	
Displaced volume	399 cc
Stroke	81.3 mm
Bore	79 mm
Connecting rod	143 mm
Compression ratio	10:1
Number of valves	4
Exhaust valve open	153 CAD ATDC
Exhaust valve close	360 CAD ATDC
Inlet valve open	357 CAD ATDC
Inlet valve close	144 CAD BTDC

ATDC, after top dead centre; BTDC, before top dead centre; CAD, crank angle degree.

in the ranges of 1,000 to 2,000 mbar and 290 to 340 K, respectively.

A quartz pressure transducer was flush-installed in the region between the intake-exhaust valves at the side of the spark plug. The transducer allowed performing in-cylinder pressure measurements in real time. An elongated engine piston was used; it was flat and its upper part was transparent since it was made of fused silica UV enhanced ($\Phi = 57$ mm). To avoid window contamination by the lubricating oil, self-lubricant Teflon-Bronze composite piston rings were used in the optical section.

To reduce the initial conditions effects, the engine was preheated by a conditioning unit, and it was maintained in motored condition by an electrical engine until the temperature reached 65°C. After the warm up, the engine worked in fired conditions for 300 consecutive cycles. The engine parameters and pressure measured in the last 200 cycles were considered in the work. Then the engine returned to motored condition for 100 cycles. This phase allowed checking a possible change in thermal and fluid dynamic status of the engine from the beginning of the measurements.

Figure 1 shows the experimental apparatus for the optical investigations and the bottom field of view of the combustion chamber. During the combustion process, the light passed through a quartz window located in the piston, and it was reflected toward the optical detection assembly by a 45°-inclined UV-visible mirror located at the bottom of the engine.

Cycle-resolved flame visualization was performed using an Optronis model CamRecord 5000 high-speed camera (Optronis GmbH, Kehl, Germany). It was a complementary metal oxide semiconductor (CMOS) monochrome image sensor; its full chip dimension was 512 × 512 pixel, and the pixel size was 16 × 16 μm. The camera A/D conversion was 8 bit, and the spectral range extended from 390 to 900 nm. The camera was equipped with a 50-mm focal Nikon lens (Nikon UK Ltd., Surrey, England); a camera region of interest was selected (480 × 480 pixel) to obtain the best match between spatial and temporal resolution. All the images are a line-of-sight average. The optical assessment allowed a spatial resolution in a 2D field of view around 120 μm/pixel. The exposure time was fixed at 10 μs, and the frame rate was 5,392 fps. In this work, the optical measurements were performed during 100 consecutive engine cycles after an engine warm up under motored conditions and 100 fired cycles. National Instruments LabVIEW acquisition system driven by an optical encoder with 0.2 crank angle degree resolution recorded the TTL signals from the high-speed camera acquisitions and the pressure transducer. In this way, it was possible to determine the crank angles at which the optical measurements were carried out. Finally, to quantify the variability of indicated work per cycle, the COV_{IMEP} was calculated.

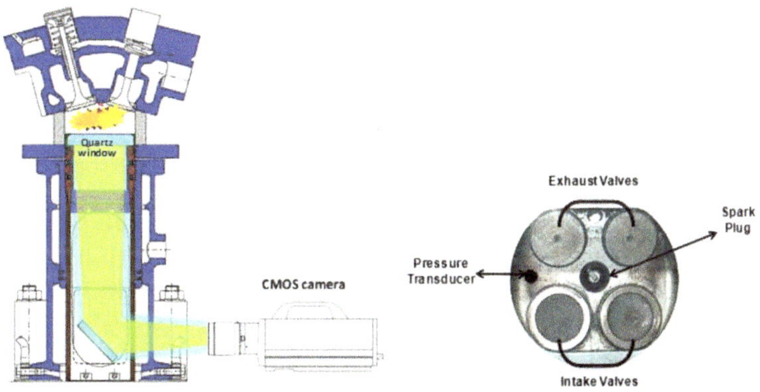

Figure 1 Experimental setup for optical measurements with bottom field of view of the combustion chamber.

In-house LabVIEW numerical procedures were applied for the retrieving of the optical data. The 8-bit gray-scale images were converted to numerical matrices. In this way, it was possible to evaluate the luminous signals locally or as integral on the whole combustion chamber. In a second procedure, a threshold was applied to each image. On the resulted binary images, a morphology function able to evaluate the mean radius of the flame front as distance between the combustion chamber center and the outline of the flame was applied.

Combustion tests were carried out using, as baseline fuel, the commercial gasoline; moreover, two blends with the volume of 20% and of 40% n-butanol with gasoline were tested. The blends were indicated in the following as BU20 and BU40, respectively. The main properties of the two fuels are reported in Table 2. Finally, steady-state measurements of HC, NO_x and soot were performed in the undiluted exhaust. Gaseous emissions were measured by an AVL Digas 4000 (AVL DiTEST, Graz, Austria) equipped with an electrochemical sensor for NO_x and nondispersive infrared analyzers for HC. Smoke was measured by a part flow opacimeter (AVL Opacimeter 439). The opacity can be directly correlated to the particulate mass concentration [15].

Results and discussion

All the tests presented in this paper were carried out at an engine speed of 2,000 rpm and wide-open throttle. Absolute intake air pressure and temperature were fixed at 1.4 bar and 338 K, respectively. The relative injection pressure was settled at 3.5 bar. The spark timing was changed in the range 12 to 20 crank angle degree before top dead centre (CAD BTDC) in order to identify the maximum brake torque and the knocking limit. To distinguish normal combustion cycles from knocking cycles, the knocking signal was evaluated through 5- to 30-kHz band-pass filtering of the pressure signals [16-18]. For all cycles, the evolution of the knock pressure was calculated using the absolute value of the knock signal. The combustion cycles were classified in the following way according to their knock intensity [19-21]:

- normal combustion (knock pressure 2% lower than the motored pressure at TDC),
- borderline knocking (knock pressure between 2% and 5% of the motored pressure),
- standard knocking (knock pressure between 5% and 25% of the motored pressure),
- heavy knocking (knock pressure was 25% higher than the motored pressure).

The duration of injection (DOI) was changed in order to set $\lambda = 1.0$ (±3%), as measured by a lambda sensor at the engine exhaust, and averaged over 200 consecutive engine cycles. The fuel injection conditions are resumed in Table 3. The injection timings were fixed at 130 CAD after top dead centre (ATDC) and 300 CAD BTDC in order to inject the fuel when the intake valves were closed (CV) and open (OV), respectively.

In a port-fuel-injected engine, the fuel is generally injected at the backside of a closed intake valve to take advantage of the warm valve and port surfaces for vaporization [22-24]. However, a large part of the injected spray is deposited on the intake manifold surfaces and forms a layer of liquid film on the valve and port surfaces. The film needs to be re-atomized by the shearing airflow as the intake valves open. If these fuel layers are not well atomized, they enter the cylinder as drops and ligaments [25-27]. These phenomena occur in varying degrees and depend upon the engine design, injector location and engine operation. Previous experiments on the same engine fuelled by gasoline and BU20 demonstrated that the injector sprayed the fuel towards the plate between the intake valves and on the intake valves stems [28-30]. The droplet impingement induced fuel layer formation on the intake manifold walls. The fuel layers were drawn by gravity to the valve head and gap, where they remained as film due to surface tension. The stripping of the squeezed fuel film created fuel

Table 2 Fuel specifications

	Gasoline	Butanol
Low heating value (MJ/kg)	43.5	32.1
Latent heat of vaporization (kJ/L)	223	474
Reid vapor pressure (kPa)	60 to 90	18.6
Stoichiometric air-to-fuel ratio	14.6	11.1
Density (kg/m^3)	720 to 775	813
Oxygen (wt.%)	<2.7	21.6
Research octane number	95	113
Adiabatic flame temperature (K)	2,370	2,340

Table 3 Fuel injection conditions

Label	Fuel	SOI	DOI
Gas_OV	gasoline	300 CAD BTDC	133 CAD
Gas_CV	gasoline	130 CAD ATDC	148 CAD
BU20_OV	BU20	300 CAD BTDC	145 CAD
BU20_CV	BU20	130 CAD ATDC	157 CAD
BU40_OV	BU40	300 CAD BTDC	153 CAD
BU40_CV	BU40	130 CAD ATDC	165 CAD

ATDC, after top dead centre; BTDC, before top dead centre; BU20, blend of 20% n-butanol with gasoline; BU40, blend of 40% n-butanol with gasoline; CAD, crank angle degree; CV, closed valve; DOI, duration of injection; Gas, gasoline; OV, open valve; SOI, start of injection.

deposits on the optical window. When the injection occurred in open-valve condition, part of the droplets was carried directly into the combustion chamber by the gas flow. These droplets, sucked in the combustion chamber, stuck on the cylinder walls. In both injection conditions, the fuel deposits on the combustion chamber walls created fuel-rich zones on the piston surfaces that influenced the composition of the mixture charge and, hence, the combustion process. During the normal combustion process, only a fraction of the fuel deposits was completely burned. Thus, more fuel should be injected to reach the selected air-fuel ratio measured at the exhaust [23].

In the open-valve injection, the fuel deposits amount and size were smaller than in the closed-valve injection; thus, the duration of injection resulted shorter [25,29]. Regarding the difference in the injection duration, as with any alcohol, gasoline-butanol blends have a lower stoichiometric air-fuel ratio. Therefore, when using gasoline blended with butanol, fuel flow must be increased to ensure the same relative air-fuel ratio as with pure gasoline.

In order to estimate the effect on engine performance of the selected fuel injection conditions, IMEP and COV_{IMEP} were evaluated as average on 100 consecutive engine cycles. For each spark timing, the IMEP variation was lower than 5% for all engine conditions and fuels. For each fuel, the IMEP of CV injection was higher than that of OV, in agreement with previous works [20,29]. For BU20, the IMEP in the OV condition was higher than gasoline in OV condition and very close to gasoline in CV condition. This occurred because butanol-gasoline blend burns faster than pure gasoline at the same conditions, making higher the indicated efficiency of the engine work cycle. For BU20_OV, the best stability was measured too [7]. This concurs with the results reported in the works by Irimescu [31] and Yang et al. [32] in which, at full load, the power drop is significant only for butanol concentrations higher than 30% to 40%. About the spark timing effect, for gasoline fuel, the knocking limit was evaluated around 16 CAD BTDC for both fuel injection conditions.

For BU20, the knocking limit was evaluated around 18 CAD BTDC, and for BU40, it advanced until around 20 CAD BTDC. It means that butanol blend allowed working in more advanced spark timing without occurrence of knocking combustion. The COV_{IMEP} increased with spark advance until reaching the highest value in knocking regime. From 16 CAD BTDC, the COV_{IMEP} increased with retarding spark timing too. This result agrees with those reported in the works by Szwaja and Naber [7] and Morey and Seers [33]. It occurs because, when the ignition is too advanced, the cylinder temperature is comparatively low. Besides, the quite low and uneven mixture concentration near the spark plug brings negative influence on the flame kernel initiation and development. When the ignition is too delayed, the low combustion efficiency does harm to combustion stability. In this work, the optical results obtained at 14 CAD BTDC spark timing are discussed. This choice was done in order to evaluate the effect of selected fuels on the normal combustion process at comparable IMEP (±1%) with satisfactory engine stability. Cycle-resolved visualization was used for detailing thermal and fluid dynamic phenomena that occur in the combustion chamber. Figures 2 and 3 report images of cycle-resolved flame front evolution for gasoline, BU20 and BU40 in CV and OV conditions. The images' brightness and contrast were changed to enhance the kernel flame luminosity. The combustion process was visualized from the spark ignition until the flame front reached the cylinder walls. As expected, for gasoline [29,31], after the evidence of the spark ignition, thanks to the plasma luminosity detected at 14 CAD BTDC, the flame front started as kernel from the spark plug, and then, it spread with radial-like behavior for around 6 to 10 CAD. Then, the flame front shape showed an asymmetry that induced the flame to reach first the cylinder wall in the exhaust valves region, around 20 CAD after the start of spark timing. This was due to the fuel film deposited on the intake valves and combustion chamber surfaces previously discussed. The fuel film develops dynamically under the effect of the gas flow, influencing mixture composition and combustion process. In fact, the flame propagation is influenced by the thermodynamic conditions, mixture composition and local turbulence intensity. When a flame propagates in the normal direction to a region with an equivalence ratio gradient, each part of the front evolves in a field with varying fuel concentration. This induces propagation speed variation along the flame front and an increase in flame wrinkling in comparison with the homogeneous case. A similar combustion process evolution was detected for both butanol blends.

As can be observed in Figure 2, the asymmetry was less evident for BU40, which showed a more regular evolution of the flame front with slight border wrinkling. This was a first marker of a lower amount of fuel deposits near the valves for the blend if compared to pure gasoline. Bright spots were observed in the burned gas before the flame front reached the chamber wall. The bright spots were due to the fuel deposits on the optical window. During the injection of fuel at closed intake valve, when the gas flow passes through the valves, the fuel droplets are stripped from the fuel film layer. After reaching the combustion chamber, the fuel droplets stuck on the piston surfaces. These fuel deposits also created fuel-rich zones that ignited when reached by the normal flame front. In CV condition, the spots are bigger

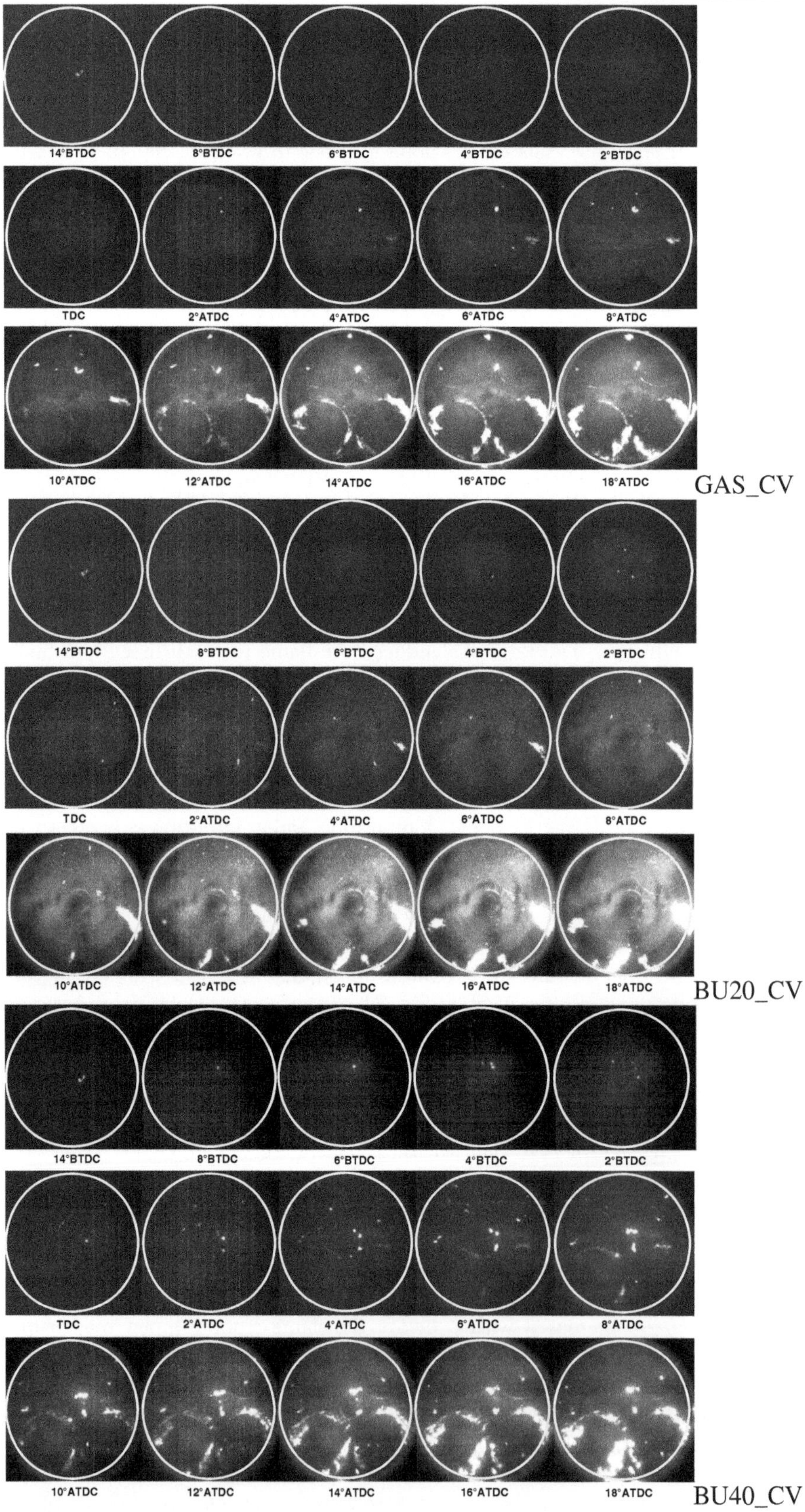

Figure 2 Cycle-resolved flame front evolution for gasoline, BU20 and BU40 in CV condition.

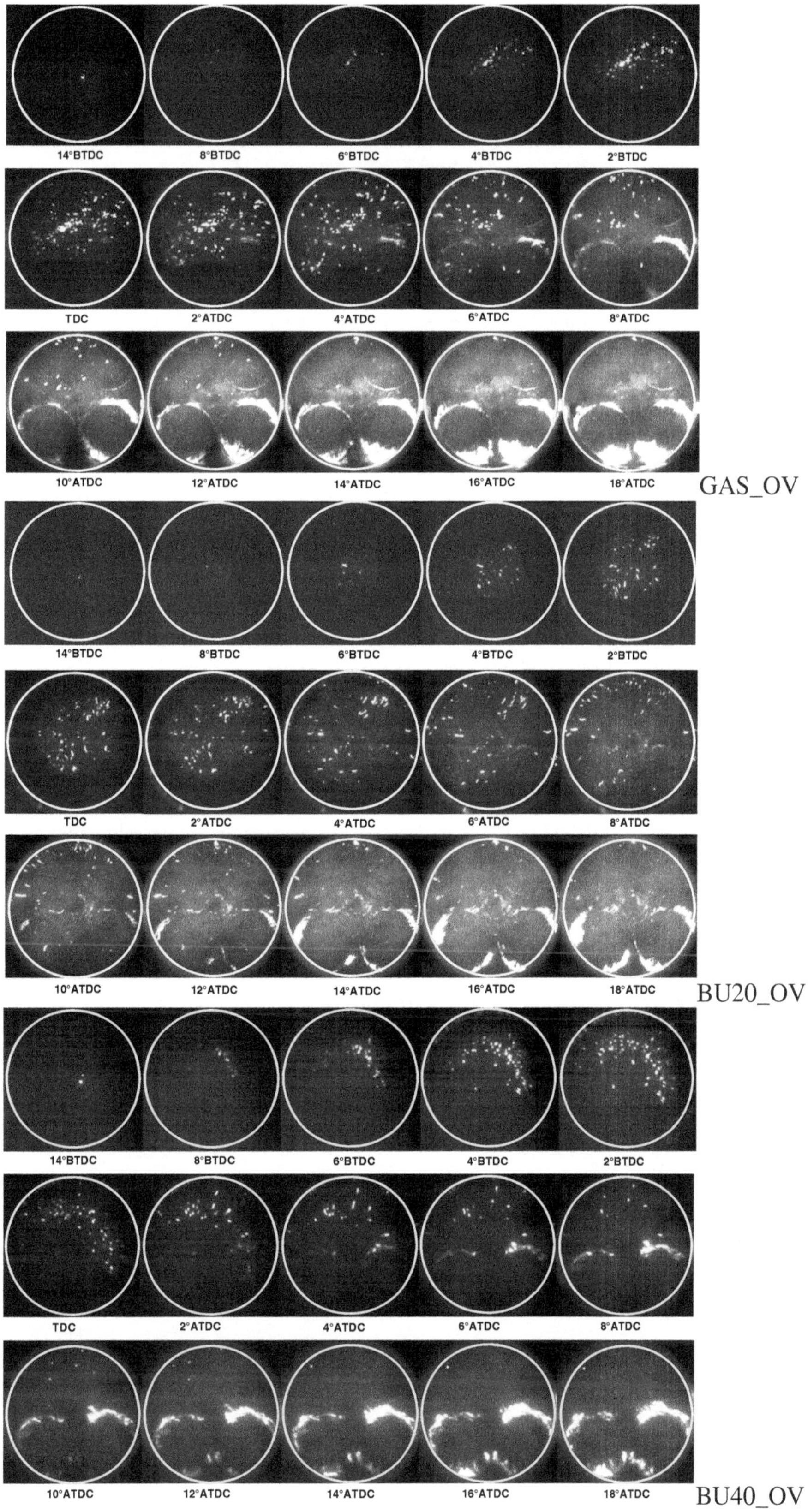

Figure 3 Cycle-resolved flame front evolution for gasoline, BU20 and BU40 in OV condition.

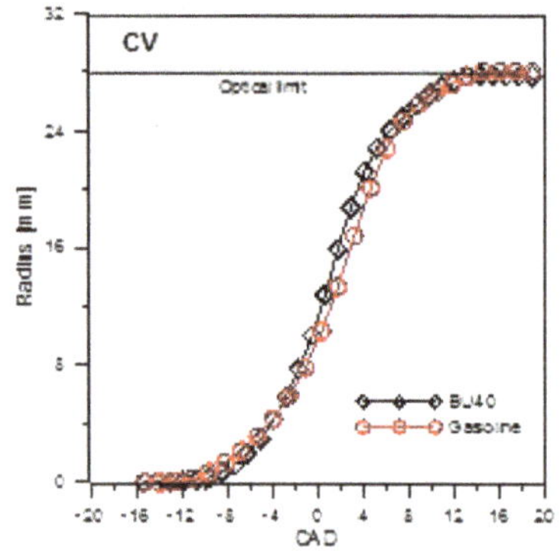

Figure 4 Evaluated flame radius of the 100 consecutive engine cycles.

but less in number than in OV condition. When the injection occurred in open-valve condition, the fuel droplets' sticking was enhanced by the partial carrying of the injected fuel droplets directly into the combustion chamber due to the gas flow. The bright spots have a random nature; during the combustion process, they decreased in size and number, and then, they disappeared before the exhaust valve opening [29]. The evidence of the bright spots decreased with the butanol percentage increasing. This means that the chemical composition of the blends helped the vaporization of the low-volatile component and then the fuel deposits burning. The presence of the fuel deposits, as squeezed film or impinged droplets, had direct effect on the flame radius evolution in terms of kernel cyclic variability and flame stability [25,34]. Figure 4 reports the trend of the flame radius for gasoline and BU40 evaluated on 100 consecutive cycles. Figure 5 shows the outline of the flame front evaluated for two CADs from the sequences of Figure 2. From its inception until around 6 CAD BTDC, the flame kernels had the same trends for both the fuels. Then, a little difference was observed; this was due to the reduced fuel amount deposited on the intake valves and on the piston surface that influenced the flame propagation. At around 4 to 6 CAD ATDC, flame radius evolution changed dramatically as shown in Figure 5 that shows the outline of the flame front evaluated from the sequences of Figure 2. This was due to approaching the intake valve region. In fact, the heat exchange between the intake ports and the surrounding gas led to the fuel film deposit evaporation. It influenced the composition of the mixture, creating locally fuel-rich zones. The higher fuel amount near the intake valves for gasoline in CV condition induced fuel-richer zones that slowed down the flame front more than in the other conditions.

When that normal flame front reached the intake valves, the fuel film layer around the valves burned, and strong-intensity flames were observed [35,36]. The

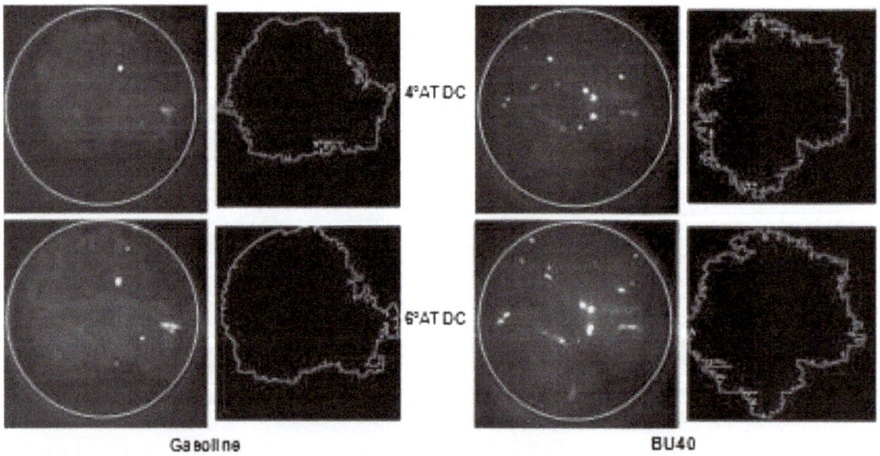

Figure 5 Flame fronts and related outlines evaluated from the sequences of Figure 2.

outlines of the valves are clearly distinguished since 8 CAD ATDC for all the tested conditions (Figures 2 and 3). Previous works demonstrated the presence of diffusion-controlled flames. Their inception was possible since the oxygen was not completely consumed after the normal flame front propagation [28,29,37]. The diffusion-controlled flames persisted in the late combustion phase, and their optical evidence could be detected until the exhaust valve opening.

The spatial distribution of diffusion-controlled flames can be analyzed through the images reported in Figure 6. For all fuels in closed valve injection condition, the resulting flames were more intense and have bigger surface than in open-valve condition due to the higher amount of fuel deposited on the valve stem for CV. Until 70 CAD ATDC, the highest intensity of the flames was observed near the intake valves, as expected, for all conditions. Then, different evolutions and spatial distributions were detected. This was in part due to the burning of fuel deposits carried by the gas motion from the intake to the exhaust valves. For BU40, the abnormal combustion was lower in intensity and duration. While for gasoline, the diffusion flame intensity decreased after 150 CAD ATDC due to the exhaust valve opening; for BU40, flame intensity were strongly reduced just around 130 CAD ATDC. Again, BU20 showed a mean behavior.

The abnormal combustion did not contribute to the engine work, and it did not influence the pressure signal [35]: it induced surface diffusion flames that warmed up the nearby in-cylinder gas by thermal diffusion. This phenomenon increased the pressure much slower than the reduction of pressure produced by the movement of the piston during the expansion stroke. A comparison between the pressure-related measurements and processed optical data was performed.

Figure 7 shows the evolution of the combustion pressure signal and the integral luminosity measured during the engine cycles of Figures 3 and 4. The combustion pressure signal was calculated by the subtraction of the motored in-cylinder pressure from the fired one. From the spark ignition to the maximum values, the luminous and pressure signals showed similar trends for all the fuels and injection conditions. The sharp increase was due to the chemical reactions occurring in the first moments of the combustion process that are exothermic and radiative in the wavelength range of the CMOS camera. Figure 8 reports typical spectra detected in the center of the combustion chamber in the early stage of the combustion process for gasoline fuel [37]. In Figure 8a, the results obtained at 7 CAD BTDC are shown. For both signals, the spectral features of OH and CH were detected [38-40]. In particular, the highest heads at 306 to 309 nm of the OH band system (250 to 320 nm) were well resolved. Excited OH radical was formed in the primary combustion zone by the chemiluminescent reaction: $CH + O2 \rightarrow CO + OH$. Moreover, the CH systems were observed near 431, 390 and 314 nm. The 431-nm band is the brightest; the 390-nm band system is very weak with closely packed heads. The 310-nm band is usually obscured by OH. In addition to the OH and CH features, a continuum on which two groups of diffuse bands were superimposed was detected. The first group was due to the Emeleus' bands of formaldehyde molecule CH_2O, and it had the highest emission in the range of 350 to 460 nm. The second band system identified the Vaidya's bands of HCO with the highest heads from 290 to 360 nm. Thus, the longer wavelength bands were overlapped by the CH_2O and continuum emission [39-41]. When the flame front overcame the spectroscopic measurement region, CH disappeared and high OH emission was measured, as shown in Figure 8b. Moreover, the burned gas is characterized by a broadband emission from UV to visible that is related to the CO_2 chemiluminescence [42]. Even if, in the flames, there is no sufficient energy to excite stable atoms or molecules to high electronic states, electronic states of CO_2 can be excited during the combustion by consecutive transitions from the ground state level to intermediate vibrationally activated levels [43,44]. The emission of CO-O appears as a continuum, which extends from 300 to 600 nm with a broad maximum around 375 nm.

The behavior of the combustion pressures and luminous intensities plotted in Figure 7 became quite different after the maximum around 20 CAD ATDC. For gasoline, the luminous signal decreased until a local minimum and then increased in the late combustion phase, while the pressure signal rapidly decreased. The diffusion-controlled flames greatly influenced the evolution of luminous signal, but their contribution to the combustion pressure was negligible. The diffusion flame intensity was higher for closed-valve injection than for open-valve due to the higher fuel amount deposition. For both conditions, the maximum was detected around 70 to 80 CAD ATDC. Previous investigations for gasoline fuel [37] showed that the diffusion-controlled flames were characterized by the optical markers of carbonaceous structures. Figure 8c reports typical spectra detected in the combustion chamber in the late combustion phase. The spectra presented a strong continuous contribution that increased with the wavelength in the visible range; this was representative of blackbody-like emission of soot precursors. Different levels of visible wavelength luminous intensity in the late combustion phase were related to different soot concentrations in the combustion chamber. The results confirmed those reported in the works of Witze and Green [36] and Kayes et al. [45] that assigned to the fuel deposition burning the cause of the volatile organic carbon compounds and ultrafine particles emission at the SI PFI exhaust. Regarding the

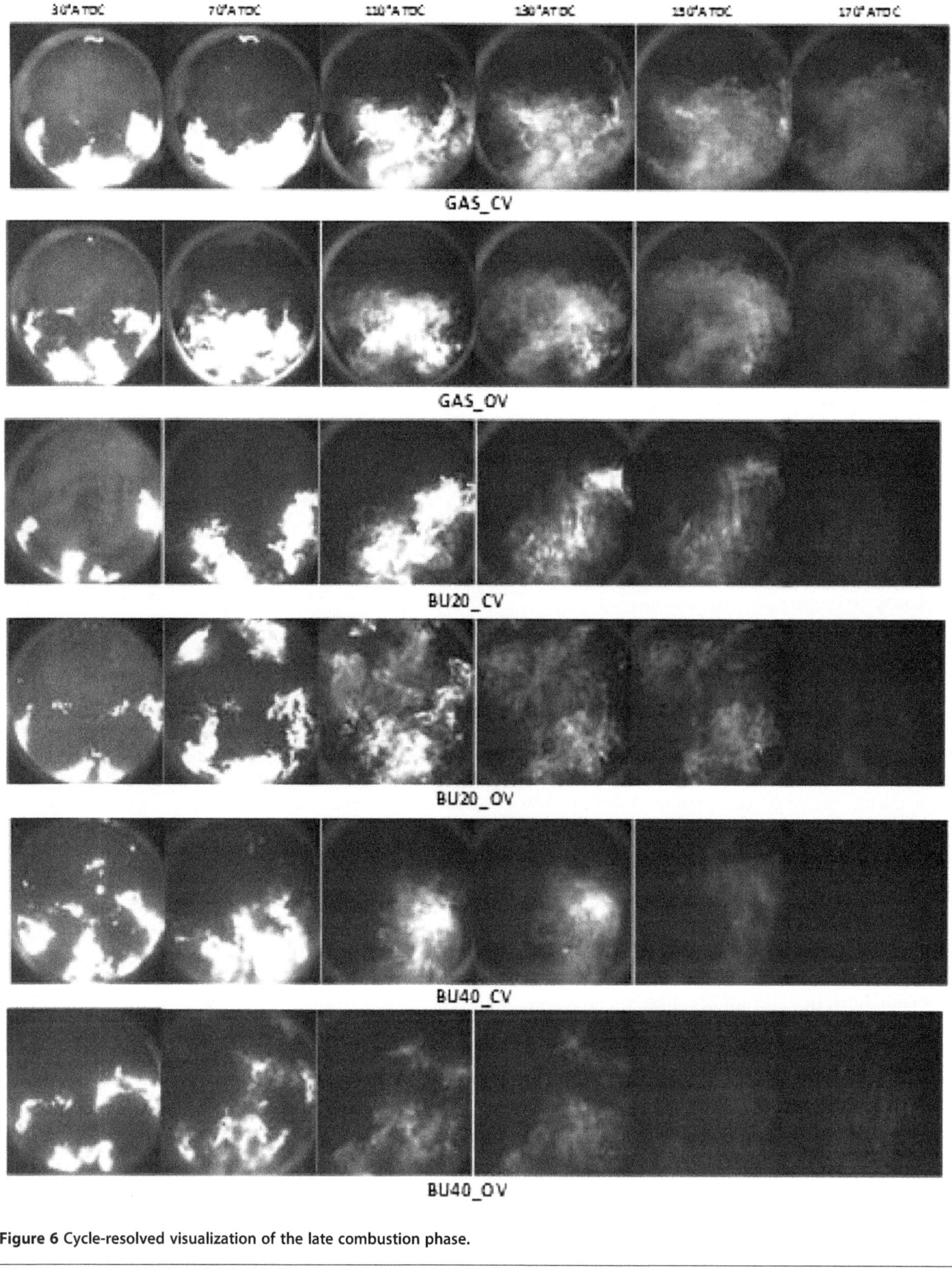

Figure 6 Cycle-resolved visualization of the late combustion phase.

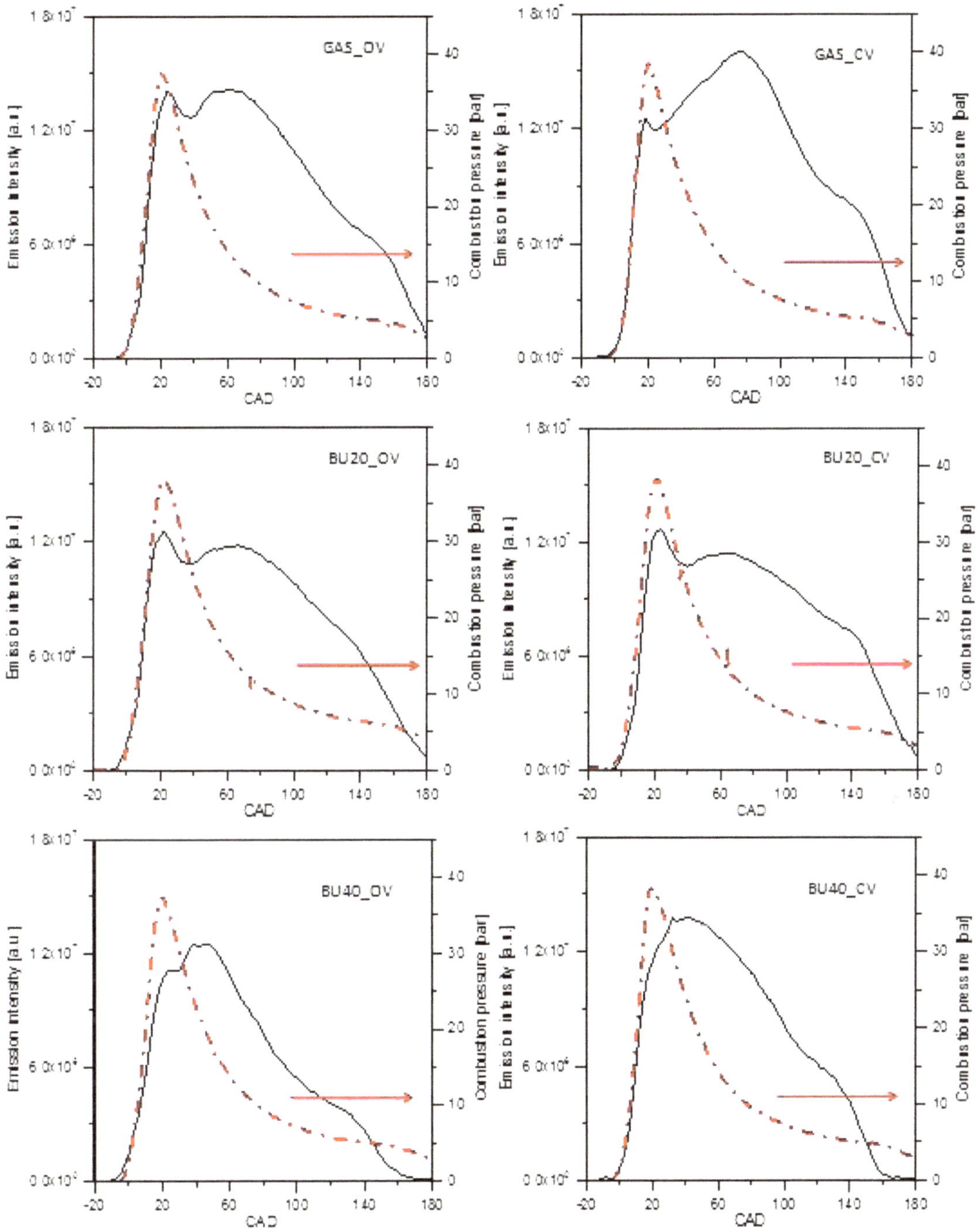

Figure 7 Evolution of the combustion pressure signal (dotted line) and the integral luminosity (solid line). Comparison between the net combustion pressure signal with the integral luminosity measured during the engine cycles of Figures 3 and 4.

temporal behavior of luminous signal, it should be noted that the spectra detected in the late combustion in Figure 8c showed a well-resolved signal due to OH radical that featured the soot oxidation phase [46]. OH emission was comparable for both fuel injection conditions. As a consequence, the open-valve condition showed not only a different spatial distribution of diffusion-controlled flame if compared to the closed-valve condition but also a more efficient soot oxidation phase. For butanol blends, the resulting integral luminosities in Figure 7 were less intense, demonstrating a lower particulate amount produced than gasoline with stronger soot reduction. This was more evident for BU40 due to higher oxygen content in the fuel. Even for butanol blends, higher particulate concentration

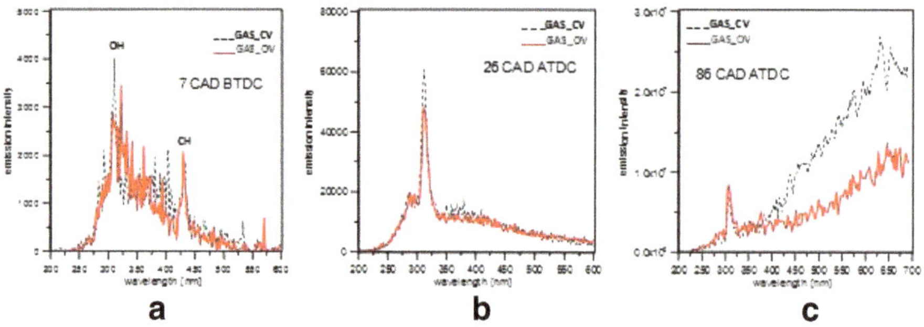

Figure 8 Typical spectra detected in the center of the combustion chamber for gasoline fuel [37]. (a) Results obtained at 7 CAD BTDC. **(b)** Measurement of high OH emission. **(c)** Typical spectra detected in the combustion chamber in the late combustion phase.

was detected for closed valve than for open one, as expected. At the exhaust valve opening, the soot reduction rate was not sufficient to complete the oxidation; thus, part of the particulate matter formed in the combustion chamber was emitted in the exhaust line. This result partially agrees with the opacity values measured at the undiluted exhaust and reported in Table 4. The discrepancy could be due to the sensitivity of the opacimeter to gaseous species such as heavy HC and NO_2 that have high absorption cross section [47,48] in the visible wavelength range that corresponds to the opacimeter working spectral region. It should be stressed that the absence of a catalyst device determined very high concentrations of HC and NO_x, and these can contribute to opacity value. Anyway, for all the tested fuels, HC in OV condition resulted higher than in CV. This occurred because, even if open-valve injection greatly reduces the amount of intake port wetting, it also induced larger in-cylinder wall wetting due to the direct fuel impingement. Experiments [49] showed that the many droplets associated with open-valve injection survived to the crank angle of ignition. This induced an inhomogeneous charge with poorer flame-front propagation that is responsible for reduction in performance and higher HC emission than closed-valve injection. Moreover, for small n-butanol blending (BU20), HC emissions were included among those of gasoline and BU40. The reduction with respect to gasoline was due to the decrease of the hydrocarbon fraction that led to the decrease of HC formation. The increase with BU40 could be due to the higher latent heat of vaporization than BU20. As reported in the work by Gu et al. [50], increasing butanol concentration in the blend with gasoline led to a decrease in the HC oxidation during expansion and exhaust processes.

Table 4 Engine out emissions

Label	HC (ppm)	NO_x (ppm)	Opacity (%)
Gas_CV	395	3,547	23.8
Gas_OV	458	3,749	15.1
BU20_CV	235	2,030	10.7
BU20_OV	408	2,905	9.3
BU40_CV	320	1,968	15.1
BU40_OV	433	2,528	11.5

BU20, blend of 20% n-butanol with gasoline; BU40, blend of 40% n-butanol with gasoline; CV, close valve; Gas, gasoline; OV, open valve.

Conclusions

The effect on the spark-ignition combustion process of n-butanol blended in volume with pure gasoline was investigated through cycle-resolved visualization applied in a single-cylinder PFI SI engine working at low speed, medium boosting and wide-open throttle. Two injection timings were fixed in order to inject the fuel at closed intake valve and open intake valve, respectively. The spark timing was changed to identify the maximum brake torque and the knocking limit. Blends of butanol up to 40% allowed working in more advanced spark timing without negative effects on performance. To work with a stoichiometric mixture for both fuels, the duration of injection was slightly increased for the blend. DOI in CV resulted longer than in OV for both fuels because, in CV injection, part of the injected spray is deposited on the intake manifold surfaces, forming a layer of liquid film. If these fuel layers are not well atomized, they enter the cylinder as drops and ligaments. During the normal combustion process, only part of the fuel deposits was completely burned. Thus, more fuel should be injected to reach the selected air-fuel ratio measured at the exhaust. When the normal flame front reached the fuel deposits, abnormal combustion was incepted. This was characterized by intense diffusion-controlled flames. Their contribution to the combustion pressure was negligible. The different levels of intensity were related to different carbonaceous structures and soot precursor concentrations. CV condition was characterized by higher fuel deposition amount and thus more intense diffusion-controlled flames than OV. Gasoline in CV condition showed the highest luminosity, and BU40 in OV condition, the lowest one. This demonstrated that

BU40_OV allowed the reduction of emission of ultrafine carbonaceous particles at the exhaust and the optimization of fuel consumption at fixed performance. Moreover, medium-low percentage of butanol in the gasoline allowed the reduction of NO_x and unburned hydrocarbon emission. Finally, even if an increase in the injected fuel amount should be considered to obtain the same air-fuel ratio for butanol-gasoline blend, if compared to pure gasoline, the better efficiency of fuel deposit burning allowed the reduction of that amount.

Acknowledgement
The authors would like to express their sincere appreciation to Mr. A. Mazzei for the precious technical support.

Competing interests
The authors declare that they have no competing interests.

Authors' contributions
SSM participated in the design of the study, performed the post processing of the data and drafted and corrected the manuscript. CT participated in the design of the study, carried out the experimental activity and helped in the organization and presentation of the data. LM participated in the design of the study, carried out the experimental activity and drafted the manuscript. GV participated in the design and coordination of the study and corrected the draft manuscript. FEC participated in the design and coordination of the study. All authors read and approved the final manuscript.

Authors' information
SSM received from the University of Naples in Italy her *Summa cum Laude* Magister Degree in Physics in 1995 and the PhD in Chemical Engineering in 2000. Since 2001, she is a full-position researcher in Istituto Motori of the Italian National Research Council. She specializes in the application of optical diagnostics to the combustion process investigation. Her research work focuses on the experimental analysis of the thermo fluid-dynamic phenomena that occur in-cylinder and at the exhaust of the internal combustion engines fuelled by conventional and alternative fuels. Since 1996, she has been involved with and has the direct responsibility for experimental activities in several national and international research projects.
CT received from the University of Naples in Italy her *Summa cum Laude* Magister Degree in Chemical Engineering in 2004 and PhD in Chemical Engineering in 2007. Her education activities were focused on the optical diagnostics of nanoparticles at internal combustion engine exhaust. From 2007 to 2011, she worked as associate researcher at Istituto Motori of the Italian National Research Council. Since 2011, she is a full-position researcher and is involved in research projects on the optical characterization of combustion process in internal combustion engines.
LM got his Mechanical Engineering MD at the University of Naples in Italy in 2006, working on fuel injection apparatuses for HD diesel engines. Since 2006, he is an associate researcher at Istituto Motori of the Italian National Research Council. His activity is mainly focused on fluid-dynamic characterization of sprays and high-injection pressure combustion by non-intrusive techniques using both mineral and biodegradable (first and second generation) fuels. He currently uses optical (Imaging, PDPA, PIV, UV–VIS chemiluminescence techniques) and non-conventional techniques (X-Rays) as well as forecast simulating codes to study mechanisms influencing the air/fuel mixture formation in internal combustion engines.
GV was born in 1957. He received his masters degree with honor in Mechanical Engineering from the University of Naples "Federico II" in 1980. He joined Istituto Motori (IM) of the National Research Council (CNR) as a researcher in 1984. In 2006, he was promoted to the rank of research manager and is currently overseeing one of the divisions of Istituto Motori focused on the mixture formation process and combustion within direct-injection engines. GV is an author of a large number of papers in qualified scientific conferences and magazines, with more than 20 years of expertise in optical diagnostics for fluid-dynamic and spray combustion research in internal combustion engines.
FEC received his degree in Electrical Engineering from the University of Naples "Federico II" in 1973. He joined Istituto Motori of the National Research Council in 1974 as a researcher. In 1991, he became research manager and, from 1997 to 2002, director of Istituto Motori. He is the chairman of the Executive Committee of International Energy Agency Implementing Agreement "Energy Conservation in Combustion and Emissions Reduction". He is the head of the Fluidynamics and Combustion Division of Istituto Motori, and he manages the research activity in thermodynamics, fluid mechanics, internal combustion engines and emission formation processes. He has published extensively in this field throughout the world (more than 270 papers). In 2000, he founded the SAE Naples Section. He is an SAE Fellow and a recipient of Distinguished Speaker of Japan Society of Automotive Engineers Award. From 2004 to 2008, he was a member of the SAE Board of Directors. In 1995 and 2008, he was a guest professor at the University of Wisconsin, Madison, USA. He was a guest professor at Wayne State University, Detriot, USA, in 2006 and at the University of Erlangen, Nurberg, Germany, in 2010.

References

1. Veloo, PS, Wang, YL, Egolfopoulos, FN, Westbrook, CK: A comparative experimental and computational study of methanol, ethanol, and n-butanol flames. Combustion and Flame **157**(10), 1989–2004 (2010)
2. Ramey, DE: Continuous two stage, dual path anaerobic fermentation of butanol and other organic solvents using two different strains of bacteria. (1998). US Patent 5753474, 19 May 1998
3. Yacoub, Y, Bara, R, Gautam, M: The performance and emission characteristics of C1–C5 alcohol–gasoline blends with matched oxygen content in a single cylinder spark ignition engine. Proc. Inst. Mech. Eng. A. – Power. Energy **212**, 363–379 (1998)
4. Gautam, M, Martin, DW: Emission characteristics of higher-alcohol/gasoline blends. Proc. Inst. Mech. Eng. A. – Power. Energy **214**, 165–182 (2000)
5. Gautam, M, Martin, DW: Combustion characteristics of higher-alcohol/ gasoline blends. Proc. Inst. Mech. Eng. **214A**, 497–511 (2000)
6. Dernotte, J, Mounaim-Rousselle, C, Halter, F, Seers, P: Evaluation of butanol–gasoline blends in a port fuel-injection, spark-ignition engine. Oil & Gas Science and Technology – Rev. IFP **65**(2), 345–351 (2010)
7. Szwaja, S., Naber, JD: Combustion of n-butanol in a spark-ignition IC engine. Fuel **89**, 1573–1582 (2010)
8. Alasfour, FN: Butanol - a single-cylinder engine study: availability analysis. Appl. Therm. Eng. Vol. **17**(6), 537–549 (1997)
9. Alasfour, FN: NOx emission from a spark-ignition engine using 30% iso-butanol – gasoline blend: Part 1: preheating inlet air. Appl. Therm. Eng. Vol. **18**(5), 245–256 (1998)
10. Alasfour, FN: NOx emission from a spark-ignition engine using 30 % iso-butanol – gasoline blend: Part 2: ignition timing. Appl. Therm. Eng. Vol. **18** (8), 609–618 (1998)
11. Broustail, G, Seers, P, Halter, F, Moréac, G, Mounaim-Rousselle, C: Experimental determination of laminar burning velocity for butanol and ethanol iso-octane blends. Fuel **90**, 1–6 (2011)
12. Yang, J, Wang, Y, Feng, R: The performance analysis of an engine fueled with butanol-gasoline blend. SAE Technical Paper (2011). doi:10.4271/2011-01-1191
13. Wigg, B, Coverdill, R, Lee, C-F, Kyritsis, D.: Emissions characteristics of neat butanol fuel using a port fuel-injected, spark-ignition engine. SAE Technical Paper (2011). doi:10.4271/2011-01-0902
14. Dagaut, P, Togbé, C: Oxidation kinetics of butanol–gasoline surrogate mixtures in a jet-stirred reactor: experimental and modeling study. Fuel **87**, 3313–3321 (2008)
15. Lapuerta, M, Martos, FJ, Cárdenas, MD: Determination of light extinction efficiency of diesel soot from smoke opacity measurements. Meas. Sci. Technol. Vol. **16**, 2048–2055 (2005)
16. Heywood, JB: Internal combustion engine fundamentals. McGraw-Hill, New York (1988)
17. Checkel, MD, Dale, JD: 1989. SAE Technical Paper (1989). doi:10.4271/ 890243
18. Brunt, MF, Pond, CR, Biundo, J: Gasoline engine knock analysis using cylinder pressure data. SAE Technical Paper (1998). doi:10.4271/980896

19. Mittal, V, Revier, BM, Heywood, JB: Phenomena that determine knock onset in spark-ignition engines. SAE Technical Paper (2007). doi:10.4271/2007-01-0007
20. Merola, SS, Sementa, P, Tornatore, C, Vaglieco, BM: Knocking diagnostics in the combustion chamber of boosted port fuel injection spark ignition optical engine. International Journal of Vehicle Design **49**(1–3), 70–90 (2009)
21. Merola, SS, Sementa, P, Tornatore, C: Experiments on knocking and abnormal combustion through optical diagnostics in a boosted spark ignition port fuel injection engine. International Journal of Automotive Technology **12**(1), 93–101 (2011)
22. Henein, NA, Tagomori, MK: Cold-start hydrocarbon emissions in port-injected gasoline engines. Progress in Energy and Combustion Science **25**, 563–593 (1999)
23. Behnia, M, Milton, BE: Fundamentals of fuel film formation and motion in SI engine induction systems. Energy Conversion and Management **42**(15–17), 1751–1768 (2001)
24. Costanzo, VS, Heywood, JB: Mixture preparation mechanisms in a port fuel injected engine. SAE Technical Paper (2005). doi:10.4271/2005-01-2080
25. Gold, MR, Arcoumanis, C, Whitelaw, JH, Gaade, J, Wallace, S: Mixture preparation strategies in an optical four-valve port-injected gasoline engine. Int. J. of Engine Research **1**(1), 41–56 (2000)
26. Nogi, T, Ohyama, Y, Yamauchi, T, Kuroiwa, H: Mixture formation of fuel injection systems in gasoline engines. SAE Technical Paper (1988). doi:10.4271/880558
27. Meyer, R, Heywood, JB: Liquid fuel transport mechanisms into the cylinder of a firing port-injected SI engine during start up. SAE Technical Paper (1997). doi:10.4271/970865
28. Merola, SS, Vaglieco, BM: Optical investigations of fuel deposition burning in ported fuel injection (PFI) spark-ignition (SI) engine. Energy **34**, 2108–2115 (2008)
29. Merola, SS, Sementa, P, Tornatore, C, Vaglieco, BM: Effect of injection phasing on valves and chamber fuel deposition burning in a PFI boosted spark-ignition engine. SAE International Journal of Fuels and Lubricants **1**(1), 192–200 (2009)
30. Merola, S, Tornatore, C, Valentino, G, Marchitto, L, Corcione, F: Optical investigation of the effect on the combustion process of butanol-gasoline blend in a PFI SI boosted engine. SAE Technical Paper (2011). doi:10.4271/2011-24-0057
31. Irimescu, A: Full load performance of a spark ignition engine fueled with gasoline -isobutanol blends. Analele Universității "Eftimie Murgu", Reşiţa **16** (1), 151–156 (2009)
32. Yang, J, Yang, X, Liu, J, Han, Z, Zhong, Z: Dyno test investigations of gasoline engine fueled with butanol-gasoline blends. SAE Paper (2009). doi:10.4271/2009-01-1891
33. Morey, F, Seers, P: Comparison of cycle-by-cycle variation of measured exhaust-gas temperature and in-cylinder pressure measurements. Applied Thermal Engineering **30**, 487–491 (2010)
34. Witze, P, Hall, M, Bennet, M: Cycle-resolved measurements of flame kernel growth and motion correlated with combustion duration. SAE Technical Paper (1990). doi:10.4271/900023
35. Zhu, GS, Reitz, RD, Xin, J, Takabayashi, T: Modelling characteristics of gasoline wall films in the intake port of port fuel injection engines. Int. J. Engine Research **2**(4), 231–248 (2001)
36. Witze, PO, Green, RM: LIF and flame-emission imaging of liquid fuel films and pool fires in an SI engine during a simulated cold start. SAE Technical Paper (1997). doi:10.4271/970866
37. Merola, SS, Sementa, P, Tornatore, C, Vaglieco, BM: Spectroscopic investigations and high resolution visualization of the combustion phenomena in a boosted PFI SI engine. SAE International Journal of Engines **2**(1), 1617–1629 (2009)
38. Gaydon, AG: The Spectroscopy of Flames. Chapman and Hall Ltd, London (1957)
39. Alkemade, CTJ, Herrmann, R: Fundamentals of Analytical Flame Spectroscopy. Hilger, Bristol (1979)
40. Dieke, GH, Crosswhite, HM: The ultraviolet bands of OH. J. Quant. Spectrosc. Radiat. Transfer **2**, 97–199 (1962)
41. Gaydon, AG, Wolfhard, HG: Mechanism of formation of CH, C2, OH and HCO radicals in flames. Symposium (International) on Combustion **4**(1), 211–218 (1953)
42. Samaniego, JM, Egolfopoulos, FN, Bowman, CT: CO2* chemiluminescence in premixed flames. Combust. Sci. and Tech. **109**, 183–203 (1995)
43. Gaydon, AG: The flame spectrum of carbon monoxide. In: (ed.) Proceedings of the Royal Society of London. Series A, Mathematical and Physical Sciences, vol. 176(967), pp. 505–521. Royal Society Publishing, London (1940)
44. Dixon, RN: In: (ed.) Proceedings of the Royal Society of London. Series A, Mathematical and Physical Sciences, vol. 275(1362), pp. 431–446. Royal Society Publishing, London (1963)
45. Kayes, D, Hochgreb, S: Mechanisms of particulate matter formation in spark-ignition engines 1. Effect of engine operating conditions. Environ. Sci. Technol. **33**, 3957–67 (1999)
46. Nagle, J, Strickland-Constable, RF: Oxidation of carbon between 1000°-2000°C. In: (ed.) Proceedings of the 5th Conference on Carbon, vol. 1, pp. 154–164. Pergamon Press, Oxford (1962)
47. Orphala, J, Chanceb, K: Ultraviolet and visible absorption cross-sections for HITRAN. Journal of Quantitative Spectroscopy & Radiative Transfer **82**, 491–504 (2003)
48. Verstraete, L, Leger, A: The visible and ultraviolet absorption of large polycyclic aromatic hydrocarbons. Astronomy and Astrophysics **266**(1), 513–519 (1992)
49. Chappuis, S, Cousyn, B, Posylkin, M, Vannobel, F, Whitelaw, JH: Effects of injection timing on performance and droplet characteristics of a sixteen-valve four cylinder engine. Experiments in Fluids **22**(4), 336–344 (1997)
50. Gu, X, Huang, Z, Cai, J, Gong, J, Wu, X, Lee, C: Emission characteristics of a spark-ignition engine fuelled with gasoline-n-butanol blends in combination with EGR. Fuel **93**, 611–617 (2012)

Advanced exergy-based analyses applied to a system including LNG regasification and electricity generation

Tatiana Morosuk*, George Tsatsaronis, Alicia Boyano and Camilo Gantiva

Abstract

Liquefied natural gas (LNG) will contribute more in the future than in the past to the overall energy supply in the world. The paper discusses the application of advanced exergy-based analyses to a recently developed LNG-based cogeneration system. These analyses include advanced exergetic, advanced exergoeconomic, and advanced exergoenvironmental analyses in which thermodynamic inefficiencies (exergy destruction), costs, and environmental impacts have been split into avoidable and unavoidable parts. With the aid of these analyses, the potentials for improving the thermodynamic efficiency and for reducing the overall cost and the overall environmental impact are revealed. The objectives of this paper are to demonstrate (a) the potential for generating electricity while regasifying LNG and (b) some of the capabilities associated with advanced exergy-based methods. The most important subsystems and components are identified, and suggestions for improving them are made.

Keywords: LNG, Exergy analysis, Advanced exergy analysis, Exergoeconomics, Advanced exergoeconomic analysis, Exergoenvironmental analysis, Advanced exergoenvironmental analysis

Background

Several concepts of a system for generating electricity while vaporizing liquefied natural gas (LNG) have been developed by Griepentrog et al. [1,2]. These concepts have some thermodynamic and economic advantages over systems proposed in the past. A detailed discussion of the advantages and disadvantages of these concepts has been given in [3].

In this paper, advanced exergy-based methods, including advanced exergetic and exergoenvironmental analyses, are applied to the base case of the LNG regasification system. The exergy destruction within components as well as the cost and environmental impact associated with each component is split into avoidable and unavoidable parts to help engineers identify the potential for improvement from the viewpoints of thermodynamics, cost, and environmental impact.

* Correspondence: morozyuk@iet.tu-berlin.de
Technische Universität Berlin, Berlin, 10587, Germany

Methods

Simulation

The base case is shown in Figure 1. The overall system consists of three subsystems with the following initial data and assumptions:

- LNG subsystem (process 1–2 – 3–4) - LNG from the storage system is (a) compressed by an LNG pump (P), (b) vaporized in heat exchanger II (HE II) using the waste heat from the nitrogen power system, and (c) expanded in expander III (EX III).
- N2 subsystem (process 11–12 – 13–14) - The N2 subsystem is a closed-cycle gas-turbine power system. After being cooled in HE II, the nitrogen is compressed in compressor III (CM III), heated in heat exchanger I (HE I) using the waste heat from an open gas-turbine power system, and expanded in expander II (EX II).
- Open gas-turbine power subsystem (process 21 through 28) - Air after compression in compressor I (CM I) is cooled in the cooler (CL) transferring thermal energy to the environment and is compressed in compressor II (CM II). After the combustion process in the combustion chamber

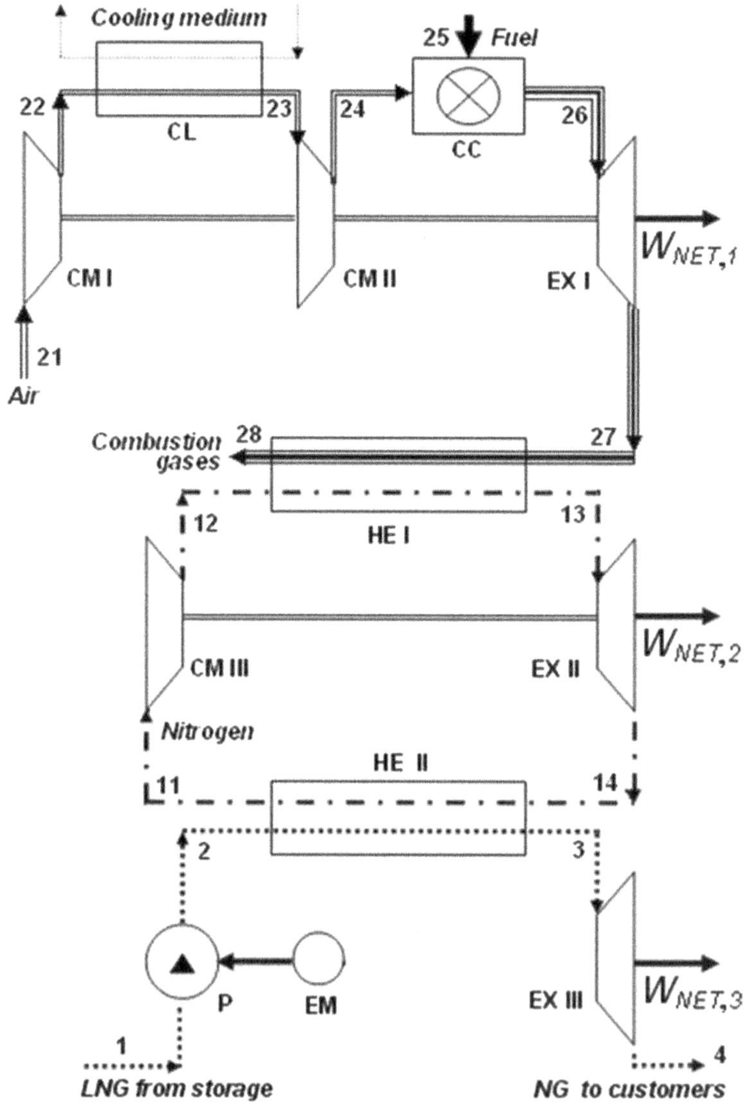

Figure 1 Schematic of the cogeneration system for vaporizing LNG.

(CC), the combustion gases are expanded in expander I (EX I) and rejected to the atmosphere after being cooled in HE I. The open gas-turbine power subsystem is based on an LMS 100 gas turbine (General Electric Company, Fairfield, CT, USA) [4].

For the simulation and the exergetic analyses, the softwares GateCycle (General Electric Company, Fairfield, CT, USA) [5], Gatex (Institut für Energietechnik, Technische Universität Berlin, Germany) [6], and EES (F-Chart Software, LLC, Madison, WI, USA) [7] were used. Table 1 presents some results obtained from the simulation [3]. The following assumptions were used: η_{CMI} = 90 %, η_{CMII} = 90 %, $p24/p21$ = 42, T_{26} = 1,290°C, $\eta_{EX\ I}$ = 94 %, $\Delta T_{HEI}^{min} = 20$ K, Δp_{HEI} = 3 %, $\eta_{CM\ III}$ = 85 %, $p12/p11$ = 15, $\eta_{EX\ II}$ = 88 %, $\Delta T_{HEI}^{min} = 15$ K, Δp_{HEII} = 3 %, η_P = 66.5 %, and $\eta_{EX\ III}$ = 85 %.

Exergy-based analyses

Exergoeconomics is a unique combination of exergy analysis and cost analysis conducted at the component level to provide the designer or operator of an energy conversion system with information crucial to the design or operation of a cost-effective system. Decisions are made, however, at the plant component level [8,9]. A complete exergoeconomic analysis consists of (a) an exergetic analysis, (b) an economic analysis, and (c) an exergoeconomic evaluation.

Table 1 Thermodynamic data for the material streams in the base case

Material stream	State	$\dot{m}$(kg/s)	T(°C)	p(bar)	e^T(kJ/kg)	e^M(kJ/kg)	e^{PH}(kJ/kg)	e^{CH}(kJ/kg)	e(kJ/kg)
LNG	1	65.03	−160	10	669.7	339.3	1,009	-	1,009
LNG	2	65.03	−144	272	250.5	778.5	1,029	-	1,029
NG	3	65.03	86	270	25.3	777.7	803	-	803
NG	4	65.03	2	80	1.0	630.0	631	-	631
N_2	11	217	−129	2.85	58.6	88.4	147	-	147
N_2	12	217	70	42.75	5.0	319.0	324	-	324
N_2	13	217	415	40.61	162.2	314.8	477	-	477
N_2	14	217	101	2.99	11.4	92.6	104	-	104
Air	21	209	15	1.013	-	-	0	1	1
Air	22	209	242	6.66	-	-	217	1	218
Air	23	209	117	6.53	-	-	170	1	171
Air	24	209	416	43.47	-	-	468	1	469
CH_4	25	5.1	15	45	-	-	566	51,534	52,100
Combustion gases	26	214.1	1,290	41.95	-	-	1,281	9	1,290
Combustion gases	27	214.1	435	1.08	-	-	190	9	199
Combustion gases	28	214.1	90	1.025	-	-	18	9	27

An *exergoenvironmental analysis* is considered as one of the most promising tools to evaluate energy conversion processes from an environmental point of view [10]. Exergoenvironmental analysis is a proper combination of exergy analysis and life cycle assessment (LCA). The exergoenvironmental analysis consists of three steps: The first step is an exergy analysis. In the second step, an LCA of (a) each relevant system component and (b) all relevant input streams to the overall system is carried out. In the last step, the environmental impact obtained from the LCA is assigned to the exergy streams in the system.

Exergy analysis

Table 1 shows the value of exergy for each material stream. The chemical exergies for material streams of the N_2 and the LNG subsystems do not need to be considered in the exergetic analysis because only the physical exergy of the working fluid is used in the corresponding subsystems. We considered the chemical exergies only in the open gas-turbine subsystem, where combustion takes place. The physical exergies of LNG, NG, and N_2 are split into their thermal and mechanical parts according to the approach presented in [11].

The exergetic analysis has been conducted at the component level using the 'exergy of the fuel' and the 'exergy of the product' [8,9]. The definitions of $\dot{E}_{F,k}$ and $\dot{E}_{P,k}$ for each system component are given in [3]. Table 2 shows some data obtained from the conventional exergetic analysis.

Exergoeconomic analysis

The exergoeconomic model for an energy conversion system consists of *cost balances* written for the *k*th component and auxiliary equations based on the P and the F rules [8,9]. The cost balances can be written as

$$\dot{C}_{P,k} = \dot{C}_{F,k} + \dot{Z}_k \tag{1a}$$

or

$$c_{P,k}\dot{E}_{P,k} = c_{F,k}\dot{E}_{F,k} + \dot{Z}_k, \tag{1b}$$

where

$$\dot{Z}_k = \dot{Z}_k^{CI} + \dot{Z}_k^{OM}. \tag{2}$$

Table 2 Results obtained from the conventional exergetic analysis [3]

Component	$\dot{E}_{F,k}$(MW)	$\dot{E}_{P,k}$(MW)	$\dot{E}_{D,k}$(MW)	ε_k(%)
CM I	48.418	45.434	2.984	93.84
CL	Dissipative component		9.993	-
CM II	65.224	62.458	2.766	95.76
CC	267.603	178.131	89.466	66.57
EX I	233.661	227.569	6.092	97.39
HE I	36.926	33.091	3.834	89.83
CM III	57.021	51.222	5.799	89.83
EX II	80.956	72.829	8.127	89.96
HE II	19.722	14.427	5.295	73.15
P	34.295	28.582	5.713	83.34
EX III	11.249	9.479	1.770	84.26
Overall system ($\dot{E}_{L,tot}$ = 5.768 MW)	311.415	163.801	141.846	52.60

To simplify the discussion, we assume that the contribution of $\dot{Z}_k^{OM}$ remains constant when the design changes, and therefore, the changes in the value of $\dot{Z}_k$ are associated only with changes in the capital investment cost $\dot{Z}_k^{CI}$.

The real cost sources in an energy conversion system are the (a) capital investment (and operating maintenance expenses) for each component, (b) cost of exergy destruction within each component, and (c) cost of exergy loss from the overall system. The last two terms can be revealed only through an exergoeconomic analysis:

- The cost rate associated with exergy destruction within the kth component is

$$\dot{C}_{D,k} = c_{F,k} \cdot \dot{E}_{D,k}. \quad (3)$$

- The cost rate associated with exergy loss from the overall system is

$$\dot{C}_{L,tot} = \dot{C}_{28}. \quad (4)$$

The exergoeconomic model for the base case has been discussed in detail in [12]. Table 3 shows selected data obtained from the conventional exergoeconomic analysis. Here, the cooler is considered together with the cooling tower ($\dot{Z}_{CT} = 3.25$ \$/h), and the pump is considered together with the electrical motor ($\dot{Z}_{EM} = 1.33$ \$/h).

For the economic analysis, the methodology presented in [8] is applied using the following assumptions and sources:

- The purchased equipment cost of turbomachinery is based on data from [8,13].
- The purchased equipment cost of heat exchangers is based on data from [8].
- The cost of LNG is equal to \$12/GJ [14].
- The average cost of money is i_{eff} = 10 %.

Table 3 Exergoeconomic variables for the LNG-based cogeneration systems

Component	$\dot{Z}_k$(\$/h)	$\dot{C}_{D,k}$(\$/h)	$\dot{Z}_k + \dot{C}_{D,k}$(\$/h)	$c_{F,k}$(\$/GJ)
CM I	64.67	554	619	51.63
CL	11.10	Dissipative component		
CM II	97.01	514	611	51.63
CC	92.39	9,493	9,585	29.47
EX I	207.90	1,097	1,305	50.04
HE I	16.04	691	707	50.04
CM III	16.23	3,007	3,023	144.00
EX II	20.49	4,133	4,154	14.13
HE II	13.76	920	934	48.26
P	7.01	783	788	38.06
EX III	2.65	432	435	67.76

- The plant economic life is n = 15 years with 7,300 h/year.
- The average general inflation rate is r_n = 2.5 %

Exergoenvironmental analysis

Exergy analysis provides a powerful tool for assessing the quality of a resource as well as the location, magnitude, and causes of thermodynamic inefficiencies. In addition, LCA supplies the environmental impacts associated with a component or an overall system during its entire useful life. In the exergoenvironmental analysis, the environmental impacts obtained by LCA are apportioned to the exergy streams pointing out the main system components with the highest environmental impact and possible improvements associated with these components. Finally, exergoenvironmental variables are calculated, and an exergoenvironmental evaluation is carried out.

Life cycle assessment is a technique for assessing the environmental aspects associated with a product over its life cycle. The LCA process consists of goal definition and scoping (defining the system under consideration), inventory analysis (identifying and quantifying the consumption and release of materials), and interpretation (evaluation of the results) [15].

In general, any of recently introduced indicators can be used for LCA. For this exergoenvironmental analysis, an impact analysis method called Eco-indicator 99 [16] has been selected because it considers many environmental aspects and uses average European data.

In order to identify the raw material inlet flows, it is first necessary to perform a sizing of the plant components and to collect information about the weights, main materials, production processes, and scrap outputs of all relevant pieces of equipment needed to build the plant. This information is usually not very widely published (compared with the corresponding cost information). In this way, only rough calculations of the employed main materials and corresponding weights can be conducted.

The data collected in [17,18] were generalized in the form of equations (Tables 4 and 5) and used for estimating the component-related environmental impact that occurs during the construction phase. If the materials of a component correspond to the data given in Table 4, then the values of $\bar{b}$ (relative environmental impact) and $\bar{m}$ (relative mass) are equal to 1. If the selected material is different, then $\bar{b} = \frac{b_{\text{given material}}}{b_{\text{new material}}}$ and $\bar{m} = \frac{\rho_{\text{given material}}}{\rho_{\text{new material}}}$ (where ρ is the density of the material, kg/m^3).

For the LCA of the system being analyzed, we assumed in analogy with the economic analysis a life time of 15 years and 7,300 working hours per year at full capacity.

The exergoenvironmental model for an energy conversion system consists of *environmental impact balances* written for the kth component and auxiliary equations

Table 4 Eco-indicator 99 values and material composition of components

Equipment	Construction material	Eco-indicator 99 (mPoints/kg)	Material composition (% *w/w*)
CM I, CM II	Steel	86	33.33
	Steel low alloy	110	44.45
	Cast iron	240	22.22
CL	Steel	86	100
CT	Concrete	3.8	91.00
	PVC	280	9.00
CC	Steel	86	33.34
	Steel high alloy	910	66.66
EX I, EX II, EX III	Steel	86	25.00
	Steel high alloy	910	75.00
HE I, HE II	Steel	86	25.00
	Steel low alloy	110	75.00
CM III	Steel	86	33.33
	Steel low alloy	110	44.45
	Cast iron	240	22.22
P	Steel	86	35.00
	Cast iron	240	65.00

based on the P and F rules [10]. The environmental impact balances can be written as

$$\dot{B}_{\mathrm{P},k} = \dot{B}_{\mathrm{F},k} + \left(\dot{Y}_k + \dot{B}_k^{\mathrm{PF}}\right) \tag{5a}$$

or

$$b_{\mathrm{P},k}\dot{E}_{\mathrm{P},k} = b_{\mathrm{F},k}\dot{E}_{\mathrm{F},k} + \left(\dot{Y}_k + \dot{B}_k^{\mathrm{PF}}\right), \tag{5b}$$

where $\dot{Y}_k$ is the environmental impacts that occur during the three life-cycle phases: Construction $\dot{Y}_k^{\mathrm{CO}}$, operation and maintenance, $\dot{Y}_k^{\mathrm{OM}}$, and disposal $\dot{Y}_k^{\mathrm{DI}}$ constitute the component-related environmental impact associated with the *k*th component $\dot{Y}_k$:

$$\dot{Y}_k = \dot{Y}_k^{\mathrm{CO}} + \dot{Y}_k^{\mathrm{OM}} + \dot{Y}_k^{\mathrm{DI}}. \tag{6}$$

To simplify the discussion, we assume in this paper that the value of $\dot{Y}_k$ is mainly associated with $\dot{Y}_k^{\mathrm{CO}}$.

To account for *pollutant formation* within the *k*th component, a new variable was recently introduced $\dot{B}_k^{\mathrm{PF}}$ [19,20]. This term $\dot{B}_k^{\mathrm{PF}}$ is zero if no pollutants are formed within a process, i.e., for processes without a chemical reaction (compression, expansion, heat transfer, etc.). For components, where chemical reactions occur (combustion, for example), the value of $\dot{B}_k^{\mathrm{PF}}$ is

$$\dot{B}_k^{\mathrm{PF}} = \sum_i b_i^{\mathrm{PF}}\left(\dot{m}_{i,\mathrm{out}} - \dot{m}_{i,\mathrm{in}}\right), \tag{7}$$

where only pollutant streams which will finally be emitted to the environment are taken into account: CO, CO_2, CH_4, N_2O, NO_x, and SO_x [10].

The environmental impact of exergy destruction $\dot{B}_{\mathrm{D},k}$ identifies the environmental impact due to the exergy destruction within the *k*th component [10]:

$$\dot{B}_{\mathrm{D},k} = b_{\mathrm{F},k}\dot{E}_{\mathrm{D},k}. \tag{8}$$

To identify the most important components from the viewpoint of formation of environmental impacts, the

Table 5 Environmental impact functions of components for the construction phase

Component	Environmental impact function, Y_k(Pts of Eco-indicator 99)	Variables
CM 1, CM 2, CM 3	$Y_{CM} = (0.0039 \cdot \dot{W}_{CM}^2 - 2,070.68 \cdot \dot{W}_{CM} + 2.62 \cdot 10^6) \times (0.3333 \cdot \bar{b}_S \cdot \bar{m}_S + 0.4445 \cdot \bar{b}_{SLA} \cdot \bar{m}_{SLA} + 0.2222 \cdot \bar{b}_{CI} \cdot \bar{m}_{CI})$	$\dot{W}_{CM}$ (kW)
CL	$Y_{CL} = (0.703 \cdot A + 12.03) \cdot \bar{b}_S \cdot \bar{m}_S$	Heat exchange area, A (m^2)
CT	$Y_{CT} = 3098 \cdot \dot{m}_{water} \cdot \left(0.09 \cdot \bar{b}_{PVC}^{fill} \cdot \bar{m}_{PVC}^{fill} + 0.91 \cdot \bar{b}_{CON} \cdot \bar{m}_{CON}\right)$	$\dot{m}_{water}$ (m^3/h)
CC	$Y_{CC} = \left(-0.2965 \cdot \dot{m}_{\underset{gas}{comb.}}^2 + 454.12 \cdot \dot{m}_{\underset{gas}{comb.}} - 60909.11\right) \times (0.3334 \cdot \bar{b}_S \cdot \bar{m}_S + 0.6666 \cdot \bar{b}_{SHA} \cdot \bar{m}_{SHA})$	$\dot{m}_{\underset{gas}{comb.}}$ (kg/s)
EX I, EX II, EX III	$Y_{EX} = (0.2457 \cdot \dot{W}_{EX} + 15,538)(0.25 \cdot \bar{b}_S \cdot \bar{m}_S + 0.75 \cdot \bar{b}_{SHA} \cdot \bar{m}_{SHA})$	$\dot{W}_{EX}$ (kW)
HE I, HE II	$Y_{HE} = (11.28 \cdot A + 119.07) \times \left(0.75 \cdot \bar{b}_{SLA}^{tubes} \cdot \bar{m}_{SLA}^{tubes} + 0.25 \cdot \bar{b}_S^{casing} \cdot \bar{m}_S^{casing}\right)$	Tubes of SLA Casing of steel Heat exchange area, A (m^2)
P	$Y_P = \left(0.0016 \cdot \dot{W}_P^2 - 0.5559 \cdot \dot{W}_P + 12.66\right) \cdot \bar{b}_{CI} \cdot \bar{m}_{CI}$ (for $p_{discharge} > 5$ bar)	$\dot{W}_P$ (kW)

sum of environmental impacts $\left(\dot{Y}_k + \dot{B}_k^{PF} + \dot{B}_{D,k}\right)$ is used.

The detailed exergoenvironmental model for the LNG-based cogeneration system (Figure 1) will be presented in a future publication. In this paper, some data obtained from the conventional exergoenvironmental analysis are given in Table 6. Here, the cooler is considered together with the cooling tower ($\dot{Y}_{CT} = 0.065\text{Pts/h}$), and the pump is considered together with the electrical motor, $\dot{Y}_{EM} = 0.363\text{Pts/h}$.

Advanced exergy-based analyses

The real potential for improving the system from a thermodynamic, economic, and environmental impact point of view can be estimated when the following are split into avoidable/unavoidable parts [17,21-23]:

- exergy destruction within each (important) system component,
- investment cost and environmental impact associated with such component, and
- cost of exergy destruction and environmental impact associated with the exergy destruction for each (important) system component.

The *unavoidable* exergy destruction cannot be further reduced due to technological limitations such as availability and cost of materials and manufacturing methods. The difference between total and unavoidable exergy destruction for a component is the *avoidable* exergy destruction. Only this value and not the total exergy destruction should be considered during the improvement procedure.

$$\dot{E}_{D,k}^{AV} = \dot{E}_{D,k} - \dot{E}_{D,k}^{UN} \tag{9}$$

The unavoidable investment cost ($\dot{Z}_k^{UN}$) and unavoidable component-related environmental impact ($\dot{Y}_k^{UN}$) for a component can be calculated by assuming the minimum values of $\left(\frac{\dot{Z}_k}{\dot{E}_{P,k}}\right)^{UN}$ and $\left(\frac{\dot{Y}_k}{\dot{E}_{P,k}}\right)^{UN}$, respectively. These values will always be exceeded as long as such a component is used in a real system. The avoidable investment cost and component-related environmental impact are the differences between the total value and unavoidable part of this variable, i.e.,

$$\dot{Z}_k^{AV} = \dot{Z}_k - \dot{Z}_k^{UN} \tag{10}$$

$$\dot{Y}_k^{AV} = \dot{Y}_k - \dot{Y}_k^{UN}. \tag{11}$$

The value of the unavoidable exergy destruction within the kth component is calculated using the ratio $\left(\frac{\dot{E}_{D,k}}{\dot{E}_{P,k}}\right)^{UN}$ that refers to the case where only unavoidable exergy destruction occurs:

$$\dot{E}_{D,k}^{UN} = \dot{E}_{P,k}^{real}\left(\frac{\dot{E}_{D,k}}{\dot{E}_{P,k}}\right)^{UN}. \tag{12}$$

The values of unavoidable capital investment cost and component-related environmental impact can be calculated using similar equations:

$$\dot{Z}_k^{UN} = \dot{E}_{P,k}^{real}\left(\frac{\dot{Z}_k}{\dot{E}_{P,k}}\right)^{UN} \tag{13}$$

$$\dot{Y}_k^{UN} = \dot{E}_{P,k}^{real}\left(\frac{\dot{Y}_k}{\dot{E}_{P,k}}\right)^{UN}. \tag{14}$$

The approaches for estimating the values of $\left(\frac{\dot{E}_D}{\dot{E}_P}\right)_k^{UN}$, $\left(\frac{\dot{Z}}{\dot{E}_P}\right)_k^{UN}$, and $\left(\frac{\dot{Y}}{\dot{E}_P}\right)_k^{UN}$ are given in [3,12].

Selected data obtained from the advanced exergy-based analyses for the LNG regasification system are given in Table 7. In this paper, we assumed that components of the open gas-turbine subsystem cannot be improved because

Table 6 Exergoenvironmental variables for the LNG-based cogeneration systems

Component	$\dot{Y}_k$(Pts/h)	$\dot{B}_{D,k}$(Pts/h)	$\dot{B}_k^{PF}$ (Pts/h)	$\dot{Y}_k + \dot{B}_{D,k} + \dot{B}_k^{PF}$ (Pts/h)	$b_{F,k}$ (Pts/GJ)
CM I	1.254	90.718	-	91.972	8.443
CL	0.090		Dissipative component		
CM II	1.054	84.085	-	85.139	8.443
CC	1.200	1,128.000	1,345.320	2,474.520	3.501
EX I	4.175	180.241	-	184.416	8.218
HE I	19.732	114.450	-	134.182	8.218
CM III	1.307	185.565	-	186.871	8.880
EX II	1.953	283.449	-	285.402	9.678
HE II	6.828	31.053	-	37.881	1.643
P	1.454	1.913	-	3.367	0.114
EX III	1.043	2.650	-	3.693	0.416

Table 7 Selected data obtained from the advanced exergy-based analyses

Component	$\left(\frac{\dot{E}_D}{\dot{E}_P}\right)_k^{UN}$	$\dot{E}_{D,k}^{UN}$ (MW)	$\dot{E}_{D,k}^{AV}$ (MW)	$\left(\frac{\dot{Z}}{\dot{E}_P}\right)_k^{UN}$ ($/MJ)	$\dot{Z}_k^{UN}$ ($/h)	$\dot{Z}_k^{AV}$ ($/h)	$\dot{C}_{D,k}^{UN}$ ($/h)	$\dot{C}_{D,k}^{AV}$ ($/h)	$\dot{Z}_k^{AV}+\dot{C}_{D,k}^{AV}$ ($/h)	$\left(\frac{\dot{Y}}{\dot{E}_P}\right)_k^{UN}$ (Pts/MJ)	$\dot{Y}_k^{UN}$ (Pts/h)	$\dot{Y}_k^{AV}$ (Pts/h)	$\dot{B}_{D,k}^{UN}$ (Pts/h)	$\dot{B}_{D,k}^{AV}$ (Pts/h)	$\dot{B}_{D,k}^{AV}+\dot{Y}_k^{AV}$ (Pts/h)
HE I	0.0451	1.492 (39%)	2.342 (61%)	0.231	7.66 (48%)	8.38 (52%)	269	422	430(61%)	0.3290	10.879 (55%)	8.835 (45%)	44.140	70.309	79.144 (58%)
CM III	0.0593	3.037 (52%)	2.762 (48%)	0.098	5.03 (31%)	11.20 (69%)	1,574	1,433	1,444(48%)	0.0063	0.323 (25%)	0.984 (75 %)	97.087	88.478	89.462 (48%)
EX II	0.0511	3.722 (46%)	4.405 (54%)	0.114	8.29 (40%)	12.20 (60%)	1,901	2,232	2,244(54%)	0.0059	0.433 (23%)	1.520 (77%)	129.677	153.772	155.292 (54%)
HE II	0.0973	1.404 (27%)	3.891 (73%)	0.469	6.77 (49%)	6.99 (51%)	244	676	623(73%)	0.0740	1.071 (16%)	5.757 (84%)	8.304	22.748	28.505 (75%)
P	0.0874	2.498 (44%)	3.125 (56%)	0.082	2.33 (33%)	3.35 (67%)	342	441	444(56 %)	0.0155	0.444 (30%)	1.010 (70%)	0.842	1.071	2.081 (62%)
EX III	0.0913	0.865 (49%)	0.905 (51%)	0.126	1.20 (45%)	1.45 (55%)	211	221	222(51%)	0.0240	0.228 (22%)	0.815 (78%)	1.298	1.351	2.166 (59%)

this subsystem represents an already commercially available unit.

Results and discussion

Conventional exergy based analyses

The conclusions which can be obtained from the conventional exergetic analysis of the N_2 and LNG subsystems are based on the values of $\dot{E}_{D,k}$ (Table 2). The components with the highest potential for improvement are EX II, CM III, and HE II.

The economic analysis (value $\dot{Z}_k$ in Table 3) shows that EX II is the most expensive component among the N2 and LNG subsystems followed by HE I and CM III.

The results from the exergoeconomic analysis (values of $\dot{Z}_k + \dot{C}_{D,k}$ in Table 3) show that EX II and CM III are by far the most important components from the economic viewpoint and that the high costs ($\dot{Z}_k + \dot{C}_{D,k}$) for these components are caused primarily by the exergy destruction ($\dot{C}_{D,k}$). This demonstrates (a) the importance of the N_2 subsystem for the economics of the overall system and (b) the necessity to keep the thermodynamic inefficiencies occurring within the N_2 subsystems to a minimum.

The results obtained from the LCA (value $\dot{Y}_k$ in Table 6) demonstrate that the component-related environmental impact associated with HE I and HE II is the highest among all components of the overall system.

The exergoenvironmental analysis (value $\dot{B}_{D,k}$ in Table 6) shows that for all components, the environmental impact associated with the exergy destruction is much higher than the component-related environmental impact ($\dot{B}_{D,k}+\dot{Y}_k$). Only for the pump and EX III are these values comparable. Based on the sum ($\dot{Y}_k + \dot{B}_{D,k} + \dot{B}_k^{PF}$), the most important components are again EX II and CM III. They can be improved by decreasing the exergy destruction and, therefore, decreasing the environmental impact associated with the exergy destruction.

The conventional exergy-based analyses suggest to initially decrease the exergy destruction within the N_2 subsystem and mainly within EX II and CM III. This decrease of exergy destruction will not only increase the overall efficiency, but will also simultaneously reduce both costs and environmental impact associated with the overall system.

Advanced exergy-based analyses

The advanced exergy-based analyses (results shown in Table 7) refine and correct the results from the conventional analyses. From the thermodynamic point of view, for example, CM III does not have the importance that the conventional exergetic analysis suggests because most of the exergy destruction in CM III is unavoidable. HE II and P are thermodynamically more important than CM III when only avoidable exergy destruction within each component is considered. Thus, improvement efforts should focus more on EX II, HE II, and P (where the potential for improvement is higher) than in CM III.

However, from the cost viewpoint, CM III is much more important than HE II or P because the cost per unit of exergy supplied to the compressor (cost of fuel) has the highest value among all components (see $c_{F,k}$ values in Table 3). The highest potential for reducing the cost of the overall product is still associated with EX II and CM III that exhibit the highest value of $\dot{C}_{D,k}^{AV} + \dot{Z}_k^{AV}$.

The values of $\dot{B}_{D,k}^{AV} + \dot{Y}_k^{AV}$ indicate that the highest potential for reducing the overall environmental impact is associated with EX II, CM III, and HE I. Thus, the advanced exergoenvironmental analysis emphasizes the importance of component HE I compared with the conventional exergoenvironmental analysis.

The advanced analyses confirm the conclusion from the conventional analyses that by decreasing the exergy destruction within the N_2 subsystem, the efficiency of the overall system would increase while the cost and the environmental impacts would decrease.

Conclusions

The present work identified the importance of the N_2 subsystem in improving the overall system and demonstrated the advantages of splitting thermodynamic inefficiencies, cost, and environmental impacts into unavoidable and avoidable parts.

Results show that efforts should focus on EXII, HEII, and P in order to improve the thermodynamic efficiency and reduce the environmental impact. To improve the cost effectiveness, effort should focus on EXII and CMIII. Thus, EX II is the most important system component regardless of the viewpoint of the analyst.

Even more accurate information is obtained when these variables are split into their endogenous and exogenous parts because, then, the interactions among components become transparent. The results from complete advanced exergy-based analyses will be presented in subsequent publications.

Nomenclature

b specific environmental impact per unit of exergy (Pts/J) or per unit of mass (Pts/kg)
$\dot{B}$ environmental impact rate associated with exergy (Pts/s)
C cost associated with an exergy stream ($)
c cost per unit of exergy ($/J)
$\dot{E}$ exergy rate (J)
e specific exergy (J/kg)
k kth component
$\dot{m}$ mass flow rate (kg/s)
p pressure (Pa)
$\dot{Q}$ heat rate (W)
T temperature (K)
$\dot{W}$ power (W)
$\dot{Y}$ component-related environmental impact rate (Pts/s)
$\dot{Z}$ cost rate associated with investment expenditures ($)

Greek symbols

ε exergetic efficiency (%)
η isentropic efficiency (%)

Superscripts

AV avoidable
CH chemical
CT cooling tower
M mechanical
PF pollutant formation
PH physical
T thermal
UN unavoidable

Subscripts

PI cast iron
CON concrete
D exergy destruction
F exergy of fuel
k kth component
L exergy loss
P exergy of product
PVC polyvinylchlorid
S steel
SHA high alloy steel
SLA low alloy steel
tot overall system
0 thermodynamic environment (reference state)

Abbreviations

CC, Combustion chamber; CL, Cooler; CM, Compressor; CT, Cooling tower; EM, Electrical motor; EX, Expander; HE, Heat exchanger; P, Pump.

Competing interests

The authors declare that they have no competing interests.

Acknowledgments

The authors would like to thank Professor Hartmut Griepentrog (Greif-Foundation, Gelsenkirchen, Germany) for developing the base case and for many helpful discussions regarding gas-turbine power systems and LNG vaporization.

Authors' contributions

TM co-supervised the work, conducted the exergy analysis, and drafted the manuscript. GT co-supervised the work, conducted the economic analysis, and corrected the manuscript. AB conducted the life cycle assessment. CG conducted the calculations according to the instructions provided by the supervisors. All authors read and approved the final manuscript.

Authors' information

Professor TM is employed at the Technische Universität Berlin. She studied refrigeration engineering at the Odessa State Academy of Refrigeration, Ukraine and received her diploma in 1990. She received her Ph.D. in 1994 and professorship in 2001, all in Ukraine. Her field of interest is the application of modern exergy-based methods to the analysis and improvement of power plants and refrigeration systems. She is the author or co-author of six books and more than 200 publications and has 10 patents.

GT is the Bewag Professor of Energy Conversion and Protection of the Environment at Technische Universität Berlin. He received a diploma in Mechanical Engineering (NTU Athens, Greece), and his MBA, Ph.D. in Combustion, and Dr. Habilitatus Degree in Thermoeconomics, all from the RWTH Aachen, Germany. His areas of interest include the design, development, analysis, and optimization of energy conversion systems. He contributed significantly to the fundamentals of exergoeconomics, a term that he coined in 1984. He co-authored the book *Thermal Design and Optimisation* (Wiley, 1996), has published more than 270 papers, received several international awards and recognitions, and has served as chairman or co-chairman of 20 international conferences.

AB received her Ph.D. in Chemical Engineering and Environmental Technologies in the University of Zaragoza (Spain). Since 2010, she is a researcher in the Joint Research Centre of Institute of Prospective Technological Studies of the European Commission in the area of sustainability. She gained experiences with the exergoeconomic and exergoenvironmetal analyses in Technische Universität Berlin (post-doc during the years 2008 to 2009). She is a co-author of more than 40 papers published in the International Journals and Proceedings of the International conferences.

CG is a Chemical Engineer (2004, Universidad Nacional de Colombia) and M. Sc. in Process Energy Environmental Systems Engineering (2011 - Technische Universität Berlin, Germany).

References

1. Griepentrog, H, Tsatsaronis, G, Morosuk, T: A novel concept for generating electricity and vaporizing LNG. In: Ziebik A, Kolenda Z., Stanek W (eds) Proceedings of the 21th International Conference on Efficiency, Cost, Optimization, Simulation and Environmental Impact of Energy Systems, Cracow-Gliwice (2008) .
2. Griepentrog, H, Tsatsaronis, G, Morosuk, T: LNG vaporization using a novel co-generation system. Proceedings of the ASME International Mechanical Engineering Congress and Exposition, Boston, In (2008)
3. Tsatsaronis, G, Morosuk, T: Advanced exergetic analysis of a novel system for generating electricity and vaporizing liquefied natural gas. Energy **35**, 820–829 (2010)
4. Reale, MJ: New high efficiency simple cycle gas turbine – GE's LMS100™, General Electric Company, USA, 2004. www.ge.com. Accessed 26 February 2007
5. GateCycle for Windows, Version 5.52.0.r., The General Electric Company, 1989–2004. http://www.gepower.com/enter. Accessed 1 March 2007
6. Gatex (Technische Universität Berlin, Institut für Energietechnik), 2002. Ref: Eisermann W, Hasberg W, Tsatsaronis G. THESIS - Ein Rechenprogramm zur Simulation und Entwicklung von Energieumwandlungsanlagen, Brennst.-Wärme-Kraft, 1984, Vol. 36(1–2), pp. 45–51.
7. EES (Engineering Equation Solver), 1992–2010, V7.847, #92: McGraw-Hill. www.fchart.com. Accessed 01 September 2011
8. Bejan, A, Tsatsaronis, G, Moran, M: Thermal Design and Optimization. Wiley, New York (1996)
9. Lazzaretto, A, Tsatsaronis, G: SPECO: a systematic and general methodology for calculating efficiencies and costs in thermal systems. Energy **31**(8–9), 1257–1289 (2006)
10. Meyer, L, Tsatsaronis, G, Buchgeister, J, Schebek, L: Exergoenvironmental analysis for evaluation of the environmental impact of energy conversion systems. Energy Int J. **34**, 75–89 (2009)
11. Morosuk, T, Tsatsaronis, G: Graphical models for splitting physical exergy. In: Kjelstrup, S, Hustad, JE, Gundersen, T, Rosjorde, A, Tsatsaronis, G (eds.) Shaping Our Future Energy Systems, pp. 377–384. Tapir academic press, Trondheim (2005)
12. Tsatsaronis, G, Morosuk, T, Cziesla, F: LNG-based cogeneration systems. Part 2. Advanced exergy-based analyses of a concept. Paper presented at the ASME 2009 International Mechanical Engineering Congress & Exposition (IMECE2009), Lake Buena Vista, Florida, 13–19 November 2009, IMECE2009-10460
13. Pequot: Gas Turbine World Handbook. Southport (2006).
14. US Department of Energy. www.eia.doe.gov. Accessed 01 September 2011
15. International Organization for Standardization (ISO): Environmental management—life cycle assessment. European Standard ENISO14040 and 14044, Geneva (2006)
16. Goedkoop, M, Spriensma, R: The Eco-indicator 99: a damage oriented method for life cycle impact assessment. Methodology Report. Amersfoort, Netherlands, 2000. www.pre.nl. Accessed 06 August 2007
17. Tsatsaronis, G, Morosuk, T: A general exergy-based method for combining a cost analysis with an environmental impact analysis. Part II—application to a cogeneration system. In: Proceedings of the ASME International Mechanical Engineering Congress and Exposition, Boston, 31 October–6 November 2008, IMECE2008-67219
18. Cabrera, M: Exergoenvironmental analysis of oxyfuel-based combined-cycle power plants including CO_2 capture. Master Thesis, Technische Universität Berlin (2010)
19. Boyano, A, Blanco-Marigorta, AM, Morosuk, T, Tsatsaronis, G: Exergoenvironmental analysis of a steam methane reforming process for hydrogen production. Energy Int J. (2010). doi:10.1016/j.energy.2010.05.020
20. Petrakopoulou, F: Comparative evaluation of power plants with CO2 capture: thermodynamic, economic and environmental performance. Ph.D, Dissertation, Technische Universität Berlin (2010)
21. Tsatsaronis, G, Morosuk, T: A general exergy-based method for combining a cost analysis with an environmental impact analysis. Part I–theoretical development. In: Proceedings of the ASME International Mechanical Engineering Congress and Exposition, Boston, 31 October–6 November 2008, IMECE2008-67218
22. Tsatsaronis, G, Park, MH: On avoidable and unavoidable exergy destructions and investment costs in thermal systems. Energy Convers Manage **43**, 1259–1270 (2002)
23. Cziesla, F, Tsatsaronis, G, Gao, Z: Avoidable thermodynamic inefficiencies and costs in an externally fired combined cycle power plant. Energy Int. J. **31** (10–11), 1472–1489 (2006)

Performance evaluation of low-cost FRP parabolic trough reflector with mild steel receiver

Atul A Sagade[1*], Satish Aher[2] and Nilkanth N Shinde[3]

Abstract

Solar collector and concentrator system can be used for industrial process heat application in various industries. Apart from the low temperature applications, there are several potential fields of application for solar thermal energy at medium-high temperatures (80°C to 300°C). This paper describes the experimental results of the prototype parabolic trough made of fiberglass-reinforced plastic with its aperture area coated by aluminum foil with a reflectivity of 0.86. From Indian conditions, there is a large potential available for low-cost solar-concentrating technologies for domestic as well as industrial process heat applications. This line-focusing parabolic trough with mild steel receiver coated with black proxy material has been tested with and without glass cover. Instantaneous efficiency of 51% and 39% has been achieved with and without glass cover, respectively. Performance evaluation of the prototype system has been done during the months of April and May 2010 at Shivaji University, Kolhapur (16.42°N latitude, 74.13°W longitude). The total cost of the prototype system developed has been calculated as Rs10,000 (US$200).

Keywords: Solar collector, Solar thermal energy, Fiberglass-reinforced plastic, Solar parabolic trough, Renewable energy technology

Background

Kalogirou et al. described a low-cost method for mass production of parabolic surfaces with fiberglass. The article indicates that the accuracy of the parabolic surface depends on the accuracy of the mold, and details of the mold production and the procedure for producing the parabolic surface are presented. The authors used fiberglass of 4-mm (mean value) thickness. The cost of the parabolic surface is US$30/$m^2$ of aperture area with 90° rim angle [1].

Valan Arasu and Sornakumar explained the design and manufacturing of a smooth 90°-rim-angle fiberglass-reinforced parabolic trough for water-heating application. The total thickness of the parabolic trough is 7 mm. The concave surface where the reflector is fixed is manufactured to a high degree of surface finish. They have found that the standard deviation of the distribution of the parabolic surface errors is 0.0066 radians from the collector performance test according to ASHRAE Standard 93 (1986), which indicates the high accuracy of the parabolic surface [2].

Martinez et al. developed a 2.37-m aperture, 1.14-m-long parabolic trough concentrator with first surface solar mirrors made over floated soda lime glasses with concave geometry and built with 16 mirrors with sizes of 0.3 × 0.6 m. This reflector yields specular reflectance around 86%. The field test of focusing such a concentrator gave a size focus of about 5.08 cm where over 90% of the reflected beam solar irradiance arrived on a simulated pipe receiver with this diameter [3].

Thomas describes a simple welded frame structure for parabolic trough concentrators suitable for developing countries. He conducted static load tests on this structure to study its deflection characteristics under various load conditions. The results show that the slope error of the reflector support corresponding to normal wind loads was found to be within the specified limits, and the structure withstood the load corresponding to extreme wind load conditions [4].

The collector's performance has been tested by Kalogirou according to the ASHRAE Standard 93(1986). The collector's efficiency and incidence angle modifier have

* Correspondence: atulsagade@gmail.com
[1]Solar Thermal Research Laboratory, New Satara College of Engineering & Management (Polytechnic), Korti-Pandharpur, Maharashtra 413304, India
Full list of author information is available at the end of the article

been measured. The test slope and intercept were found to be 0.387 and 0.638, respectively. The author explained that the collector's time constant is less than 1 min and the collector's acceptance angle obtained from the test is ± 0.5°, which in combination with the tracking mechanism maximum error (± 0.2°) implies that the system works continuously at almost maximum possible efficiency [5].

Thomas and Guven have explored and reviewed the design aspects of the structural, optical, and thermal subsystems of parabolic trough concentrators. Existing methods of performance evaluation and techniques to improve their performance were also discussed by the authors [6].

Gong et al. explains that the vacuum solar receiver is the key component of a parabolic trough solar plant, which plays a prominent role in the gross system efficiency. This paper first establishes and optimizes a one-dimensional (1-D) theoretical model using MATLAB program to compute the receiver's major heat loss through glass envelope and then systematically analyzes the major influence factors of heat loss. With the laboratorial steady state test stand, the heat losses of both good vacuum and non-vacuum Sanle-3 receivers were surveyed by the authors. The authors' comparison shows that the original 1-D model agrees with the end-covered test while remarkably deviating from the end-exposed test. The authors also developed a 3-D model by CFD software to further investigate the different heat transfer processes of the receiver's end components [7].

Skeiker explains that the important parameter affecting the performance of a solar collector is its tilt angle with the horizon. This is because the variation of tilt angle changes the amount of solar radiation reaching the collector surface. Skeiker has developed a mathematical model for estimating the solar radiation on a tilted surface and to determine the optimum tilt angle and orientation (surface azimuth angle) for the solar collector in the main Syrian zones on a daily basis as well as for a specific period. The results expose that changing the tilt angle 12 times in a year maintains approximately the total amount of solar radiation near the maximum value that is found by changing the tilt angle daily to its optimum value. This achieves a yearly gain in solar radiation of approximately 30% more than in the case of a solar collector fixed on a horizontal surface [8].

System description

A schematic sketch of the test setup of the constructed parabolic trough concentrator for domestic hot water application is shown in Figures 1 and 2. The test setup was set at the Department of Technology, Shivaji University. It consists of a solar collector, storage tank of 100-L capacity, non-return valves fitted in the pipe line to define the flow direction, and control valve used to regulate the flow rate through the circuit. The necessary instruments are attached to the apparatus and then connected to the data acquisition system. For performance evaluation of a system, data collection is important, and for the data collection, measuring instrument is needed. The following instruments were used.

Figure 1 The fiberglass-reinforced plastic parabolic trough.

1. Solar radiation: pyranometer (CM6b, Kipp and Zonen, Delft, The Netherlands)
2. Wind Velocity: digital anemometer (AM-4201, Lutron, Taipei, Taiwan)
3. Temperature measurements: set of thermocouple with digital displays.
4. Mass flow rate: flow meter

Fabrication of the fiberglass-reinforced parabolic trough

For construction of the parabolic trough, a composite material is used. E-glass fiberglass-reinforced plastic (FRP) has been selected as material for collector. There are a number of benefits obtained from using FRP composite; these benefits and characteristics were considered during the design and development process. The parabolic trough was constructed by the hand lay-up method. First, the mold was made of a plywood-type wooden material. The accuracy of the fiberglass-reinforced parabolic trough surface depends on the accuracy of the mold; therefore, extreme care has been taken during the construction of the wooden mold. The wooden mold has been constructed with plywood ISI BWR IS 303, which has been manufactured using high-density timber from selected hardwood layers of uniform thickness for good strength and stability. The thickness of the plywood was 20 mm. Firstly, a parabolic rib having a width of 110.23 cm was manufactured. Two parabolic pieces of the same type of were manufactured using a saw machine; these two parabolic pieces were positioned in such a way that the

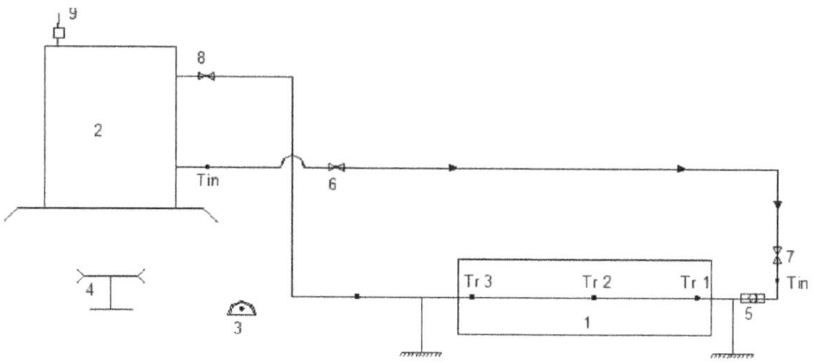

1. Parabolic trough
2. Storage Tank
3. Pyranometer4. Anemometer
5. Flow meter

6, 7 Flow meter
8. Non return valve
9. Pressure relief valve
Tr_1, Tr_2, Tr_3:- K type Thermocouples

Figure 2 Schematic diagram of the test setup.

space between the adjacent forms on either side is equal. Small recesses were cut on the mold frame to accommodate the parabolic forms, and the parabolic forms were nailed perpendicularly on the mold frame and then connected by means of a wooden strip.

On top of these parabolic forms, a 6-mm smooth-surface plywood was fixed with nails. Wood putty was used to eliminate any marks left by the nails that were polish by sandpaper after the wood putty was set; plane smooth sunmica (a type of smooth plywood) was stuck on the plywood surface with sticky materials (Fevicol, Pidilite, Mumbai, India). A thin film of plastic, 200-μm thick, with size equal to the parabola mold was laid on the wooden parabolic mold. The thin film was used as a release agent and to get a smooth surface inside the parabola. Then, 450-μm chopped strand fiberglass was placed on top of the thin film. The catalyst or Methyl Ethyl Ketone Peroxide Hardener and accelerator cobalt were mixed in polyester resin in a separate bowl. The mixed resin was applied by brushes to the surface of the chopped strand mat, and then, another layer of chopped strand mat was placed on the mixed resin. The fiberglass mat was saturated with the resin by rolling the surface with a brush. The resin was applied on the mat by brushing. A brush was used to work the resin in to the fiber. During the process, extreme care was taken so that the material is securely attached to the mold, and no air is trapped in between the fiberglass and mold. The rolling action of the squeeze assists in removing air bubbles that could prove detrimental to the laminate performance. This process of laying alternate layers of polyester resin and chopped strand fiberglass mat was repeated until a 2-mm-thick fiberglass-reinforced laminate was obtained. Table 1 shows the parameters and dimensions of the terms used in the experimentation.

Methods

Experimental methods

The experiment procedure was started by flushing the system. Then, the system was filled with water and the flow rate was adjusted to the required value. The solar collector was allowed to run for over 20 min to achieve quasi-steady-state conditions before starting the experiments, and the proper working of all measuring instruments was checked, including the connection of the temperature indicator to the temperature sensor, pyranometer, anemometer, and tracking of parabolic trough. Cold water from the storage tank enters the mild steel receiver of the parabolic trough collector. The tank was located above the level of the collector to assure the natural flow of water. As water in the receiver tube, which is located at the focal axis of

Table 1 Parameters and dimensions of terms used in the experimentation

Parameter	Dimension
Aperture of the concentrator (W)	1.10 m
Inner diameter of absorber tube (D_i)	0.020 m
Outer diameter of absorber tube (D_o)	0.025 m
Inner diameter of glass tube	0.050 m
Outer diameter of glass tube	0.056 m
Length of parabolic trough	1.21 m
Concentration ratio	13.69
Collector aperture area	1.33 m^2
Specular reflectivity of concentrator (ρ)	0.85
Glass cover transitivity for solar radiation(τ)	0.85
Absorber tube emissivity/emissivity (α)	0.82
Intercept factor	0.95
Emissivity of absorber tube surface (ε_p)	0.08
Emissivity of glass (ε_c)	0.82

the trough, is heated by solar energy, heated water flows automatically to the top of the water tank and is replaced by cold water from the bottom of the tank. When the water gets heated upon rising to the collector, its density will decrease and the lighter-density water will move up and be stored on top of the storage tank. Higher-density water from the bottom of the tank again enters the parabolic trough and gets heated and moves up and stored in the top of the storage tank. Data of all readings of ambient, fluid, receiver body, and storage tank temperatures and total solar radiations with wind speed every half an hour were collected. The experiment has been performed for 7 h over the day from 0930 hours to 1530 hours. The experiment has been continued with changing input parameters, such as receiver materials and mass flow rate of water.

During the experimentation, a cylindrical parabolic collector has been oriented with its focal axis pointed in the east–west (E-W) orientation so that the focal axis is horizontal. Sun tracking was provided to the parabolic trough collector. Manual tracking was provided to the parabolic trough collector. The trough was rotated about a horizontal (E-W) axis and adjusted manually so that solar beam makes minimum angle of incidence with the aperture plane at all times.

Experimental data collection and calculations

Tables 2 and 3 present the experimental measurements and calculations for the mild steel receiver with and without glass cover, respectively.

Thermal performance calculations

The useful energy delivered from the concentrator can be given by Equations 1 and 2 [9,10]:

$$q_u = mc_p(T_o - T_{in}) \tag{1}$$

$$q_u = mc_p\left[\frac{CS}{U_l} + T_a - T_{in}\right]\left[1 - \exp\left\{-\frac{F'\pi D_o U_l L}{mc_p}\right\}\right] \tag{2}$$

where q_u is the useful energy delivered from the concentrator (W); m, mass flow rate(kg/s); T_o, outlet fluid temperature (°C); T_{in}, inlet fluid temperature (°C); C_p, specific heat of water (kJ/kg°C); C, concentration ratio; S, incident solar flux absorbed in the absorber plate (W/m^2); U_l, overall heat loss coefficient (W/m^{2}°C); T_a, ambient temperature (°C); F', collector efficiency factor; D_o, outer diameter of the tube (m); and L is the length of the concentrator (m).

The useful energy gain per unit of the collector length can be expressed by Equations 3 and 4 in terms of the local receiver temperature T_r [1,9,10].

$$q_{u'} = \frac{q_u}{L} \tag{3}$$

$$q_{u'} = \frac{q_u}{L} = F'\left[S - \frac{U_l}{C}(T_r - T_a)\right](W - D_o) \tag{4}$$

where $q_{u'}$ is the useful energy gain per unit of the collector length; T_r, mean receiver surface temperature (°C); W, width of the parabolic reflector (m), and where F' is the collector efficiency factor defined by Equations 5 and 6 [1,5,9,10].

$$F' = 1/U_l\left[\frac{1}{U_l} + \frac{D_o}{D_i h_f}\right] \tag{5}$$

$$q_u = F_R(W - D_o)L\left[S - \frac{U_l}{C}(T_{in} - T_a)\right] \tag{6}$$

where D_i is the inner diameter of the tube (m); h_f, heat transfer coefficient on the inside surface of the tube (W/m^{2}°C); and F_R is the collector heat removal factor.

Heat removal factor is given by Equation 7 [4,9,10],

$$F_R = \frac{mc_p}{\pi D_o L U}\left[1 - \exp\left\{-\frac{F'\pi D_o U_l L}{mc_p}\right\}\right] \tag{7}$$

and collector efficiency can be obtained by dividing q_u by I_b WL. The instantaneous collection efficiency can also be calculated by Equation 8 [9,10]

$$\eta_{in} = \frac{q_u}{IbrbWL} \tag{8}$$

where n_{in} is the instantaneous collection efficiency.

Table 2 Experimental measurements and calculation for mild steel receiver without glass cover

Time (hours)	Direct solar radiation on the collector (W/m^2)	Wind speed (m/s)	T_{amb} (°C)	T_{in} (°C)	T_{out} (°C)	ΔT (°C)	T_r (°C)	Flux absorbed by the receiver (W/m^2)	Useful heat gained by water (W)	Overall heat loss coefficient (W/m^{2}°C)	Instantaneous collector efficiency (%)
0930	667	3.1	29	26	28	02	33	405.5	362.7	44.0	39
1030	861	5.1	31	27	30	03	38	523.4	458.9	44.4	40
1130	977	3.8	34	28	36	08	45	594.0	529.7	49.1	40
1230	1004	6.2	36	29	41	12	49	610.4	457.2	44.1	41
1330	949	5.3	36	30	40	10	54	576.9	493	50.5	39
1430	841	3.5	35	30	39	09	53	511.3	431.1	52.8	37
1530	783	5.2	34	30	32	02	48	476.0	378.0	58.2	36

Mass flow rate, 0.023 kg/s. T_{amb}, ambient temperature; T_{in}, inlet fluid temperature; T_{out}, outlet fluid temperature; ΔT, temperature gradient; T_r, receiver temperature.

Table 3 Experimental measurements and calculation for mild steel receiver with glass cover

Time (hours)	Direct solar radiation on the collector (W/m^2)	Wind speed (m/s)	T_{amb} (°C)	T_{in} (°C)	T_{out} (°C)	ΔT (°C)	T_r (°C)	Flux absorbed by the receiver (W/m^2)	Useful heat gained by water (W)	Overall heat loss coefficient (W/m^2°C)	Instantaneous collector efficiency (%)
0930	696	3.2	29	27	34	07	45	400.8	468.2	18.4	50
1030	764	2.8	32	28	42	14	52	440.0	530.9	12.2	52
1130	793	5.3	36	28	48	20	56	456.7	544.5	16.2	51
1230	928	3.2	38	29	60	31	69	534.5	649.0	11.6	52
1330	976	2.1	39	29	67	38	75	562.1	704.6	8.9	54
1430	899	5.2	39	30	57	27	68	517.8	629.7	12.3	52
1530	812	6.3	34	30	52	22	61	467.7	558.6	14.1	51
1600	764	5.9	33	31	39	08	49	410.9	487.0	20.5	47

Mass flow rate, 0.023 kg/s. T_{amb}, ambient temperature; T_{in}, inlet fluid temperature; T_{out}, outlet fluid temperature; ΔT, temperature gradient; T_r, receiver temperature.

Overall loss coefficient and heat correlations

The calculations of overall loss coefficient were based on convection and re-radiation losses. Heat loss rate per unit length can be given by Equations 9 and 10 [1,5,9,10]:

$$\frac{q_l}{L} = h_{pc}(T_r - T_c)\pi D_o + \sigma\pi D_o\left(T_r^4 - T_c^4\right)/\left\{\frac{D_o}{\frac{1}{\varepsilon_p} + D_{ci}\left(\frac{1}{\varepsilon_c} - 1\right)}\right\} \tag{9}$$

$$\frac{q_l}{L} = h_w\,(T_c - T_a)\pi Dco + \sigma\pi Dco\varepsilon c\left(T_r^4 - T_a^4\right) \tag{10}$$

where T_c is the temperature of the cover; h_w, wind heat transfer coefficient(W/m^2-K); ε_p, emissivity of absorber surface for long-wavelength radiation; and ε_c, emissivity of the cover for long-wavelength radiation.

Heat transfer coefficient between the absorber tube and the cover

The heat transfer coefficient h_{pc} for the enclosed annular space between a horizontal absorber tube and a concentric cover is calculated by Equations 11, 12, and 13 [9,10]:

$$\frac{K_{eff}}{k} = 0.317(Ra^*)^{0.25} \tag{11}$$

$$(Ra^*)^{0.25} = \frac{\ln\left(\frac{D_{ci}}{D_o}\right)}{b^{0.75}\left(\frac{1}{D_o^{0.6}} + \frac{1}{D_{ci}^{0.6}}\right)} Ra^{0.25} \tag{12}$$

where Ra^* is Rayleigh's number.

Thus,

$$h_{pc} = \frac{2K_{eff}}{D_o \times ln\left(\frac{D_{ci}}{D_o}\right)}. \tag{13}$$

Heat transfer coefficient on the inside surface of the absorber tube

The convective heat transfer coefficient h_f on the inside surface of the absorber tube can be calculated. For Reynolds number greater than 2,000, the flow is turbulent and the heat transfer coefficient may be calculated from Equation 14 [9,10]:

$$Nu = 0.023\mathrm{Re}^{0.8}P_r^{0.4} \tag{14}$$

Results and discussion

Figure 3 shows the variation of inlet and outlet temperatures of the water. The trend explains that, as solar radiation on the receiver enhances, outlet water temperature increases. In comparison, the receiver with glass cover yields much higher outlet water temperature than the receiver without glass cover. The glass cover over the mild steel receiver acts as an insulator, and outlet water temperature and temperature gradient between inlet and outlet water rise sharply. It has been observed that, with glass cover, outlet water temperature and temperature gradient increase by 29% and 68%, respectively.

Figure 4 explores the variation of receiver temperature. From Figure 4, it is clear that the receiver temperature increases with increase in solar radiation. When the receiver is exposed to atmosphere, there are heavier convective heat losses from the receiver as compared to the receiver with glass cover. As said earlier, the glass cover acts as an insulator and also helps in concentrating more solar radiation at the receiver. It has been observed that, with glass cover, receiver temperature has been increased by 23%, which assures hotter water temperature at the outlet.

Figure 5 indicates useful heat gained by the water flowing through the receiver throughout the day. Useful heat gained by the water is affected by various parameters such as wind speed, receiver temperature, and solar radiation. Receiver temperature is the average of three

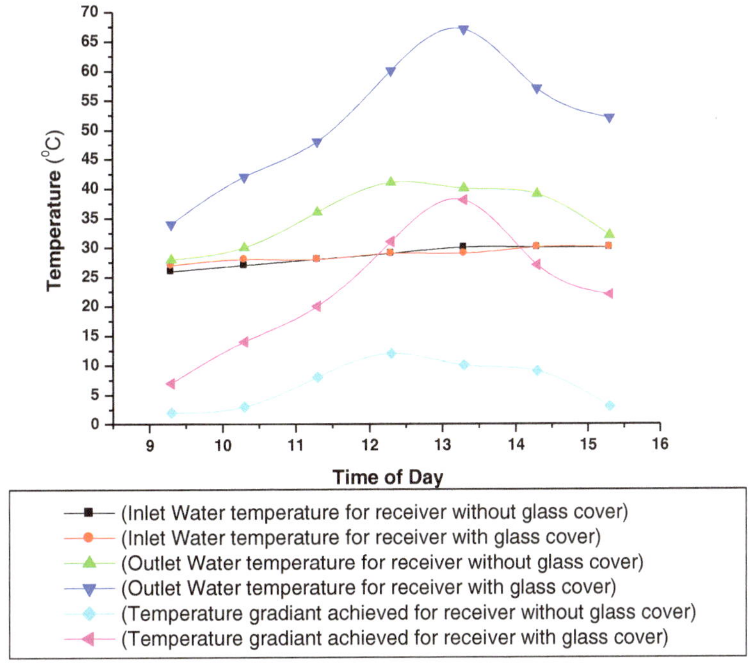

Figure 3 **Variation of temperatures throughout the day.**

values over the space and not over the time. Temperature was measured at four locations over the length of the receiver. This plotted variation shows that as the receiver surface temperature increases, it increases the thermal conductivity of the air surrounding the receiver. When air conductivity increases, heat losses also increase, but this is not the only cause of heat loss. Figure 5 shows that, with glass-covered receiver, it is possible to achieve more useful heat gain. It has been observed that as the beam solar radiation incident on the collector increases, more optical energy is captured by the receiver. Compared to the decrease in solar radiation and wind velocity by 4% and 7%, respectively, useful heat gained by the receiver with glass cover increases by 22%

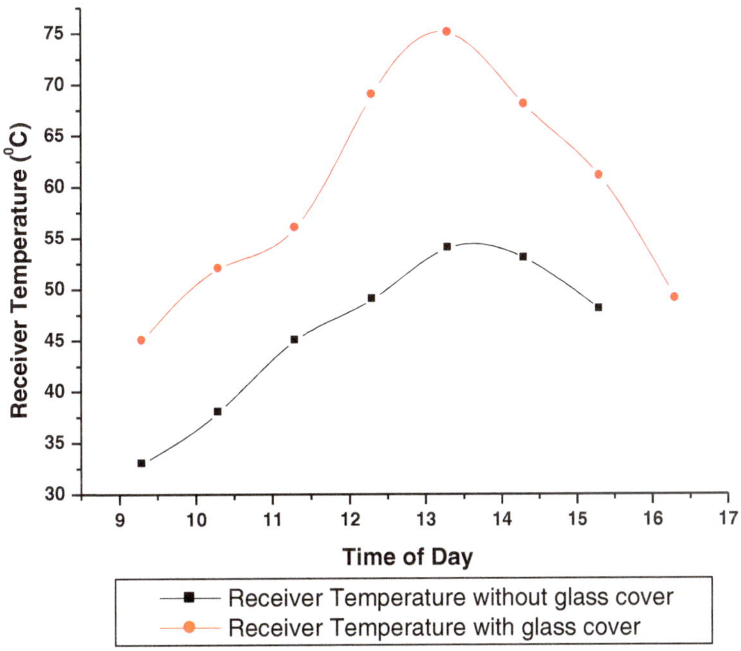

Figure 4 **Variation of receiver temperature throughout the day.**

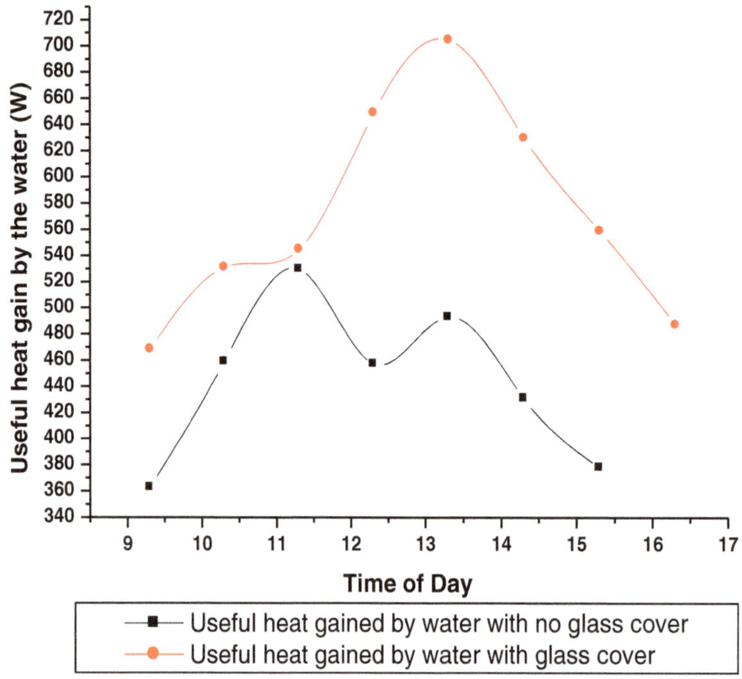

Figure 5 Variation of useful heat gained by water throughout the day.

throughout the day, and average receiver temperature increased by 23%.

Figure 6 explains the variation of collector efficiency and convective heat loss coefficient. The higher the wind speed, the higher will be the convective heat loss and the lower will be the collector efficiency at an instant. It has been observed that the average decrease in heat loss coefficient is 70% when the receiver was covered with glass. When the heat loss coefficient decreased by 70%, the instantaneous efficiency of the collector still increased

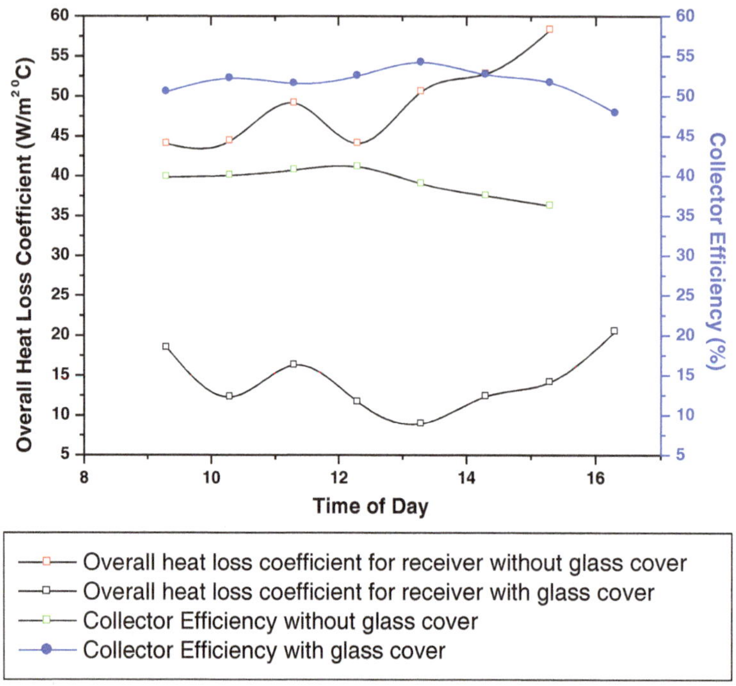

Figure 6 Variation of overall heat loss coefficient and collector efficiency throughout the day.

by 13% with the glass-covered receiver. Instantaneous efficiency of 51% has been achieved with the glass-covered receiver. Radiative heat losses are very small, so their effect on collector efficiency is very small. The main cause for heat loss is temperature gradient between the receiver and surrounding and wind velocity at an instant which therefore increase convective and radiative heat losses from the receiver. An average receiver temperature has been determined to provide accurate receiver loss predictions. Also, wind velocity affects the variation of convective heat loss coefficient. This heat loss also depends upon solar radiation at an instant.

Conclusions

From Indian conditions, low-cost FRP parabolic trough system can prove to be beneficial for industrial heating applications as well as domestic heating. With the system described in this paper, the following conclusions are drawn:

1. Instantaneous efficiency of the collector has been increased by 13%. Instantaneous efficiency of 51.67% has been achieved with the glass-covered receiver.
2. Useful heat gained by the receiver increases by 22% with the glass-covered receiver throughout the day, and the average receiver temperature increased by 23%.
3. It has been observed that, with glass-covered receiver, outlet water temperature and temperature gradient increase by 29% and 68%, respectively.
4. It has been observed that the average decrease in heat loss coefficient is 70% when the receiver was covered with glass.

Competing interests

The authors declare that they have no competing interests.

Authors' contributions

AS and SA carried out all the computations and system designing and drafted the manuscript. NS conceived of the study and participated in its design and coordination. All authors read and approved the final manuscript.

Authors' information

AS, M. Tech, Principal at New Satara College of Engineering & Management, Head of Solar Thermal Research Lab, is involved in research and development programs of solar thermal energy technology in Maharashtra, India, with more than 6 years of experience in the field of solar thermal energy. He was also invited as a solar expert for IEA workshops on the road map of solar heating and cooling at Germany, China, and Australia. SA, M. Tech in Energy, is working as a lecturer at the Amrutvahini College of Engineering and is involved in research activities in the field of thermal energy technology at Shivaji University, Kolhapur, Maharashtra, India. NS, a coordinator and associate professor at the Department of Energy Technology, has 30 years of experience in the field of energy-related issues. He is involved in various energy research projects of the Department of Technology at Shivaji University, Kolhapur, Maharashtra, India. He has more than 150 publications under his name.

Acknowledgments

The authors want to thank the Department of Energy Technology, Shivaji University, Kolhapur, for providing lab support and testing equipment.

Author details

[1]Solar Thermal Research Laboratory, New Satara College of Engineering & Management (Polytechnic), Korti-Pandharpur, Maharashtra 413304, India. [2]Amrutvahini College of Engineering, Sangamner, Maharashtra 422608, India. [3]Department of Energy Technology, Shivaji University, Kolhapur, Maharashtra 416004, India.

References

1. Kalogirou, S, Eleftheriou, P, Lloyd, S, Ward, J: Low cost high accuracy parabolic troughs construction and evaluation. Renew. Energy. **5**(4), 384–386 (1994)
2. Valan Arasu, A, Sornakumar, T: Design, manufacture and testing of fiberglass reinforced parabola trough for parabolic trough solar collectors. Sol. Energy. **81**(10), 1273–1279 (2007)
3. Martinez, I, Almanza, R, Mazari, M, Correa, G: Parabolic trough reflector manufactured with aluminum first surface mirrors thermally sagged. Sol. Energ. Mat. Sol. C. **64**(1), 85–96 (2000)
4. Thomas, A: Simple structure for parabolic trough concentrator. Energ. Convers. Manage. **35**(7), 569–573 (1994)
5. Kalogirou, S: Parabolic trough collector system for low temperature steam generation: design and performance characteristics. Appl. Energ. **55**(1), 1–19 (1996)
6. Thomas, A, Guven, HM: Parabolic trough concentrators—design, construction and evaluation. Energ. Convers. Manage. **34**(5), 401–416 (1993)
7. Gong, G, Huang, X, Wang, J, Hao, M: An optimized model and test of the China's first high temperature parabolic trough solar receiver. Sol. Energy. **84**(12), 2230–2245 (2010)
8. Skeiker, K: Optimum tilt angle and orientation for solar collectors in Syria. Energ. Convers. Manage. **50**(9), 2439–2448 (2009)
9. Duffie, J, Beckman, W: Solar Engineering of Thermal Processes. Wiley, New York (2006)
10. Yogi Goswami, D, Kreith, F, Kreider, JF: Principals of Solar Engineering. Taylor and Fancis, Philadelphia (2003)

Rapid thermodynamic simulation model for optimum performance of a four-stroke, direct-injection, and variable-compression-ratio diesel engine

Maher M Abou Al-Sood*, Mahmoud Ahmed and Yousef M Abdel-Rahim

Abstract

A thermodynamic simulation model for the performance of a four-stroke, direct-injection diesel engine is developed. The simulation model includes detailed sub-models for fuel burning rate, combustion products, thermodynamic properties of working fluid, heat transfer, fluid flow, and both soot and oxides of nitrogen (NO_x) formation mechanisms. To validate the model, comparisons between experimental and predicted results for different engines, operating under different conditions, were conducted. The comparisons show that there is a good concurrence between measured and predicted values. An optimization analysis is conducted for seeking an optimum variation of compression ratio to achieve pre-set objective targets such as constant minimum brake-specific fuel consumption and constant maximum torque. The optimization analysis is performed under the constraint that the maximum pressure and temperature inside the cylinder do not exceed the maximum allowable pressure and temperature of the conventional engine (constant compression ratio).The varying compression ratio is optimized with each of the previous conditions separately. Results indicated that varying the compression ratio to achieve previous targets leads to saving fuel consumption, higher brake efficiency and power, and reduction in soot emission from the engine. Also, an increase in NO_x is noticed at low speed. This drawback is considerable and can be overcome by reducing the operation speed range.

Keywords: Thermodynamic, Combustion, Compression ratio, Optimum performance, Diesel engine

Background

Thermodynamic models (zero-dimensional) and turbulent models (multi-dimensional) are the two types of models that have been used in internal combustion engine simulation modeling. Nowadays, trends in combustion engine simulations are towards the development of comprehensive multi-dimensional models that accurately describe the performance of engines at a very high level of details. However, these models need a precise experimental input and substantial computational power, which make the process significantly complicated and time-consuming [1-6]. On the other hand, zero-dimensional models, which are mainly based on energy conservation (first law of thermodynamics) with a limited set of experimental data, are preferred due to their simplicity, of being less time-consuming, and their relatively accurate results [7-10].

Requirements to provide environmentally friendly vehicles (i.e., low fuel consumption with less emission) have resulted in using the most complicated and intellectual means of internal combustion engine development. Nowadays, new ideas, which could not be discussed two decades ago, are being considered by automotive manufacturers. In particular, many leading automotive companies have approached practically the very complicated design ideas with different aspects of diesel/petrol engine design. These aspects have been under extensive theoretical and experimental investigations. The most important aspect of design is aimed to vary the engine compression ratio (r_c) depending on load, speed, or both. Several trials have been done in

* Correspondence: m_aboualsood@hotmail.com
Department of Mechanical Engineering, Assiut University, Assiut 71516, Egypt

that respect with extensive design, experimentation, and measurements [11-19]. All attempts to change the r_c are achieved by one or more of the following concepts [11]: (1) moving the cylinder head, (2) variation of combustion chamber volume, (3) variation of piston deck height, (4) modification of connecting rod geometry (usually by means of some intermediate member), (5) moving the crankpin within the crankshaft (effectively varying the stroke), and (6) moving the crankshaft axis. Benefits and challenges of variable-compression-ratio (VCR) spark-ignition engines had been illustrated, examined, and critically reviewed by Roberts [11]. Also, the implications for volume manufacture and the strategy for VCR implementation in order to produce the maximum benefit had been discussed.

Ladommatos and Balian [12] studied the effects of an increased clearance volume on the performance of a direct-injection diesel engine. This study was conducted to overcome problems with combustion at standard r_c for high-speed, direct-injection (DI) diesel engine. Modifications were done in the injection system and operating conditions to improve the power output and fuel consumption. Sobotowski et al. [13] modified a prototype DI diesel engine to incorporate BP Oil's VCR engine concept. Their objective was achieved by altering the phase relation between two pistons linked to separate crankshafts and operating in two cylinders arranged in an opposed engine or interconnected through a transfer port. According to their concept, a novel crankshaft phasing mechanism was employed to achieve r_c variation with a number of significant advantages over other VCR engine designs. These advantages include (a) mechanical simplicity and compact design, (b) capability to externally control r_c during engine operation, (c) automatic variation of valve timing, and (d) compatibility with current production technology.

Rychter et al. [14] evaluated the concept of varying r_c to give different expansion and compression ratios by means of a theoretical simulation of a turbocharged diesel engine. They varied the ratio of connecting rod length to crank throw to study the effect of r_c on engine performance at a fixed load. The principal benefits, at 3/4 load and r_c of 20, were a reduction in fuel consumption by about 2% and a reduction in ignition delay that leads to an estimated 6-dB reduction in combustion noise. Variable-stroke engine mechanism patented by Freudenstein and Maki [15] was used by Yamin and Dado [16] as it gives availability to change the stroke length and the corresponding r_c in the range of 6.82 to 10. Their results showed a significant improvement in the engine's power and reduction in the fuel consumption, in addition to an increase in carbon monoxide and nitric oxides compared with the actual engine. Finally, they concluded that the engine performance can be optimized for a full range of driving conditions, such as acceleration, speed, and load. Such optimized performance was obtained by incorporating some means to suppress the increase in emission rate to be applicable in the automotive field due to the strict anti-pollution laws.

The effect of varying compression ratio and other parameters such as injection timing, engine load, and fuel on the performance of diesel engine was recently studied experimentally [17-20]. Raheman and Ghadge [17] studied the performance of a Ricardo EG engine using biofuel and its blend with high-speed diesel fuel at varying compression ratio, injection timing, and engine load. They concluded that the performance of the engine reduced with the increase in the percentage of biofuel in the blend above 20% compared to that of conventional engine. This reduction in engine performance can be overcome by increasing the compression ratio and injection timing. Jindal et al. [18] investigated the effect of compression ratio and injection pressure on the performance of DI diesel engine running on Jatropha methyl ester. They demonstrated that an increase in compression ratio associated with an increase in injection pressure improves the performance of the engine in terms of brake-specific fuel consumption, brake thermal efficiency, and emissions (i.e., hydrocarbon, NO_x, and smoke opacity). They concluded that for fuelling the engine with bio-diesel, one should go for higher compression ratio associated with higher injection pressure. Muralidharan et al. [19] estimated the performance, emission, and combustion characteristics of a single-cylinder, four-stroke, VCR, multi-fuel engine fuelled with waste cooking oil methyl ester and its blends with standard diesel. The performance parameters include brake thermal efficiency, brake-specific fuel consumption, brake power, indicated mean effective pressure, mechanical efficiency, and exhaust gas temperature. They compared and analyzed their results with standard diesel, and a considerable improvement in the performance parameters as well as exhaust emissions was observed especially at full load. Gandure and Ketlogetswe [20] carried out their work to compare the performance of variable compression engine fuelled with native marula seed oil and petrodiesel fuels. The results indicated that engine performance when powered with marula oil as fuel was very close to that when powered with petrodiesel and at an 80% load, the compression ratio of 16:1 yields optimum performance in terms of engine torque and brake power for both petrodiesel and marula oil fuels.

The purpose of this study is to develop a simple, rapid, and accurate simulation model based on thermodynamic analysis without the need for a great deal of computational power or knowledge of precise engine geometrical data to predict the range of compression ratio corresponding to the optimum engine performance (minimum brake-specific fuel consumption (bsfc) and maximum brake torque) without an increase in the maximum cylinder pressure and temperature of the conventional engine. The developed model is based on mass and energy conservation laws, equation of state, heat transfer between the working fluid and cylinder walls, fuel energy release rate, and engine friction losses. Single-zone combustion is assumed, and thermodynamic properties are calculated from chemical equilibrium principles for six and/or eleven species. A stepwise solution is carried out for the varying compression ratio and the engine speed over its entire range (i.e., 1,500 to 2,800 rpm). In calculations, the temperature and pressure cannot exceed the maximum allowable temperature and pressure of the conventional engine, i.e., constant compression ratio, which are 2,100 K and 10.7 MPa, respectively.

Methods

Mathematical model

A thermodynamic analysis of a four-stroke, direct-injection diesel engine is conducted. It is based on a single-zone combustion modeling with a rapid premixed burning phase followed by a slower mixing-controlled burning phase. Although a single-zone model is an old one, it was selected in this study for the sake of its simplicity and of being less time-consuming for the program execution. Works with two- and multi-zone models are in progress. The analysis includes detailed sub-models representing different engine aspects integrated in the main engine simulation model. These sub-models are engine geometry, combustion products, thermodynamic properties, engine friction and heat transfer, mass exchange, combustion rate, temperature-pressure, and emission mechanisms. The analysis presented in this article is based on the following assumptions:

1. The content of the cylinder is a homogeneous gas mixture of air (79% N_2 and 21% O_2 by volume), where fuel vapor, combustion products, and its thermodynamic properties are calculated using ideal gas law with temperature-dependent specific heats.
2. Pressure and temperature inside the cylinder are uniform and vary with the crank angle.
3. The gas motion inside the cylinder (created by pressure gradients due to piston motion) is neglected.
4. A single-zone combustion process starts with a rapid premixed burning phase, followed by a slower mixing-controlled burning phase.
5. Temperatures of the cylinder head, cylinder walls, and piston crown are assigned constant values.

In the present work, the detailed sub-models are formulated as follows:

Engine geometry

The rate of change of the volume with respect to time is given as

$$dot; V = V_c\left[\frac{r_c - 1}{2}\sin\theta\frac{1+\cos\theta}{\sqrt{\lambda^2 - \sin^2\theta}}dot; \theta\right]. \tag{1}$$

Combustion product model

For temperature <1,600 K, the equilibrium composition of gases produced by the combustion of a general hydrocarbon fuel with air (having a general fuel-air molecule in the form $C_wH_xO_yN_z$) is calculated based on the assumption that only gaseous phases are considered. The general combustion equation of the fuel in air can be written as

$$C_wH_xO_yN_z \rightarrow n_1CO_2 + n_2H_2O + n_3CO + n_4H_2 + n_5O_2 + n_6N_2. \tag{2}$$

For temperature ≥1,600 K, 11 combustion products are assumed. The chemical reaction can be written as

$$C_wH_xO_yN_z \rightarrow n_1CO_2 + n_2H_2O + n_3CO + n_4H_2 + n_5O_2 + n_6N_2 + n_7OH + n_8H + n_9NO + n_{10}N + n_{11}O. \tag{3}$$

Obtaining the moles of the 11 combustion products, 11 equations are needed (i.e., four equations from the mass balance of C, H, O, and N in Equation 3, and seven equations from the chemical equilibrium). These equations have been solved by the procedure mentioned in [21].

Thermodynamic properties of cylinder content model

During each thermal cycle, the cylinder contains, in general, air, fuel vapor, and combustion products with different proportions depending on the engine conditions under consideration. The specific heat under constant pressure, enthalpy, and entropy of the cylinder content are given by

$$\bar{c}_p = \sum_{i=1}^{zz} X_i\bar{C}_{p_i},\ \bar{h} = \sum_{i=1}^{zz} X_i\bar{h}_i, \tag{4}$$

where $X_i = n_i/\sum_{i=1}^{zz} n_i$, and zz (number of different gases) is equal to 6 and 11 for $T < 1{,}600$ K, and $T \geq 1{,}600$ K, respectively.

Engine friction model

Empirical correlations of losses are in terms of compression ratio, engine speed, bore, stroke, intake and exhaust manifold pressures, and valve geometry [22] are used in this work. These correlations were concluded to the following friction categories as follows:

1. Piston losses:

$$(fmep)_{\mathrm{p}} = 6.2n_{\mathrm{r}} + \left[0.606 \;\; r_{\mathrm{c}} + 1.254\, r_{\mathrm{c}}^{1.37-0.0238S_{\mathrm{p}}}\right] + 9.3S_{\mathrm{p}}^{1.03} \tag{5}$$

2. Blowby losses:

$$(fmep)_{\mathrm{b}} = 11.86r_{\mathrm{c}}^{0.4} - (3.38 + 0.103r_{\mathrm{c}}) + 6.89[N/1,000]^{1.185} \tag{6}$$

3. Exhaust and inlet system throttling losses:

$$(fmep)_{\mathrm{ei}} = P_{\mathrm{im}} + P_{\mathrm{em}}/2.75 \tag{7}$$

4. Crankcase mechanical losses:

$$\begin{aligned}(fmep)_{\mathrm{c}} = {} & 12.122\left(\frac{dN}{1,000L}\right) \\ & + 0.07\left[30 - \left(\frac{4N}{1,000}\right)\right]\left[\frac{n_{\mathrm{iv}}D_{\mathrm{iv}}^{1.75}}{d^2L}\right] \\ & + 2.688\left(\frac{N}{1,000}\right)^{1.5}\end{aligned} \tag{8}$$

5. Combustion chamber and valve pumping losses:

$$(fmep)_{\mathrm{cc}} = 8.96(N/1,000)^{1.7} \times \left[2.98\, V_{\mathrm{d}}/n_{\mathrm{iv}}\, n_{\mathrm{c}}\, D_{\mathrm{iv}}^2\right]^{1.298} \tag{9}$$

The intake and exhaust manifold gage pressures (p_{im}, p_{em}) in Equation 7 are taken as reported by Eichelberg [23] as $p_{\mathrm{im}} = 0$, $p_{\mathrm{em}} = 0.062(N/1,000)^2$.

The overall engine frictional power losses are obtained by

$$(fmep)_{\mathrm{overall}} = V_{\mathrm{d}}(N/120)n_{\mathrm{c}}\sum_{i=1}^{5}(fmep)_i. \tag{10}$$

Engine heat transfer model

Eichelberg [23] derived a simple equation for calculating the instantaneous heat flux out of the engine $dot; q_{\mathrm{w}}$ (in Watts per square meter) in terms of cylinder temperature, pressure, and piston mean velocity. Rakopoulos and Hountalas [24] modified this equation into the form

$$dot; q_{\mathrm{w}} = 1.99(Sp)^{1/3}(pT)^{1/2}(T - Tw). \tag{11}$$

Radiation heat transfer was neglected in some studies (e.g., [25-28]), but it had been considered in others (e.g., [29-31]). In the present simulation, the Eichelberg model [23] as modified by Rakopoulos and Hountalas [24] is used with assumed specified values for piston crown, cylinder head, and cylinder wall temperatures.

Mass exchange model

Air intake and exhaust exchange. The mass flow rates through the valves are usually described by the equation for compressible, one-dimensional, isentropic flow. The real flow effects are included by means of an experimentally determined discharge coefficient. In the present simulation, the discharge coefficient is taken as reported by Campbell [29] and is equal to $\sin\left[\pi(\theta - \theta_{\mathrm{vo}})/\Delta\theta_{\mathrm{open}}\right]^{1/3}$. The gas flow rate is related to the upstream stagnation pressure, stagnation temperature, and static pressure just downstream of the flow restriction. For the flow into the cylinder through intake valve, the upstream stagnation pressure and temperature are the intake manifold pressure and temperature (p_{im}, T_{im}), respectively, and the downstream static pressure is the cylinder pressure (p). For the flow out of the cylinder through exhaust valve, the upstream stagnation pressure and temperature are the cylinder pressure and temperature (p, T), respectively, and the downstream static pressure is the exhaust manifold pressure (p_{em}) [29,32]. The mass flow rates entering and leaving the cylinder are defined as follows [33]:

For subsonic inflow where $\frac{p}{p_{\mathrm{in}}} < \left(\frac{2}{k+1}\right)^{\frac{k}{k-1}}$,

$$dot; m_{\mathrm{in}} = A_{\mathrm{in}}p_{\mathrm{im}}\sqrt{\left(\frac{2k}{(k-1)RT_{\mathrm{im}}}\right)\left(\frac{p}{p_{\mathrm{im}}}\right)^{\frac{2}{k}}\left(1 - \left(\frac{p}{p_{\mathrm{im}}}\right)^{\frac{k-1}{k}}\right)}. \tag{12}$$

For sonic and supersonic inflow where $\frac{p}{p_{\mathrm{in}}} \geq \left(\frac{2}{k+1}\right)^{\frac{k}{k-1}}$,

$$dot; m_{\mathrm{in}} = A_{\mathrm{in}}p_{\mathrm{im}}\sqrt{\left(\frac{k}{RT_{\mathrm{im}}}\right)\left(\left(\frac{2}{k+1}\right)^{\frac{k+1}{k-1}}\right)}. \tag{13}$$

For subsonic outflow where $\frac{p}{p_{\mathrm{em}}} < \left(\frac{2}{k+1}\right)^{\frac{k}{k-1}}$,

$$dot; m_{\mathrm{ex}} = A_{\mathrm{ex}}p\sqrt{\left(\frac{2k}{(k-1)RT}\right)\left(\frac{p_{\mathrm{em}}}{p}\right)^{\frac{2}{k}}\left(1 - \left(p_{em}{}^{\frac{em}{p}}\right)^{\frac{k-1}{k}}\right)}. \tag{14}$$

For sonic and supersonic outflow where $\frac{p_{\mathrm{em}}}{p} \geq \left(\frac{2}{k+1}\right)^{\frac{k}{k-1}}$,

$$dot; m_{\mathrm{ex}} = A_{\mathrm{ex}}p\sqrt{\left(\frac{k}{RT}\right)\left(\left(\frac{2}{k+1}\right)^{\frac{k+1}{k-1}}\right)}. \tag{15}$$

Mass of injected fuel. The mass of fuel injected into the cylinder m_{f} (relative to total injected mass m_{ft}) is written (according to Zelenik [34] and Gu and Ball [35]) in terms of crank angle (θ) as follows:

$$\left(\frac{m_{\mathrm{f}}}{m_{\mathrm{ft}}}\right) = \frac{1}{2}\left\{1 - \cos\left(\frac{\pi(\theta - \theta_o)}{\Delta\theta_{\mathrm{injection}}}\right)\right\}. \tag{16}$$

Differentiation of the above equation with respect to time leads to the following equation:

$$\left(\frac{dot; m_{\mathrm{f}}}{m_{\mathrm{ft}}}\right) = \left(\frac{\pi}{2\Delta\theta_{\mathrm{injection}}}\right)\sin\left(\frac{\pi(\theta - \theta_o)}{\Delta\theta_{\mathrm{injection}}}\right) dot; \theta. \tag{17}$$

Combustion rate model

Watson et al. [36] developed equations for fuel energy release appropriate for diesel engine simulations. In their development, the combustion process starts from a rapid premixed burning phase (represented by function f_1), followed by a slower mixing-controlled burning phase (represented by function f_2), with both functions empirically linked to the duration of ignition delay ($\tilde{\tau}_{\mathrm{id}}$) and the duration of combustion ($\Delta\theta_{\mathrm{comb}}$). According to their model, the burned mass of fuel at any angle (θ) related to total injected mass (m_{ft}) is

$$m_{\mathrm{fb}}\left(\hat{\theta}\right)/m_{\mathrm{ft}} = \beta f_1 + (1 - \beta) f_2, \tag{18}$$

where $f_1 = 1 - \left(1 - \hat{\theta}^{k}\right)^{k}$, $\hat{\theta} = (\theta - \theta_{\mathrm{ig}})/\Delta\theta_{\mathrm{comb}}$, and $\beta = 1 - 0.92\phi^{0.37}\tau_{\mathrm{id}}^{-0.26}$.

The ignition delay ($\tilde{\tau}_{\mathrm{id}}$) is calculated in degrees of crank angle by empirical correlations mentioned by Heywood [32] and Mehta et al. [37]. These correlations are in terms of intake manifold temperature (T_{im}, in Kelvin) and pressure (p_{im}, in bars), compression ratio (r_c), fuel cetane number (CN), and mean piston velocity (S_p, in meters per second) as follows:

$$\tilde{\tau}_{\mathrm{id}} = \left(0.36 + 0.622S_{\mathrm{p}}\right)\exp\left[\left(\frac{618,840}{\mathrm{CN} + 25}\right) \times \left(\frac{1}{RT_{\mathrm{im}}r_{\mathrm{c}}^{k-1}} - \frac{1}{17,190}\right) \times \left(\frac{21.2}{P_{\mathrm{im}}r_{\mathrm{c}}^{k} - 12.4}\right)^{0.63}\right]. \tag{19}$$

The ignition delay period can be expressed in milliseconds according to $\tau_{\mathrm{id}} = [\tilde{\tau}_{\mathrm{id}}/0.006N]$. Values of the empirical coefficients k_1, k_2, k_3, and k_4 are reported by Heywood [32] and Rossini et al. [38] as follows: $k_1 = 2 + 1.25 \times 10^{-8}(\tau_{\mathrm{id}}/N)^{2.4}$, $k_2 = 5000$, $k_3 = (14.2/\phi^{0.644})$, and $k_4 = 0.79k_3^{0.25}$. The fuel burning rate with respect to the crank angle (θ) can be obtained by differentiating Equation 18 as follows:

$$\frac{\left(\frac{dm_{\mathrm{fb}}}{d\theta}\right) = m_{\mathrm{ft}}\beta k_1 k_2\left(1 - \hat{\theta}^{k_1}\right)^{(k_2-1)}\hat{\theta}^{(k_1-1)} + k_3 k_4(1-\beta)\hat{\theta}^{(k_4-1)}\exp\left(-k_3\hat{\theta}^{k_4}\right)}{\Delta\theta_{\mathrm{comb}}}. \tag{20}$$

Temperature-pressure relationships

The changes of pressure and temperature with respect to time are defined as follows:

$$dot; P = \frac{\rho}{\partial\rho/\partial P}\left(-\frac{dot; V}{V} - \frac{dot; T}{\rho}\frac{\partial\rho}{\partial T} - \frac{dot; \phi}{\rho}\frac{\partial\rho}{\partial\phi} + \frac{dot; m}{m}\right), \tag{21}$$

$$dot; T = \frac{B^\circ}{A^\circ}\left\{-\frac{dot; V}{V} + \left(1 - \frac{h}{B^\circ}\right)\frac{dot; m}{m} - \frac{C^\circ}{B^\circ}dot; \phi + \frac{1}{mB^\circ}\left(\sum_j dot; m_j h_j + dot; Q_w\right)\right\}, \tag{22}$$

where $A^\circ = \left(\frac{\partial h}{\partial T}\right) + \frac{(\partial\rho/\partial T)}{(\partial\rho/\partial P)}\left(\frac{1}{\rho} - \frac{\partial h}{\partial P}\right)$, $B^\circ = \frac{1-\rho(\partial h/\partial P)}{(\partial\rho/\partial P)}$, $C^\circ = \frac{\partial h}{\partial\phi} + \frac{(\partial\rho/\partial\phi)}{(\partial\rho/\partial P)}\left(\frac{1}{\rho} - \frac{\partial h}{\partial P}\right)$.

Soot mechanism

Hiroyasu and Kadota [39] developed a simple soot model to predict the production rate of soot mass ($dot; m_s$) using a single-step competition between the soot mass formation rate $dot; m_{\mathrm{sf}}$ and the soot mass oxidation rate $dot; m_{\mathrm{so}}$ according to $dot; m_s = dot; m_{\mathrm{sf}} - dot; m_{\mathrm{so}}$. Morel and Keribar [40] and Wahiduzzaman et al. [41] suggested models for soot formation and their oxidation in diesel engines. In their models, the time rates of soot formation and oxidation, respectively, are given by

$$dot; m_{\mathrm{sf}} = 0.3\ m_{\mathrm{fb}}\exp\left(-\frac{3,000}{T_{\mathrm{af}}}\right) dot; \theta, \tag{23}$$

$$dot; m_{\mathrm{so}} = \left(\frac{0.4m_s\exp\left(-\frac{10,000}{T}\right)\sqrt{p_{O_2}}}{6\rho_s d_s N}\right) dot; \theta, \tag{24}$$

where m_{fb} is the mass of fuel burned, T_{af} is the adiabatic flame temperature at an equivalence ratio of 1.1, d_s (the soot particle diameter) = 0.022×10^{-6} m, ρ_s (soot density) = 1,800 kg/m^3 [42].

NO_x mechanism

The extended Zeldovich mechanism for NO_x formation rate is formed as follows [29]:

$$\frac{d}{dt}[NO] = 2k_{1f}[O]_e[N_2]_e\left(\frac{1 - [NO]^2/K_{12}[O_2]_e[N_2]_e}{1 + k_{1b}[NO]/\left(k_{2f}[O_2]_e + k_{3f}[OH]\right)_e}\right), \tag{25}$$

where $K_{12} = (k_{1f}/k_{1b})/(k_{2f}/k_{2b})$. The constants above are taken as follows: $k_{1f} = 7.6 \times 10^{13}\exp\left(-\frac{38,000}{T}\right)$, $k_{2f} =$

$6.4 \times 10^9 T \exp\left(-\frac{3{,}150}{T}\right)$, $k_{3f} = 4.1 \times 10^{13}$, $k_{1b} = 1.6 \times 10^{13}$, and $k_{2b} = 1.5 \times 10^9 T \exp\left(-\frac{19{,}500}{T}\right)$.

Solution procedure

The governing equations (Equations 1 to 25) of the comprehensive model subjected to the assumptions previously mentioned in the mathematical model are solved for different processes of the engine cycle. The fourth-order Runge-Kutta method is used to simulate the comprehensive model equations at the prescribed initial conditions. The engine cycle starts from the moment the exhaust valve closes. The solution is carried out with end results of each process taken as the starting conditions for the following process and end values of the completed thermal cycle taken as the starting values for the subsequent cycle. The solution is iterated and considered satisfactory when the end values of the temperature and pressure of the cycle are within ±0.001% from the starting values of the temperature and pressure of the same cycle. Inputs to the code are the engine specifications, operating conditions, and fuel data listed in Table 1. The numerical output results are the instantaneous pressure, temperature, volume, heat transfer, work, mass exchange, fuel heat release, and engine performance variables such as bsfc, indicated mean effective pressure, brake mean effective pressure, indicated power, brake power, and volumetric, indicated, brake, and mechanical efficiencies.

Table 1 Engine specifications and operating conditions

Specifications/operating conditions	Inputs/values
Engine type	HELWAAN M114
Cylinder bore	112 mm
Stroke	115 mm
Number of cylinders	4
Intake valve open angle	9° BTDC
Intake valve close angle	43° ABDC
Exhaust valve open angle	40° BBDC
Exhaust valve close angle	12° ATDC
Start of injection	22° BTDC
Number of injector nozzles	1
Diameter of injector nozzles	0.45 mm
Conventional compression ratio	16.4
Engine speed range	1,500 to 2,800 rpm
Overall equivalence ratio	0.47
Fuel type	$C_{10.8}H_{18.7}$
Lower latent heat of fuel	42 MJ/kg

BTDC before top dead center, *ABDC* after bottom dead center, *BBDC* before bottom dead center, *ATDC* after top dead center.

Results and discussions

Simulation results and its validation

The numerical simulation results are performed for a conventional DI diesel engine with constant compression ratio (specifications and operating conditions are given in Table 1). Samples of these results represent the engine thermodynamic cycle, the temperature and pressure variation throughout the cycle, the rates of mass entering and leaving the cylinder, and the in-cylinder NO_x and soot history along the cycle which are displayed in Figures 1, 2, and 3, respectively.

In order to validate the present developed model, comparisons between the numerical results and the available measurements are conducted. Two sets of experimental data reported by Han and Reitz [3] and Wahiduzzaman et al. [41] are used to validate the developed model as shown in Figure 4. Figure 4 shows the comparison between the predicted and measured pressures and heat release versus the crank angle. Based on this figure, one can observe that there is a good concurrence between measured and predicted results for different engines under different operating conditions. More details about the simulation results and validation can be found in a previous work of the authors [43].

Variation of the compression ratio

The numerical simulation results are used to determine the range of r_c corresponding to the optimum engine performance for two cases: (1) minimum bsfc and (2) maximum brake torque. This was computed using optimization techniques under the restriction of maximum allowable pressure and temperature. The numerical results of the optimization process, for engine

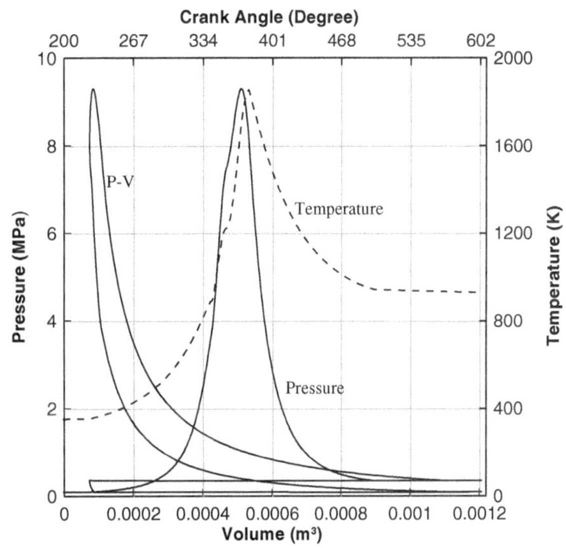

Figure 1 Pressure-volume diagram along engine cycle; cylinder pressure and temperature versus crank angle at 2,100 rpm.

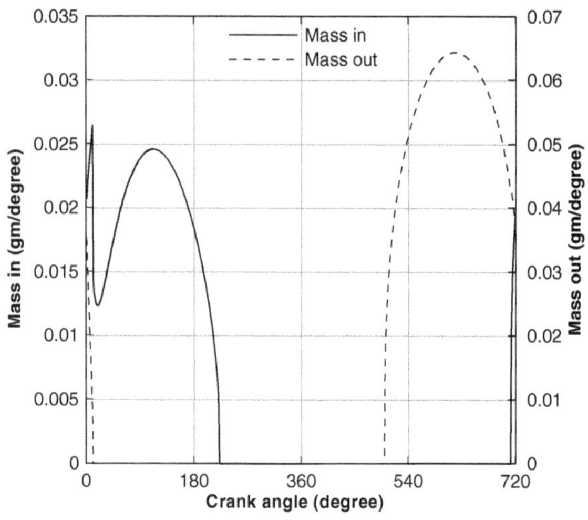

Figure 2 **Instantaneous mass flow rates through intake and exhaust valves versus crank angle at 2,100 rpm.**

specifications previously mentioned as shown in Table 1, are presented in Figures 5,6,7,8,9, Figures 10,11,12,13,14. It is worth mentioning that for all figures, the solid lines represent the performance of optimum variable r_c, while dashed lines represent the performance of conventional constant r_c engine.

Case 1: minimum brake-specific fuel consumption

Figures 5, 6, 7, 8, and 9 show the results at minimum bsfc. Figure 5 shows the comparison between the operation under constant compression ratio ($r_c = 16.4$) and the hypothetical operation under varying compression ratio to achieve working with minimum bsfc (230 g/kW.h) as possible. The figure shows that the bsfc for the constant-compression-ratio engine changes between

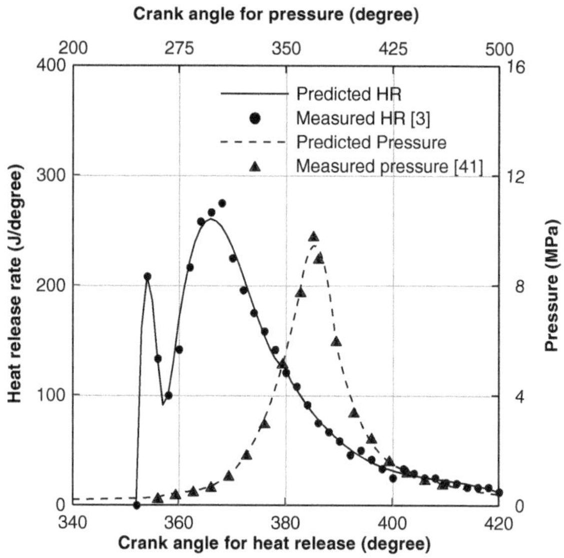

Figure 4 **Predicted and measured pressures and heat release versus crank angle.**

238.5 and 261 g/kW.h passing through a minimum value of 230 g/kW.h at an engine speed of about 1,900 rpm. In addition, for the variable-compression-ratio engine, the bsfc is kept constant at its minimum value of 230 g/kW.h by varying the compression ratio between 16.4 and 17.8. The shown variation trend in compression ratio emphasizes that the variation in compression ratio should be considered to decrease the bsfc to its minimum possible value. Since the increase in compression ratio is accompanied by an increase in both cylinder pressure and brake power, consequently, a decrease in bsfc is achieved. The percentage reduction in bsfc ranges

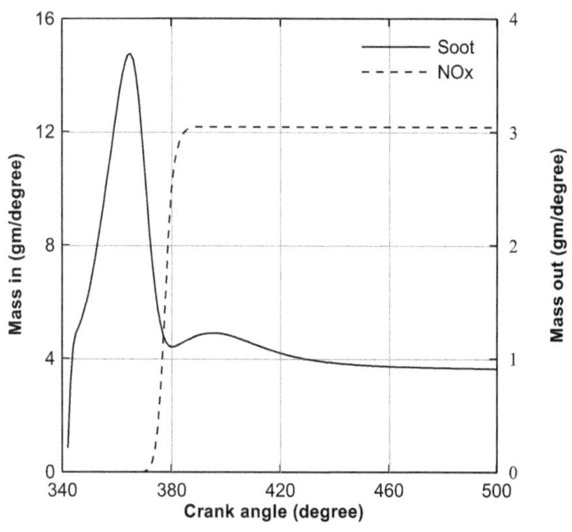

Figure 3 **Soot and NO_x inside the cylinder of the engine versus crank angle at 2,100 rpm.**

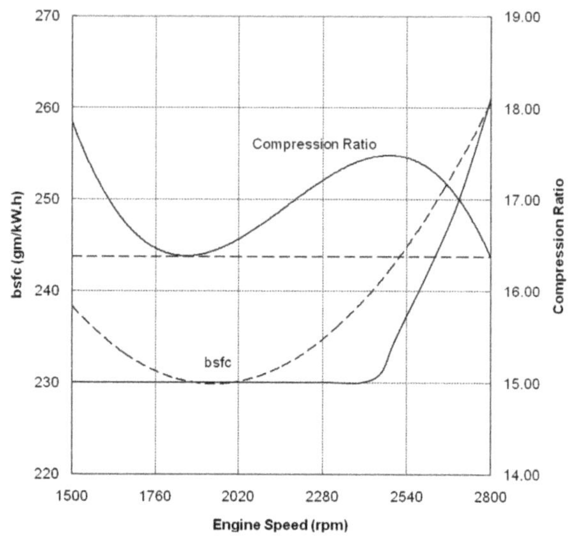

Figure 5 **bsfc and compression ratio versus speed for constant- and variable-compression-ratio engines.**

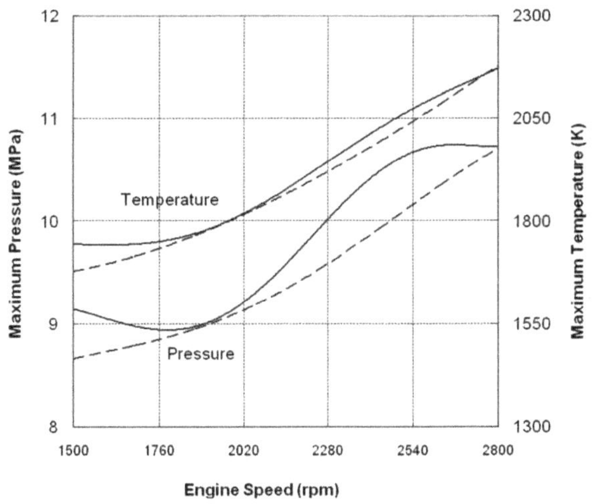

Figure 6 Maximum pressure and temperature versus speed for constant- and variable-compression-ratio engines.

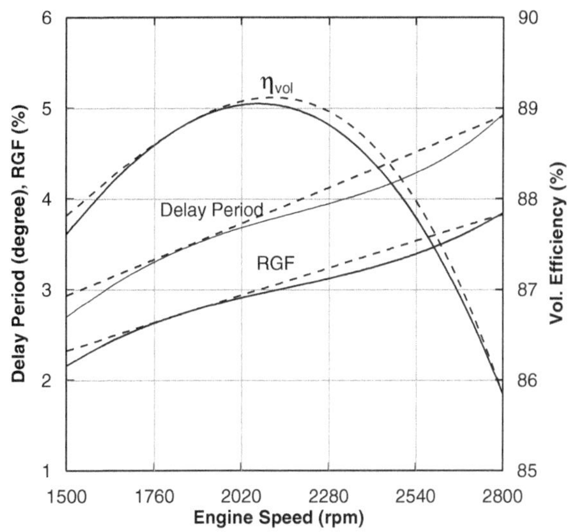

Figure 8 Delay period, RGF, and volumetric efficiency versus speed for constant- and variable-r_c engines for minimum bsfc.

between 3.7% at 1,500 rpm and 4.4% at 2,460 rpm passing through 0% at 1,900 and 2,800 rpm. Changing the compression ratio for an operating engine is desirable and could be attained. Its effect on maximum cycle pressure and temperature must be studied as they represent important parameters in the design of engine's different parts.

Figure 6 shows the increase in maximum pressure and temperature when operating under variable compression ratio compared to the operation under constant compression ratio. Compared to the case of a constant compression ratio, the increase in maximum cycle pressure is about 0.53 MPa at 1,500 rpm and 0.55 MPa at 2,460 rpm passing through 0 MPa at both 1,900 and 2,800 rpm with variable compression ratio. The noticeable increase in maximum temperature is about 72 K at 1,500 rpm and 40 K at 2,460 rpm passing through 0 K at both 1,900 and 2,800 rpm. These increases still do not exceed the engine design's maximum pressure and temperature of the conventional engine. Changes in brake power and torque with speed for the variable compression ratio at constant minimum bsfc are shown in Figure 7. It shows an increase in brake power by about 1.67 kW at 1,500 rpm and 1.6 kW at 2,460 rpm. Moreover, the brake torque increases by 9 N.m at 1,500 rpm and 8 N.m at 2,460 rpm.

Figure 8 shows that varying the compression ratio results in a decrease in the delay period, volumetric efficiency, and residual gas fraction (RGF) over the engine speed range (1,500 to 2,460 rpm). It is seen that this

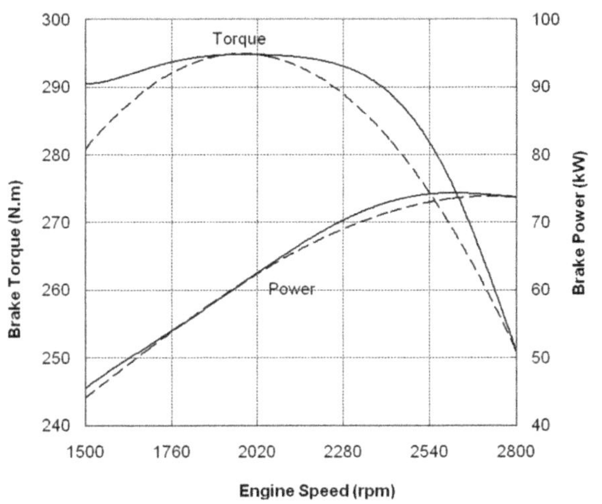

Figure 7 Brake power and torque versus speed for constant- and variable-compression-ratio engines.

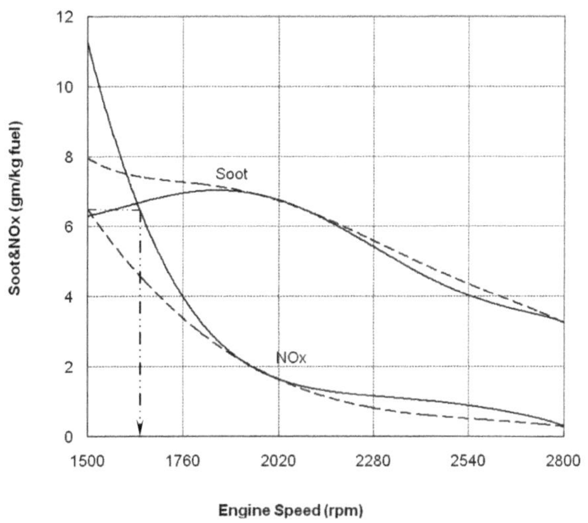

Figure 9 Soot and NO_x emission versus speed for constant- and variable-r_c engines for minimum bsfc.

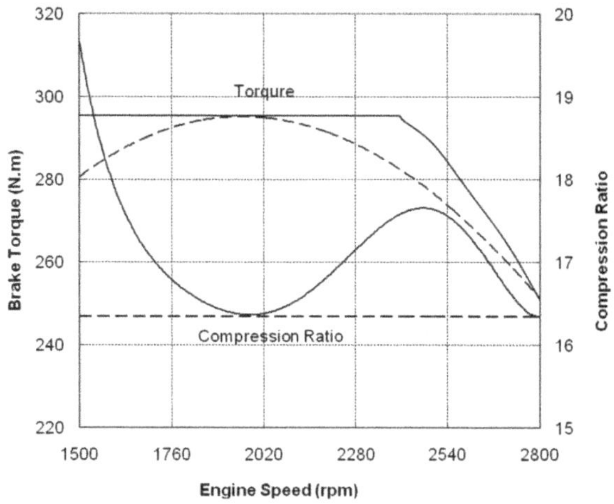

Figure 10 Brake torque and power versus speed for constant- and variable-compression-ratio engines.

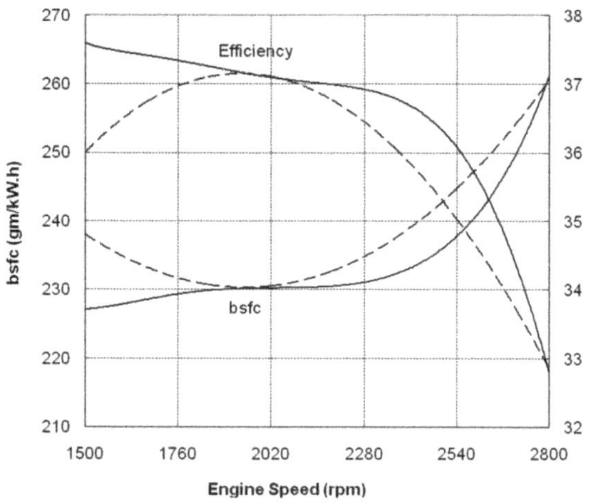

Figure 12 bsfc and thermal efficiency versus speed for constant- and variable-r_c engines.

decrease changes between 0.2° of crank angle at 1,500 rpm and 0.2° of crank angle at 2,460 rpm. The decrease occurs due to the fact that the increase in compression ratio results in an increase of the air pressure and temperature at the end of the compression process which are inversely proportional to ignition delay. The decrease in the ignition delay is desirable as it causes ignition to occur before greater amounts of fuel is injected in the cylinder. Consequently, the rates of heat release and pressure rise are then controlled primarily by the rate of injection and air-fuel mixing, resulting in a smoother engine operation. Long ignition delay allows for more fuel to be injected before auto-ignition occurs, which results in a very rapid pressure rise as the combustion starts and reaches high-peak pressures. Under extreme conditions, when auto-ignition of most of the injected fuel occurs, an audible knocking sound is noticed, often referred to as a 'diesel knock.' Also, if long ignition delay occurs sufficiently late and reaches the expansion process, the burning process may be quenched. This leads to incomplete combustion, reduced power output, and poor fuel-power conversion efficiency. Also, the figure shows that the volumetric efficiency and RGF are reduced by about 4.3% and 0.3%, respectively, at a speed of 1,500 rpm. These reductions continue to decrease until reaching 0% at 1,900 rpm. These reductions, then, increase again until reaching about 4.3% and 0.1% at 2,460 rpm. This variation was attributed to the increase in compression ratio. It means that a decrease in clearance volume leads to a decrease in charging-up the

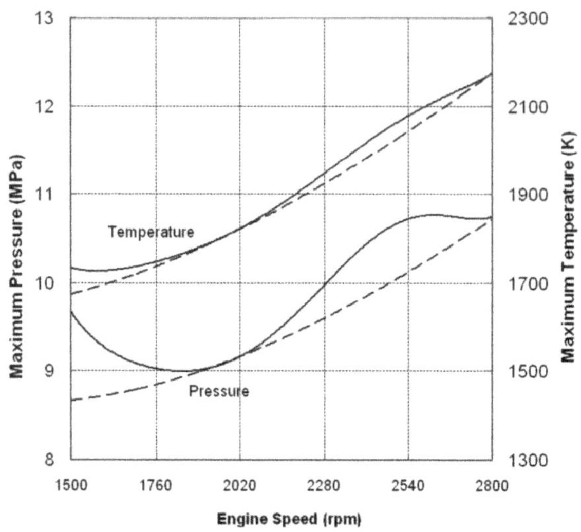

Figure 11 Maximum cylinder pressure and temperature versus speed for constant- and variable-compression-ratio engines.

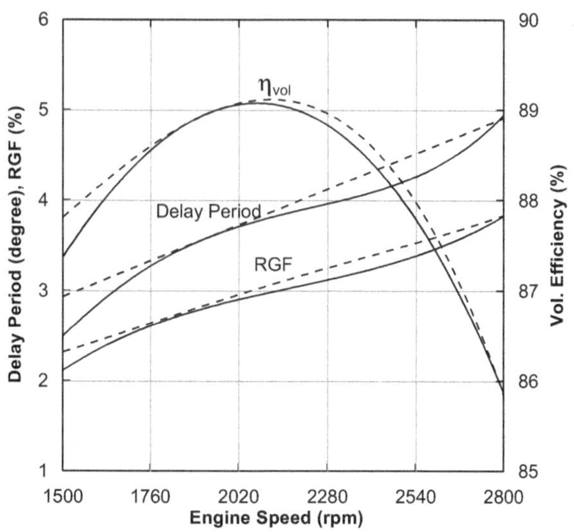

Figure 13 Delay period, RGF, and volumetric efficiency versus speed for constant- and variable-r_c engines for maximum torque.

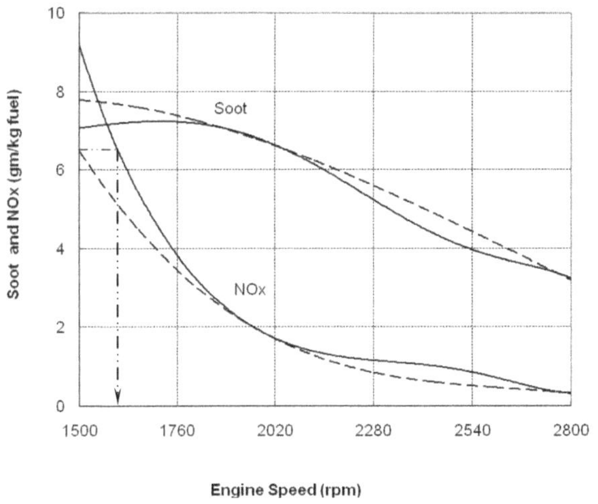

Figure 14 Soot and No_x emission versus speed for constant- and variable-compression-ratio engines.

cylinder. The general decrease in residual gas fraction is desirable because it decreases the dilution of the charge in the cylinder and increases the scavenging efficiency of the engine. However, these reductions are small and can be neglected.

The effect of the optimal variation of the compression ratio corresponding to the minimum bsfc on soot and NO_x emissions is shown in Figure 9. For all reported speed range, the soot emission from the engine with variable compression is less than that of the conventional engine. The decrease in soot emission is found to be about 21% at 1,500 rpm and decreases until it reaches 0% at 1,900 rpm. Then, it increases again to 9% at 2,460 rpm and returns to 0% at 2,800 rpm. On the other hand, the formed NO_x inside the variable-compression-ratio engine is much higher compared to that inside the constant-compression-ratio engine. The maximum increase reaches about 11.2 g/kg of fuel instead of 6.4 g/kg of fuel (r_c = constant) at 1,500 rpm. Beyond 1,900 rpm, the NO_x emission from the engine with variable compression is also greater than that of conventional engine and is about 0.97 g/kg of fuel at 2,460 rpm. The increase of NO_x is a drawback to the advantages mentioned when using variable-compression-ratio engines to achieve minimum bsfc (maximum efficiency). This drawback can be overcome by reducing the engine speed range from 1,500 to 2,800 rpm to 1,665 to 2,800 rpm as illustrated on the figure.

To conclude based on Figures 6, 7, 8, and 9, varying the compression ratio, within the entire range of engine speed of 1,500 to 2,800 rpm, to attain a minimum bsfc results in significant variations in the predicted performance variables compared to those predicted at constant compression ratio ($r_c = 16.4$). These variations can be summarized as follows: (1) at an engine speed of 2,460 and 2,800 rpm, there is no variation in all investigated performance variables; (2) for the other values of engine speed, the maximum pressure, maximum temperature, brake torque, brake power, and unfavorable NO_x are increasing. On the other hand, the delay period, volumetric efficiency, RGF, and soot formation are decreasing. Table 2 summarizes the predicted results of performance variables at different values of engine speed.

Case 2: maximum brake torque

The predicted performance variables corresponding to maximum brake torque compared to those predicted at a conventional constant compression ratio ($r_c = 16.4$) are displayed in Figures 10, 11, 12, 13, and 14. Figure 10 shows that the brake torque, in the case of a constant compression ratio, changes between 281 and 252 N.m passing through a maximum of 295 N.m at about 1,900 rpm. For the variable compression ratio, the brake torque is kept constant at its maximum value of 295 N. m over a speed ranging from 1,500 to 2,460 rpm by

Table 2 Results of varying engine r_c compared to constant r_c ($r_c = 16.3$) to attain the minimum bsfc

Engine performance variables	$N = 1{,}500$ rpm		$N = 1{,}900$ rpm		$N = 2{,}460$ rpm	
	Increase (%)	Decrease (%)	Increase (%)	Decrease (%)	Increase (%)	Decrease (%)
bsfc	-	3.7	0	0	-	4.4
Maximum pressure	6	-	0	0	5.5	-
Maximum temperature	4.3	-	0	0	1.9	-
Brake torque	3.3	-	0	0	2.8	-
Brake power	3.8	-	0	0	2.2	-
Delay period	-	6.9	0	0	-	4.5
Volumetric efficiency	-	0.3	0	0	-	0.1
RGF	-	4.3	0	0	-	4.3
Soot	-	21	0	0	-	8.5
NO_x	75	-	0	0	76	-

bsfc brake-specific fuel consumption, *RGF* residual gas fraction.

Table 3 Results of varying engine r_c compared to constant r_c ($r_c = 16.4$) to attain maximum brake torque

Engine performance variables	$N = 1{,}500$ rpm		$N = 1{,}900$ rpm		$N = 2{,}460$ rpm	
	Increase (%)	Decrease (%)	Increase (%)	Decrease (%)	Increase (%)	Decrease (%)
bsfc	-	4.6%	0	0	-	2.7%
Maximum pressure	10.3	-	0	0	6%	-
Maximum temperature	3.6	-	0	0	2	-
Brake torque	5.4	-	0	0	4.6	-
Thermal efficiency	4.4	-	0	0	2.8	-
Delay period	-	15.3	0	0	-	6.8
Volumetric efficiency	-	0.48	0	0	-	0.25
RGF	-	6.5	0	0	-	5
Soot	-	9.7	0	0	-	12
NO_x	21.7	-	0	0	90.7	-

bsfc brake-specific fuel consumption, *RGF* residual gas fraction.

varying the compression ratio between 16.4 and 19.3. This means that the brake torque increases by 5.4% at 1,500 rpm and 4.6% at 2,460 rpm. The shown variation trend in compression ratio urges that actual variation in compression ratio should be considered to increase the brake torque to its constant maximum value.

Figure 11 shows the increase in p_{max} and T_{max} when operating under variable compression ratio compared to the operation under constant compression ratio. The increase in p_{max} is about 0.9 MPa at 1,500 rpm and 0.6 MPa at 2,460 rpm passing through 0 MPa both at about 1,900 and 2,800 rpm. The noticeable increase in T_{max} is about 60 K at 1,500 rpm and 40 K at 2,460 rpm passing through 0 K both at 1,900 and 2,800 rpm. These increases in p_{max} and T_{max} are tolerable, as they are still below the allowable pressure and temperature of the conventional engine. Similar observations for the changes of brake efficiency and bsfc versus engine speed for operation under constant and variable compression ratios are shown in Figure 12. If the compression ratio is constant, the brake thermal efficiency varies between 36% and 32.9% passing through a maximum value of 37.3% at an engine speed of 1,900 rpm. In case of varying compression ratio (between 16.4 and 19.3 for maximum torque), the efficiency changes between 37.6% at 1,500 rpm and 36.5% at 2,460 rpm. In addition to gains in brake torque and efficiency, the figure shows a decrease in the bsfc over an engine speed ranging from 4.6% at 1,500 rpm to 2.7% at 2,460 rpm passing through 0% both at 1,900 and 2,800 rpm.

Figure 13 shows a significant decrease in delay period, volumetric efficiency, and RGF over an engine speed ranging from 1,500 to 2,460 rpm with values between 15.3% (0.45° of crank angle) at 1,500 rpm and 6.8% (0.3° of crank angle) at 2,460 rpm. Also, the figure shows that the volumetric efficiency is reduced by about 0.48% at a speed of 1,500 rpm and 0.25% at 2,460 rpm passing through 0% both at 1,900 and 2,800 rpm. Also, this figure shows that variation of the compression ratio for optimum brake torque results in a general decrease in RGF by about 6.5% at 1,500 rpm and 5% at 2,460 rpm passing through 0% both at 1,900 and 2,800 rpm.

The effect of the optimal variation of compression ratio for maximum brake torque on soot and NO_x emission is shown in Figure 14. The soot emission from the engine with variable compression ratio is less than that of the conventional engine by about 9.7% at 1,500 rpm, 0% both at 1,900 and 2,800 rpm, and 12% at 2,460 rpm. The formed NO_x inside the variable-compression-ratio engine is very high compared to that inside the constant-compression-ratio engine by about 21.7% (2.65 g/kg of fuel) at 1,500 rpm, 90.7% (0.461 g/kg of fuel) at 2,460 rpm, and 0% at both 1,800 and 2,800 rpm. This increase is due to the increase of cylinder temperature with increasing compression ratio. The increase in NO_x at low speed, i.e., 1,500 rpm, can be avoided by avoiding the engine to run in the speed range of 1,500 to 1,620 rpm as illustrated in Figure 14.

The conclusion based on Figures 10, 11, 12, 13, and 14 is that varying the compression ratio of the diesel engine, within the entire range of engine speed of 1,500 to 2,800 rpm, to achieve a maximum engine torque leads to variations in the predicted performance variables compared to those predicted at constant compression ratio ($r_c = 16.4$). These variations can be summarized as follows: (1) at an engine speed of 1,500 and 2,460 rpm, there is no variation in all studied variables; (2) for the other values of engine speed, the maximum pressure, maximum temperature, brake torque, thermal efficiency, and unfavorable NO_x are increasing. On the contrary, the delay period, volumetric efficiency, residual gas fraction, and soot formation are decreasing. Table 3 summarizes the predicted results of performance variables at different values of engine speed.

Conclusions

The present rapid thermodynamic simulation for the direct-injection diesel engine is based on a single-zone combustion modeling with a rapid premixed burning phase followed by a slower mixing-controlled burning phase. Comparison with other experimental results (under constant compression ratio) verifies very well the simulation predictions of cycle performance. The main conclusions of this study indicate that varying the compression ratio to achieve minimum bsfc or maximum brake torque for moderate diesel engine under the restriction that the maximum cylinder pressure and temperature do not exceed those of the conventional engine results in saving fuel consumption, higher brake efficiency and power, and reduction in soot emission from the engine. Also, an increase in NO_x is noticed at low speed. This drawback is considerable and can be overcome by reducing the operation speed range from 1,500 to 2,800 rpm to 1,665 to 2,800 rpm in the case of minimum bsfc and from 1,500 to 1,800 rpm to 1,620 to 2,800 rpm in the case of maximum brake torque.

Abbreviations

CN: fuel cetane number; c_p, c_v: constant pressure and volume specific heats; D_{iv}: intake valve head diameter; *D*: cylinder bore; *H*, *h*: enthalpy and specific enthalpy; *K*: specific heat ratio; *M*: mass; *N*: crank speed; *N*: number of moles; n_c: number of cylinders per engine; n_{iv}, n_r: number of intake valves and rings per cylinder; *p*: pressure; *Q*, *q*: heat transfer and heat flux; r_c: compression ratio; *R*: gas constant; S_p: mean piston velocity; *T*: cylinder temperature; *V*, V_c, V_d: swept, clearance, and maximum displacement volumes; *X*: mole fraction; Θ: crank position angle; Λ: connecting rod length to crank radius ratio; θ_{evo}, θ_{ivo}: exhaust and intake valve opening angles; $\Delta\theta_{eopen}$: opening duration angle of exhaust valve; $\Delta\theta_{iopen}$: opening duration angle of intake valve; ρ: density; φ: equivalence ratio; τ, T: ignition delay period and ignition delay angle; B: backward; comb: combustion; e: exit, equilibrium; em: exhaust manifold; ex: exhaust; f: fuel, forward; fb: fuel burning; ft: total fuel; i: indicated, inlet, injection start; ig: ignition; im: intake manifold; in: intake; o: ambient condition; p: piston; s: stoichiometry, soot; sf: soot formation; so: soot oxidation; w: wall, atoms of carbon in fuel-air molecule; x: hydrogen atoms in fuel-air molecule; y: oxygen atoms in fuel-air molecule; z: nitrogen atoms in fuel-air molecule.

Competing interests

The authors declare that they have no competing interests.

Authors' contributions

MM conceived the concept and procedures of the present work, developed the model, and carried out the computations and analysis. MA checked the equations and analysis, and reviewed the revised manuscript. YM reviewed the graphs and results. All authors read and approved the final manuscript.

Authors' information

MM earned his Ph.D. degree in Mechanical Engineering from the University of Manitoba, Canada, in 2007. He is currently an assistant professor in the Department of Mechanical Engineering at the Assiut University, Egypt. He has been interested in heat and mass transfer relating to droplets and spray modeling, turbulence, combustion, and modeling and simulation of internal combustion engines. MA received his Ph.D. degree in Mechanical Engineering from Mississippi State University, USA, in 1994. He is currently a professor in the Department of Mechanical Engineering at the Assiut University, Egypt. His current research interests focus on sprays and liquid atomization, fuel cells, solar hydrogen generation, and two-phase flows. YM received his Ph.D. degree in Mechanical Engineering from Kansas State University, USA, in 1984. Now, he is an emeritus professor in the Department of Mechanical Engineering at the Assiut University, Egypt. His research interests focus on combustion and management and optimization of thermal systems.

Acknowledgment

This work has been fully supported by Assiut University and the Mechanical Engineering Department.

References

1. Kaario, O, Lendormy, E, Sarjovaara, T, Larmi, M, Rantanen, P: In-cylinder flow field of a diesel engine. SAE Paper No. 2007-01-4046., (2007). doi:10.4271/2007-01-4046
2. Lee, C-FF, Zhao, J: Modeling of blowby in a small-bore high-speed direct injection optically accessible diesel engine. SAE Paper No. 2006-01-0649., (2006). doi:10.4271/2006-01-0649
3. Han, Z, Reitz, RD: Turbulence modeling of internal combustion engines using RNG k-ϵ models. Combust. Sci. Technol. **106**, 267–295 (1995)
4. Hessel, RP, Rutland, CJ: Intake flow modeling in a four-stroke diesel engine using KIVA-3. J. Propulsion Power **11**, 378–384 (1995)
5. Reitz, RD, Rutland, CJ: Development and testing of diesel engine CFD models. Prog. Energy Sci. **21**, 173–196 (1995)
6. Kopa, RD, Hollander, BR, Hollander, FH, Kimura, H: Combustion temperature, pressure and products at chemical equilibrium. Report No. 64-1 Dept. of Mech. Eng., Univ. of California, (1964)
7. Arrègle, J, Lòpez, JJ, Garcia, JM, Fenollosa, C: Development of of a zero-dimensional diesel combustion model. Part 1: analysis of the quasi-steady diffusion combustion phases. Appl. Therm. Eng. **23**, 1301–1317 (2003)
8. Kouremenous, DA, Rakopoulos, CD, Hountalas, D: Thermodynamic analysis of indirect injection diesel engine by using two-zone modeling of combustion. Trans. ASME, J. Eng. Gas Turbine Power **112**, 138–149 (1990)
9. Chmela, FG, Pirker, GH, Wimmer, A: Zero-dimensional ROHR simulation for DI diesel engines–a generic approach. Energy Convers. Manage. **48**, 2942–2950 (2007)
10. Basha, SA, Gopal, KR: In-cylinder fluid flow, turbulence and spray models—a review. Renew. Sustain. Energy Rev. **13**, 1620–1627 (2009)
11. Roberts, M: Benefits and challenges of variable compression ratio (VCR). SAE Paper No. 2003-01-0398., (2003). doi:10.4271/2003-01-0398
12. Ladommatos, N, Balian, RA: Combustion in a direct injection diesel engine with increased clearance volume. Proc. Instn. Mech. Engr. **204**, 187–197 (1990)
13. Sobotowski, RA, Potrer, BC, Pilley, AD: The development of novel variable compression ratio, direct injection diesel engine. SAE Paper No. 910484., (1991). doi:10.4271/910484
14. Rychter, TJ, Teodorczyk, A, Stone, CR, Leonard, HJ, Ladommatos, N, Charlton, SJA: Theoretical study of a variable compression ratio turbocharged diesel engine. Proc. Instn. Mech. Engr. **206**, 227–238 (1992)
15. Freudenstein, F, Maki, ER: Variable-displacement piston engine. US Patent 4,270,495, 2 June 1981
16. Yamin, JA, Dado, MH: Performance simulation of a four-stroke engine with variable stroke-length and compression ratio. Appl. Energy **77**, 447–463 (2004)
17. Raheman, H, Ghadge, SV: Performance of diesel engine with biodiesel at varying compression ratio and ignition timing. Fuel **87**, 2659–2666 (2008)
18. Jindal, S, Nandwana, BP, Rathore, NS, Vashistha, V: Experimental investigation of the effect of compression ratio and injection pressure in a direct injection diesel engine running on Jatropha methyl ester. Appl. Therm. Eng. **30**, 442–448 (2010)
19. Muralidharan, K, Vasudevan, D, Sheeba, KN: Performance, emission and combustion characteristics of biodiesel fuelled variable compression ratio engine. Energy **36**, 5385–5393 (2011)
20. Gandure, J, Ketlogetswe, C: Comparative performance analysis of marula oil and petrodiesel fuels on a variable compression ratio engine, pp. 13–15. AFRICON, 2011, Victoria Falls, Livingstone (2011)
21. Bishop, IN: Effect of design variables on friction and economy. SAE Trans. **3**, 334–358 (1965)

22. Heywood, JB: Modeling combustion and performance characteristics of internal combustion engines. In: Proceedings of the 1976 Heat Transfer and Fluid Mechanics Institute, pp. 180–195. Stanford University Press, Palo Alto (1976)
23. Eichelberg, G: Some new investigations on old combustion-engine problems. Engineering **149**, 463–547 (1939)
24. Rakopoulos, CD, Hountalas, DT: Net and gross heat release rate calculation in DI diesel engine using various heat transfer models, pp. 13–19. Paper presented at the 1994 American Society of Mechanical Engineers-winter annual meeting, Chicago (1994)
25. Annand, WJD: Heat transfer in cylinders of reciprocating internal combustion engines. Proc. Instn. Mech. Engr. **177**, 973–990 (1963)
26. Annand, WJD, Ma, TH: Instantaneous heat transfer rates to cylinder head surface of small compression ignition engine. Proc. Instn. Mech. Engr. **185**, 967–987 (1970)
27. Woschni, GA: Universally applicable equation for the instantaneous heat transfer coefficient in internal combustion engine. SAE Trans. **76**, 3065–3083 (1967)
28. Blawton, B: Effect of compression and expansion on instantaneous heat transfer in reciprocating internal combustion engines. Proc. Instn. Mech. Engr. **201**, 175–186 (1987)
29. Campbell, AS: Thermodynamic Analysis of Combustion Engines. John Wiley & Sons Press, New York (1979)
30. Ogurt, T, Takeda, A, Torii, K, Inaba, S: Radiation heat transfer of combustion flames in a diesel engine. Bull. JSME **28**, 638–646 (1985)
31. Cheung, CG, Leung, CW, Leung, TP: Modelling spatial radiative heat flux distribution in a direct injection diesel engine. Proc. Instn. Mech. Engr. **208**, 275–283 (1994)
32. Heywood, JB: Internal Combustion Engine Fundamentals, 1st edn. McGraw-Hill Science/Engineering/Math, New York (1988)
33. Ramos, JL: Internal Combustion Engine Modeling, 1st edn. Taylor & Francis, London (1989)
34. Zelenik, FJ: Combustion modeling in internal combustion engines. Combust. Sci. Technol. **12**, 159–164 (1976)
35. Gu, F, Ball, AD: Diesel injector dynamic modelling and estimation of injection parameters from impact response. Part 1: modelling and analysis of injector impacts. Proc. Instn. Mech. Engr **210**, 293–302 (1996)
36. Watson, N, Pilley, AD, Marzouk, MA: Combustion correlation for diesel engine simulation. SAE Paper No. 800029., (1980). doi:10.4271/800029
37. Mehta, PS, Singal, SK, Pundir, BPA: Comprehensive simulation model for mixing and combustion characteristics of small direct injection diesel engine. Proc. Instn. Mech. Engr. **209**, 117–126 (1995)
38. Rossini, FD, Pitzer, KS, Arnett, RL, Braun, RM, Primentel, GC: Selected Values of Physical and Thermodynamic Properties of Hydrocarbons and Related Compounds. Carnegie Press, Pittsburgh (1953)
39. Hiroyasu, H, Kadota, T: Models of combustion and formation of nitric oxide and soot in DI diesel engine. SAE Paper No. 760129., (1976). doi:10.4271/760129
40. Morel, T, Keribar, R: Heat radiation in D.I. diesel engines. SAE Paper No. 860445., (1986). doi:10.4271/860445
41. Wahiduzzaman, S, Morel, T, Timar, J, DeWitt, DP: Experimental and analytical study of heat radiation in a diesel engine. SAE Paper No. 870571., (1987). doi:10.4271/870571
42. Han, Z, Uludogan, A, Hampson, GJ, Reitz, RD: Mechanism of soot and NOx emission reduction using multiple-injection in a diesel engine. SAE Paper No. 960633., (1996). doi:10.4271/960633
43. Al-Sood, M, Abdel-Latif, A, Abdel-Rahim, Y, Ibrahim, A: Optimum compression ratio variation of a 4-stroke, direct injection diesel engine for minimum bsfc. SAE Paper No. 1999-01-2519. (1999). doi:10.4271/1999-01-2519

Permissions

All chapters in this book were first published in IJEEE, by Springer; hereby published with permission under the Creative Commons Attribution License or equivalent. Every chapter published in this book has been scrutinized by our experts. Their significance has been extensively debated. The topics covered herein carry significant findings which will fuel the growth of the discipline. They may even be implemented as practical applications or may be referred to as a beginning point for another development.

The contributors of this book come from diverse backgrounds, making this book a truly international effort. This book will bring forth new frontiers with its revolutionizing research information and detailed analysis of the nascent developments around the world.

We would like to thank all the contributing authors for lending their expertise to make the book truly unique. They have played a crucial role in the development of this book. Without their invaluable contributions this book wouldn't have been possible. They have made vital efforts to compile up to date information on the varied aspects of this subject to make this book a valuable addition to the collection of many professionals and students.

This book was conceptualized with the vision of imparting up-to-date information and advanced data in this field. To ensure the same, a matchless editorial board was set up. Every individual on the board went through rigorous rounds of assessment to prove their worth. After which they invested a large part of their time researching and compiling the most relevant data for our readers.

The editorial board has been involved in producing this book since its inception. They have spent rigorous hours researching and exploring the diverse topics which have resulted in the successful publishing of this book. They have passed on their knowledge of decades through this book. To expedite this challenging task, the publisher supported the team at every step. A small team of assistant editors was also appointed to further simplify the editing procedure and attain best results for the readers.

Apart from the editorial board, the designing team has also invested a significant amount of their time in understanding the subject and creating the most relevant covers. They scrutinized every image to scout for the most suitable representation of the subject and create an appropriate cover for the book.

The publishing team has been an ardent support to the editorial, designing and production team. Their endless efforts to recruit the best for this project, has resulted in the accomplishment of this book. They are a veteran in the field of academics and their pool of knowledge is as vast as their experience in printing. Their expertise and guidance has proved useful at every step. Their uncompromising quality standards have made this book an exceptional effort. Their encouragement from time to time has been an inspiration for everyone.

The publisher and the editorial board hope that this book will prove to be a valuable piece of knowledge for researchers, students, practitioners and scholars across the globe.

List of Contributors

Katarzyna Stępczyńska
Institute of Power Engineering and Turbomachinery, Silesian University of Technology, Konarskiego 18, Gliwice, 44-100, Poland

Henryk Łukowicz
Institute of Power Engineering and Turbomachinery, Silesian University of Technology, Konarskiego 18, Gliwice, 44-100, Poland

Sławomir Dykas
Institute of Power Engineering and Turbomachinery, Silesian University of Technology, Konarskiego 18, Gliwice, 44-100, Poland

Maatouk Khoukhi
Civil and Architectural Engineering Department, College of Engineering, University of Sultan Qaboos, P.O. Box 33, Al-Khoudh, Muscat 123, Oman

Naïma Fezzioui
Department of Sciences, University of Bechar, P.O. Box 417, Bechar 08000, Algeria

Pragya Sharma
GERMI RIIC (Research and Innovation Centre), Gandhinagar, Gujarat 382007, India

Tirumalachetty Harinarayana
GERMI RIIC (Research and Innovation Centre), Gandhinagar, Gujarat 382007, India
CSIR-NGRI (National Geophysical Research Institute), Hyderabad 500007, India

Mohammad Ghadimi
Department of Energy Engineering, Science and Research Branch, Islamic Azad University, P.O. Box 147789385, Tehran, Iran

Hossein Ghadamian
Department of Energy Engineering, Science and Research Branch, Islamic Azad University, P.O. Box 147789385, Tehran, Iran

Aliasghar A Hamidi
Department of Chemical Engineering, Tehran University, P.O. Box 147789385, Tehran, Iran

Farivar Fazelpour
Department of Energy System Engineering, Islamic Azad University of South Tehran Branch, P.O. Box 189, Tehran, Iran

Mehdi A Behghadam
Department of Mechanical Engineering, Islamic Azad University of Roudehen branch, P.O. Box 189, Tehran, Iran

Mohan T Raj
Department of Mechanical Engineering, SASTRA University, Thanjavur, Tamilnadu, India

Murugumohan Kumar K Kandasamy
Department of Mechanical Engineering, Pavendar Bharathidasan College of Engineering and Technology, Trichy, Tamilnadu, India

Yosephine Yulia Margaretha
Department of Chemical Engineering Widya Mandala Surabaya Catholic University, Kalijudan 37, Surabaya 60114, Indonesia

Aning Ayucitra
Department of Chemical Engineering Widya Mandala Surabaya Catholic University, Kalijudan 37, Surabaya 60114, Indonesia

Suryadi Ismadji
Department of Chemical Engineering Widya Mandala Surabaya Catholic University, Kalijudan 37, Surabaya 60114, Indonesia

Hifjur Raheman
Agricultural & Food Engineering Department, Indian Institute of Technology, Kharagpur 721302, India

Prakash C Jena
Agricultural & Food Engineering Department, Indian Institute of Technology, Kharagpur 721302, India

Snehal S Jadav
Agricultural & Food Engineering Department, Indian Institute of Technology, Kharagpur 721302, India

Walfrido Alonso Pippo
UNICAMP/IFGW/DFA/Grupo Combustíveis Alternativos, Cidade Universitária 'Zeferino Vaz', Campinas, Sao Paulo CEP 13083-859, Brazil

Carlos A Luengo
UNICAMP/IFGW/DFA/Grupo Combustíveis Alternativos, Cidade Universitária 'Zeferino Vaz', Campinas, Sao Paulo CEP 13083-859, Brazil

Pragya Sharma
GERMI-RIIC, Gandhinagar, Gujarat 382007, India

Tirumalachetty Harinarayana
GERMI-RIIC, Gandhinagar, Gujarat 382007, India
CSRI-NGRI, Hyderabad 500007, India. population-based mortality survey in Iraq. Int. J. Health Geographics 11, 12 (2012)

Yau Chwan Kun
Faculty of Environmental Studies, Universiti Putra Malaysia, Serdang, Selangor 43400, Malaysia

Ahmad Makmom Haji Abdullah
Faculty of Environmental Studies, Universiti Putra Malaysia, Serdang, Selangor 43400, Malaysia

Stephen J Mitchual
Department of Design and Technology Education, University of Education, Winneba, Kumasi Campus, Kumasi, Ghana

Kwasi Frimpong-Mensah
Department of Wood Science and Technology, Kwame Nkrumah University of Science and Technology, Kumasi, Ghana

Nicholas A Darkwa
Department of Wood Science and Technology, Kwame Nkrumah University of Science and Technology, Kumasi, Ghana

Olaleye M Amoo
Stevens Institute of Technology, Hoboken, NJ 07030, USA

Richard Layi Fagbenle
Mechanical Engineering, Obafemi Awolowo University, Ile-Ife, Osun State, Nigeria

Abdul Waheed Badar
Institut für Energietechnik, KT 2, FG Maschinen- und Energieanlagentechnik,Technische Universität Berlin, Marchstraße 18, Berlin 10587, Germany

Reiner Buchholz
Institut für Energietechnik, KT 2, FG Maschinen- und Energieanlagentechnik,Technische Universität Berlin, Marchstraße 18, Berlin 10587, Germany

Yongsheng Lou
Institut für Energietechnik, KT 2, FG Maschinen- und Energieanlagentechnik,Technische Universität Berlin, Marchstraße 18, Berlin 10587, Germany

Felix Ziegler
Institut für Energietechnik, KT 2, FG Maschinen- und Energieanlagentechnik,Technische Universität Berlin, Marchstraße 18, Berlin 10587, Germany

Eriola Betiku
Biochemical Engineering Laboratory, Department of Chemical Engineering, Obafemi Awolowo University, Ile-Ife, Osun State 220005, Nigeria

Tunde Folorunsho Adepoju
Department of Chemical Engineering, Landmark University, P.M.B 1001,Omu-Aran, Kwara State, Nigeria

Khandakar Niaz Morshed
Mechanical Engineering Department, Colorado State University, Fort Collins, CO 80523, USA

Mosfequr Rahman
Mechanical Engineering Department, Georgia Southern University, Statesboro, GA 30460, USA

Gustavo Molina
Mechanical Engineering Department, Georgia Southern University, Statesboro, GA 30460, USA

Mahbub Ahmed
Biochemistry, Chemistry & Physics, Southern Arkansas University, Magnolia, AR 71753, USA

Aditya Rachman
Department of Mechanical Engineering of Haluoleo University, Anduonohu Kendari 93232 Indonesia

Ridway Balaka
Department of Mechanical Engineering of Haluoleo University, Anduonohu Kendari 93232 Indonesia

Jenny Delly
Department of Mechanical Engineering of Haluoleo University, Anduonohu Kendari 93232 Indonesia

Yuspian Gunawan
Department of Mechanical Engineering of Haluoleo University, Anduonohu Kendari 93232 Indonesia

Peiwen Li
Department of Aerospace and Mechanical Engineering, University of Arizona, Tucson, AZ, 85721, USA

Jeong-Pill Ki
Department of Aerospace and Mechanical Engineering, University of Arizona, Tucson, AZ, 85721, USA
Department of Mechanical and Aerospace Engineering, University of Texas at Arlington, Arlington, TX, 76019, USA

Hong Liu
Department of Aerospace and Mechanical Engineering, University of Arizona, Tucson, AZ, 85721, USA

Mi Hyung Kim
Department of Environmental Planning, Graduate School of Environmental Studies, Seoul National University, San 56-1, Sillim-dong, Gwanak-gu, Seoul 151-742, Republic of Korea

Han Byul Song
Department of Chemical Engineering, Soongsil University, Sangdo-ro 369, Dongjak-gu, Seoul 156-743, Republic of Korea

Yuleum Song
Department of Molecular Microbiology and Immunology, School of Medicine, University of Missouri, Columbia, MO 65201, USA

In Tae Jeong
Carbon Management Office, Korea Environmental Industry and Technology Institute, 215 Jinheung-ro, Bulgwang-dong, Eunpyeong-gu, Seoul 122-706, Republic of Korea

Jung Wk Kim
Department of Environmental Planning, Graduate School of Environmental Studies, Seoul National University, San 56-1, Sillim-dong, Gwanak-gu, Seoul 151-742, Republic of Korea

Purabi Mazumdar
Center for Energy, Indian Institute of Technology Guwahati, Guwahati, Assam 781039, India

Venu Babu Borugadda
Department of Chemical Engineering, Indian Institute of Technology Guwahati, Guwahati, Assam 781039, India

Vaibhav V Goud
Center for Energy, Indian Institute of Technology Guwahati, Guwahati, Assam 781039, India
Department of Chemical Engineering, Indian Institute of Technology Guwahati, Guwahati, Assam 781039, India

Lingaraj Sahoo
Center for Energy, Indian Institute of Technology Guwahati, Guwahati, Assam 781039, India
Department of Biotechnology, Indian Institute of Technology Guwahati, Guwahati, Assam 781039, India

Oluseyi O Ajayi
Mechanical Engineering Department, Covenant University, P. M. B. 1023, Ota, Nigeria

Richard O Fagbenle
Mechanical Engineering Department, Obafemi Awolowo University, Ile Ife, Nigeria

James Katende
College of Engineering and Technology, Botswana International University of Science and Technology, Gaborone, Botswana

Samson A Aasa
Mechanical Engineering Department, Covenant University, P. M. B. 1023, Ota, Nigeria

Joshua O Okeniyi
Mechanical Engineering Department, Covenant University, P. M. B. 1023, Ota, Nigeria

Simona Silvia Merola
Istituto Motori-CNR, Via Marconi 8, Naples, 80125, Italy

Cinzia Tornatore
Istituto Motori-CNR, Via Marconi 8, Naples, 80125, Italy

Luca Marchitto
Istituto Motori-CNR, Via Marconi 8, Naples, 80125, Italy

Gerardo Valentino
Istituto Motori-CNR, Via Marconi 8, Naples, 80125, Italy

Felice Esposito Corcione
Istituto Motori-CNR, Via Marconi 8, Naples, 80125, Italy

Tatiana Morosuk
Technische Universität Berlin, Berlin, 10587, Germany

George Tsatsaronis
Technische Universität Berlin, Berlin, 10587, Germany

Alicia Boyano
Technische Universität Berlin, Berlin, 10587, Germany

Camilo Gantiva
Technische Universität Berlin, Berlin, 10587, Germany

Atul A Sagade
Solar Thermal Research Laboratory, New Satara College of Engineering & Management (Polytechnic), Korti-Pandharpur, Maharashtra 413304, India.

Satish Aher
Amrutvahini College of Engineering, Sangamner, Maharashtra 422608, India

Nilkanth N Shinde
Department of Energy Technology, Shivaji University, Kolhapur, Maharashtra 416004, India

Maher M Abou Al-Sood
Department of Mechanical Engineering, Assiut University, Assiut 71516, Egypt

Mahmoud Ahmed
Department of Mechanical Engineering, Assiut University, Assiut 71516, Egypt

Yousef M Abdel-Rahim
Department of Mechanical Engineering, Assiut University, Assiut 71516, Egypt